WAVE PROPAGATION,SCATTERING AND EMISSION IN COMPLEX MEDIA

WAVE PROPAGATION, SCATTERING AND EMISSION IN COMPLEX MEDIA

Editor: Ya-Qiu Jin

Science Press

World Scientific

Responsible Editors: Bin PENG and Lu QIU

Wave Propagation: Scattering and Emission in Complex Media

Copyright © 2004 by Science Press and World Scientific Publishing Co. Pte. Ltd.
Published by Science Press, and World Scientific

Science Press
16 Donghuangchenggen North Street
Beijing, 100717
P.R. China

World Scientific Publishing Co. Pte. Ltd.
5 Toh Tuck Link
Singapore, 596224

Printed in Beijing.

ISBN 7-03-012464-2 (Beijing)
ISBN 981-238-771-4(Singapore)

Preface

Electromagnetic wave propagation, scattering and emission are the physics basis for modern information technology. Great advances of satellite-borne remote sensing, wireless communication, and other information high technology during recent decades have promoted extensive studies on electromagnetic fields and waves, such as theoretical modeling, experiments and data validation, numerical simulation and inversion. To exchange most recent research achievements and discuss the future topics, the International Workshop on Wave Propagation, Scattering and Emission (WPSE2003) as an international forum was scheduled to be held in Shanghai, China, 1-4 June 2003.

This workshop was sponsored by the IEEE Geoscience and Remote Sensing Society, the USA Electromagnetic Academy, PIERS, the Fudan University, and the City University of Hong Kong. Our many colleagues, especially many distinguished scientists in the world, enthusiastically prepared their presentations for the WPSE. However, unfortunately, due to the SARS infection in the spring season 2003 in some places of Asia, this workshop had to be cancelled.

To meet the goal of our international forum, we edit this book to enclose some full papers, which cover seven areas:

1. Polarimetric scattering and SAR imagery,
2. Scattering from randomly rough surfaces,
3. Electromagnetics of complex materials,
4. Scattering from complex targets,
5. Radiative transfer and remote sensing,
6. Wave propagation and wireless communication,
7. Computational electromagnetics.

I hope this book as a good reference would benefit to our colleagues to learn and exchange the research progress in this rapidly developing area of information technology.

We are very grateful to that all authors for their excellent contributions and make this book publication possible.

This book is financially supported by Ministry of Science and Technology of China via the China State Key Basic Research Project 2001CB309400, the National Natural Science Foundation of China, and Shanghai Magnolia Foundation.

Ya-Qiu Jin

WPSE2003 Chairman
Center for Wave Scattering and Remote Sensing
Fudan University
Shanghai

Contents

Preface

III. Electromagnetics of Complex Materials

IV. Scattering from Complex Targets

V. Radiative Transfer and Remote Sensing

VI. Wave Propagation and Wireless Communication

VII. Computational Electromagnetics

Author Index

I. Polarimetric Scattering and SAR Imagery

Wave Propagation,Scattering and Emission in Complex Media
Edited by Ya-Qiu Jin
Science Press and World Scientific,2004

EM Wave Propagation and Scattering in Polarimetric SAR Interferometry

Shane R Cloude,

AEL Consultants,

Granary Business Centre, Cupar, KY15 5YQ, Scotland, UK

Tel/Fax : (44) 1334 652919/654192, Email: scloude@aelc.demon.co.uk

Abstract   In this paper we develop a multi-layer coherent polarimetric vegetation scattering model to investigate physical parameter estimation using fully polarimetric multi-baseline multi-frequency radar interferometry. It is shown that 2-layer model inversion for single baseline/single frequency sensors requires regularisation to remove multiple solutions. Traditionally this is achieved by using assumptions about polarimetric ground scattering ratios. By employing dual frequency or multi-baseline data these multiple solutions can be removed. However, the model inversion becomes ill-conditioned unless the correct baseline and frequency ratios are employed. In this paper we show how wave propagation and scattering models can be used to help devise robust inversion methods for land surface parameter estimation.

1. Introduction

In previous studies we have shown that by using single baseline polarimetric interferometry (SBPI), estimates of vegetation height and ground topography can be obtained without the need for external reference DEMs or data specific regression formulae [1,2,3,4]. However, the robustness of this inversion process is based on the assumption that in at least one (not all) of the observed polarisation channels, the ratio of ground to volume scattering is small (typically less than −10dB for 10% height accuracy). We have found two main limitations to this SBPI approach. The first is that the polarisation response of the sub-canopy ground cannot be properly estimated, as by definition the ground scattering is assumed zero in one of the channels. Hence, if we could devise a sensor capable of estimating the full polarisation response of the underlying surface then this would lead to the following improvements:
  a) free the technique from the need to assume a directly observable volume coherence
  b) enable several interesting extensions of the method such as to sub-canopy moisture and surface roughness estimation.
A second limitation of SBPI is the inability of single baseline techniques to determine vertical structure. The model assumes a vertically uniform spatial density of scatterers which maps into an exponentially weighted integral to determine the coherence. However, many trees show important variations in canopy density and if we could devise a sensor configuration capable of estimating this structure, then we could augment the height information for improved species and biomass related studies. In either case we need to increase the number of observations in the data. There are several possible ways to do this. Dual frequency or multiple angles of incidence are important examples.   The use of dual baseline regularisation has already been treated in [5] and the use of an extra frequency channel in [6]. Here we review the background to this important topic and highlight the key

contribution made by electromagnetic wave propagation and scattering models to the development of quantitative parameter estimation algorithms.

2. COHERENT VECTOR INSAR MODEL

The key observable of interest in interferometry is phase. The phase difference between signals from positions 1 and 2 and is non-zero due to the slightly different propagation path lengths Δr. This phase has the form shown in equation 1 (where B_n is the normal component of the baseline)

$$\exp(i2k\Delta r) \approx \exp(\frac{4\pi}{\lambda}\delta\theta m) \approx \exp(\frac{4\pi B_n}{\lambda R}m) \tag{1}$$

where $\theta \approx B_n/R$ if $R \gg B$ and the co-ordinate m is defined as normal to the slant range direction. Transforming to the surface y,z co-ordinates using the mean angle of incidence θ we can also express equation 1 in the modified form $\exp(i\ \phi(y, z))$ where

$$\phi(y,z) = y\left(\frac{2kB_n \cos\theta}{R} - 2\Delta k \sin\theta\right) + z\left(\frac{2kB_n \sin\theta}{R} + 2\Delta k \cos\theta\right) \quad \Delta k = \frac{kB_n}{R \tan\theta} \tag{2}$$

In equation 2 we have further included the possibility of making a wavenumber shift Δk between the two images. This shift can best be derived using a k-space representation of interferometry as shown in figure 1 [12, 13]. Here the radial co-ordinate is the wavenumber k = $\tilde{\pi\lambda}$ and the polar co-ordinate the angle of incidence. By making a frequency shift Δf to one of the signals then we see that we can equalise the k_y components of the wavenumber.

As is apparent from equation 2, we can then always remove the 'y' dependence of the phase ϕ by choosing Δk based on the geometry of the system. In this case the interferometric phase depends only on the height of the scatterers above the reference plane (the z co-ordinate) i.e. we need consider only the volume scattering contributions.

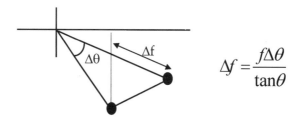

$$\Delta f = \frac{f\Delta\theta}{\tan\theta}$$

Figure 1 : K-space representation of radar interferometry

To study decorrelation in the 'z' direction, we then define an effective vertical propagation constant using 1 and 2 so that

To study decorrelation in the 'z' direction, we then define an effective vertical propagation constant using 1 and 2 so that

$$k_z = \frac{4\pi\delta\theta}{\lambda\sin\theta} \approx \frac{4\pi B_n}{\lambda R \sin\theta} \tag{3}$$

A scatterer that changes in height by Δz will then have an associated phase change of $k_z\Delta$? . In foliage examples there will be a random distribution of scatterers in the vertical direction. This will cause fluctuations in the phase which are manifest as a drop in the interferometric coherence γ [7,8]. In polarimetric systems we have 3 channels of complex data at positions 1 and 2 (HH, VV and HV). Hence to generate the appropriate coherence we need first to project the 3 channels onto a unitary weight vector $\underline{w}_1$ to generate a complex scalar s_1 as shown in equation 4. Similarly we can define a different weight vector $\underline{w}_2$ and scalar s_2 at position 2. The interferometric coherence is then defined from the normalised product of the scalar projections as shown in equation 4

$$\left. \begin{array}{l} s_1 = \underline{w}_1^{*T} \underline{k}_1 \\ s_2 = \underline{w}_2^{*T} \underline{k}_2 \end{array} \right\}$$

$$\Rightarrow \left\{ \begin{array}{l} \gamma = \dfrac{\left|\langle s_1 s_2^* \rangle\right|}{\sqrt{\langle s_1 s_1^* \rangle \cdot \langle s_2 s_2^* \rangle}} = \dfrac{\left|\underline{w}_1^{*T} \Omega_{12} \underline{w}_2\right|}{\sqrt{\underline{w}_1^{*T} T_{11} \underline{w}_1 \underline{w}_2^{*T} T_{22} \underline{w}_2}} \quad 0 \le \gamma \le 1 \\[3ex] \phi = \arg(s_1 s_2^*) = \arg\left(\underline{w}_1^{*T} \Omega_{12} \underline{w}_2\right) \end{array} \right. \tag{4}$$

We see that the coherence and phase can be expressed in terms of a vector of scattering coefficients $\underline{k}$ and 3x3 block elements of a 6 x 6 coherency matrix [P] as defined in equation 5

$$[P] = \left\langle \begin{bmatrix} \underline{k}_1 \\ \underline{k}_2 \end{bmatrix} \begin{bmatrix} \underline{k}_1^{*T} & \underline{k}_2^{*T} \end{bmatrix} \right\rangle = \begin{bmatrix} [T_{11}] & [\Omega_{12}] \\ [\Omega_{12}]^{*T} & [T_{22}] \end{bmatrix} \tag{5}$$

In the presence of speckle [11,24], the 6 dimensional coherent polarimetric signal can then be modelled as a multi-variate Gaussian distribution of the form shown in equation 6

$$\underline{u} = \begin{bmatrix} \underline{k}_1 \\ \underline{k}_2 \end{bmatrix} \Rightarrow p(\underline{u}) = \frac{1}{\pi^6 \det([P])} e^{-\underline{u}^T [P]^{-1} \underline{u}} \tag{6}$$

Here again the matrix [P] determines the fluctuation statistics of the signals. Hence to evaluate the coherence [14] for arbitrary polarisation, we need to estimate the block matrices [T] and [Ω]. It is here that use can be made of EM wave propagation and scattering models.

For example, in many vegetation problems the scatterers in a volume have some residual orientation correlation due to their natural structure (branches in a tree canopy) or due to agriculture (oriented corn stalks). In this case the volume has two eigen-propagation states a and b (which we assume are unknown but orthogonal linear polarisations). Only along these eigenpolarisations is the propagation simple, in the sense that the polarisation state does not change with depth into the volume. If there is some mismatch between the radar co-ordinates and the medium's eigenstates then a very complicated situation arises where the polarisation of the incident field changes as a function of distance into the volume [9,10]. By assuming that the medium has reflection symmetry about the (unknown) axis of its eigenpolarisations, we obtain a polarimetric coherency matrix [T] and corresponding covariance matrix [C] for backscatter from the volume as shown in equation 7 [2,9]

$$[T] = \begin{bmatrix} t_{11} & t_{12} & 0 \\ t_{12}^* & t_{22} & 0 \\ 0 & 0 & t_{33} \end{bmatrix} \quad \Leftrightarrow \quad [C] = \begin{bmatrix} c_{11} & 0 & c_{13} \\ 0 & c_{22} & 0 \\ c_{13}^* & 0 & c_{33} \end{bmatrix} \tag{7}$$

We can now obtain an expression for the matrices $[T_{11}]$ and $[\Omega_{12}]$ for an oriented volume extending from $z = z_0$ to $z = z_0 + h_v$ as vector volume integrals as shown in equation 8

$$[\Omega_{12}] = e^{-\frac{(\sigma_a + \sigma_b) h_v}{\cos \theta_o}} e^{i\phi(z_o)} R(2\beta) \left\{ \int_0^{h_v} e^{ik_z z'} e^{\frac{(\sigma_a + \sigma_b) z'}{\cos \theta_o}} P(\tau) \, T \, P(\tau^*) dz \right\} R(-2\beta)$$

$$[T_{11}] = e^{-\frac{(\sigma_a + \sigma_b) h_v}{\cos \theta_o}} R(2\beta) \left\{ \int_0^{h_v} e^{\frac{(\sigma_a + \sigma_b) z'}{\cos \theta_o}} P(\tau) \, T \, P(\tau^*) dz \right\} R(-2\beta) \tag{8}$$

where for clarity we have dropped the brackets around matrices inside the integrals and defined R as a rotation matrix to allow for mismatch between the radar co-ordinates and the projection of the eigenstates into the polarisation plane (equation 9).

$$R(\beta) = \begin{bmatrix} 1 & 0 & 0 \\ 0 & \cos \beta & \sin \beta \\ 0 & -\sin \beta & \cos \beta \end{bmatrix} \tag{9}$$

Differential propagation through the medium is modelled as a matrix transformation as shown in equation 10, with the complex differential propagation constant shown in equation 11

$$P(\tau) \, T \, P(\tau^*) =$$
$$\begin{bmatrix} \cosh \tau & \sinh \tau & 0 \\ \sinh \tau & \cosh \tau & 0 \\ 0 & 0 & 1 \end{bmatrix} \begin{bmatrix} t_{11} & t_{12} & 0 \\ t_{12}^* & t_{22} & 0 \\ 0 & 0 & t_{33} \end{bmatrix} \begin{bmatrix} \cosh \tau^* & \sinh \tau^* & 0 \\ \sinh \tau^* & \cosh \tau^* & 0 \\ 0 & 0 & 1 \end{bmatrix} \tag{10}$$

$$\tau = vz = \left(\frac{\sigma_a - \sigma_b}{2} + ik(\chi_a - \chi_b) \right) \frac{z'}{\cos \theta_o} \tag{11}$$

This now enables us to generate the coherence for any choice of weight vector $\underline{w}$. in equation 4. However we are often more interested in the maximum variability of coherence with changes in $\underline{w}$ and so we employ the coherence maximiser, which requires solution of an eigenvalue problem as shown in equation 12 [1,3]

$$\left.\begin{array}{c} [T_{22}^{-1}] \, [\Omega_{12}]^{*T} \, [T_{11}^{-1}] \, [\Omega_{12}] \, \underline{w}_2 = \lambda \underline{w}_2 \\ [T_{11}^{-1}] \, [\Omega_{12}] \, [T_{22}^{-1}] \, [\Omega_{12}]^{*T} \, \underline{w}_1 = \lambda \underline{w}_1 \end{array}\right\} \quad 0 \leq \lambda = \gamma_{opt}^2 \leq 1 \qquad (12)$$

The eigenvalues of this matrix then indicate the maximum change of coherence with polarisation. As the eigenvalues are invariant to unitary transformations of the vector $\underline{k}$ we can replace [T] by [C] in equation 12. To account for the effects of propagation on the polarimetric response of an oriented volume, it is simpler to employ the covariance [C] rather than the coherency matrix [T]. For a general oriented volume we then have from 6,7 and 8

$$C_{11} = \begin{bmatrix} c_{11} I_1 & 0 & c_{13} I_2 \\ 0 & c_{22} I_3 & 0 \\ c_{13}^* I_2^* & 0 & c_{33} I_4 \end{bmatrix}$$

$$\Rightarrow C_{11}^{-1} = \frac{1}{f} \begin{bmatrix} c_{33} I_4 & 0 & -c_{13} I_2 \\ 0 & \dfrac{f}{c_{22} I_3} & 0 \\ -c_{13}^* I_2^* & 0 & c_{11} I_1 \end{bmatrix} \qquad (13)$$

where $f = (c_{11} c_{33} - c_{13} c_{13}^*) I_1 I_4$ and similarly for the Ω matrix we can write

$$\Omega_{12} = e^{i\phi(z_o)} \begin{bmatrix} c_{11} I_5 & 0 & c_{13} I_6 \\ 0 & c_{22} I_7 & 0 \\ c_{13}^* I_8 & 0 & c_{33} I_9 \end{bmatrix} \qquad (14)$$

Note that Ω_{12} is neither symmetric nor Hermitian. The integrals $I_1 - I_9$ are defined as [5]

$$I_1 = \int_0^{h_v} e^{2\sigma_a z} dz \quad I_2 = \int_0^{h_v} e^{2(k_a + k_b^*) z} dz \quad I_3 = \int_0^{h_v} e^{(\sigma_a + \sigma_b) z} dz \quad I_4 = \int_0^{h_v} e^{2\sigma_b z} dz \quad I_5 = \int_0^{h_v} e^{ik_z} e^{2\sigma_a z} dz \qquad (15)$$

$$I_6 = \int_0^{h_v} e^{ik_z} e^{2(k_a + k_b^*) z} dz \quad I_7 = \int_0^{h_v} e^{ik_z} e^{(\sigma_a + \sigma_b) z} dz \quad I_8 = \int_0^{h_v} e^{ik_z} e^{2(k_b + k_a^*) z} dz \quad I_9 = \int_0^{h_v} e^{ik_z} e^{2\sigma_b z} dz$$

and k_z is defined in equation 3, k_a, k_b are the complex propagation constants of the eigenpolarisation states and σ are the real extinction rates in the medium. Hence the first part of the optimisation matrix (equation 12) has the form

$$C_{11}^{-1}\Omega_{12} = \frac{e^{i\phi(z_o)}}{f}\begin{bmatrix} c_{33}I_4 & 0 & -c_{13}I_2 \\ 0 & \dfrac{f}{c_{22}I_3} & 0 \\ -c_{13}^*I_2^* & 0 & c_{11}I_1 \end{bmatrix}\begin{bmatrix} c_{11}I_5 & 0 & c_{13}I_6 \\ 0 & c_{22}I_7 & 0 \\ c_{13}^*I_8 & 0 & c_{33}I_9 \end{bmatrix}$$

(16)

which is diagonal if $I_4I_6 - I_2I_9 = I_8I_1 - I_2^*I_5 = 0$. From equation 15 we can easily show that both equations are satisfied for arbitrary medium parameters as we have

$$I_4I_6 = \int_0^{h_v} e^{2\sigma_b z} e^{ik_z z} e^{2(k_a+k_b^*)z}dz = I_2I_9$$

$$I_8I_1 = \int_0^{h_v} e^{2\sigma_a z} e^{ik_z z} e^{2(k_b+k_a^*)z}dz = I_2^*I_5$$

(17)

$$\tilde{\gamma}_1 = f(\sigma_a) = \frac{2\sigma_a e^{i\phi(z_o)}}{\cos\theta_o(e^{2\sigma_a h_v/\cos\theta_o}-1)}\int_0^{h_v} e^{ik_z z} e^{\frac{2\sigma_a z}{\cos\theta_o}}dz'$$

$$\tilde{\gamma}_2 = f(\sigma_a,\sigma_b) = \frac{(\sigma_a+\sigma_b)e^{i\phi(z_o)}}{\cos\theta_o(e^{(\sigma_a+\sigma_b)h_v/\cos\theta_o}-1)}\int_0^{h_v} e^{ik_z z} e^{\frac{(\sigma_a+\sigma_b)z}{\cos\theta_o}}dz'$$

$$\tilde{\gamma}_3 = f(\sigma_b) = \frac{2\sigma_b e^{i\phi(z_o)}}{\cos\theta_o(e^{2\sigma_b h_v/\cos\theta_o}-1)}\int_0^{h_v} e^{ik_z z} e^{\frac{2\sigma_b z}{\cos\theta_o}}dz'$$

(18)

The optimum coherence values can then be calculated as shown in equation 18. It follows from the above that $K_c = C_{11}^{-1}\Omega_{12}C_{11}^{-1}\Omega_{12}^{*T}$ is also diagonal and hence by using the relationship between [T] and [C] we can show that the eigenvectors of $K = T_{11}^{-1}\Omega_{12}T_{11}^{-1}\Omega_{12}^{*T}$ are functions of β as

$$\underline{w}_1 = \frac{1}{\sqrt{2}}\begin{bmatrix} 1 \\ -\cos2\beta \\ \sin2\beta \end{bmatrix} \quad \underline{w}_2 = \begin{bmatrix} 0 \\ \sin2\beta \\ \cos2\beta \end{bmatrix} \quad \underline{w}_3 = \frac{1}{\sqrt{2}}\begin{bmatrix} 1 \\ \cos2\beta \\ -\sin2\beta \end{bmatrix}$$

(19)

Equation 18 shows that the maximum coherence is obtained for the medium eigenpolarisation with the highest extinction. The lowest coherence is then obtained for the orthogonal polarisation that has lower extinction and hence better penetration into the vegetation. The cross polar channel which propagates into the volume on one eigenpolarisation and out on the other has a coherence between these two extremes.

However, in real applications forest cover will often be random and any orientation effects are likely to be weak. For this reason we must consider a special case of equation 18 when the vegetation shows full azimuth symmetry in the plane of polarisation. In this case there can, by definition, be no preferred β angle in equation 19 and hence the coherency matrix for the volume must be diagonal with 2 degenerate eigenvalues of the form [3]

$$[T_v] = m_v \begin{bmatrix} 1 & 0 & 0 \\ 0 & \eta & 0 \\ 0 & 0 & \eta \end{bmatrix} \quad 0 \le \eta \le 1 \tag{20}$$

where η depends on the particle shape in the volume and on the presence of multiple scattering. If there is no distinction between the eigenvectors in equation 19 then the eigenvalues in equation 18 must become equal. This arises when $\sigma_? = \sigma_b = \sigma$ i.e. when the extinction in the medium becomes independent of polarisation. In this case volume scattering alone leads to equal eigenvalues in 12 and hence the coherence is no longer a function of polarisation. The situation is practice is made more complicated because some penetration of the vegetation occurs and combined surface and volume scattering occurs. It is well known that surface scattering is strongly polarisation dependent and hence in practice we need to extend the above argument to a multi-layer model.

3. MULTI-LAYER COHERENT SCATTERING MODEL

We consider a general multi-layer oriented random media problem as shown schematically in figure 2. We assume that each layer is composed of a cloud of identical spheroidal particles, where the density of each layer is tenous enough so that we can ignore reflections at the boundaries between layers. We further assume that we can ignore any refraction at the boundaries between the layers, as justified in [4]. We assume that the propagation eigenstates are orthogonal linear polarisations, but allow for the fact that they may not be aligned with the radar h and v axes.

In order to assess the complexity of the inverse problem consider the following argument. According to our model, each layer is characterised by a set of 9 parameters,

β -   orientation of the eigen-propagation states with respect to the radar co-ordinate system

h_V –   Layer Thickness in metres

ϕz_o - Interferometric Phase of the bottom of the layer relative to reference $z = 0$

χ_a –   Refractivity (index of refraction-1) for eigenstate a

χ_b –   Refractivity (index of refraction-1) for eigenstate b

σ_a -   Extinction coefficient for the a eigenpolarisation

σ_b -   Extinction coefficient for the b eigenpolarisation

γ -   Shape parameter of particles in the volume

m_c -   the backscatter coefficient from the volume

Hence we face a set of at least 9n physical parameters to totally characterise the multi-layer problem. A fixed frequency, single baseline polarimetric interferometer measures up to 36 parameters (the elements of P). Hence we see that, using this simplified argument, we might consider inversion up to 4-layer structures with such a single baseline system.

In practice, for small angle baselines in the absence of temporal decorrelation, we can assume that $T_{11} = T_{22}$ and hence only 27 parameters are available from P. Hence we are limited to consider up to 3-layer structures. Nonetheless, many physical vegetation structures can be characterised in terms of a small number of distinct layers and hence are suitable candidates for this single base-line technology.

The most general formulation of polarimetric interferometry for an N-layer problem can then be written in compact matrix notation as

$$T_{11} = R_N\left(I_1^N + P_N R_{N(N-1)}(I_1^{N-1} + P_{(N-1)} R_{(N-1)(N-2)}(I_1^{N-2} + \ldots\ldots) R_{(N-2)(N-1)} P_{N-1}^*)\ldots\ldots\right)R_{-N} \quad (21)$$

$$\Omega_{12} = R_N\left(e^{i\phi_N} I_2^N + P_N R_{N(N-1)}(e^{i\phi_{N-1}} I_2^{N-1} + P_{(N-1)} R_{(N-1)(N-2)}(e^{i\phi_{N-2}} I_2^{N-2} + \ldots\ldots) R_{(N-2)(N-1)} P_{N-1}^*)\ldots\ldots\right)R_{-N}$$
$$(22)$$

$$I_1^i = e^{-\frac{\sigma_{ai}+\sigma_{bi}}{\cos\theta_o}h_{vi}} \int_0^{h_{vi}} e^{\frac{\sigma_{ai}+\sigma_{bi}}{\cos\theta_o}z'} P(\tau_i)T_i P(\tau_i^*)dz' \qquad I_2^i = e^{-\frac{\sigma_{ai}+\sigma_{bi}}{\cos\theta_o}h_{vi}} \int_0^{h_{vi}} e^{ik_z z'} e^{\frac{\sigma_{ai}+\sigma_{bi}}{\cos\theta_o}z'} P(\tau_i)T_i P(\tau_i^*)dz'$$

$$(23)$$

$$\tau = vz = \left(\frac{\sigma_a - \sigma_b}{2} + ik(\chi_a - \chi_b)\right)\frac{z'}{\cos\theta_o} \quad (24)$$

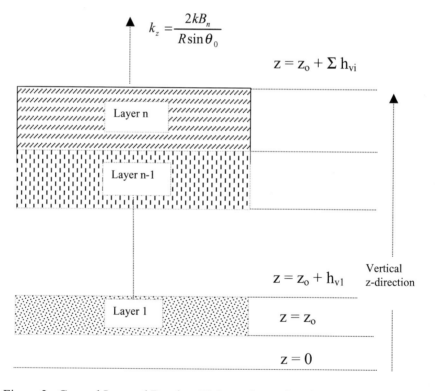

$$k_z = \frac{2kB_n}{R\sin\theta_0}$$

$z = z_0 + \Sigma\, h_{vi}$

Layer n

Layer n-1

$z = z_0 + h_{v1}$ Vertical z-direction

Layer 1

$z = z_0$

$z = 0$

Figure 2 : General Layered Random Volume Scattering Geometry
For Polarimetric Interferometry

where the 3x3 matrices P_N, R_N are defined as

$$P_N = e^{-\frac{\sigma_{aN}+\sigma_{bN}}{2\cos\theta_o}h_{vN}} \begin{bmatrix} \cosh\varepsilon_N & \sinh\varepsilon_N & 0 \\ \sinh\varepsilon_N & \cosh\varepsilon_N & 0 \\ 0 & 0 & 1 \end{bmatrix} \quad \varepsilon_N = \left(\frac{\sigma_{aN}-\sigma_{bN}}{2} + ik(\chi_{aN}-\chi_{bN})\right)\frac{h_{vN}}{\cos\theta_o}$$

$$R_N = \begin{bmatrix} 1 & 0 & 0 \\ 0 & \cos\beta_N & -\sin\beta_N \\ 0 & \sin\beta_N & \cos\beta_N \end{bmatrix} \quad R_{N(N-1)} = \begin{bmatrix} 1 & 0 & 0 \\ 0 & \cos(\beta_N-\beta_{N-1}) & -\sin(\beta_N-\beta_{N-1}) \\ 0 & \sin(\beta_N-\beta_{N-1}) & \cos(\beta_N-\beta_{N-1}) \end{bmatrix}$$

$$(25)$$

and T_j is the 3 x 3 polarimetric coherency matrix for the jth layer. To simplify matters, we now assume that the eigenpolarisations for each layer are aligned so that $R_{N(N-1)} = I$, the identity matrix. By considering the case of a 3-layer structure suitable for single baseline polarimetric interferometry we then obtain the specific form of the polarimetric interferometric equations as

$$T_{11} = R_3\left(I_1^3 + P_3(I_1^2 + P_2 I_1^1 P_2^*)P_3^*\right)R_{-3}$$

$$\Omega_{12} = R_3\left(e^{i\phi_3}I_2^3 + P_3(e^{i\phi_2}I_2^2 + P_2 e^{i\phi_1}I_2^1 P_2^*)P_3^*\right)R_{-3}$$

$$(26)$$

where the three phase centres are defined as

$$\phi_j = \phi(\sum_{i=0}^{j-1}h_{vi}) \quad \text{where} \quad h_{v0} = z_o \quad\quad (27)$$

Hence we see that information about the position of each layer lies in the phase terms of Ω_{12}, while the extent of each layer can be found from the ratio of integrals I_2/I_1 from Ω_{12} and T_{11}.

The 2-layer problem takes on the following further simplified form

$$T_{11} = R_2\left(I_1^2 + P_2 I_1^1 P_2^*\right)R_{-2}$$
$$\Omega_{12} = R_2\left(e^{i\phi_2}I_2^2 + e^{i\phi_1}P_2 I_2^1 P_2^*\right)R_{-2}$$

$$(28)$$

Finally for a single layer problem, the equations can be written as

$$T_{11} = R_1 I_1^1 R_{-1}$$
$$\Omega_{12} = e^{i\phi_1}R_1 I_2^1 R_{-1}$$

$$(29)$$

The key remaining problem is how to best employ polarimetry and interferometry to separate each layer. In this paper we consider the analytical structure of three important examples of these layered media problems, namely

- Scattering by a single Random Volume
- Scattering by an single Oriented Volume
- Scattering from a 2-layer Random Volume over a ground surface

The first two have already been dealt with in equation 18. The last problem is of particular importance in radar remote sensing, as microwaves can penetrate vegetation cover and are backscattered from the underlying ground surface. Any successful inversion scheme must then account for this mixture of volume and surface scattering. To do this we require a parametric model for surface scattering.

4.POLARIMETRIC SURFACE SCATTERING MODELS FOR INSAR

In order to assess the role of surface scattering in polarimetric interferometry, we need to first understand the polarimetric properties of surfaces. Figure 3 shows the geometry for surface scattering problems, where by convention horizontal polarisation (H) is perpendicular to the plane of incidence (S) and vertical polarisation (V) is in the plane (P). The total observed scattering coherency matrix for a non-vegetated rough surface must satisfy the following general constraints, based on experimental observations made in the literature:

- The backscatter coherency matrix [T] has the general form of a reflection symmetric medium, since the mean surface normal imposes a strong symmetry axis on the problem.
- The HH and VV back-scattering coefficients are generally not equal, except for very rough surfaces
- The surface generates non-zero HV back-scattered power. Some of this can be correlated with surface slope effects but some is due to depolarisation by the surface.
- Surfaces depolarise in backscatter so that the polarimetric coherences are less than unity

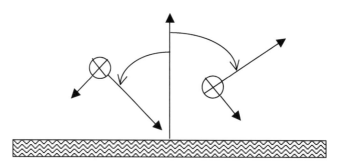

Figure 3 : General Surface Scattering Geometry

There is no general solution for surface scattering that can account for all of these features in a fully quantitative manner and hence an approximate model must be used. We have developed a simple surface backscatter model based on these observations and satisfying all of the above constraints. The model may be motivated as follows. A perfectly smooth surface has zero backscatter and acts as a limiting case in our analysis. However we are interested in considering the problem of slight roughness parameters, particularly the case of ks < 1, where s is the rms roughness. In this situation the roughness is seen as a perturbation of the smooth surface problem and we can generate a model for the polarimetric structure of the backscattered signal as follows.

We obtain the backscatter coefficients for a slightly rough surface using a small perturbation model from Maxwell's equations. According to this model the backscatter from a surface is non-zero and depends on the component of the power spectrum of the surface that matches the incident wavelength and angle of incidence (Bragg scattering). In this model the scattering matrix for the surface has the form shown in equation 30

$$[S] = \begin{bmatrix} R_S(\theta, \ \varepsilon_r) & 0 \\ 0 & R_P(\theta, \ \varepsilon_r) \end{bmatrix} \left\{ \begin{array}{l} R_S = \dfrac{\cos\theta - \sqrt{\varepsilon_r - \sin^2\theta}}{\cos\theta + \sqrt{\varepsilon_r - \sin^2\theta}} \\[3ex] R_P = \dfrac{(\varepsilon_r - 1)(\sin^2\theta - \varepsilon_r(1 + \sin^2\theta))}{(\varepsilon_r \cos\theta + \sqrt{\varepsilon_r - \sin^2\theta})^2} \end{array} \right. \tag{30}$$

This model yields zero cross-polarization and zero depolarization and so is inadequate as it stands. Further, although this model has different HH (SS) and VV (PP) scattering coefficients, it tends to overestimate the ratio of VV/HH scattered power. In order to widen the applicability of this model, we propose a configurational average of the above solution, to obtain a coherence less than unity at the same time as generating cross polarized energy and a reduced VV/HH ratio. The configurational averaging is taken over a uniform distribution about zero of the surface slope in the plane perpendicular to the scattering plane. If this slope is termed β then we propose a uniform distribution of half-width β_1 as shown in figure 4. Assuming now a non-coherent summation of energy across the distribution of β?then the coherency matrix for the surface has the analytic form shown in equation 31 (where sinc(x) =sin(x)/x)

$$[T] = \begin{bmatrix} A & B\text{sinc}(2\beta_1) & 0 \\ B^*\text{sinc}(2\beta_1) & C(1 + \text{sinc}(4\beta_1)) & 0 \\ 0 & 0 & C(1 - \text{sinc}(4\beta_1)) \end{bmatrix} \quad 0 \le \beta_1 \le \frac{\pi}{2} \tag{31}$$

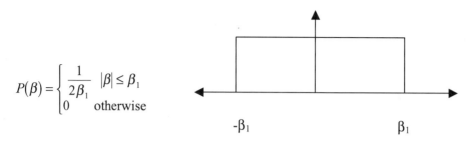

$$P(\beta) = \begin{cases} \dfrac{1}{2\beta_1} & |\beta| \le \beta_1 \\ 0 & \text{otherwise} \end{cases}$$

Figure 4 : Uniform Distribution of Surface Slope and the Bragg model

where the coefficients A B and C are given in terms of the smooth surface solution from equation 30 as:

$$A = |R_S + R_P|^2 \qquad B = (R_S + R_P)(R_S^* - R_P^*) \qquad C = \frac{1}{2}|R_S - R_P|^2 \tag{32}$$

In this way both the polarimetric coherence and the level of cross-polarized power is controlled by a single parameter (β_1). We see that for $\beta_1 = 0$ the coherence is unity and the HV power is zero, as expected for the limiting smooth Bragg surface. As β_1 increases so the HV power increases. We see that the coherence also reduces monotonically from 1 for a smooth surface to zero for $\beta_1 = 90$ degrees. This limiting case obtains complete azimuthal symmetry and the surface scattering shows zero dependence on polarization angle.

According to this model all rough surfaces are characterised by a triplet of real eigenvalues $\lambda_1 \ge \lambda_2 \ge \lambda_3 \ge 0$. A key parameter in characterising the surface roughness is the scattering anisotropy defined as

$$A = \frac{\lambda_3 - \lambda_2}{\lambda_3 + \lambda_2} \qquad 0 \le A \le 1$$

(33)

Note that the increase in HV power in 31 is faster than the fall-off in coherence and so for small β_1 the Scattering Anisotropy will be high (close to 1). As β_1 increases so the anisotropy falls monotonically to zero. Such behaviour has been confirmed in radar observations of natural surfaces.

We see from the above analysis that direct surface polarimetric backscatter must be considered a rank 3 coherency matrix process. For slightly rough surfaces, when the anisotropy is high, a rank 2 model may suffice but in any case we see that we cannot effectively model a surface with a rank 1 matrix, corresponding to a single scattering mechanism. From the point of view of polarimetric interferometry this is very important, as we cannot then easily separate surface from volume scattering using polarimetry alone.

There are two other aspects of surface scattering to be discussed before proceeding to an analysis of polarimetric interferometry. The first is the presence of specular dihedral scattering between ground and volume. This can cause large amplitude surface returns. The presence of such double scattering can in extreme cases lead to an effective rank 1 ground coherency matrix (with a scattering mechanism with $45° < \alpha < 90°$). Hence any inverse model we propose must be able to account for arbitrary rank 1, 2 or 3 ground components depending on circumstances, and permit the full range of ? values to occur. We shall use this result in the design of our inversion algorithm and hence achieve robust performance. However, a second key feature of surface scattering is the fact that it is localised in the z-direction and hence represents a fixed phase centre in interferometry. Even the roughest surfaces have a (range-filtered) coherence so close to unity that they can be assumed a fixed phase centre in the image. Note that this is also true for dihedral scattering, where the phase centre is along the apex of the dihedral and hence remains on the surface.

We now use these observations to consider 2-layer polarimetric interferometric scattering problems where one of the layers is the surface and the other a random volume. We begin by assuming the simplest case of a rank-1 surface coherency matrix as this simplifies the equations and demonstrates some important general features of the solution method. We then consider the more general case of a rank 3 ground coherency matrix before finally accounting for ground/volume dihedral interactions.

There are 8 physical parameters characterising this random volume/rank-1ground scattering model. They are defined as follows:

$\phi(z_0)$ - Local Topographic Phase relative to reference $z = 0$

h_V –   Volume Thickness in metres

σ -   extinction coefficient in the volume in dB/m

κ -   shape parameter of particles in the volume

m_C -   the scattering coefficient from the volume

α -   the polarimetric scattering mechanism of the rank one underlying surface

ψ -   Polarisation Phase shift due to the ground scatter

m_g -   the reflection coefficient of the underlying surface

The optimum coherence values are given as the square root of the eigenvalues of the 3 x 3 matrix K which can be expressed directly in terms of these parameters in the form

$$K = K_1K_2 \Rightarrow \begin{cases} K_1 = \dfrac{1}{d}\begin{bmatrix} \dfrac{I_2}{I_1}\left(1+\dfrac{m_g}{I_1\kappa m_c}\sin^2\alpha\right)+\dfrac{m_g}{I_1 m_c}\cos^2\alpha & \dfrac{m_g}{I_1 m_c}\sin\alpha\cos\alpha e^{i\psi}(1-\dfrac{I_2}{I_1}) & 0 \\[3mm] \dfrac{m_g}{I_1\kappa m_c}\sin\alpha\cos\alpha e^{-i\psi}(1-\dfrac{I_2}{I_1}) & \dfrac{I_2}{I_1}\left(1+\dfrac{m_g}{I_1 m_c}\cos^2\alpha\right)+\dfrac{m_g}{I_1\kappa m_c}\sin^2\alpha & 0 \\[3mm] 0 & 0 & \dfrac{I_2 d}{I_1} \end{bmatrix} \\[14mm] K_2 = \dfrac{1}{d}\begin{bmatrix} \dfrac{I_2^*}{I_1}\left(1+\dfrac{m_g}{I_1\kappa m_c}\sin^2\alpha\right)+\dfrac{m_g}{I_1 m_c}\cos^2\alpha & \dfrac{m_g}{I_1 m_c}\sin\alpha\cos\alpha e^{i\psi}(1-\dfrac{I_2^*}{I_1}) & 0 \\[3mm] \dfrac{m_g}{I_1\kappa m_c}\sin\alpha\cos\alpha e^{-i\psi}(1-\dfrac{I_2^*}{I_1}) & \dfrac{I_2^*}{I_1}\left(1+\dfrac{m_g}{I_1 m_c}\cos^2\alpha\right)+\dfrac{m_g}{I_1\kappa m_c}\sin^2\alpha & 0 \\[3mm] 0 & 0 & \dfrac{I_2^* d}{I_1} \end{bmatrix} \end{cases} \tag{34}$$

where

$$d = 1+\frac{m_g}{m_c I_1}\left(\cos^2\alpha+\frac{\sin^2\alpha}{\kappa}\right) \tag{35}$$

and the two integrals I_1 and I_2 have the explicit form

$$I_1 = \frac{\cos\theta_o}{2\sigma}(e^{\frac{2\sigma h_v}{\cos\theta_o}}-1) \quad I_2 = \frac{\dfrac{2\sigma}{\cos\theta_o}-ik_z}{k_z^2+\dfrac{4\sigma^2}{\cos^2\theta_o}}(e^{\frac{2\sigma h_v}{\cos\theta_o}}e^{ik_z h_v}-1) \tag{36}$$

The eigenvalues of the 3 x 3 complex matrix K are all real and $1\geq\lambda_1\geq\lambda_2\geq\lambda_3\geq 0$. We can see from equation 33 that one of the eigenvalues is directly obtained as

$$\begin{matrix} \text{weak ground} & \lambda_1 \\ \text{strong ground} & \lambda_3 \end{matrix}\Bigg\} = \left|\frac{I_2}{I_1}\right|^2 = \gamma_v^2 \tag{37}$$

In any case, γ_v^2 is *always* an eigenvalue of the optimiser matrix K. The other eigenvalues can be found by expressing equation 33 in the matrix diagonal form

$$K_1 = \frac{I_2}{dI_1}\begin{bmatrix}1&0&0\\0&1&0\\0&0&d\end{bmatrix}+\frac{m_g}{dm_c I_1}\begin{bmatrix}\cos\alpha & -\sin\alpha & 0\\ \kappa^{-1}\sin\alpha & \cos\alpha & 0\\ 0 & 0 & 1\end{bmatrix}\begin{bmatrix}1&0&0\\0&\dfrac{I_2}{I_1}&0\\0&0&0\end{bmatrix}\begin{bmatrix}\cos\alpha & \sin\alpha & 0\\ -\kappa^{-1}\sin\alpha & \cos\alpha & 0\\ 0 & 0 & 1\end{bmatrix} \tag{38}$$

from which the optimum coherence values are directly obtained as

$$\tilde{\gamma}_1 = \frac{I_2}{dI_1} + \frac{m_g}{dm_c I_1}(\cos^2\alpha + \kappa^{-1}\sin^2\alpha) = \frac{\dfrac{I_2}{I_1} + \dfrac{m_g}{m_c I_1}(\cos^2\alpha + \kappa^{-1}\sin^2\alpha)}{1 + \dfrac{m_g}{m_c I_1}(\cos^2\alpha + \kappa^{-1}\sin^2\alpha)} \qquad = \frac{\tilde{\gamma}_v + \mu}{1 + \mu}$$

$$\tilde{\gamma}_2 = \frac{I_2}{dI_1} + \frac{I_2 m_g}{dm_c I_1^2}(\cos^2\alpha + \kappa^{-1}\sin^2\alpha) = \frac{I_2}{I_1} \qquad\qquad\qquad\qquad = \tilde{\gamma}_v \qquad (39)$$

$$\tilde{\gamma}_3 = \frac{I_2}{I_1} \qquad\qquad\qquad\qquad\qquad\qquad\qquad\qquad\qquad\qquad = \tilde{\gamma}_v$$

where we see that we have obtained two equal eigenvalues. This is a very important result. It means that for a random volume over a *single* component ground, two of the optimum coherence values are always equal. These equal eigenvalues are also the squared coherence values for the volume alone i.e. the optimiser is 'switching-off' the ground component completely.

Based on the above physical argument, the eigenvectors of K should formally be equivalent to those obtained in the polarimetric contrast optimisation problem between the ground and volume, namely obtained from the three eigenvectors $\underline{w}_i$ such that

$$T_{vol}^{-1} T_{ground}\, \underline{w} = \mu \underline{w} \qquad (40)$$

This is an important observation as it permits the use of a simplified eigenvalue problem (equation 45) to find the eigenvectors of K. Further, the eigenvalues of the contrast optimisation problem may be related to those of K, as long as we restrict attention to random volumes. Hence we can solve the coherence optimisation problem for a random volume over a ground using equation 40, as we now show.

The contrast optimisation can be written for the case of a single ground component as

$$T_v^{-1} T_g = \frac{m_g}{I_1 m_c}\begin{bmatrix} 1 & 0 & 0 \\ 0 & \kappa^{-1} & 0 \\ 0 & 0 & \kappa^{-1} \end{bmatrix} \cdot \begin{bmatrix} \cos^2\alpha_g & \cos\alpha_g\sin\alpha_g & 0 \\ \cos\alpha_g\sin\alpha_g & \sin^2\alpha_g & 0 \\ 0 & 0 & 0 \end{bmatrix} = \frac{m_g}{I_1 m_c}\begin{bmatrix} \cos^2\alpha_g & \cos\alpha_g\sin\alpha_g & 0 \\ \kappa^{-1}\cos\alpha_g\sin\alpha_g & \kappa^{-1}\sin^2\alpha_g & 0 \\ 0 & 0 & 0 \end{bmatrix}$$

$$(41)$$

The eigenvalues of this matrix can be analytically determined as

$$\mu_1 = \frac{m_g}{I_1 m_c}\left(\cos^2\alpha_g + \kappa^{-1}\sin^2\alpha_g\right) \qquad \mu_2 = 0 \qquad \mu_3 = 0 \qquad (42)$$

from which we see that two eigenvalues are zero as expected from our assumption of a single ground scattering mechanism. The corresponding optimum coherence values are then given as

$$\tilde{\gamma}_i = \frac{\tilde{\gamma}_v + \mu_i}{1 + \mu_i} \tag{43}$$

which we see agrees with equation 39. Hence we have demonstrated that the eigenvalues of the contrast optimisation problem may be used to generate expressions for the optimum coherence values. We now use this result to deal with the more complicated problem of a rank 3 surface coherency matrix.

The results for a random volume over a rank-1 ground surface display how analytical solutions can be obtained (and provide a test for experimental data). However they are not too practical for data inversion studies, as surface scattering often has a coherency matrix with rank larger than one. However, we can use the contrast optimisation formulation to obtain an analytical solution to the problem as follows. The polarimetric coherency matrices for ground and volume can be written as

$$T_V = I_1 m_c \begin{bmatrix} 1 & 0 & 0 \\ 0 & \kappa & 0 \\ 0 & 0 & \kappa \end{bmatrix} \Rightarrow T_v^{-1} = \frac{1}{I_1 m_c} \begin{bmatrix} 1 & 0 & 0 \\ 0 & \kappa^{-1} & 0 \\ 0 & 0 & \kappa^{-1} \end{bmatrix}$$

$$T_g = m_g \begin{bmatrix} A & B\sin c2\beta & 0 \\ B\sin c2\beta & \dfrac{C}{2}(1+\sin c4\beta) & 0 \\ 0 & 0 & \dfrac{C}{2}(1-\sin c4\beta) \end{bmatrix} \tag{44}$$

where $A = \cos^2\alpha$, $B = \sin\alpha\cos\alpha$ and $C = \sin^2\alpha$ and α is the mean surface scattering mechanism. The contrast optimisation matrix is then

$$T_v^{-1}T_g = \frac{m_g}{I_1 m_c} \begin{bmatrix} A & B\sin c2\beta & 0 \\ \dfrac{B}{\kappa}\sin c2\beta & \dfrac{C}{2\kappa}(1+\sin c4\beta) & 0 \\ 0 & 0 & \dfrac{C}{2\kappa}(1-\sin c4\beta) \end{bmatrix} \tag{45}$$

The eigenvalues of this matrix can then be obtained by direct calculation as

$$\mu_1 = \frac{m_g}{2I_1 m_c}\left(A + \frac{C}{2\kappa}(1+\sin c4\beta) + \sqrt{\left(A - \frac{C}{2\kappa}(1+\sin c4\beta)\right)^2 + \frac{4B^2}{\kappa}\sin c^2 2\beta} \right)$$

$$\mu_2 = \frac{m_g}{2I_1 m_c}\left(A + \frac{C}{2\kappa}(1+\sin c4\beta) - \sqrt{\left(A - \frac{C}{2\kappa}(1+\sin c4\beta)\right)^2 + \frac{4B^2}{\kappa}\sin c^2 2\beta} \right) \tag{46}$$

$$\mu_3 = \frac{m_g}{I_1 m_c}\left(\frac{C}{2\kappa}(1-\sin c4\beta) \right)$$

and finally the optimum complex coherences can be written as

$$\tilde{\gamma}_1 = e^{i\phi}\frac{\tilde{\gamma}_v + \mu_1}{1 + \mu_1} \qquad \tilde{\gamma}_2 = e^{i\phi}\frac{\tilde{\gamma}_v + \mu_2}{1 + \mu_2} \qquad \tilde{\gamma}_3 = e^{i\phi}\frac{\tilde{\gamma}_v + \mu_3}{1 + \mu_3} \qquad (47)$$

We note the following important consequences of this result:

- The optimum coherence values are distinct from each other. We have no equality conditions as for the single surface mechanism.
- There is no longer a clean separation of ground and volume components. Hence there is no separability of volume from ground, even in the HV cross polarisation channel (μ_3)
- To obtain estimates of the volume and surface parameters, we need to use all three optimum coherence values in a model based inversion using equation 47

The above formulation can be easily extended to the case where there is a specular interaction between the ground reflection and volume scattering. This is sometimes called a 'dihedral' or double bounce response and occurs generically in radar images of vegetated terrain. The basic geometry is shown in figure 5. Due to its specular nature, such a response can usually be represented by a single polarimetric scattering mechanism. However, because the dielectric constant of the surface and volume particles is generally unknown, the HH/VV ratio is also unknown for this mechanism. For this reason we must add 3 new parameters to the coherency matrix.

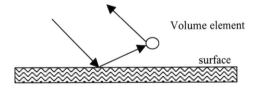

Volume element

surface

Figure 5 : Dihedral Surface/Volume Interaction geometry

The dihedral component then has a rank-1 coherency matrix of the form

$$T_d = m_d \begin{bmatrix} \cos\alpha_d \\ \sin\alpha_d e^{i\chi} \\ 0 \end{bmatrix} \begin{bmatrix} \cos\alpha_d & \sin\alpha_d e^{-i\chi} & 0 \end{bmatrix} =$$

$$m_d \begin{bmatrix} \cos^2\alpha_d & \cos\alpha_d \sin\alpha_d e^{-i\chi} & 0 \\ \cos\alpha_d \sin\alpha_d e^{i\chi} & \sin^2\alpha_d & 0 \\ 0 & 0 & 0 \end{bmatrix} = \begin{bmatrix} D & E & 0 \\ E^* & F & 0 \\ 0 & 0 & 0 \end{bmatrix} \qquad \frac{\pi}{4} \le \alpha_d \le \frac{\pi}{2}$$

$$(48)$$

From an interferometric point of view, the important aspect of such a dihedral response is that the phase centre remains fixed on the surface, even though physically such scattering occurs from the ground and volume. As a consequence of this, the coherence optimisation of equation 45 can be easily modified to accommodate such dihedral effects. The contrast optimisation then becomes

$$T_v^{-1}(T_g+T_d)=\frac{1}{I_1 m_c}\begin{bmatrix} A+D & E+B\operatorname{sinc}2\beta & 0 \\ \frac{1}{\kappa}(E^*+B\operatorname{sinc}2\beta) & \frac{1}{\kappa}(F+\frac{C}{2}(1+\operatorname{sinc}4\beta)) & 0 \\ 0 & 0 & \frac{C}{2\kappa}(1-\operatorname{sinc}4\beta) \end{bmatrix}$$

(49)

The eigenvalues μ of this new matrix can be obtained as a straightforward modification of those in equation 46. We note only that the presence of the dihedral component changes two of the   values and hence has an effect on the observable coherence through the effective ground-to-volume scattering ratio.

5. TWO-LAYER COHERENCE MODEL FOR VEGETATED TERRAIN

According to the 2-layer vegetation model derived above and shown schematically in figure 6, the complex interferometric coherence for a *random* volume over a ground can be written as a function of the polarisation scattering mechanism $\underline{w}$ as shown in equation 50

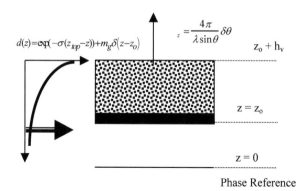

Figure 6: 2-layer Vegetation Coherence Model

$$\hat{\gamma}(\underline{w})=e^{i\phi_1}\frac{\hat{\gamma}_v+\mu(\underline{w})}{1+\mu(\underline{w})}=e^{i\phi_1}\left(\hat{\gamma}_v+\frac{\mu(\underline{w})}{1+\mu(\underline{w})}(1-\hat{\gamma}_v)\right)$$
$$=e^{i\phi_1}\left(\hat{\gamma}_v+L(\underline{w})(1-\hat{\gamma}_v)\right) \qquad 0\leqslant L(\underline{w})\leqslant 1$$

(50)

Interestingly, this is the equation of a straight line in the complex coherence plane, where the parameter L dictates the length of the line. This makes inversion very straightforward. This is such an important observation that several experimental tests have been derived to investigate under what conditions this model is valid [15, 21,22]. Furthermore, in order to find a unique solution when inverting 50, SBPI must be regularised, as there exist a family of height/extinction products that satisfy the observed coherence variations [15,16,18,19,20,29]. One way to do this is to assume that we can always find a $\underline{w}$ for which $L(\underline{w}) = 0$. In this case the coherence is just a function of height and extinction as shown in equation 51

$$\hat{\gamma}_v = \frac{2\sigma \, e^{i\phi}}{\cos\theta_o (e^{2\sigma \, h_v / \cos\theta_o} - 1)} \int_0^{h_v} e^{ik_z z n} e^{\frac{2\sigma \, z n}{\cos\theta_o}} \, dz \, n \tag{51}$$

From our discussion of surface scattering in section IV, we know that this is not true, as surfaces have a rank-3 coherency matrix. However for low entropy surfaces (low frequencies) [30] the ratio of surface to volume scattering can be sufficiently small to validate this assumption.

The ground phase parameter ϕ can then be obtained by using a line fit to the coherence data and extrapolating the line to get the unit circle intersection points. Hence equation 51 is a function of only two unknowns, h_v and σ. Avoiding height ambiguities through appropriate choice of baseline, we then have a unique solution as the intersection of equation 51 with the line (50). [29]

Figure 7 shows a simulation across 3 polarisation channels. The topographic phase is zero in this case. The white circles correspond to the observed optimum coherences. The blue line corresponds to a set of height/extinction pairs that satisfy the model. The reference values are $h_v = 10$m , $\sigma = 0.2$ dB/m and $\mu = [-10$dB -5dB 0dB]. SBPI would try and set $\mu = -\infty$dB and hence overestimate the tree height by around 10% as shown.

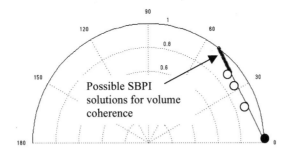

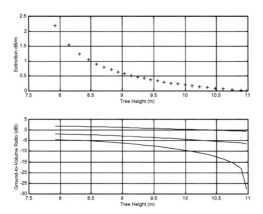

Figure 7 : Ambiguous Height/extinction and ground components in SBPI

6.DUAL BASELINE PROCESSING

The addition of a second baseline provides a change in equation 51 through diversity in the k_z value. All other parameters remain unchanged [5]. Hence our unknown volume coherence term must now satisfy equation 52. One problem is that the ground phase terms are now different, although we can use the fact that the ratio is related to the ratio of baselines.

$$
\hat{\gamma}_{v1} = \frac{2\sigma\, e^{i\phi_1}}{\cos\theta_o(e^{2\sigma\, h_v/\cos\theta_o}-1)} \int_0^{h_v} e^{ik_{z1}z\,\mathrm{n}} e^{\frac{2\sigma\, z\,\mathrm{n}}{\cos\theta_o}}\, dz\,\mathrm{n}
$$

$$
\hat{\gamma}_{v2} = \frac{2\sigma\, e^{i\phi_2}}{\cos\theta_o(e^{2\sigma\, h_v/\cos\theta_o}-1)} \int_0^{h_v} e^{ik_{z2}z\,\mathrm{n}} e^{\frac{2\sigma\, z\,\mathrm{n}}{\cos\theta_o}}\, dz\,\mathrm{n}
$$

(52)

We can interpret the dual baseline solution geometrically with reference to figure 7. By stepping along the line of possible volume coherence solutions for the first baseline, evaluating each of the solution pairs h_v,σ and then calculating the corresponding coherence for the second baseline, we can verify that both equations in 52 are satisfied. Note that this process is independent of μ and hence we are free of the assumptions required for SBPI.

Figure 8 shows a simulation with the addition of a second baseline twice as large as the first (to mimic a 'ping-pong' interferometer configuration) and 4 times as large. For clarity μ is now increased by 5dB to [−5dB 0dB 5dB] and we have offset the ground phase points. The line fits are shown, together with the 6 complex coherence points. In the noise-free case we achieve the correct solution of $h_v = 10$m, $\sigma = 0.2$ dB/m and $\mu_{min} = $ -5dB. However, we see that the mapped second baseline solution is nearly parallel to the line fit, especially around the intersection point. This is not a well-conditioned situation. To secure reliable estimation we therefore need to use a much larger second baseline. This can lead to operational difficulties, especially in single pass interferometer designs. An alternative approach is to consider use of a second frequency as we now consider.

7.DUAL FREQUENCY REGULARISATION

We have seen that dual baseline techniques can be used to help resolve several problems with SBPI. However large baseline ratios are required to avoid ill-conditioning. An alternative approach is to employ an extra frequency channel of interferometry. For example SBPI at L-band may be augmented by an X-band single polarisation channel. Dual wavelength interferometers are now operational, an important example being the GeoSAR airborne sensor, which operates at X/P bands simultaneously. In space applications the SRTM data base provides X-band coherence data for a large part of the globe. With the future launch of L-band polarimetric satellites such dual frequency POLInSAR data may then become available over large areas. Here we examine the possibility of using this extra channel (2 observables, being the phase and coherence in one polarisation channel) to help regularise SBPI.

A common assumption is that a high frequency such as X-band has negligible penetration and sees the top of the canopy. If this is the case then with SBPI we can find the ground phase centre from the extrapolated line and by using a difference of DEMs simply obtain an estimate of height without the need for model inversion. However such an assumption is not robust, as there can still be significant penetration even at X-band depending on canopy structure and tree species. Hence in general we must consider the impact of a change of frequency on the 2-layer model. We make the following assumptions:

1) We assume that for forest scattering at the higher frequency the canopy is random so that the extinction,although unknown, can be assumed scalar.
2) We assume that the frequency is high enough so that ground scattering can be ignored. Hence by combining with assumption 1 and equation 18 we are led to the conclusion that only 1 polarisation channel need be measured at the high frequency.
3) Parameter estimation then requires inversion of the following set of equations

$$\hat{\gamma}(f_1) = e^{i\phi_1}\frac{\hat{\gamma}_v(f_1) + \mu(w)}{1 + \mu(w)} \quad \hat{\gamma}_v(f_1) = \frac{2\sigma_1\,e^{ik_{z1}z_o}}{\cos\theta_o(e^{2\sigma_1\,h_v/\cos\theta_o} - 1)}\int_0^{h_v} e^{ik_{z1}z\,n}e^{\frac{2\sigma_1\,z\,n}{\cos\theta_o}}\,dz\,n$$

$$\hat{\gamma}(f_2) = \frac{2\sigma_2\,e^{ik_{z2}z_o}}{\cos\theta_o(e^{2\sigma_2\,h_v/\cos\theta_o} - 1)}\int_0^{h_v} e^{ik_{z2}z\,n}e^{\frac{2\sigma_2\,z\,n}{\cos\theta_o}}\,dz\,n \tag{53}$$

The first is just SBPI at the lower frequency, where combined ground and volume scattering is observed. There are 2 observables and 4 unknowns, h_v, σ_1, μ and z_o. By SBPI line fitting we can obtain an estimate of z_o and hence reduce to 3 unknowns.

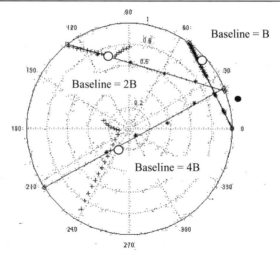

Figure 8 : Multi- Baseline Volume Coherence Estimation
O = volume coherence

By adding a second frequency we add two observables and only one unknown (σ_2). Hence we have 4 observables and 4 unknowns. Further information can be used to solve these equations. For example it would seem sensible to assume that $\sigma_2 > \sigma_1$ and in general we might try and employ some dispersion relation for the change in extinction with frequency. In general the extinction is composed of absorption and scattering contributions and we can summarise the frequency dependence as

$$\sigma_e = \sigma_a + \sigma_s \quad \begin{cases} \sigma_a \propto f \\ \sigma_a \propto f^m \quad 2 \leq m \leq 4 \end{cases} \Rightarrow \quad \sigma_e \propto f^x \quad 1 \leq x \leq 2 \tag{54}$$

For Rayleigh scattering the extinction increases approximately linearly with frequency. However at high frequencies such as X-band the Rayleigh approximation is no longer valid and so the extinction increases at a faster rate. In dual frequency POLInSAR it may become possible to obtain estimates of the ratio of these extinctions and hence important information about the structure of the vegetation canopy.

8. CONCLUSIONS

In this paper we have developed a multi-layer scattering model for the interferometric coherence of vegetated terrain. We have shown that single baseline polarimetric interferometry can be used for parameter estimation using a 2-layer model but that it needs regularisation by assuming a small ground component in one polarisation channel. We have considered two methods for overcoming this. Two baselines can be used to solve for all polarimetric ground components. However, careful choice of the baseline ratio is required to avoid poorly conditioned inversion. As an alternative we have considered the use of a second frequency to regularise the inversion process. By assuming that the frequency is high enough to see only volume decorrelation we have shown that regularisation is possible. These results have important implications for the design of dual wavelength/multi-baseline polarimetric interferometers for vegetation parameter mapping applications. The most mature current application of POlInSAR techniques is in tree height measurement for forestry and biomass estimation [17,26,27,28]. Here, its main advantage over conventional amplitude radar is the lack of saturation at high biomass levels. This makes it suitable for consideration in continental and global scale carbon missions. Studies have also shown that recent developments in low cost satellite technology could make it economically viable to deploy such advanced space borne interferometer designs in the near future [26,28].

REFERENCES

[1] CLOUDE S. R., PAPATHANASSIOU K.P., "Polarimetric Optimisation in Radar Interferometry", *Electronics Letters*, Vol. 33, N0. 13, June 1997, pp. 1176-1178

[2] CLOUDE S R, K P PAPATHANASSIOU, E POTTIER, "Radar Polarimetry and Polarimetric Interferometry", IEICE Transactions on Electronics, VOL.E84-C, No.12, 2001, pp1814-1822, December 2001

[3] CLOUDE S. R., PAPATHANASSIOU K.P, "Polarimetric SAR Interferometry", *IEEE Transactions Geoscience and Remote Sensing.*, Vol. GRS-36. No. 5, September 1998, pp. 1551-1565

[4] PAPATHANASSIOU K.P., CLOUDE S. R., "Single Baseline Polarimetric SAR Interferometry", *IEEE Transactions Geoscience and Remote Sensing*, Vol 39/11, November 2001, pp 2352-2363

[5] CLOUDE, S R, "Robust Parameter Estimation using dual baseline polarimetric SAR Interferometry", Proceedings of IEEE International Geoscience and Remote Sensing Symposium (IGARSS 2002), Toronto, Canada, Vol.2, pp 838-840, July 2002

[6] CLOUDE S.R. "POlInSAR Regularisation using Dual Frequency Interferometry", submitted to International Radar Conference, Radar2003, Adelaide, Australia, 3-5 September, 2003

[7] TREUHAFT R. N., MADSEN S., MOGHADDAM M., VAN ZYL J.J.,"Vegetation Characteristics and Underlying Topography from Interferometric Data", *Radio Science*, Vol. 31, Dec, 1996, pp. 1449-1495

[8] TREUHAFT R.N., SIQUERIA P., "Vertical Structure of Vegetated Land Surfaces from Interferometric and Polarimetric Radar", *Radio Science*, Vol. 35(1), January 2000, pp 141-177

[9] TREUHAFT R.N., CLOUDE, S.R., "The Structure of Oriented Vegetation from Polarimetric Interferometry", *IEEE Transactions Geoscience and Remote Sensing*, Vol. 37/2, No. 5, September 1999, pp 2620-2624

[10] CLOUDE S. R., PAPATHANASSIOU K.P,BOERNER W.M., "The Remote Sensing of Oriented Volume Scattering Using Polarimetric Radar Interferometry", Proceedings of ISAP 2000, Fukuoka, Japan, August 2000, pp 549-552

[11] LEE J.S.,HOPPEL K.W., MANGO S.A., MILLER A., "Intensity and Phase Statistics of Multi-Look Polarimetric and Interferometric SAR Imagery", *IEEE Transactions Geoscience and Remote Sensing.* GE-32, 1994, pp. 1017-1028

[12] REIGBER A., MOREIRA A., "First Demonstration of Airborne SAR Tomography Using Multi-Baseline LBand Data", *IEEE Transactions on Geoscience and Remote Sensing*, Vol. 38/5, September 2000, pp 2142-2152

[13] REIGBER A., "Airborne Polarimetric SAR Tomography", DLR report ISRN DLR-FB-2002-02, 2002

[14] TOUZI R, LOPES A., BRUNIQUEL J, VACHON P.W., "Coherence Estimation for SAR Imagery", *IEEE Transactions Geoscience and Remote Sensing,* VOL. GRS-37, Jan 1999, pp 135-149

[15] ISOLA M., CLOUDE, S.R., "Forest Height Mapping using Space Borne Polarimetric SAR Interferometry", CD Proceedings of IEEE International Geoscience and Remote Sensing Symposium (IGARSS 2001), Sydney, Australia, July 2001

[16] TABB M., CARANDE R., "Robust Inversion of Vegetation Structure Parameters from Low-Frequency, Polarimetric Interferometric SAR", CD Proceedings of IEEE-IGARSS 2001, Sydney Australia

[17] CLOUDE S.R., WOODHOUSE I.H., HOPE J., SUAREZ-MINGUEZ J.C., OSBORNE P., WRIGHT G., "The Glen Affric Radar Project: Forest Mapping using dual baseline polarimetric radar interferometry", Proceedings of Symposium on "Retrieval of Bio and Geophysical Parameters from SAR for Land Applications", University of Sheffield, England, September 11-14, 2001, ESA publication SP-475, pp 333-338

[18] ULBRICHT A., FABREGAS X., SAGUES L., "Applying Polarimetric Interferometric Methods to Invert Vegetation Parameters from SAR Data", CD Proceedings of the IEEE International Geoscience and Remote Sensing Symposium (IGARSS), 9-13 July, Sydney, Australia, 2001

[19] ULBRICHT A., FRABREGAS X., CASAL M., "Experimental and Theoretical Aspects of Inversion of Polarimetric Interferometric Data", Proceedings of the Open Symposium on Propagation and Remote Sensing, URSI, Commission F, 12-15 February, Garmisch-Partenkirchen, Germany, 2002

[20] BRANDFASS M., HOFMANN C., MURA J.C., MOREIRA J., PAPATHANASSIOU K.P., "Parameter estimation of Rain Forest Vegetation via Polarimetric Radar Interferometric Data", Proceedings of SPIE 2001, Toulouse, France, August 2001

[21] FLYNN T., TABB M., CARANDE R., "Coherence region Shape Estimation for Vegetation Parameter Estimation in POLINSAR", Proceedings of IGARSS 2002, Toronto, Canada, pp V 2596-2598

[22] TABB M., FLYNN T., CARANDE R., "Direct Estimation of Vegetation Parameters from Covariance Data in POLINSAR", Proceedings of IGARSS 2002, Toronto, Canada, pp III 1908-1910

[23] YAMADA H., SATO K., YAMAGUCHI Y., BOERNER W.M. "Interferometric Phase and Coherence of Forest Estimated by ESPRIT based POLINSAR", Proceedings of IGARSS 2002, Toronto, Canada, pp II 832-834

[24] LEE J.S., CLOUDE S.R., PAPATHANASSIOU K.P., GRUNES M.R., AINSWORTH T., " Speckle Filtering of POLINSAR Data", Proceedings of IGARSS 2002, Toronto, Canada, pp II 829-831

[25] HAJNSEK I. , "Inversion of Surface Parameters Using Polarimetric SAR", DLR-Science Report, vol. 30, ISSN 1434-8454, 2001, p. 224

[26] PAPATHANASSIOU K.P., HAJNSEK I., MOREIRA A., CLOUDE S.R., "Forest Parameter Estimation using a Passive Polarimetric Micro-Satellite Concept", Proceedings of European Conference on Synthetic Aperture Radar, EUSAR'02, pp. 357-360, Cologne, Germany, 4-6 June 2002.

[27] METTE T., PAPATHANASSIOU K.P., HAJNSEK I., and ZIMMERMANN R., "Forest Biomass Estimation using Polarimetric SAR Interferometry", Proceedings IGARSS'02 (CD-ROM), Toronto, Canada, 22-26 June 2002

[28] PAPATHANASSIOU K.P., METTE T., HAJNSEK I, KRIEGER G. and MOREIRA A., "A Passive Polarimetric Micro-Satellite Concept for Global Biomass Mapping", Proceedings of the PI-SAR Workshop (CD-ROM), Tokyo, Japan, 29-30 August 2002

[29] CLOUDE S R, K.P. PAPATHANASSIOU, A 3-Stage Inversion Process for Polarimetric SAR Interferometry, *Proceedings of European Conference on Synthetic Aperture Radar*, EUSAR'02, pp. 279-282, Cologne, Germany, 4-6 June 2002

[30] CLOUDE S R, POTTIER E, "An Entropy Based Classification Scheme for Land Applications of Polarimetric SAR", *IEEE Transactions on Geoscience and Remote Sensing*, Vol. 35, No. 1, pp 68-78, January 1997

Wave Propagation,Scattering and Emission in Complex Media
Edited by Ya-Qiu Jin
Science Press and World Scientific,2004

29

Terrain Topographic Inversion from Single-Pass Polarimetric SAR Image Data by Using Polarimetric Stokes Parameters and Morphological Algorithm

Ya-Qiu Jin, Lin Luo

Center for Wave Scattering and Remote Sensing

Fudan University, Shanghai 200433, China

Abstract   It has been studied that the shift of polarization orientation angle ψ at the maximum of co-polarized or cross-polarized back-scattering signature can be converted to the tilted-surface slopes. It is then utilized to generate the digital elevation mapping (DEM) and terrain topography using two-pass fully polarimetric SAR or interferometric SAR (INSAR) image data.

This paper, using the Mueller matrix solution, newly derives the ψ shift as a function of three Stokes parameters, I_{vs}, I_{hs}, U_s, which are measured by polarimetric SAR image data. Using the Euler angles transformation, the orientation angle ψ is related to both the range and azimuth angles of the tilted surface and radar viewing geometry. When only a single-pass SAR data is available, the adaptive thresholding method and image morphological thinning algorithm for linear textures are proposed to first determine the azimuth angle. Then, making use of full multi-grid algorithm, both the range and azimuth angles are utilized to solve the Poisson equation of DEM to produce the terrain topography.

Key Words   fully polarimetric Stokes parameter, the ψ shift, azimuth angle and range angle, terrain topography DEM

1. INTRODUCTION

Fully polarimetric SAR imagery has been extensively studied for terrain surface classification. The 2×2-D (Dimensional) complex scattering amplitude functions F_{pq} $(p,q=v,h)$, and 4×4-D real Mueller matrix M_{ij} $(i,j=1,...,4)$ can be measured [1]. Co-polarised or cross-polarised back-scattering signature is the function of the incidence wave with the ellipticity angle χ and orientation angle ψ. Recently, polarimetric SAR image data has been utilized to generate digital surface elevation and to invert terrain topography. When the terrain surface is flat, polarimetric scattering signature has the maximum largely at the orientation angle $\psi = 0$. However, it has been shown that as the surface is tilted, the orienattion angle ψ at the maximum of co-polarized (co-pol) or cross-polarized (cross-pol) signature can shift from $\psi = 0$. This shift can be applied to convert the surface slopes. Making use of the assumption of real co-pol and zero cross-pol scattering amplitude functions [2], the ψ shift is expressed by the real scattering amplitude functions. Since both the range and azimuth angles are coupled, two- or multi-pass SAR image data are required for solving two unknowns of the surface slopes. This approach has been well demonstrated for inversion of digital elevation mapping (DEM) and terrain topography by using airborne SAR data [3-5].

However, scattering signature is an ensemble average of echo power from random scatter media. Measurable Stokes parameters as the polarized scattering intensity should be directly related to the ψ shift. This paper, using the Mueller matrix solution, newly derives the ψ shift as a function of three Stokes parameters, I_{vs}, I_{hs}, U_s, which are measurable by the polarimetric SAR imagery. Using the Euler angles transformation between the principle and local coordinates, the orientation angle ψ is related with both the range and azimuth angles, β and γ, of the tilted surface pixel and radar viewing geometry. These results are consistent with [2] and [3], but are more general. When only the single-pass SAR data are available, the linear texture of tilted surface alignment may specify the azimuth angle γ. The adaptive threshold method and image morphological thinning algorithm are proposed to determine the azimuth angle γ from image linear textures. Thus, the range angle β is then solved, and both β and γ are utilized to obtain the azimuth slope and range slope. Then, the full multi-grid algorithm is employed to solve the Poisson equation of DEM and produce the terrain topography.

2. THE SHIFT OF ORIENTATION ANGLE AS A FUNCTION OF STOKES PARAMETERS

Consider a polarized electromagnetic wave incident upon the terrain surface at $(\pi - \theta_i, \varphi_i)$. Incidence polarization is defined by the ellipticity angle χ and orientation angle ψ [1]. The 2×2-D complex scattering amplitude functions are obtained from polarimetric measurement as

$$\overline{E}_s(\theta_s,\varphi_s) = \frac{e^{ikr}}{4\pi r}\overline{\overline{F}}(\theta_s,\varphi_s;\pi-\theta_i,\varphi_i)\cdot\overline{E}_i(\pi-\theta_i,\varphi_i;\chi,\psi) \tag{1}$$

where the elements of scattering amplitude matrix, F_{pq} ($p,q=v,h$), take account for volumetric and surface scattering from layered scatter media. From Eq.(1), the 4×4-D real Mueller matrix solution for the Stokes vector can be obtained as [1]

$$\overline{I}_s(\theta_s,\varphi_s) = \overline{\overline{M}}(\theta_s,\varphi_s;\pi-\theta_i,\varphi_i)\cdot\overline{I}_i(\pi-\theta_i,\varphi_i;\chi,\psi) \tag{2}$$

where the normalized incident Stokes vector is written as

$$\overline{I}_i(\chi,\psi) = [I_{vi}, I_{hi}, U_i, V_i]^T$$
$$= [0.5(1-\cos 2\chi\cos 2\psi), 0.5(1+\cos 2\chi\cos 2\psi), -\cos 2\chi\sin 2\psi, \sin 2\chi]^T \tag{3}$$

The elements of the Mueller matrix, M_{ij} ($i,j=1,...,4$), are expressed by the ensemble averages of real or imagery parts of $<F_{pq}F_{st}^*>$ ($p,q,s,t=v,h$). The formulations of M_{ij} can be found in [1]. It is noted that co-pol terms $<|F_{pp}|^2>$ are always much larger than cross-pol terms, such as $<F_{pq}F_{st}^*>$ ($s\neq t$).

The co-pol backscattering coefficient $\sigma_c(\theta_i,\pi-\varphi_i;\pi-\theta_i,\varphi_i)$ can be obtained as [1]

$$\sigma_c = 4\pi\cos\theta_i P_n \tag{4}$$

where

$$P_n = 0.5[I_{vs}(1-\cos 2\chi\cos 2\psi)+I_{hs}(1+\cos 2\chi\cos 2\psi)+U_s\cos 2\chi\sin 2\psi+V_s\sin 2\chi] \tag{5}$$

When the terrain surface is flat, co-pol back-scattering σ_c versus the incidence polarization (χ,ψ) has the maximum at $\psi=0$ [2]. However, it has been shown that as the surface is tilted, the orientation angle ψ for the σ_c maximum is shifted from $\psi=0$ [2,3,4,5].

Let $\partial P_n/\partial\psi=0$ at the maximum σ_c and $\chi=0$ of symmetric case, from Eq.(5), it yields

$$0 = -(I_{hs}-I_{vs})\sin 2\psi + U_s\cos 2\psi + 0.5(I_{hs}+I_{vs})'+0.5U_s'\sin 2\psi + 0.5(I_{hs}-I_{vs})'\cos 2\psi \tag{6}$$

where the superscript of prime denote $\partial/\partial\psi$. Calculations of Eq. (2) and comparison with the terms on RHS (right hand side) of Eq. (6) can show that

$$0.5(I_{hs}+I_{vs})' \sim (M_{13}+M_{23})\cos 2\psi \sim (\text{Re}<F_{vv}F_{vh}^*>+\text{Re}<F_{hv}F_{hh}^*>)\cos 2\psi \tag{7a}$$

$$0.5U'_s \sim M_{33}\cos 2\psi \sim \text{Re} < F_{vv}F_{hh}^* + F_{vh}F_{hv}^* > \cos 2\psi \tag{7b}$$

$$(I_{hs} - I_{vs}) \sim 0.5(M_{11} + M_{22})\cos 2\psi \sim 0.5(<|F_{vv}|^2 > + <|F_{hh}|^2 >)\cos 2\psi \tag{7c}$$

$$0.5(I_{hs} - I_{vs})' \sim 0.5(M_{13} - M_{23})\cos 2\psi \sim 0.5(\text{Re} < F_{vv}F_{vh}^* > - \text{Re} < F_{hv}F_{hh}^* >)\cos 2\psi \tag{8a}$$

$$U_s \sim 0.5(M_{31} + M_{32}) \sim \text{Re} < F_{vv}F_{hv}^* > + \text{Re} < F_{vh}F_{hh}^* > \tag{8b}$$

It can be seen that (7a) and (7b) are much less than (7c), and (8a) is much less than (8b), so the last three terms on RHS of Eq. (6) are now neglected. Thus, it yields the ψ shift at the σ_c maximum expressed by the Stokes parameters as follows

$$\tan 2\psi = \frac{U_s}{I_{hs} - I_{vs}} \tag{9}$$

It can be seen that the third Stokes parameter $U_s \neq 0$ does cause the ψ shift.

By the way, if both U_s and $I_{hs} - I_{vs}$ approach zero, e.g. scattering from uniformly oriented scatterers or isotropic scatter media such as thick vegetation canopy, ψ cannot be well defined by Eq.(9).

Furthermore, fom Eq.(2), it yields

$$U_s \sim 0.5(M_{31} + M_{32}) = \text{Re}(< F_{vv}F_{hv}^* > + < F_{vh}F_{hh}^* >) \tag{10a}$$

$$I_{hs} - I_{vs} \sim 0.5(M_{22} - M_{11}) = 0.5(<|F_{hh}|^2 > - <|F_{vv}|^2 >) \tag{10b}$$

It is interesting to see that if let

$$\text{Re}(< F_{vv}F_{hv}^* > + < F_{vh}F_{hh}^* >) = 2F_{hv}F_0 \text{ and } 0.5(<|F_{hh}|^2 > - <|F_{vv}|^2 >) \sim F_0(F_{hh} - F_{vv}),$$

where $F_0 = F_{vv} = F_{hh}$ is assumed, Eq. (9) can be reduced to the result of [3],

$$\tan 2\psi = \frac{2F_{vh}}{F_{hh} - F_{vv}} \tag{11}$$

Using the cross-pol backscattering coefficient σ_x, the same result can be obtained.

3. THE RANGE AND AZIMUTH ANGLES FROM EULER ANGLES TRANSFORMATION

The polarization vectors $\hat{h}_i, \hat{v}_i$ of the incident wave at $(\pi - \theta_i, \phi_i)$ in the principle coordinate $(\hat{x}, \hat{y}, \hat{z})$ are defined as [1]

$$\hat{h}_i = \frac{\hat{z} \times \hat{k}_i}{|\hat{z} \times \hat{k}_i|} = -\sin\phi_i \hat{x} + \cos\phi_i \hat{y} \text{ and } \hat{v}_i = \hat{h}_i \times \hat{k}_i \tag{12}$$

where the incident wave vector

$$\hat{k}_i = \sin\theta_i \cos\phi_i \hat{x} + \sin\theta_i \sin\phi_i \hat{y} - \cos\theta_i \hat{z} \tag{13}$$

As the pixel surface has local slope as shown in Figure 1, the polarization vectors should be re-defined following the local normal vector $\hat{z}_b$ as follows

$$\hat{h}_b = \frac{\hat{z}_b \times \hat{k}_{ib}}{|\hat{z}_b \times \hat{k}_{ib}|} \text{ and } \hat{v}_b = \hat{h}_b \times \hat{k}_{ib} \tag{14}$$

By using the transformation of the Euler angles (β, γ) between two coordinates $(\hat{x}, \hat{y}, \hat{z})$ and $(\hat{x}_b, \hat{y}_b, \hat{z}_b)$ [1], it has

$$\hat{x} = \cos\gamma \cos\beta \hat{x}_b + \sin\gamma \hat{y}_b - \cos\gamma \sin\beta \hat{z}_b \tag{15a}$$

$$\hat{y} = -\sin\gamma \cos\beta \hat{x}_b + \cos\gamma \hat{y}_b + \sin\gamma \sin\beta \hat{z}_b \tag{15b}$$

$$\hat{z} = \sin\beta \hat{x}_b + \cos\beta \hat{z}_b \tag{15c}$$

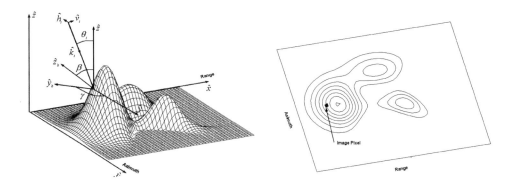

Figure 1. Incidence upon a pixel of tilted surface and the contour to be inverted

Substituting above relations into Eqs. (12,13), it yields

$$\hat{h}_i = -\cos\beta \sin(\gamma+\phi_i)\hat{x}_b + \cos(\gamma+\phi_i)\hat{y}_b + \sin\beta \sin(\gamma+\phi)\hat{z}_b \tag{16}$$

$$\hat{k}_{ib} = (\cos\beta \sin\theta_i \cos(\gamma+\phi_i) - \sin\beta \cos\theta_i)\hat{x}_b + \sin\theta_i \sin(\gamma+\phi_i)\hat{y}_b \\ -[\sin\beta \sin\theta_i \cos(\gamma+\phi_i) + \cos\beta \cos\theta_i]\hat{z}_b \tag{17}$$

Using Eq.(17) to Eq.(14), the polarization vector $\hat{h}_b$ for local surface pixel is written as

$$\hat{h}_b = \frac{a\hat{x}_b + b\hat{y}_b}{\sqrt{a^2+b^2}} \tag{18}$$

where

$$a = -\sin\theta_i \sin(\gamma+\phi_i) \\ b = \sin\theta_i \cos\beta \cos(\gamma+\phi_i) - \sin\beta \cos\theta_i \tag{19}$$

Thus, Eqs.(16,18) yield the orientation angle as

$$\cos\psi = \hat{h}_b \cdot \hat{h}_i = \frac{\cos\beta \sin\theta_i - \cos(\gamma+\phi_i)\sin\beta \cos\theta_i}{\sqrt{a^2+b^2}} \tag{20a}$$

$$\tan\psi = \frac{\tan\beta \sin(\gamma+\phi_i)}{\sin\theta_i - \cos\theta_i \tan\beta \cos(\gamma+\phi_i)} \tag{20b}$$

Taking $\phi_i = 0$, i.e. $\hat{x}$ is the range direction and $\hat{y}$ is the azimuth direction, Eq.(20b) is reduced to [3]. Note that the azimuth angle in [2] was defined by the angle between $\hat{\rho}$ and $\hat{y}$. Thus, the range and azimuth slopes of the pixel surface can be defined as [2]

$$S_R = \tan\beta \cos\gamma \\ S_A = \tan\beta \sin\gamma \tag{21}$$

Since a single ψ shift cannot simultaneously determine two unknowns of β and γ in Eq. (20b), two- or multi-pass SAR image data are usually needed. It has been discussed in [2].

4. THE AZIMUTH ANGLE OF EVERY PIXEL IN SAR IMAGE

If only the single-pass SAR image data are available, one of two unknown angles, β or γ, should be first determined. The azimuth alignment of tilted surface pixels can be visualized as a good indicator of the azimuth direction. We apply the adaptive threshold method and image morphological thinning algorithm [6] to specify the azimuth angle in all SAR image pixels. The algorithm contains the following steps:

    (a) Make speckle filtering over the entire image.

(b) Apply the adaptive threshold method [7] to produce a binary image (see Fig.2(a)). The global threshold value is not adopted because of the heterogeneity of the image pixels.

(c) Apply the image morphological processing for the binary image, remove those isolated pixels and fill small holes. Referring to the part of binary's "1" as the foreground and the part of binary's "0" as the background, the edges from the foreground are extracted (see Fig.2(b)).

(d) Each pixel on the edge is set as the center of a square 21×21 window (see Fig.2(c)), and a curve segment through the centered pixel is then obtained. Then, applying the polynomial algorithm for fitting curve segment in the least-squares sense, the tangential slope of the centered pixel is obtained (see Fig.2(d)). It yields the azimuth angle of the centered pixel. Make a mark on that pixel so that it won't be calculated in the next turn.

(e) Removing the edge in Step (d) from the foreground, a new foreground is formed. Repeat Step (d) until the azimuth angle of every pixel in the initial foreground is determined.

(f) Make the complementary binary image, i.e. the initial background now becomes the foreground (see the gray line in Fig.2(e)). Then, the Steps (d) and (e) are repeated to this complementary image until the azimuth angle of every pixel in the initial background is determined.

This approach provides a supplementary information to firstly determine the angle γ over whole image areaif there is no other information available.

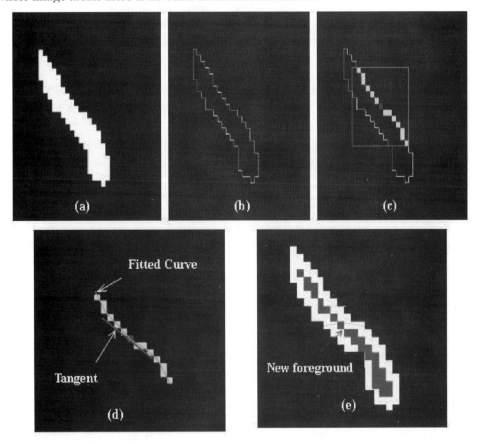

Figure 2. Morphological thinning algorithm to determine the azimuth angle

5. AN EXAMPLE OF ALGORITHM IMPLEMETATION

As an example for DEM inversion, the L-band polarimetric SAR data on October 4, 1994 over Taiwan, China measured by SIR-C of the Cal-Tech JPL is investigated. The center location is about ($120.964° E$, $23.487° N$) under $\theta_i = 32.85°$, and the pixel size is 12.5×12.5 m^2. Figure 3 is a 512px × 512px part from the original SAR image. The center location of Figure 3 is ($120.8626° E$, $23.6659° N$) situated in the Ali Mountain of Taiwan.

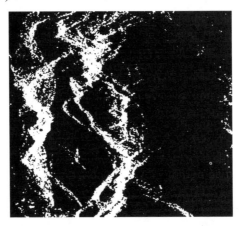

Figure 3. SAR image of total power *hh, vv, hv* at L band

To focus on the prominent topographic structure for DEM, some trivial undulations are omitted. Note that the azimuth angle γ is the angle between the slope direction $\hat{y}_b$ and the azimuth direction $\hat{y}$, as shown in Figure 1.

As the azimuth angle γ of each pixel is obtained by the adaptive threshold method and thinning algorithm as described in the above steps, the β angle of each pixel can be determined by Eq. (20b). Taking $\phi_i = 0$, it yields

$$\tan \beta = \frac{\tan \psi \sin \theta_i}{\sin \gamma + \tan \psi \cos \theta_i \cos \gamma} \tag{22}$$

where the orientation angle ψ is calculated using Eq. (9), while the incident angle θ_i is determined by the SAR view geometry.

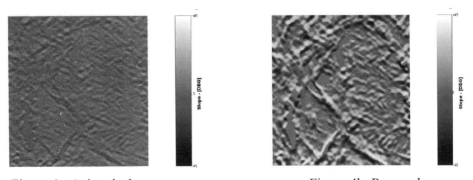

Figure 4a. Azimuth slope   Figure 4b. Range slope

Substituting β and γ into Eq. (21), the azimuth slope S_A and range slope S_R are obtained Following the approach of Refs. [10-13], we can utilize the two slopes to invert the terrain shown in Figures 4a and 4b.

The DEM can be generated by solving the Poisson equation for a $M \times N$ rectangular grid area [2,5]. The approach is similar to the phase unwrapping of INSAR [8,9,10]. The Poisson equation can be written as

$$\nabla^2 \phi(x, y) = \rho(x, y) \tag{23}$$

where ∇^2 is the Laplace operator $\partial^2 / \partial x^2 + \partial^2 / \partial y^2$. The source function $\rho(x, y)$ consists of the surface curvature calculated by the slopes $S(x, y)$. Taking the $\hat{x}$-direction as the range direction and $\hat{y}$-direction as the azimuth direction, the discrete values of the source function can be derived (the superscript A and R represent the azimuth and range directions, respectively) as follows

$$\rho_{ij}^R = S_{ij}^R - S_{i-1,j}^R, \quad \rho_{ij}^A = S_{ij}^A - S_{i,j-1}^A \quad (0 \le i \le M, 0 \le j \le N) \tag{24}$$

Assuming that the errors in the input slopes are Gaussian random number, the integrated elevation values also have Gaussian errors. Therefore, a χ^2 statistics is a good measure of matching the solution to the exact value [5]. This is also the standard assumption of the phase unwrapping algorithm of INSAR. χ^2 is given by

$$\chi^2 = \sum_{i,j} (\phi_{ij} - \phi_{i-1,j} - S_{ij}^R)^2 + \sum_{i,j} (\phi_{ij} - \phi_{i,j-1} - S_{ij}^A)^2 \tag{25}$$

The differences between the partial derivatives of the solution ϕ_{ij} and partial derivatives defined by Eq. (24) should be minimized. Differentiating χ^2 with respect to ϕ_{ij} and setting the results equal to zeros, it yields

$$(\phi_{i+1,j} - 2\phi_{ij} + \phi_{i-1,j}) + (\phi_{i,j+1} - 2\phi_{ij} + \phi_{i,j-1}) = \rho_{ij} \tag{26}$$

Eq. (26) presents the Poisson equation to be solved.

The discrete source function ρ_{ij} is defined by

$$\rho_{ij} = (S_{i+1,j}^R - S_{ij}^R) + (S_{i,j+1}^A - S_{ij}^A) \tag{27}$$

In [2,5] of INSAR approach, the ADI (Alternating Direction Iteration) algorithm was applied to generate the DEM. In this paper, the FMG (full multi-grid) algorithm [11,12] is employed to solve the Poisson equation. The benefits of FMG are due to its rapid-convergence, robustness and low computation load. For a $M \times N$ rectangular grid, the FMG method has computations on the order of $(MN)\ln(MN)$ [11]. The detailed steps of FMG are referred to Ref. [11].

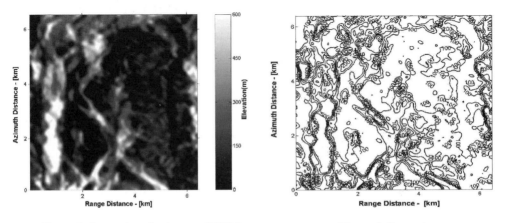

Figure 5. the psedo-color image of DEM   Figure 6. the contour map

Figure 7. a picture of inverted DEM

Figure 5 is a picture of the inverted DEM. Figure 6 is the contour map, and Figure 7 presents the pesudocolor image of DEM.

It won't be surprised that there is a similarity between the inverted topography (Fig.5) and the SAR image (Fig.3). Also, the inverted contour map has been validated in comparison to the map of Ref. [13]. But, because more accurate DEM as ground truth is not yet available, the inverted topography has been only roughly verified.

6. CONCLUSIONS

The shift of orientation angle at the maximum of polarimetric backscattering signature is derived as a function of the Stokes parameters, without the assumption of real co-pol and zero cross-pol scattering amplitude functions. It shows that the third Stokes parameter $U_s \neq 0$ causes the ψ shift.

Using the Euler angles transformation, the orientation angle is related with the range and azimuth angles of the tilted surface pixels. It bridges the polarized echo power with the polarimetric ψ shift and terrain topography. When only the single-pass SAR image data are available, the adaptive thresholding method and image morphological thinning algorithm are developed to specify the linear texture, which determine the azimuth angle of each pixel. Two unknowns of the azimuth and range angles are obtained.

Using the full multi-grid algorithm, both the range and azimuth angles are utilized to solve the Poisson equation of DEM and produce the terrain topography. An example of DEM inverted from a single-pass SAR SIR-C data is demonstrated.

ACKNOWLEDGMENTS

This work was supported by the China National Key Basic Research Program 2001CB309401, -5 and NSFC, 60171009.

REFERENCES

1. Jin Y. Q., *Electromagnetic scattering modeling for quantitative remote sensing*, 1994, (Singapore: World Scientific)
2. Lee J.S., Jansen, R.W., Schuler, D.L., Polarimetric analysis and modeling of multi-frequency SAR signatures from Gulf stream fronts, *IEEE Journal of Oceanic Engineering*, 1998, 23: 322-333
3. Schuler, D.L., Lee, J.S., Answorth, T.L., and Grunes, M.R., Terrain topography measurements using multipass polarimetric synthetic aperture radar data, *Radio Science*, 2000, 35: 813-832

4. Lee, J.S., Schuler, D.L. and Answorth, T.L., Polarimetric SAR data compression for terrain azimuthal slope variation, *IEEE Transaction on Geoscience Remote Sensing*, 2000, 38(5): 2153-2163

5. Schuler, D.L., Answorth, T.L., and Lee J.S., Topographic mapping using polarimetric SAR data, *Int. J. Remote Sensing*, 1998, 19(1): 141-160

6. Castleman, K.R., *Digital image processing*, 1996, (Printice Hall).

7. Francis H.Y.Chan, F.K.Lam and Hui Zhu, Adaptive Thresholding by Variational Method, *IEEE Transaction on Geoscience Remote Sensing*, 1998, 7(3): 468-473

8. H.Takajo and T.Takahashi, Least-squares phase estimation from phase difference, *J. Opt. Soc. Amer. A*, 1998, 5: 416-425

9. D. C. Ghiglia and L. A. Romero, Robust two-dimensional weighted and unweighted phase unwrapping that uses fast transforms and iterative methods, *J. Opt. Soc. Amer. A*, 1994, 11(1): 107-117

10. M. D. Pritt and J. S. Shipman, Least-squares two-dimensional phase unwrapping using FFT's, *IEEE Transaction on Geoscience Remote Sensing*, 1994, 32(3): 706-708

11. M. D. Pritt, Phase unwrapping by means of multi-grid techniques for interferometric SAR, *IEEE Transaction on Geoscience Remote Sensing*, 1996, 34: 728-738

12. J. Demmel, *Applied Numerical Linear Algebra*, 1997, Society for Industrial and Applied Mathematics

13. China Integrative Atlas, 1990, Beijing: China Map Publishing Company, p151

Wave Propagation,Scattering and Emission in Complex Media
Edited by Ya-Qiu Jin
Science Press and World Scientific,2004

38

Road Detection in Forested Area Using Polarimetric SAR

Guiwei Dong, Jian Yang, Yingning Peng
Dept. of Electronic Engineering, Tsinghua University, Beijing 100084, China

Chao Wang, Hong Zhang
Institute of Remote Sensing Applications, Beijing 100101, China

Abstract   This paper focuses on the use of polarimetric SAR to detect roads in forested area. After analyzing some polarimetric parameters in forested area, such as the total span of a scattering matrix, polarimetric entropy and similarity parameters, we propose an expression to combine the above polarimetric parameters with the optimal coefficients. The optimal expression is sensitive to road characteristics. By using SIR-C/X-SAR L-Band data of Tian Shan, China, we detect line segments representing roads on both the total span image and the proposed expression image. It is demonstrated that the proposed method is effective.

1.Introduction

Road detection is of great importance on SAR image process, for instance for classification and for forest monitoring. In forested area, the presence of new roads might provide an indication of active logging. So, finding the new roads in time is very important in forest monitoring. For this problem, SAR has a great predominance, in that its working for all-weathers, day and night imaging capability.

In the past, methods used for road detection were concentrated on the retrieve of features of radar image [1]. For example, many algorithms were developed for automatically detecting the linear features on radar image [1], [2]. Most of the algorithms focused on the detection of roads in cities or highways. Touzi and Sasitiwarih studied the potential of the C-band Radarsat for road detection in rainforests [1]. They modeled the SAR imaging of a road in rain forests as a function of some SAR parameters, road width and orientation, and tree height. However, this method has several disadvantages. For example, the model is very complex, and it is difficult in the detection of narrow road in tall forests.

In this paper, we focus on the applications of polarimetric SAR to detect roads in forested area. Polarimetric radars are sensitive to the shape and orientation of a target. The information contained in polarimetric SAR image is valuable for road detection in a forested area. In forested areas, backscattering wave from trees contains more dihedral component than that from space ground. In addition, the span of a scattering matrix and the entropy of several considered pixels also vary with target classes. In Section 2, some polarimetric parameters, including the span of a scattering matrix, the polarimetric entropy [3] and similarity parameters [4] are summarized. Then, an expression is proposed which combines these polarimetric parameters with the optimal coefficients. The optimal expression is sensitive to road characteristics. By using SIR-C/X-SAR L-Band data of Tian Shan, China, we detect line segments on both the total span image and the proposed expression image. The detection results demonstrate the proposed method is effective.

2.   Polarimetric Parameters

In this section, we summarize some polarimetric parameters which will be used in Section 3.

For the problems of target classification and target recognition in radar polarimetry, one important problem is how to analyze characteristics of a radar target. The similarity parameter [4] may be employed for extracting single reflected and double reflected contributions of a target. Now we give a brief introduction to this parameter. Let

$$[S] = \begin{bmatrix} s_{HH} & s_{HV} \\ s_{VH} & s_{VV} \end{bmatrix} \tag{1}$$

denote a target scattering matrix in a linear horizontal (H) and vertical (V) polarization base, and let ψ denote the orientation angle of the target, which is easy to calculate due to Huynen's decomposition [5], we may rotate the target to a special position, where its orientation angle equals zero:

$$[S^0] = [J(-\Psi)] \cdot [S] \cdot [J(\Psi)] = \begin{bmatrix} s_{HH}^0 & s_{HV}^0 \\ s_{VH}^0 & s_{HH}^0 \end{bmatrix}, \tag{2}$$

where

$$[J(\Psi)] = \begin{bmatrix} \cos \Psi & -\sin \Psi \\ \sin \Psi & \cos \Psi \end{bmatrix}. \tag{3}$$

For the reciprocal backscattering case, $s_{HV} = s_{VH}$ and $s_{HV}^0 = s_{VH}^0$. So the polarimetric scattering information may be represented by a modified Pauli-scattering vector:

$$\vec{k} = \frac{1}{\sqrt{2}} \begin{bmatrix} s_{HH} & \sqrt{2} s_{HV} & s_{VV} \end{bmatrix}^T, \tag{4}$$

where the superscript T denotes the transpose of a vector. The similarity parameter between two scattering matrices [S$_1$] and [S$_2$] is defined as [4]

$$r([S_1],[S_2]) = \frac{\left| \left(\vec{k}_1 \right)^H \left(\vec{k}_2 \right) \right|^2}{\left\| \vec{k}_1 \right\|_2^2 \left\| \vec{k}_2 \right\|_2^2}, \tag{5}$$

where the superscript H denotes the conjugate transpose. Vectors $\vec{k}_1$ and $\vec{k}_2$ are related to [S$_1$] and [S$_2$], respectively. Symbol $\| \ \|_2^2$ denotes the square sum of the absolute values of the three components in a complex scattering vector. The similarity parameter has following properties:

1) $r([J(\theta_1)][S_1][J(-\theta_1)],[J(\theta_2)][S_2][J(-\theta_2)]) = r([S_1],[S_2])$, where θ_1 and θ_2 are two arbitrary angles.

2) $r(a_1[S_1],a_2[S_2]) = r([S_1],[S_2])$, where a_1 and a_2 are two arbitrary non-zero complex numbers.

3) $0 \le r([S_1],[S_2]) \le 1$, $r([S_1],[S_2]) = 1$ if and only if $[S_2] = a[J(\theta)][S_1][J(-\theta)]$, where a is an arbitrary non-zero complex number and θ is an arbitrary angle.

4) If [S$_1$], [S$_2$] and [S$_3$] are three non-zero scattering matrices which satisfy

$$r([S_1],[S_2]) = r([S_1],[S_3]) = r([S_2],[S_3]) = 0, \tag{6}$$

then for an arbitrary non-zero scattering matrix [S], one has

$$r([S],[S_1]) + r([S],[S_2]) + r([S],[S_3]) = 1. \tag{7}$$

Now we give some interpretations of the above properties. As we know, if the scattering matrix of a target is $[S]$, and if the target is rotated an angle θ about the sight line of the monostatic radar, the scattering matrix of the target in new position is $[J(\theta)][S][J(-\theta)]$. The first property demonstrates that the similarity parameter between two scattering matrices does not vary with the orientation angles.

The second property shows that the similarity parameter is independent of spans of two scattering matrices. For some targets, i.e. spheres, plates, diplanes, etc., this property implies that the similarity parameter does not vary with target sizes.

The third property gives a range of the similarity parameter. Two scattering matrices are completely similar if and only if they have a constant difference after one is rotated an angle.

The forth property shows that the sum of the similarity parameters between an arbitrary non-zero scattering matrix and three pairwise non-similar scattering matrices equals one.

As an application, the similarity parameter can be employed to extract characteristics of a target. According to (5), one easily calculates some typical similarity parameters as follows.

i) The similarity parameter between an arbitrary scattering matrix [S] and a plane is given by:

$$r_1 = r([S], diag(1,1)) = \frac{\left| s_{HH}^0 + s_{VV}^0 \right|^2}{2 \left(\left| s_{HH}^0 \right|^2 + \left| s_{VV}^0 \right|^2 + 2 \left| s_{HV}^0 \right|^2 \right)}. \tag{8}$$

This parameter is related to the measurement of single reflections from a target.

ii) The similarity parameter between an arbitrary scattering matrix [S] and a dihedral is given by:

$$r_2 = r([S], diag(1,-1)) = \frac{\left| s_{HH}^0 - s_{VV}^0 \right|^2}{2 \left(\left| s_{HH}^0 \right|^2 + \left| s_{VV}^0 \right|^2 + 2 \left| s_{HV}^0 \right|^2 \right)}. \tag{9}$$

This parameter is related to the measurement of double reflections from a target.

2.2 Polarimetric Entropy

Entropy is originally used to describe the chaos extent in nature. To study target scattering characteristics, Jin and Cloude introduced this concept in polarimetric SAR.

Consider the following polarimetric coherent matrix

$$[T] = \left\langle \vec{k} \vec{k}^{*T} \right\rangle \tag{10}$$

This matrix can be decomposed as

$$[T] = \left\langle \vec{k} \vec{k}^{*T} \right\rangle = v_1 \left(\vec{e}_1 \vec{e}_1^{*T} \right) + v_2 \left(\vec{e}_2 \vec{e}_2^{*T} \right) + v_3 \left(\vec{e}_3 \vec{e}_3^{*T} \right), \tag{11}$$

where $v_i (i = 1,2,3)$ are the eigenvalues of matrix $[T]$, $\vec{e}_i (i = 1,2,3)$ are the corresponding eigenvectors. Then the polarimetric entropy is defined by

$$H = -\sum_{i=1}^{3} P_i \log_3 P_i, \tag{12}$$

where $P_i = \dfrac{v_i}{\sum_{j=1}^{3} v_j}$.

3. The Proposed Method

To use polarimetric information of target sufficiently, we combine the above parameters with a linear combination

$$f = C_1 \xi_1 + C_2 \xi_2 + C_3 \xi_3 + C_4 \xi_4 = \vec{C}^T \vec{\xi}, \tag{13}$$

where $\vec{C} = [C_1, C_2, C_3, C_4]^T$, $\vec{\xi} = [\xi_1, \xi_2, \xi_3, \xi_4]^T$. ξ_1 denotes the total span of polarimetric scattering matrix, ξ_2 denotes the polarimetric entropy, ξ_3 denotes the plane similarity parameter and ξ_4 denotes the dihedral similarity parameter. $C_i\ (i=1,2,3,4)$ are the corresponding coefficients. Let f_f denote the above linear combination of forested area and f_g denote that of space areas. To get the optimal coefficient, we solve the following optimization problem

$$\text{maximize} \quad F = \frac{f_g^2}{f_f^2} = \frac{\vec{C}^T\left(\vec{\xi}_g\vec{\xi}_g^T\right)\vec{C}}{\vec{C}^T\left(\vec{\xi}_f\vec{\xi}_f^T\right)\vec{C}}, \tag{14}$$

s.t. $C_1^2 + C_2^2 + C_3^2 + C_4^2 = 1$

It is equivalent to solving the following eigenvalue problem

$$\left(\vec{\xi}_g\vec{\xi}_g^T\right)\vec{C} = \lambda\left(\vec{\xi}_f\vec{\xi}_f^T\right)\vec{C} \tag{15}$$

After obtaining the optimal $\vec{C}$, f_f of every pixel can be calculated. In the image of f_f, the forested area will become darker and some roads will be detected.

4. Calculation Result and Conclusion

The used experimental polarimetric SAR image data is L-band scattering image of the Tian Shan, China, collected with the SIR-C/X-SAR system in October, 1994. The total span image of the analyzed forested region is shown in Fig.1. In the darker areas of the image, there are no trees, a road should have similar characteristics. The brighter areas denote forested area in Fig.1.

Fig.1   The total span image of a forested area, Tian Shan, China

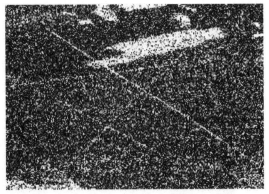

Fig.2   The image of $f = C_1\xi_1 + C_2\xi_2 + C_3\xi_3 + C_4\xi_4$

Solving (14), we obtain $\vec{C} = [-0.1956, -0.6491, 0.7351, 0.0081]^{\mathrm{T}}$. After calculating $f = C_1\xi_1 + C_2\xi_2 + C_3\xi_3 + C_4\xi_4$ for all pixels, we obtain a new image shown in Fig.2. In this image, the space areas become much brighter than those in the total span image. We can see four bright lines in this image, it is of great possibility that these bright lines are roads, but these roads are vague in the total span image.

The advantage of the proposed method can also be demonstrated from another point of view. In general, local detection of line structure is the first step of road detection process on a SAR image. The result of line segment detection affects road detection a lot. By using ratio of average (RoA) method [6], we detect local line segments on both the total span image and the proposed expression image, respectively. In the detector, we have the following two assumptions: 1) line segments are thin and elongated with a maximum width 3 pixels and minimum length 15 pixels, respectively, 2) the differences of mean values between a line segment and its two neighbor regions are no less than a given threshold (in this paper, we let the threshold T be 0.21). For any given pixel, the region centered at the pixel and satisfied the first assumption for some direction can be determined. We denote this middle region as R2 and its two neighbors as R1 and R3, respectively. Then the mean values of the optimal expression f corresponding to R1, R2 and R3 are calculated, denoted as u_1, u_2 and u_3, respectively (Fig.3). Usually, R1, R2 and R3 have the same length, but widths of R1 and R3 are bigger than that of R2.

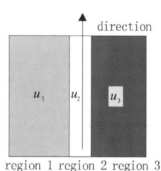

Fig.3   Line segment detector for vertical direction case

Now, we use the following RoA detector

$$\mathrm{RoA} = \min\{|u_2 - u_1|, |u_2 - u_3|\}. \tag{16}$$

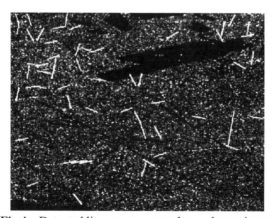

Fig.4   Detected line segments on the total span image

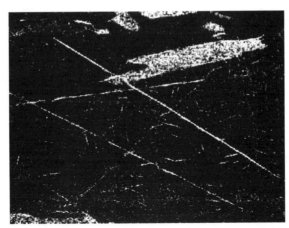

Fig.5a   Detected line segments on the proposed expression image (before filtering)

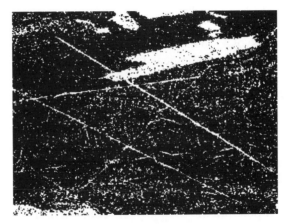

Fig.5b   Detected line segments on the proposed expression image (after filtering)

If RoA>T, we consider that the pixel belongs to a line. Fig.4 and Fig.5a show the results of line segment detection on both the total span image and the proposed expression image, respectively. To highlight the main roads in Fig.5a, we process the image based on the assumption that roads in forested area are long enough. So, a number of line segments located in some direction will be considered as belonging to the same road and they will be remained, whereas other segments will be removed from the image. This can be regarded as filtering undesired line segments. The image after filtering is shown in Fig.5b. It is clear that line segment detection on the proposed expression image is much better than that on the total span image. We therefore conclude the proposed method is effective for road detection in forested area.

Acknowledgement

This work was supported by the National High Technology Research and Development Program (863 Program 2001AA135050), the National Natural Science Foundation of China (40271077), and by the Excellent Young Teacher Program of MOE, China.

References

[1] Touzi R, Sasitiwarih A., "On the use of Radarsat and JERS-1 satellite SARs for trail and road detection in tropical rainforests", International Geoscience and Remote Sensing Symposium, 2465-2467, 1998

[2] Jeon B.-K., Jang J.-H., and Hong K.-S., "Road detection in spaceborne SAR images using a genetic algorithm", IEEE Trans Geosci Remote Sensing, vol.40, no.1, 22-29, 2002

[3] Jin Y.-Q., and Cloude S. R., "Numerical eigenanalysis of the coherency matrix for a layer of random nonspherical scatterers", IEEE Trans Geosci Remote Sensing, vol.32, no.6, 1179-1185, 1994

[4] Yang J., Peng Y.-N., Lin S.-M., "Similarity between two scattering matrices", Electron Letters, vol.37, no.3, 193-194, 2001

[5] Huynen J.R., "Phenomenological theory of radar targets", Ph.D thesis, Technical University, Delft, The Netherlands, 1970

[6] Li S.-D, Zhang C., Wang Z.-Z., "A study of road network detection method for SAR images", Journal of Astronautics, vol.23, no.1, 17-24, 2002

[7] Yang J., "On theoretical problems in radar polarimetry", Ph.D thesis, Niigata University, Niigata-Shi, Japan, 1999

[8] Geman D. and Jedynak B., "Detection of roads in satellite images", In Proceeding of International Geoscience and Remote Sensing Symposium, 2437-2447, 1991

[9] Cloude S. R., Pottier E., "An entropy based classification scheme for land applications of polarimetric SAR", IEEE Trans Geosci Remote Sensing, vol.35, no.1, 68-78, 1998

Wave Propagation,Scattering and Emission in Complex Media
Edited by Ya-Qiu Jin
Science Press and World Scientific,2004

Research on Some Problems about SAR Radiometric Resolution

Ge Dong, Minhui Zhu, Xinyuan Tian, Jingyun Fan
The National Key Laboratory of Microwave Imaging Technology,
Institute of Electronics, Chinese Academy of Sciences, Beijing, 100080, China
Email: gdong@mail.ie.ac.cn

Abstract   This paper presents some problems in SAR radiometric resolution definition, and proposes a definition formula for engineering calculation after analyzing the SAR radiometric performances, then discusses the effects of sample coherence on SAR radiometric resolution, the relation between spatial resolution and radiometric resolution. Finally, the paper makes requests for the SAR system design parameters to improve SAR radiometric resolution.
Key words: Synthetic Aperture Radar (SAR), radiometric resolution.

1. Preface

Microwave imaging technology realizes target images through the target microwave scattering performances that radar echo signals reflect. SAR (Synthetic Aperture Radar) can achieve all-weather and all-day large-scale images with high spatial resolution, thus used widely in the fields of national economy and national defense. Radiometry- calibrated   SAR images should be descriptions of the microwave scattering performances of the target scene, and the more accurately the microwave scattering performances are reflected, the higher image quality can be obtained in a SAR system. As a measure parameter of microwave system imaging capability, radiometric resolution plays a key in SAR system design definitely. Because a SAR system has a higher spatial resolution than a traditional microwave imaging system, we can interpret SAR images according to the target spatial shape. Scattering ability difference between natural background and artificial targets such as vehicles, roadways, factories and cities, even between different natural targets such as vegetation, deserts, sea and mountains is very distinct, so the application of radiometric resolution in SAR system is often ignored. However in the research of crop mapping, soil moisture discrimination, especially sea phenomena, the ability to distinguish weak scattering difference of terrain targets is particularly important. Improvement of SAR radiometric resolution benefits cognition of imaging target characteristics and the development of SAR application in new fields.

2. Definitions of radiometric resolution

At present there isn't a uniform definition for radiometric resolution of SAR. In some literature[1], background contrast sensitivity is used. Some regular definitions[1,2,3] are as follows:

$$\text{Definition 1.} \quad \delta_d = (\sqrt{M} + 1.282)/(\sqrt{M} - 1.282) \tag{1}$$

$$\text{Definition 2.} \quad \delta_d = (\sqrt{M} + 1)/(\sqrt{M} - 1) \tag{2}$$

$$\text{Definition 3.} \quad \delta_d = 1 + 1/\sqrt{M} \tag{3}$$

$$\text{Definition 4.} \quad \delta_d = 1 + (1 + 1/SNR)/\sqrt{M} \tag{4}$$

where δ_d -SAR radiometric resolution, M-accumulated independent sample number (multi-look), SNR -Signal-to-Noise Ratio of SAR system.

These definitions are based on the statistic characteristics of SAR echoes. If the area covered by SAR antenna pattern is large enough, in other words, if N_s-the total of scattering units is large enough, and there isn't any large scattering target whose contribution can predominate echo characteristics in the area, u -the instant voltage amplitude accumulated by the N_s scattering units, is independent of φ -the phase, and they are random variables in Rayleigh distribution and even distribution on $[-\pi,\pi]$ respectively. Presume that the envelope power detected is attained on a 1Ω resistor, then $P = u^2$, and we can get to know that the density function of output power level of radiodetector is in an exponential distribution with a probability density function as follows:

$$p(u)=\begin{cases}u/\sigma^2 \exp\left(-u^2/(2\sigma^2)\right) & u \geqslant 0 \\ 0 & u < 0\end{cases} \tag{5}$$

$$p(\varphi)=\begin{cases}1/(2\pi) & \varphi \in [-\pi,\pi] \\ 0 & \varphi \notin [-\pi,\pi]\end{cases} \tag{6}$$

$$p(P)=\begin{cases}1/\sigma^2 \exp\left(-P/\sigma^2\right) & P \geqslant 0 \\ 0 & P < 0\end{cases} \tag{7}$$

mean: $E[u]=\sqrt{2/\pi}\sigma$, $E[P]=\sigma^2$, standard variance: $D[u]=(2-\pi/2)\sigma^2$, $D[P]=\sigma^2$.

These two kinds of distribution have a considerable dynamic fading scope, and the probability of fading signal at 17.7dB of this dynamic scope is 90%, up to 7.68 times voltage value and 58.9 times power level value. If we sample only one from the aggregate distributed as Rayleigh distribution or exponential distribution, it is very difficult to get the mean. However, just like what stated before, estimation of the mean of the detected signal intensity is very important to remote-sensing application. The random fluctuation effect of pixel amplitude in SAR images caused by signal fading is what we say-coherent speckle or coherent speckle noise, which is a significant factor affecting SAR image radiometric resolution.

SAR system solves this problem through multi-look technology. As a rule we presume that target images in different looks are irrelative. Because the exponential distribution of radar signals, which the square-law filter outputs, is χ^2 distribution with two-order free dimensions, noncoherent averaging M independent samples in exponential distribution results in the probability density function of signal z, which is χ^2 distribution with 2M free dimensions, that is:

$$p(z)=\frac{1}{(\sigma^2/M)\Gamma(N)}\left(\frac{z}{\sigma^2/M}\right)^{M-1}\exp\left(-\frac{z}{\sigma^2/M}\right) \tag{8}$$

Here M is a positive integer, Gamma function $\Gamma(M)=(M-1)!$

$$p(z)=\frac{1}{(\sigma^2/M)(M-1)!}\left(\frac{z}{\sigma^2/M}\right)^{M-1}\exp\left(-\frac{z}{\sigma^2/M}\right) \tag{9}$$

mean $E[z]=\sigma^2$; standard variance $\sigma[z]=\sqrt{D[z]}=\sigma^2/\sqrt{M}$.

We define the ratio of two boundaries of 80% of the intensity scope $[0.1,0.9]$ in which the signal power level falls as radiometric resolution:
Definition 5:

$$\delta_d = \frac{z_1(P(z > z_1) = 0.1)}{z_2(P(z < z_2) = 0.1)} \tag{10}$$

According to this definition, conclusions can be drawn as follows in Table 1:

Table 1

Sample number M	1	2	4	10	20	100
$z_1(P(z > z_1) = 0.1)$，dB	3.63	2.89	2.23	1.53	1.13	0.53
$z_2(P(z < z_2) = 0.1)$，dB	-9.79	-5.76	-3.61	-2.16	-1.39	-0.58
dynamic scope (80%), dB	13.42	8.65	5.84	4.14	2.52	1.11

Definitions 1-4 are all derived from primitive definition 5. If the number of independent samples M is large enough e.g. $M \geqslant 10$, χ^2 distribution with 2M free dimensions approximates normal distribution, and the scope definition 5 requires is $[E[z] - 1.282\sigma[z], E[z] + 1.282\sigma[z]]$. So $\delta_d = \dfrac{E[z] + 1.282\sigma[z]}{E[z] - 1.282\sigma[z]} = \dfrac{\sqrt{M} + 1.282}{\sqrt{M} - 1.282}$, which is just definition 1, and table 2 presents the radiometric resolution drawn according to this definition. In the brackets is the probability of signal fading scope. This shows that the radiometric resolution definition 1 presents is slightly larger than that of definition 5, and the probability when the signal falls into the dynamic scope is slightly larger than 80%. Definition 2 is the simplified form of definition 1, in which the radiometric resolution is less than that of definition 5, and the probability when the signal falls into the dynamic scope is slightly less than 80% (refer to table 2). Definition 3 and definition 4 derived considering the presence of system noises are results of shortening signal fading dynamic scope to $[E[z], E[z] + \sigma[z]]$. As direct results of this, the radiometric resolution decreases distinctly, and the probability when the signal falls into the dynamic scope also decreases to 26%-33%, as can be indicated in Table 2 as follows:

Table 2

sample number	2	3	4	8	10	20	100
definition 1	13.1 (88%)	8.3	6.6	4.2	3.7	2.6	1.1 (80%)
definition 2	7.7 (74%)	5.7	4.8	3.2	2.8	2.0	0.9 (68%)
definition 3	2.3 (33%)	2.0	1.8	1.3	1.2	0.9	0.4 (26%)
definition 4 (SNR=10)	2.5	2.1	1.9	1.4	1.3	1.0	0.45
definition 5	8.65 (80%)	6.84	5.84	4.03	3.59	2.52	1.11 (80%)

Definition 4 is often used in engineering for its simple form. But the discussion above shows us that the probability when the signal falls into the dynamic scope for definition 4 which is derived from definition 3 is about 30%, much less than 80% for the primitive definition, and this can not meet practical need obviously. Here we propose a more accurate formula considering the presence of system noises for engineering estimation, and the probability when the signal falls into the dynamic scope is around 70%.

The relation between $\delta_d(n)$-SAR radiometric resolution considering the presence of system noises and δ_d-that not considering the presence of system noises is as follows[2]:

$$\delta_d(n) = \frac{\delta_d(SNR + 1) - 1}{SNR} \tag{11}$$

take definition 2 into this formula, and we can get after simplification:

$$\delta_d(n) = 1 + \left(1 + \frac{\sqrt{M}+1}{M-1}\right)\left(1 + \frac{1}{SNR}\right)$$ （12）

3. The influence of the correlation of samples to radiometric resolution

From above discussion, we can know that multi-look technology is an efficient means that can improve the radiometric resolution. The accumulated samples are supposed independent in discussion above. If there is some correlation in the samples, what will happen to radiometric resolution using multi-look noncoherent accumulation?

In short, the noncoherent accumulation of the samples that have some correlation corresponds to the coherent accumulation of the independent samples. Its result can't get the effect of the noncoherent accumulation of the independent samples by any means. We analyze this quantitatively using the knowledge of probability as follows.

The echoes of radar are $V_i = u_i \exp(j\varphi_i)$, $i = 1,2,...N$, and the united density function of noncoherent accumulation is[4]:

$$p(V_1,V_2,...V_N) = \frac{1}{\pi^N|\vec{J}|}\exp\left(-\vec{V}^\tau \vec{J}^{-1}\vec{V}\right)$$ （13）

where $\vec{V} = (V_1,V_2,...V_N)^\tau$, $\vec{J} = E\left[\vec{V},\vec{V}^\tau\right]$ - correlated matrix

$$E\left[V_i,\overline{V}_k^\tau\right] = \begin{cases} \sigma_i^2 & i = k \\ (\alpha_{ik} + j\beta_{ik})\sigma_i\sigma_k & i \neq k \end{cases}$$

to the square-law filter radar $z = u_i^2$, according to the equation above and the eigenfunction distributed as multivariate Gauss probability //of multivariate Gauss probability distribution, we can get:

the mean $E(z) = \sum_{i=1}^{N}\sigma_i^2$, and the variance $D(z) = \sum_{i=1}^{N}\sum_{k=1}^{N}P_{ik}^2\sigma_i^2\sigma_k^2$, $P_{ii} = 1$, $0 \leqslant P_{ik} \leqslant 1$

From this we know when the samples are not correlated, viz. when $P_{ik} = 0$, $D(z)$ gets minimal, and the radiometric resolution defined by definitions 1-5 gets best. Oppositely, when the samples are absolutely correlated, viz. when $P_{ik} = 1$, $D(z)$ gets maximal, and the radiometric resolution defined by definitions 1-5 gets worst.

Take two independent samples as an example to explain this problem. The curve that the radiometric resolution δ_d (dB) defined in definition 2 follows the related coefficient P_{12} between two samples is drew in the plot below (fig.1. The real line is the curve δ_d follows P_{12} when $\sigma_1 = \sigma_2$, and the broken line is the value of the radiometric resolution δ_d defined by definition 2).

Based on the plot, when P_{12} equals zero, the noncoherent accumulation of the samples gets the best effect, which is the value that defined. As the correlation in the samples increases, the effect that the noncoherent accumulation of the samples improves on the radiometric resolution drops off.

The discussion above explains the influence that the correlation in the samples improves on the radiometric resolution. We can ensure the validity that the noncoherent accumulation of the samples improves on the radiometric resolution only if we can ensure to get the independence of the samples. In fact, there is correlation in the samples obtained by SAR. The most typical example is the imaging process of the moving target. The phase difference hidden in the correlation of two-look image is the gist to determine the information of the moving target.

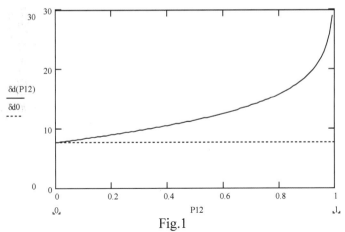

$$\frac{\delta d(P12)}{\delta d0}$$

Fig.1

To land imaging, because of the relative stillness of the land, two-look images are viewed relatively independent（$P_{ik} > 0.9$）. But the problem will be more complex to the imaging of ocean surface. For the variety of ocean surface, the correlation in the samples lies on the selection of the time of synthetic aperture. Particularly when a lot of samples accumulation is needed, the independence of the samples is especially important.

4. The relation between spatial resolution and radiometric resolution

Based on the working theory of SAR, we know that if we adopt sampling in frequency domain to improve the radiometric resolution of the system, the bandwidth of single sample will be reduced when the bandwidth of the system is definite. The signal bandwidth determines the spatial resolution of SAR on range direction (perpendicular to track direction) $\delta y = c /(2\Delta f \sin\theta)$ [5], where c is velocity of light, Δf is bandwidth of signal, and θ is angle of incidence. So the result that we adopt sampling in frequency domain to improve the radiometric resolution of the system is that spatial resolution on range direction has been reduced.

If we adopt sampling in time domain to improve the radiometric resolution of the system, the time of synthetic aperture T_s will be reduced. While the time of synthetic aperture T_s determines the spatial resolution of SAR on azimuth direction (along track direction) $\delta x = \lambda R /(2VT_s)$ [5], where λ is the wavelength of signal, R is inclined distance, and V is the flight rate of carrier. So the result that we adopt sampling in time domain to improve the radiometric resolution of the system is that spatial resolution on azimuth direction has been reduced.

Finally, if we adopt sampling in space, in other words, we make some resolving units smooth, the result will be more obvious, that is, the radiometric resolution has been improved at the cost of losing the spatial resolution.

It's not difficult to get the conclusion that it's ambivalent each other between improving the spatial resolution and radiometric resolution.

5. Conclusions

When we design the system parameter of SAR, first of all we must consider the geometric character and scattering character of the target to ensure reasonable spatial resolution and radiometric resolution. Thereout the mode and quantity of samples can be ensured, thus the bandwidth and the time of synthetic aperture are ensured. Then the width of antenna pattern can be ensured. The dynamic character of the target should be considered sufficiently to ensure the independency of the samples when we select the time of synthetic aperture. Except of this, to improve radiometric resolution the SNR and sampling precision of the system should be tried to reduce, which includes reducing the noise of receiver, and improving the power of transmitter, the stability of signal, the stability of flat roof of antenna, and the precision and dynamic domain of A/D switch, etc.

Acknowledgment

This work was sponsored by the China State High Technology Research and Development 863 Program.

References

1   Соколов Г. А. Разрешение дискретных случайных сигналов при ограниенной априорной статистике. Радиотехника, 1989, (4)
2   Frost V. S. Probability of error and radiometric resolution for target discrimination in radar images. IEEE Trans. Rem. Sens., 1984, GE-22(4): 121-125
3   Ulaby F.T., Moore R.K, Fung A.K., Microwave Remote Sensing (vol.2), 1980, Mass: Artech
4   Goodman N. R. Statistical analysis based on a certain multivariable complex guassian distribution (an introduction). Ann Math. Statist., 1963, 34(1): 152-177
5   C. Zhang, Synthetic Aperture Radar, 1980, Beijing: Science Press

Wave Propagation,Scattering and Emission in Complex Media
Edited by Ya-Qiu Jin
Science Press and World Scientific,2004

A Fast Image Matching Algorithm
for Remote Sensing Applications

Zhiqiang Hou, Chongzhao Han, Lin Zheng, Xin Kang
School of Electronics and Information Engineering,
Xi'an Jiaotong University, Xi'an 710049, China.

Abstract   Aiming at the weakness of the high computational cost of correlation-based matching method, a fast matching algorithm is presented in this paper. A coarse-to-fine strategy is adopted in the new algorithm, i.e., the circle template is used in matching area to complete the coarse matching such that the matching windows can be determined firstly and then the whole template is used to decide the final correct window. This algorithm has been applied to match the remote sensing images successfully. The simulation results show that the new algorithm can not only match images correctly, but also decrease the computational cost greatly. Compared with the correlation-based matching method, the computational cost of the new algorithm is only its 10%.

Key Words   Image Matching, Remote Sensing, Correlation-based, Circle Template Matching

1. Introduction

In image matching, the correlation-based method[1] is a classical one, and it suits for matching those images which have complex texture especially. Because the optical remote sensing images have complex texture, the correlation-based method is used for matching remote sensing images frequently. But there is very high computational cost in the correlation-based method. In order to improve its performance, the special hardware system should be designed to increase the computing efficiency sometimes. At the same time, some improved algorithms have been proposed to decrease the computational cost. One of the improved methods is the wavelet-based algorithm [2], the images are compressed by using the wavelet transformation at first, and then the coarse-to-fine strategy is adopted, i.e., the coarse matching is performed at the low resolution level and the fine matching is completed at the high resolution level. However, the computing complexity of this method increases because the wavelet transformation is joined into the algorithm. Another is the feature-based matching method. The image features such as edges, points, lines and inflexions etc should be extracted to be the matching gist, and furthermore, a little computational cost is needed for the feature matching. The general feature-extracting methods such as the LoG Operator[3] or Canny Operator[4] can be used to detect the edges or lines, and the wavelet transformation [5] or mathematics morphology [6] can be used to extract the points or inflexions. However, there is a great deal of computation in these pixel-level operations. Meanwhile, the features-extracting by using the above methods are not always good enough.

Based on the correlation-based method, a fast algorithm called to be the *Circle Template Matching* (CTM) is presented in this paper. The coarse-to-fine strategy is adopted in the new algorithm, i.e., the circle template is used in matching area to complete the coarse matching such that the matching windows can be determined firstly, and then the whole template is used to decide the final correct window. Simulation results show that the computational cost of the new algorithm is only 10% of cost of the original one.

The remainder of this paper is organized as follows: the algorithm is described in detail in Section 2; Section 3 gives the simulation results, and the comparative research with the original correlation-based method; the conclusion appears in Section 4.

2. Algorithm Description

2. 1 Circle template selection

When the remote sensing images are matched under assumption that the matching template is known, a part of the template pixels can be used as feature pixels to perform the coarse matching for their textures are complex, and then the full template can be used to complete the fine matching. In order to simplify problem, only the template translation is considered in this paper.

Because human vision cells are circle distributed in retina, the circle pixels are adopted as the feature pixels in this paper. The circle pixels are a set of points, which are all on a circle edge, the center of the circle is the template center, and the circle is not out of the template. The feature points set is composed of all these circle pixels and called to be the *circle template* in this paper.

In general, the image pixels are arrayed in matrix form. Therefore, the positions that the circle pixels occupied, its center in the image is fixed, must be defined. These positions can be calculated by using the Chamfer transformation [7]. There are two templates to be used, one is so called the 3-4 template, and another is the 5-7-11 template. In this paper, the 5-7-11 template is adopted to transform the distances because it has only $\pm 2\%$ error with the real Euclid distance, but the 3-4 template has $\pm 8\%$ error.

2.2 Algorithm steps

There are 2 steps in the new algorithm. Step 1 is the coarse matching, and the obvious non-matching points can be eliminated and the preparative matching points can be chosen by using circle template in this step. Step 2 is the fine matching, and the correct matching position from the preparative matching points is decided to be the current position of the template by using full template in the step.

2.2.1 Coarse matching

In order to obtain the point (m, n) in the remote sensing image, which will be matching with the circle template, *Pel Difference Classification* [8] (PDL) is chosen to define the similarity between two matching points, it is

$$A(m,n) = \sum_{i=1}^{p} D(f_1^{i}, f_2^{i}(m,n)), \qquad (1)$$

$$D(f_1^{i}, f_2^{i}(m,n)) = \begin{cases} 1, & |f_1^{i} - f_2^{i}(m,n)| \leq T \\ 0, & |f_1^{i} - f_2^{i}(m,n)| > T \end{cases}, \qquad (2)$$

where $f_1^{i} = f_1(x_i, y_i)$ is the ith circle pixel in the template, and the sum of the circle pixels is p; meanwhile $f_2^{i}(m,n) = f_2(x_i + m, y_i + n)$ represents all circle pixels with the center (m, n) in the searching area, which are corresponding to the ith circle pixel in the template. The sum of these pixels is also p. T is a threshold. If the gray absolute difference between the prepared match pixel and the corresponding circle template pixel is less than T, these two pixels are considered to be matched and the value is denoted by 1; meanwhile if it is not less than T, these two pixels are considered not to be matched and the value is denoted by 0.

The more pixels the circle template pixels are matching with, the larger $A(m,n)$ is, and therefore the more possibility the window corresponding to point $A(m,n)$ is the correct window including target. Because of the noise effect and the target state changing, the maximum $A(m,n)$ is not always the correct match point. Despite this, in the matching area, $A(m,n)$ always has a larger value, so S, a threshold, is selected as follows:

If $A(m,n) \geq S$, the point (m, n) is reserved as a preparative matching pixel at the next step.

If $A(m,n) < S$, the point (m, n) is deleted as a non-matching pixel and does not process any more.

So, the computational consumption is decreased obviously.

2.2.2 Fine matching

After coarse matching, the number of match pixels is calculated by using full template in the preparative matching pixels, and the maximum point position is the correct match position. The formula to calculate the match point number by using full template is the same as the formula by using circle pixels. The formula is as the following:

$$B(m,n) = \sum_{x=1}^{M}\sum_{y=1}^{N} D(f_1(x,y), f_2(x+m,y+n)) \tag{3}$$

$$D(f_1(x,y),f_2(x+m,y+n)) = \begin{cases} 1, & |f_1(x,y)-f_2(x+m,y+n)| \leq T \\ 0, & |f_1(x,y)-f_2(x+m,y+n)| > T \end{cases} \tag{4}$$

where, $f_1(x,y)$ is template pixel gray-level, template size is $M \times N$; $f_2(x+m,y+n)$ is match window pixel gray-level at the point (m, n) in the searching area ; T is the same in Eq. (2).

By means of calculating, the window with the maximum $B(m,n)$ can be chosen to be the best match area.

3. Simulation results and Comparative Research

Both the correlation-based algorithm and the new algorithm proposed in this paper are respectively used to match the real optical remote sensing images. Three sets of simulation results are shown in figure 1, figure 2 and figure3 respectively.

Figure 1 is a village remote sensing image, the image size is 150×150, 256 gray level. Here, figure 1(a) is the initial remote sensing image, and figure 1(b) is the full template, and its size is 20×20; meanwhile figure 1(c) is the circle template used to coarse matching, the diameter of circle is 17 pixels, and figure 1(d) is the matching position showed by the white rectangle in the image.

The comparison analysis in computational cost between correlation-based method and the new algorithm is given in Table 1, where NCCF means *Normalize Cross-Correlation Function*, and CTM means *Circle Template Matching*.

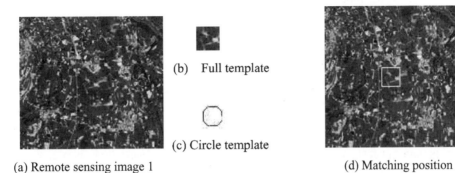

(b)   Full template

(c) Circle template

(a) Remote sensing image 1   (d) Matching position

Fig 1. Village remote sensing image

Figure 2 is a lake remote sensing image, the image size is 200×200, 256 gray level. Here figure 2(a) is the initial remote sensing image, and figure 2(b) is the full template, and its size is 20×20; meanwhile figure 2(c) is the circle template used to coarse matching, the diameter of circle is 17 pixels, and figure 2(d) is the matching position showed by the white rectangle in the image. The comparison analysis is also given in Table 1.

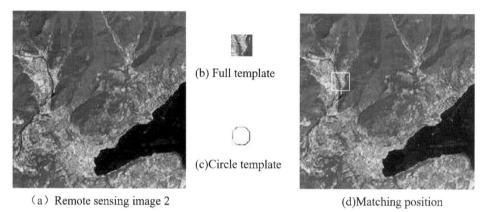

(b) Full template

(c)Circle template

（a）Remote sensing image 2

(d)Matching position

Fig 2. Lake remote sensing image

Table 1. Comparison of the computational costs between NCCF and CTM

	NCCF	CTM
Village Remote sensing image	566 s	57 s
Lake Remote sensing image	1324 s	116 s
Mountain Remote sensing image	3452 s	323 s

Figure 3 is a mountain remote sensing image, the image size is 300×300, 256 gray level. Here figure 3(a) is the initial remote sensing image, and figure 3(b) is the full template, and its size is 20×20; meanwhile figure 3(c) is the circle template used to coarse matching, the diameter of circle is 17 pixels, and figure 3(d) is the matching position showed by the white rectangle in the image. The comparison analysis is also given in Table 1.

The simulation results indicate that both the correlation-based algorithm and the new algorithm can be used to match images correctly, but the computational cost of the new algorithm is only 10% of that of the correlation-based algorithm. The programs are performed by using Matlab 5.3.

4. Conclusions

In this paper, a fast matching algorithm, *circle template matching*, is presented, and is used to match remote sensing images. Simulation results show that this algorithm can not only match correctly, but also decrease computational cost greatly. Compared with correlation-based matching method, this algorithm has superiority in calculation cost.

Meanwhile, this method can be used with the wavelet transformation and perform at the different remote sensing image scale, so this method can match more large remote sensing image correctly and quickly.

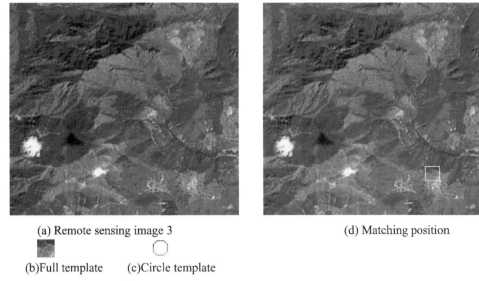

(a) Remote sensing image 3   (d) Matching position

(b)Full template   (c)Circle template

Fig 3. Mountain remote sensing image

Acknowledgments

This work was supported by the China State Key Basic Research Project 2001CB309403.

Reference

[1] R.C. Gonzalez, P. Wintz, *Digital Image Processing*, Addison-Wesley Publishing Company, Inc., 1977

[2] J. Deubler, J. Olivo, A Wavelet-Based Multiresolution Method to Automatically Register Images, *Journal of Mathematical Imaging and Vision 7*, 1997, 199–209

[3] D.Marr and B. Hildreth, Theory of edge detection, *Proc. Royal Society of London, Series B*, 1980, 207:187-217

[4] J. Canny, A computational approach to detection, *IEEE. Trans. on PAMI*, 1986, 8(6): 679-698

[5] A. Quddus, M.M. Fahmy, Fast wavelet-based corner detection technique, *Electronics Letters*, 1999, 35(4):287-288

[6] R. Laganiere, A morphological operator for corner detection, *Pattern Recognition*, 1998, 31(11): 1643-1652

[7] G. Borgefors, Distance transformations in digital images, *Comput. Vision, Graphics, Image Processing*, 1986, 34:344-371

[8] H. Gharavi and M. Mills, Blockmatching Motion Estimation Algorithms – New Results, *IEEE Trans. Circuits and System*, 1990, 37(5):649-651

A New Algorithm of Noised Remote Sensing Image Fusion Based on Steerable Filters

Xin Kang, Lin Zheng, Chongzhao Han, Qian Ma, Chenhao Zhao
School of Electronics and Information Engineering
Xi'an Jiaotong University, Xi'an 710049, China

Abstract   A fusion algorithm of noised remote sensing images based on steerable filters is presented in this paper. A quadrature pair that consists of a steerable filter and its Hilbert transform is used for analyzing the dominant orientation and the local energy in images. By using this algorithm, the original remote sensing images can be denoised, and then their features can be fused successfully. The simulation results indicate that this new remote sensing image fusion algorithm performs well in denoising, preserving, and enhancing edge or texture information. Therefore its comprehensive visual effect is very good.

Key Words   remote sensing; image denoising; dominant orientation; local oriented energy; image fusion

The image fusion combines different aspects of information from the same imaging modality or from two distinct image modalities [1]. The fused image can provide the robust operational performance, i.e., confidence increasing, ambiguity reducing, improved reliability or a deeper insight about the nature of observed data. As a powerful tool, the image fusion is widely used in the digital image processing and understanding, such as image enhancing, automatic object detection, 3D image reconstruction, and the image sharpening, especially in the remote sensing image processing.

In fact, the original remote sensing images are always polluted. A preprocessing stage is needed to denoise the images. Because the noised pixel is odd, so it can be considered to be a feature in the image. The essential of the remote sensing image fusion is combining the complementary features in different remote sensing images. So it is the basic of denoising and fusing images to find out the features in original images and to determine their type and characteristic.

According to the theory of Venkatesh and Owen, the phase congruency is proportional to local energy, which is defined as a sum of the squared responses of a quadrature pair of symmetric and anti-symmetric filters. Therefore the maximum in phase congruency at features is corresponding to the maximum of local energy at features [2]. In order to extract the local energy, many kinds of quadrature pair have been proposed [3]. One simple and suitable quadrature pair used in this paper is composed of a steerable filter and its Hilbert transform, which not only detect the local energy, called to be the *local oriented energy* here, but also the local dominant orientation along the feature orientation [4]. Then two remote sensing images are used to do experiment such that the features in the images can be enhanced by means of denoising and fusion.

1. Transform with Steerable Filters

Steerable filters are filters whose arbitrary orientation can be synthesized as a linear combination of a set of basis filters [4]. When they are used as a quadrature pair, they have a good performance in edge detection, texture analyses and oddity point detection.

1.1 Design of quadrature pair of steerable filter

The basis filters in a steerable filter are a set of filters whose responses have some overlapped area in frequency domain, and anyone of which can be represented a rotation form when the steerable filter is fixed in some orientation. In this paper, the second derivative of the Gaussian function G_2 is selected to be a steerable filter, which has the following form

$$G_2(x, y) = \frac{\partial^2}{\partial x^2} e^{-(x^2+y^2)} = (4x^2 - 2)e^{-(x^2+y^2)} \qquad (1)$$

Making $x = r\cos\varphi, y = r\sin\varphi$, then the Fourier series of $G_2(x, y)$ can be written as

$$G_2(r, \varphi) = r^2 e^{-r^2}(e^{i2\varphi} + e^{-i2\varphi}) + (2r^2 - 2)e^{-r^2}, \qquad (2)$$

which has three nonzero coefficients. So three basis filters can be selected [5]. In practice, for reasons of symmetry and robustness against noise, the three basis filters can be distributed between 0 and π uniformly, i.e., the three angles are

$$\theta_1 = 0; \quad \theta_2 = \pi/3; \quad \theta_3 = 2\pi/3,$$

and three basis filters are the rotation forms $G_2^0, G_2^{\pi/3}, G_2^{2\pi/3}$ of G_2 at θ_1, θ_2, θ_3 respectively. Therefore, the arbitrary rotation of G_2 can be described as

$$G_2^\theta(x, y) = k_1(\theta)G_2^0(x, y) + k_2(\theta)G_2^{\pi/3}(x, y) + k_3(\theta)G_2^{2\pi/3}(x, y), \qquad (3)$$

where $k_i(\theta), (i = 1,2,3)$ are three interpolation functions, θ is the steering angle.

Figure 1 Three basis filters of $G_2(x, y)$

For analyzing the local energy of the features in the image, a quadrature pair of filters should be established. The quadrature filter of G_2 is its Hilbert transform, which is unable to steer. So a product of a third-order polynomial with a Gaussian function can be used to approximate it, i.e.,

$$H_2(x, y) = (-4.7867x + 2.123x^3)e^{-(x^2+y^2)}. \qquad (4)$$

Using above method, four corresponding basis filters of $H_2(x, y)$ and its interpolation functions can be worked out.

1.2 Local oriented energy analysis

For each feature pixel in an image, its energy in arbitrary direction can be calculated by using the quadrature filter combined by steerable filter and its Hilbert transform

$$E(x, y, \theta) = |I(x, y) * G_2^\theta(x, y)|^2 + |I(x, y) * H_2^\theta(x, y)|^2, \qquad (5)$$

where "$*$" indicates the convolution, $I(x, y)$ is the original image, and θ is the rotation angle. This energy is called to be the *oriented energy* of the pixel (x, y) at θ angle.

At a feature pixel, the oriented energy is changing with rotation of the steerable filter. When orientation of the quadrature pair of filters fits with the feature orientation, the oriented energy takes its maximum, is called to be the *local oriented energy*, and therefore the corresponding orientation is defined to be the *local dominant orientation* of the pixel. According to the theory in [3], if the local oriented energy of a pixel is the maximum in its vicinity, the feature point can be easily distinguished.

The value of the maximum local oriented energy is corresponding to the quality of feature point. If a clear image $I(x, y)$ is blurred, the blurry image can be express as the convolution of the original image and a Gaussian function with a variance σ. Therefore, the blurry image is

$$I^{'}(x,y) = I(x,y) * G_{\sigma}(x,y).$$

Since $G_{\sigma}(x,y)$ is a low-pass filter along every spatial direction, some high frequency information of the original image is filtered, including those on features. Using the quadrature pair of steerable filters $G_2(x,y)$ and $H_2(x,y)$, the local oriented energies on $I(x,y)$ and $I^{'}(x,y)$ can be analyzed respectively, and the response on $I^{'}(x,y)$ must be weak than that on $I(x,y)$ along the dominant orientation. According to the definition of local oriented energy, it is clear that the local oriented energy on $I^{'}(x,y)$ must be less than that on $I(x,y)$. Because the image has more high frequency components in the feature point, and more energy on it may be filtered. Therefore, its local oriented energy is remarkably reduced. According to this, the higher quality the feature point has, the larger the local oriented energy is.

2. Fusion Algorithm of Noised Remote Sensing Images

There are many image fusion methods used before. The simplest one is to calculate the average of two images, and some complicated algorithms such as Laplacian pyramid [5], wavelet transform [6], etc. are used to get more comprehensive information. In general, before fusion of remote sensing images, it is necessary to denoise the images because the remote sensing images are always polluted.

2.1 Image denoising based on steerable filters

Assume the original remote sensing images have additive isolated noises, experiment results approve that the noise is generally to be an odd point in single pixel. To denoise image, the simplest method is accomplished by averaging two original images. But there is an obvious weakness in the simplest method that the features in original image are reduced significantly when the noised points are filtered.

If a preprocessing algorithm is adopted to find out the noise points in the original images, then the local average filter working only on these points can be used to obtain most features in the original image and most noise can be eliminated.

According to the definition of steerable filters, the isolated point can be determined by the following two steps:

1) Noise point which similar to feature point is an odd point, so it has larger local oriented energy.

2) The isolated noise point has not direction, so the energy in different direction has little difference. It means that the difference of the maximum and the minimum of the local oriented energy is small.

As mentioned above, the local oriented energy on each image is calculated by using the quandrature pair of steerable filters. If the steerable filters designed in section 1.1 are used to analyse the oriented energy in each image, it is can be found that the orientation in which the local oriented energy takes its minimum is perpendicular to the dominated orientation.

Expending the oriented energy expression (5) by using trigonometric series and omitting high level polynomial, it can be obtained as

$$E(x,y,\theta) \approx C_1 + C_2 \cos(2\theta) + C_3 \sin(2\theta) = C_1 + S\cos(2(\theta - \theta_d))$$

where $C_1, S = \sqrt{C_2^{~2} + C_3^{~2}} >= 0$ and $\theta_d = \frac{1}{2} arctg(C_3 / C_2)$.

It is easily proven [7] that the oriented energy of the pixel takes its maximum value $C_1 + S$ when $\theta = \theta_d$ and takes its minimum value $C_1 - S$, when $\theta = \theta_d + \pi/2$.

So, the result can be obtained that the orientation in which the noised pixel has the minimum oriented energy is perpendicular to the dominated orientation. It means that the noised pixel can be found by comparing the oriented energy in dominated direction and its perpendicular orientation. Then, the following denoising procedures can be used to eliminate noised pixels in the image.

Algorithm 1 can be described as follows:

(1) the dominated orientation $\theta_d(x, y)$ and the local oriented energy $E_{\theta_d}(x, y)$ for each pixel can be analyzed;

(2) it is obtained that the local oriented energy γ in perpendicular orientation to the dominated orientation;

(3) the $ratio = \left| \ln(E_{\theta_d} / E_{\theta_v}) \right|$ is calculated to determine the pixel is a noised pixel or not;

(4) if $ratio <= T1$ and $E_{\theta_d} >= T2$, the pixel is a noised pixel; and if $ratio > T1$ or $E_{\theta_d} < T2$, the pixel is no a noised pixel.

Where parameter T1 means that there is no orientation feature in the noised pixel, and takes value $1.3 \sim 1.8$ in usual case. Parameter T2 means the amount of the odd feature in the noised pixel, and can be calculated by using the mean value of the local oriented energy in the image, i.e.,

$$T2 = k \frac{1}{MN} \sum_{i=1}^{M} \sum_{j=1}^{N} E_{\theta_d}(i, j),$$

where $k = 0.8 \sim 1$.

(5) the image denoising at the noise pixels can be done by using the median-value filtering algorithm.

Simulation Example: The example is used to explain the denoising effect of the remote sensing image.

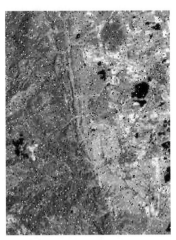
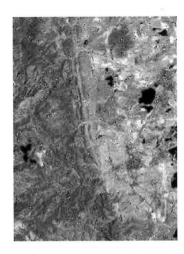

(a) TM band 4 noised image   (b) denoised image
Figure 2 denoising effect of remote sensing image

The signal-to-noise ratios of the noised image and the denoised image are shown in Table 1 respectively.

Table 1   signal-to-noise ratios

	Noised image	Denoised image
SNR	12.7215	16.0676

2.2 Remote sensing image fusion base on steerable filters

A remote sensing image fusion algorithm base on steerable filters is proposed in this paper. It can be used to eliminate isolated noised pixels, and at same time, it can preserve the compensative features in the original images and can fuse them to a final image.

According to the properties of the local oriented energy, the higher the feature quality the pixel is of, the larger the local oriented energy is. And the blur point has less energy. Therefore, the local oriented energy can be used as the weighting coefficients when fusing two or more images.

As a result, the procedure of **Algorithm 2**, i.e. the fusion algorithm of remote sensing images can be described as:

(1) registration of original remote sensing images by using the least squares algorithm based on the features;

(2) the local oriented energy of two original images is analyzed, and isolated noised pixels are eliminated according to Algorithm1;

(3) the local oriented energy and dominant orientation in the denoised images are re-calculated;

(4) because the two images are registered, the difference between dominant rotation angles of pixel (x, y) in two images is not big enough, so it can be defined as

$$\theta_d(x, y) = \frac{1}{2}(\theta_d^1(x, y) + \theta_d^2(x, y)).$$

To rotate the steerable filters to angle $\theta_d(x, y)$, and to calculate and obtain the modified local oriented energy as

$$E_j(x, y, \theta_d), (j = 1, 2).$$

(5) to fuse the two images by using the weighting average, and the fused image is

$$I(x, y) = E_1(x, y, \theta_d)I_1(x, y) + E_2(x, y, \theta_d)I_2(x, y).$$

As the result, a fused image cam be obtained, which has higher quality than two original images.

3. Experiment Results

Wapplied the above methodology to fusion two remote sensing images for Boulder, Colorado, USA, showed in figure 3(a) and 3(b) respectively. The figure 3(a) and 3(b) are band 3 and band 4 image of the TM data. Because band 3 and band 4 use different wave lengths and reflect the different land features, so there has something clearly in one image but not in another one. These two images can be fused to obtain much more land feature information.

Comparing the result with average and 3-level SIDWT with Haar fusion method, it can be found that the steerable filters preserve the features in each image and get a global legible fusion image. And comparing with other methods, the effect of the new algorithm is the same or a little better than those.

4. Conclusion

A fusion algorithm of noised remote sensing images based on steerable filters is proposed in this paper. By analyzing the oriented energy and dominated orientation of the image, it can be used to denoise and fuse two or more polluted complementary remote sensing images. The experiment example can be used to prove that this new algorithm has capability in remote sensing image denoising and fusing. It can obtains and enhance the features for the remote sensing images.

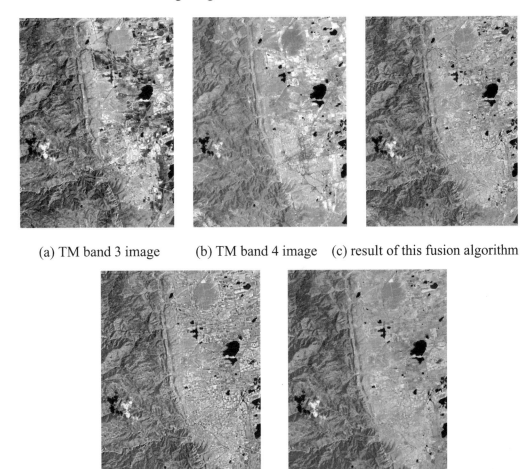

(a) TM band 3 image   (b) TM band 4 image   (c) result of this fusion algorithm

(d) result of average fusion   (e) result of 3 level Haar wavelet fusion

Figure 3 Fusion experiment for TM band 3 and band 4 remote sensing images

Table 2 Evaluation on effect of fusion results

	Stdev	Ave. Gradient	Entropy
This method	48.1095	18.1968	7.650410
Average	48.9774	12.7420	7.571796
3-level Haar wavelet	51.7002	22.5396	7.618907

Acknowledgment
This work was supported by the China State Basic Research Project 2001CB309403.

References
[1] Aggarwal J K. Multisensor fusion for computer vision [C]. Berlin Heidelberg: Springer-Verlag, 1993
[2] Venkatesh S, Owens R. On the classification of image features [J]. Pattern

Recognition Letters. 1990, 11:339-349

[3] Robins M J. Local energy feature tracing in digital images and volumes [D]. http://citeseer.nj.nec.com/ 1999-06-15/2002-01-12：37-38

[4] Freeman W T, Adelson E H. The design and use of steerable filters [J]. IEEE Trans, Pattern Anal. Machine Intell., 1991,13(9): 891-906

[5] Akerman, A. Pyramid techniques for multisensor fusion [A].Proc. SPIE, sensor fusion.1992,1828:124- 131

[6] Yocky David A. Image merging and data fusion by means of the discrete two-dimensional wavelet transform [J]. Opt. Soc. Am. A. 1995,12(9): 1834~1841

[7] Lin Zheng, Chongzhao Han, Dongguang Zuo and Xin Kang. Fusion Algorithm on Noised Images Based on Steerable Filters (in Chinese), [J]. J. of Xi'an Jiaotong University, 2002,Vol. 36 , No.12, 1236-1239

[8] Lin Zheng, Chongzhao Han, Dongguang Zuo and Xin Kang. Different Focus Points Image Fusion Based on Steerable Filters (in Chinese), [J]. J. of Computer Engineering and Application, 2002, No.7, 23-25

Wave Propagation,Scattering and Emission in Complex Media
Edited by Ya-Qiu Jin
Science Press and World Scientific,2004

Adaptive Noise Reduction of InSAR Data Based on Anisotropic Diffusion Models and Their Applications to Phase Unwrapping[†]

Chao Wang, Xin Gao, Hong Zhang

(Institute of Remote Sensing Application, CAS, 100101, Beijing, China)

*Email: cwang@public.bta.net.cn

Abstract   In this paper, the relationship between noise and phase unwrapping is analyzed. Advanced global and local denoising techniques have been exploited to serve the key step in InSAR application-phase unwrapping. Phase noise, abrupt slope, layover and shadow are main factors to degrade the quality of phase patterns. Some noise reduction techniques are presented, and their comparisons are discussed. Two main phase unwrapping algorithm are especially described. We present the improved minimum norm models with nonlinear regularization techniques. Although, higher computation load has to be paid currently, high quality unwrapped phase insensitive to noises makes the model useful. The vector digital filtering has been employed to the interferogram images, which aim at reducing the residues. Numerical examples show significantly lower computation cost.

Key Words   vector digital TV filtering; nonlinear diffusion equation; phase unwrapping

1. Introduction

Remote sensing image and medical image are two important application fields. Phase unwrapping is a key step for the two aspects. In this paper, we focus on InSAR techniques, which have a wide and fine application in Measuring Topography, Deformation of the Earth Surface, Object detection and identification etc. The advent of interferometric SAR for geophysical studies, in particular, has resulted in the need for accurate, efficient methods of two-dimensional phase unwrapping. Radar methods for fast and precise measurement of topographic data, determination of cm--level surface deformation fields, and surface velocity fields all require that the absolute phase which should be unwrapped from the relative phase value. However, phase presented by SAR sensors is very sensible to the baseline errors, thermal noise, atmospheric disturbances, and various types decorrelation, that is called "noises". In company with layovers, shadows and denser fringes, these noises make trouble, which cause a lot of residues to affect the accuracy and efficiency of phase unwrapping. Many filtering algorithms have been employed to preprocess the SLC data information, interferogram, even phase itself information. Here, we employ anisotropic diffusion models [1,2] and give a local method--the vector digital total variation (TV) filtering [3] with robust, accuracy and efficiency. In the second section, we simply describe interferometric SAR and classical phase noise treatment, and review our research on phase unwrapping techniques based on minimum norm method prefiltering by anisotropic diffusion. The third section, we show the principle of anisotropic diffusion filtering. As the high computation cost, in the forth section we present a local method --the vector digital TV filtering for InSAR interferogram to serve phase unwrapping.

The last section we give the numerical result for optical image filtering and interferogram data filtering.

[†] Research supported by National Key Basic Research Program of China (Grant No. 2001CB309406) and National Hi-tech program of China (Grant No. 2001AA135050)

2. Phase Unwrapping with phase noise treatment

A. Phase unwrapping concept

The SAR interferometric technique combines two complex images recorded with two antennas separated spatially or with one antenna separated in time. The phase and magnitude information of the two complex images is used to produce an interferogram. To get interferometric phase ϕ, two complex value SLC data for the same scene from two different positions are needed. After co-registering the SLC data, we gain two complex images: $z_1 = |z_1|.\exp(j\phi_1)$ and $z_2 = |z_2|.\exp(j\phi_2)$. Phase pattern can be obtained by taking the argument of the product of the first image with the complex conjugate of the second:

$$z_{\text{int}} = z_1.z_2^* = |z_1||z_2|\exp(j(\phi_1 - \phi_2)).\tag{1}$$

The phase difference $(\phi_1 - \phi_2)$ is determined from the real and imaginary part of z_{int} by arctan function. We only get the principal value ψ of phase:

$$\psi = W(\phi_1 - \phi_2),\tag{2}$$

where W is an operator that wraps the phase into interval $[-\pi, \pi)$. Phase can only be detected between $-\pi$ and π, but the actual phase shift between two waves is often more than this. Phase unwrapping is the process of reconstructing the original phase shift from this "wrapped" representation. It consists of adding or subtracting multiples of 2π in the appropriate places to make the phase image as smooth as possible. To convert interferometric phase into elevation, you must perform phase unwrapping. The direct linear equations that relate ψ to ϕ are

$$\psi(x, y) = \kappa\phi(x, y) + \varepsilon(x, y) + 2\pi k(x, y),\tag{3}$$

for $(x, y) \in \Omega$, a set for which valid observations are available, and κ is a linear operator, $\varepsilon(x, y)$ is a random variable that models the measurement noise, and the integer--valued field $k(x, y)$ is such that $-\pi < \psi(x, y) \leq \pi$. It is an inverse problem to reconstruct ϕ from the measurements ψ, which is ill--posed in a mathematical sense.

In next subsection, we analyze the traits of InSAR phase noise and last subsection we detail our phase unwrapping models with anisotropic diffusion filtering based on minimum norm.

B. Phase noise traits

The accuracy of the DEM is strongly influenced by phase noise. If the phase noise is too strong, some fringes will be completely lost which will result in great loss of the accuracy of the DEM. So keeping phase information and removing phase noise in an interferogram is a key point in order to obtain a highly accurate topographic map.

Apart from general noise characters, coherence interference due to reflection by random scatters degrades the complex--valued image. The amplitude is corrupted by multiplicative noise, while the phase is corrupted by additive noise. Actually, by using the complex-amplitude model, we may treat the noise consistently as complex-valued multiplicative noise. By using exponential expression of a complex-valued image, it is obvious that the noise effect on the phase will be additive. Because speckle noise is inherent trouble, multilook processing is usually applied. Actually, multilook processing does not improve coherence, but reduces the standard deviation of ψ. After reducing some noise by the multilook process, the remaining noise has to be reduced by a filtering process, such as boxcar filtering, median filtering, Goldstein—Werner (G--W) filtering, or the adaptive Lee filtering [4]. The presence of noise may cause phase inconsistency called residue or singular point, that will become an obstacle in unwrapping process because the integration depending on the path.

One of the major sources of phase noise is decorrelation, which is an inherent property of repeat pass interferometry. If we now assume that the decorrelation depends on the ground cover, we can say that the variance of the phase noise depends on the type of terrain that is being spotted. This assumption makes it possible to apply local adaptive filters (e.g. speckle filters) for reducing the phase noise in interferograms. Speckle arises because the spatial resolution of the sensor is not sufficient to resolve all the individual scatterers. Due to this lack of resolution the resulting complex backscatter of a single resolution cell is the phasor sum of randomly distributed scatterers. It can be shown that the phase of a SAR radar echo, backscattered from a single resolution cell follows a uniform distribution. Speckle is believed to be multiplicative and can be reduced by averaging independent values with multi-look processing. However, the drawback of multilook processing is that it will result in a loss of spatial resolution since the images used for averaging must be obtained from the same aperture. In order to remedy speckle more efficiently as well as to preserve spatial resolution of the images, several specklefiltering techniques have been developed from the image processing field. One of the most representative techniques of speckle reduction in SAR post-processing is the adaptive filtering technique, which was first proposed by Lee in 1980. Since then many adaptive filters have been proposed to reduce speckle whilst at the same time preserving the SAR information. However, the full removal of speckle without losing any information is still a long way off. Any filtering technique will more or less damage useful information and leave some residual speckle as it is implemented. The reason for this is that the modern adaptive filters, such as the Lee, Enhanced Lee, Frost, Enhanced Frost, Kuan and Gmap speckle filters, use the local statistical information related to the filtered pixel. The pixel that is filtered is replaced by a value, which depends on the statistical information in the filter window. This will only result in efficient speckle removal and information preservation when the filter window covers a homogenous area. When the filter window for example covers an edge the filtered pixel value is replaced by statistical information from both sides of the edge, or from two different phase/intensity distributions. In 1994 Wilhelm Hagg proposed the EPOS filter of a variable window size and sub-windows. The basic approach of this filter is to search for the biggest homogenous area around the central pixel and use only the statistical information present in this area to filter the central pixel. The EPOS speckle filter is very efficient for removing speckle whilst preserving edges. However, it is not very computationally efficient. Recently ITC proposed a new adaptive speckle filter \cite (ITC). It uses a common weighting function as the filter kernel, which is modified by an edge detector to tackle the problem of using statistical information from unwanted areas. It uses not only the local statistical information related to the central pixel but also that from the central pixels neighborhood in the filter window size. The value of this ITC filter for enhancing ERS SAR images has been clearly demonstrated.

C. The algorithms of phase unwrapping

Many algorithms serve phase unwrapping for InSAR. Two essential approaches are residue-cut "tree" procedure proposed by Goldstein et al [6] and least-squares procedure in common use today presented by Ghiglia and Romero [7], who applied mathematical variation optimization models. The former is a local method and the latter is a global method. For the residue-cut, there are two possible types of errors in the unwrapped sequence. The first are local errors in which only a few points are corrupted by noise; the second are global errors in which the local error may be propagated down the entire sequence. The global errors arise from the residues. Residues are local errors in the measured phase caused by the noise in the signals or by actual discontinuities, i.e., layovers, in the data. One major difference between the residue-cut and least-squares solutions is that in the residue-cut approach only integral numbers of cycles are added to the measurements to produce the result.

Unlike local PhU, LS approach is based on global integration criteria: they minimize the distance between the gradient estimated from the wrapped data and the gradient of the solution. Any value may be added to ensure smoothness and continuity in the solution. These techniques are generally more robust than local integration methods. Thus the spatial error distribution may differ between the approaches, and the relative merits of each method must be determined depending on the application. One advantage of the LS algorithms over residue-cut is that results may be obtained more readily in the residue-rich regions, permitting use on noisy data that would have been difficult or impossible to unwrap because of the dense tree network in the Goldstein et al. algorithm.

Other existing algorithms proposed for InSAR phase unwrapping applications include Green's function approaches, multigrid algorithms, network program techniques , and neural network or genetic algorithms, especially, some algorithms evolved from least-squares, preconditioned conjugate gradient(PCG) algorithm, multigrid algorithm, and region growing LS algorithm. More detail can be available in [8,9,10] and our anisotropic diffusion regularization model[2].

3.   The Filtering analysis of anisotropic diffusion models

In this section, we focus on image processing by using Partial Different Equation (PDE).

A. The evolution of PDE image application

The excellent work is attributed to Lions et al [11], who first bridged the gap between the approximation of filter and PDE and proposed *Scale Space*. Perona and Malik [12], Osher and Rudin [19] proposed many novel PDE models for image smoothing and denoising. Especially, in [12], Perona and Malik proposed to enhance a given noisy image I by solving the following anisotropic diffusion equation:

$$\partial_t u = div(g(|\nabla u|)\nabla u), (u,t) \in \Omega \times R_+ \tag{4}$$

with initial condition $u(x,y,0) = I(x,y)$ and boundary condition $\partial u / \partial n \big|_{\partial\Omega \times R_+} = 0$. The diffusion coefficient $g(\cdot)$ is a nonnegative non-increasing function of the magnitude of local image gradient with $g(0) = 1, g(s) \geq 0, \text{and}\}$ $g(s) \to 0$ as $s \to \infty$. The desirable diffusion coefficient should be such that (1) diffuses more in smooth areas and less around large intensity transitions, so that small variations in image intensity such as noise and unwanted texture are smoothed and edges are preserved. Another objective for the selection of $g(\cdot)$ is to incur backward diffusion around large intensity transitions so that edges are sharpened, and to assure forward diffusion in smooth areas for noise removal. The often choice for $g(\cdot)$ is of the form

$$g_k(s) = 1/(1 + ks^2) \tag{5}$$

If $g \equiv \text{constant } t$, the model equation (1) reduces to the traditional isotropic Gaussian filtering, which tends to over-smoothing. But in computer vision and pattern recognition literature, preserving important triangular points, corners and T-junctions besides general edges is our pursuit. Formula (2) makes model equation (1) have the abilities to perform selective smoothing depending on the magnitude of the image gradient at the point. Since edges are characterized by large values of $|\nabla u|$, so that $g_k(|\nabla u|)$ is small, as like performing simulated annealing there.

For mitigating the large oscillations of $|\nabla u|$ in model equation (1), Catte et al. [23] improved the above model as follows

$$\partial_t u = div(g(|\nabla v|)\nabla u), (u,t) \in \Omega \times R_+ \tag{6}$$

where $v = G_\sigma * u$, $G_\sigma(x) = \exp(-|x|^2 / 4\sigma^2)/\sigma\sqrt{4\pi}$. This change results in a much improved denoising procedure, but important edges, especially triagonal points, corners and T-junctions, are still not well preserved due to the presence of Laplacian operator $\Delta u = div(grad\ u)$.

Alvarez, Lions and Morel [24] introduced a new method of curvature-based evolution of iso-intensity contours, based on the equation:

$$\partial_t u = g(\nabla G_\sigma * u)|\nabla u|div(\nabla u / |\nabla u|) \tag{7}$$

This model diffuses the level curves of u in the direction orthogonal to ∇u with the speed $g(|\nabla G_\sigma * u|)k$, while $k = div(\nabla u / |\nabla u|)$ is the local curvature of the iso-intensity contour. Therefore, the image is smoothed on both sides of the edges with minimal smoothing of the edge itself, and at speeds which are lower in neighborhoods of edges, and higher interior to smooth regions [14].

In [15], basing on the Alvarez-Lions-Modrel model (4), they presented the model

$$\partial_t u = |\nabla u|div(g_k(|\nabla G_\sigma * u|)\frac{\nabla u}{|\nabla u|}) - \lambda(u - I)|\nabla u| \tag{8}$$

and in [16] the more general model was proposed as follows

$$\partial_t u = \alpha g_k(|\nabla v|)|\nabla u|div(\frac{\nabla u}{|\nabla u|}) + \alpha\nabla(g_k(|\nabla v|)) \cdot \nabla u - \lambda(u - I)|\nabla u| \tag{9}$$

Considering the importance of v in image restoration, many choice, for examples, $v = u$, $v = G_\sigma * I$, $v = G_\sigma * u$, have been developed as the preceding models by many scholars. For stability and simplicity, they also determines v by the following nonlinear diffusion equation (7) coupled with equation (6) based on variation method thoughts

$$\partial_t v = a(t)div(\frac{\nabla v}{\nabla v}) - b(v - u) \tag{10}$$

with insulated boundary conditions and initial conditions

$$u(x,y,0) = I(x,y), v(x,y,0) = I(x,y), (x,y) \in \Omega.$$

and a(t) is a parameter changing with time, while b remains constant. Although the couple equations (9) and (10) can reduce more computational load than direct convolution with Gaussian function, the computational load still is extremely heavy. Therefore, after detail numerical schemes analysis, we will consider a derived digital filtering version to serve phase unwrapping.

B. Numerical schemes

Several schemes have been employed to solve the PDE's, as above mentioned, e.g. finite difference method, level-set method, and finite element method and multigrid method.   Here we introduce a numerical scheme together with finite difference and level-set method to solve the couple equations (2.14), (2.15) with its insulated boundary and initial conditions (11)) originating from [13,17,18,20,21]. For our interesting image $u(x,y)$ to be reconstructed, we let $u_{i,j}$ denote u(i,j), the evolution equations obtain images at times $t_n = n\Delta t$, we denote $u(i,j,t_n)$ by u_{ij}^n.

The time derivative u_t at (i,j,t_n) is approximated by the forward difference $(u_{ij}^{n+1} - u_{ij}^n)/\Delta t$. The diffusion terms

$$|\nabla u|(div(\frac{\nabla u}{|\nabla u|})) = \frac{u_x^2 u_{yy} - 2u_x u_y u_{xy} + u_y^2 u_{xx}}{u_x^2 + u_y^2} \qquad (12)$$

$$div(\frac{\nabla u}{|\nabla u|}) = \frac{u_x^2 u_{yy} - 2u_x u_y u_{xy} + u_y^2 u_{xx}}{(u_x^2 + u_y^2)^{3/2}} \qquad (13)$$

in (9), (10) are approximated by using the center differences, where

$$(u_x)_{ij}^n = \frac{u_{i+1,j}^n - u_{i-1,j}^n}{2h}, (u_y)_{ij}^n = \frac{u_{i,j+1}^n - u_{i,j-1}^n}{2h}, \quad (u_{xx})_{ij}^n = \frac{u_{i+1,j}^n - 2u_{i,j}^n + u_{i-1,j}^n}{h^2},$$

$$(u_{yy})_{ij}^n = \frac{u_{i,j+1}^n - 2u_{i,j}^n + u_{i,j-1}^n}{h^2}, \quad (u_{xy})_{ij}^n = \frac{u_{i+1,j+1}^n - u_{i-1,j+1}^n - u_{i+1,j-1}^n + u_{i-1,j-1}^n}{4h^2}.$$

The term $\nabla g \cdot \nabla u$ permits the development of discontinuities, which indicate the presence of object boundaries. The main idea here is to use forward or backward finite difference in computing the spatial derivatives of u in a manner that is consistent with the development of the shock, and it dose not reflect the smoothness that may occur near the shock region. This feature is achieved by the following scheme developed by Osher and Sethian [13]

$$(\nabla g \cdot \nabla u)_{ij} = \max(\Delta_x g_{i,j}, 0)\Delta_x^+ u_{i,j} + \min(\Delta_x g_{i,j}, 0)\Delta_x^- u_{i,j}$$
$$+ \max(\Delta_y g_{i,j}, 0)\Delta_y^+ u_{i,j} + \max(\Delta_x g_{i,j}, 0)\Delta_y^+ u_{i,j} \qquad (14)$$

where $\Delta_x^- u_{i,j} = u_{i,j} - u_{i-1,j}, \Delta_x^+ u_{i,j} = u_{i+1,j} - u_{i,j}, \Delta_y^- u_{i,j} = u_{i,j} - u_{i,j-1}, \Delta_y^+ u_{i,j} = u_{i,j+1} - u_{i,j},$

and $\Delta_x u_{i,j} = \frac{u_{i+1,j} - u_{i-1,j}}{2h}, \Delta_y u_{i,j} = \frac{u_{i,j+1} - u_{i,j-1}}{2h}.$

The gradients of u and v presented in the error term and in the term $g_k(|\nabla v|)$ are approximated by the upwind scheme developed in [13]

$$|\nabla v| = [(\max(\Delta_x^- u_{ij}^n, 0))^2 + (\max(\Delta_x^+ u_{ij}^n, 0))^2 + (\max(\Delta_y^- u_{ij}^n, 0))^2 + (\max(\Delta_y^+ u_{ij}^n, 0))^2]^{1/2} \quad (15)$$

For Neumann boundary conditions (11), we can use the iteration formula

$$u_{ij}^{n+1} = u_{ij}^n + \Delta t L(u_{ij}^n), n = 1, 2, \cdots, N \qquad (16)$$

with $u_{ij}^0 = z(x_i, y_j)$, and $L(u) = g_k(|\nabla u|)div(\nabla u/|\nabla u|) + \nabla g \cdot \nabla u - \lambda |\nabla u|(u-z)$. The term v in g(v) is solved by the semi-implicit scheme[1].

Another, Level-set numerical method and Multigrid numerical method are tow very influential as the above applications.

C.   Digital TV filtering

We aimed at the Digital TV Filtering (DTVF)[22] techniques to serve the InSAR phase preprocessing, and phase unwrapping based on Goldstein's branch-cut [6]. The papers [1,17] represent the evolution history of anisotropic diffusion models and show robust and reliable filtering algorithms for high noisy data with discontinuation. For operation simplified and computation reduced, we here use the DTVF for nonlinear denoising, which applied in the denoising of two-dimensional data with irregular structures, gray and color images, and nonflat image features such as chromaticity.

In the following, we introduce DTVF algorithm basing on a better anisotropic diffusion model [17]

$$\partial_t u = |\nabla u| div(g_k(|\nabla G_\sigma * u|)\nabla u/|\nabla u|) - \lambda(u - I)|\nabla u| \qquad (17)$$

where $g_k(\cdot)$ is often chosen as $g_k(s) = 1/(1 + ks^2)$, $G_\sigma(x) = \exp(-|x|^2/4\sigma^2)/(\sigma\sqrt{4\pi})$

is Gaussian kernel. Positive parameter λ is called *fitting parameter*-Lagrange multiplier, which is the trade-off between approximation and regularization. To avoid singularities in flat regions or at extreme, $|\nabla u|$ in model (17) is regularized to

$$|\nabla u|_a = \sqrt{|\nabla u|^2 + a^2} \tag{18}$$

for some small positive parameter a, which is mainly for numerical purposes, and it appears in denominators later. The performance of the DTVF is insensitive to the parameter a as long as it keeps small. Considering the node stencil in Fig.1, we denote the coordinate of center node as $\alpha = (m, n)$, and other nodes can be expressed by matrix structure. For the center node α, $T(\alpha)$ is employed as the neighbors set. The {local variation} $|\nabla_\alpha u|$ at any node α is defined by

$$|\nabla_\alpha u| = \sqrt{\sum_{\beta \in T(\alpha)} (u_\beta - u_\alpha)^2} \tag{19}$$

For a given noisy signal u, the digital TV filter $F^{\lambda, \alpha}$ is a nonlinear data-dependent filter $F^{\lambda, \alpha} : u \to v$, where u is any existing signal on Ω and v output. For any node $\alpha \in \Omega$

$$v_\alpha = F_\alpha(u) = \sum_{\beta \in T(\alpha)} h_{\alpha\beta}(u) u_\beta + h_{\alpha\alpha}(u) u_\alpha \tag{20}$$

where the important filtering coefficients are given as follows

$$h_{\alpha\beta}(u) = \frac{w_{\alpha\beta}(u)}{\lambda + \sum_{\gamma \in T(\alpha)} w_{\alpha\gamma}(u)} \tag{21}$$

$$h_{\alpha\alpha}(u) = \frac{\lambda}{\lambda + \sum_{\gamma \in T(\alpha)} w_{\alpha\gamma}(u)} \tag{22}$$

$$w_{\alpha\beta}(u) = \begin{cases} 0, & |\nabla_\beta u| \geq T \\ \dfrac{1}{|\nabla_\alpha u|_a + |\nabla_\beta u|_a}, & |\nabla_\beta u| < T. \end{cases} \tag{23}$$

The node α and β can be seen in Fig.1, α is central node and β as neighborhood nodes, T is a threshold parameter. In our designed algorithm, we toggle a-type and b-type in Fig.1 as an iteration filtering. Here, 5-node stencil is employed, and 9-node stencil suggested.

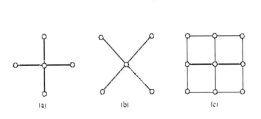

Fig.1 5-node stencil and 9-node stencil

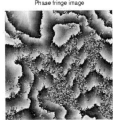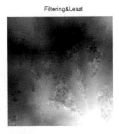

Fig.2 Phase fringes graph & LS phase unwraping

More attention must be paid for phase fringe image (scale 0-1), not only important edges but also abrupt phase fringes caused by atan-function transform. So key protected techniques have developed while filtering phase images for residues reduction. For complex-value interferogram images, we should use vector digital TV filtering (VDTVF) instead of general DTVF. Defining a map $\Gamma : \Omega \to \Re^2$, and denoting Γ_α as the value of interferogram at a node α, we have local variation of TV-norms which is well defined as $|\nabla_\alpha \Gamma| = \sum_{\beta \in T(\alpha)} |\Gamma_\beta - \Gamma_\alpha|_{TV}$. However, in this paper, we mainly focus on phase image using DTVF.

4. Numerical examples and conclusions

We take the representative phase image data ifsar512.pha (512×512) (Fig. 2: left) presented by Ghiglia and Pritt [9], which is of some better phase fringes and some worse noise or discontinuation. Our numerical experiments operate on PC with Pentium 1.0G CPU and 512MB memory, Visual C++ as developed language. Some key processing schemes is as follows. Firstly, we adopt three-node stencil scheme, toggling between a-type and b-type, as well as residues map upgrading. Secondly, the residue map is employed as a filter-guide, wishing zero-weight for α points from β points of high gradient value. Finally, Branch-cut phase wrapping proceeds after profiteering. In our reported numerical results, the trade-off parameter λ is 0.9 in model (21, 22), the threshold parameter T is 1.5 for 2-norm and 0-1 phase scale, the number of iteration filtering is 4 by alternative a-type or b-type.

Fig.3 shows the linking residues distribution and branch-cut map. For whole image ifsar512.pha, the number of residues goes down from 5885 to 2875, and more load generating branch-cut reduced, which saving about 1/3 time-consuming contrast to the unfiltering.

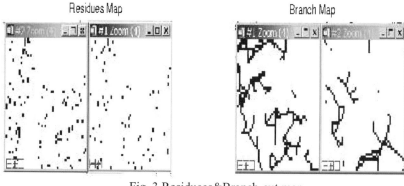

Fig. 3 Residuces&Branch-cut map

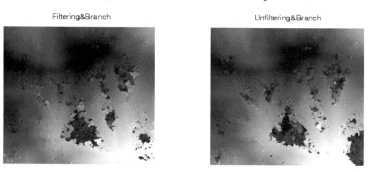

Fig. 4 Branch-cut phase wrapping&prefiltering

Another, Fig.2 (right) shows Least-square phase unwrapping result whose time-cost is 66 seconds. While, both results in Fig. 4 cost less than 1 second (not including filtering), and filtering time cost less 2 seconds. Actually, with the increase of image size, prefiltering cost becomes more and more valuable. Besides computation, prefiltering is also important for quality improved.

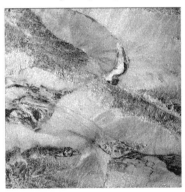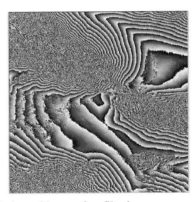

Fig. 5 Ifsar4096 interferogram and phase fringes after filtering

Fig. 5 is the interferogram of ifsar4096.corr and interferometric phase fringes after filtering. Fig.6 is the phase unwrapping result. The numerical experiment shows the ratio of residues number being 109755:53063 between before and after filtering. Further, after filtering the phase fringes data, we gain a new ratio 53063:38263, which greatly reduces the computation load.

Fig. 6 Phase Unwrapped result after filtering Ifsar4096

In summary, the improved DTVF algorithm for phase unwrapping has less computation load than anisotropic diffusion filtering and provides reliable accuracy for phase unwrapping. Our numerical results provide phase fringes protected and residues largely reduced, which lay the foundation for efficient branch-cut phase wrapping.

References

[1] Gao X, Liu L. and Huang H., A Survey - image processing and analysis using PDE and geometry-driven diffusion, *Advanced Mathematics ACTA*, 2003, 32(3): 285-294

[2] Gao X., Wang C. and Zhang H., Anisotropic diffusion filtering and phase unwrapping for Interferometric SAR, *Proceedings of IGARSS'2002*, Tolando, Canada

[3] Wang C., Gao X., Zhang H., The vector digital TV filtering and phase unwrapping, *Proceedings of IGARSS 2003*, Toulouse, France

[4] Lee J. S., Note: Digital image smoothing and the sigma filter, *Computer Vision, Graphics, and Image Processing*, 1983, 24: 255-269

[5] van Genderen J. L., Huang Y., The ITC filter for improving ERS SAR interferograms, http://earth.esrin.esa.it

[6] Goldstein R. M., Zebker H.A. and Werner C., Satellite interferometry: 2-D phase unwrapping, (\sl Radio science), 1988, 23(4): 713-720

[7] Ghiglia D. C., and Romero L.A., Robust two-dimensional weighted and unweighed phase unwrapping that uses fast transforms and iterative methods, *J. Optical Society of America A*, 1994, 11(1): 107-117

[8] Gens R. and Van Genderen J.L., Review article SAR interferometry-issues, techniques, applications, *Int. J. Remote Sensing*, 1996, 17(10): 1803-1835

[9] Ghiglia, D. C. and Pritt M.D., Two-dimensional phase unwrapping: theory, algorithms, and software. New York: Wiley, 1998

[10] Zebker H. A., and Lu Y., Phase unwrapping algorithms for radar interferometry: residue-cut, least-squares, and synthesis algorithms, *J. Opt. Soc. Am. A.*, 1998, 15(3): 586-598

[11] Alvarez L. et al, "Axioms and fundamental equation of image processing", Arch Ration Mechan, 1993, 123: 199-257

[12] Perona P. and Malik J., Scale-space and edge detection using anisotropic diffusion, *IEEE Trans. Pattern Analysis and Machine Intelligence*, 1990, 12(7): 629-639

[13] Osher S. and Sethian J., Fronts propagating with curvature dependent speed, algorithms based on the Hamilton-Jacobi formulation, *J. Comp. Physics*, 1988, 79: 12-49

[14] You Y. L. and Kaveh M., Blind image restoration by anisotropic regularization, *IEEE Trans. Image Processing*, 1999, 8(3)

[15] Chen Y., Vemari B.C. and Wang L., Image denoising and segmentation via nonlinear diffusion, *SIAM J. Appl. Math*, to be published

[16] Chen Y., Barcelos C.A. and Mair B.A., `Selective smoothing and segmentation by time-dependent penalized total variation, preprint

[17] Barcelos C.A., Chen Y. Heat flows and related minimization problem in image restoration, *Computers and Mathematics with Application*, 2000, 39: 81-97

[18] Malladi R, Sethian J, Vemuri B.C., A fast level set based logarithm for topology-independent shape modeling, *J. Mathematical Imaging and Vision*, 1996, 6: 269-289

[19] S. Osher and L. I. Rudin, Feature-Oriented Image Enhancement Using Shock Filters, *SIAM J. Numerical Analysis*, 1990, 27(4): 919-940

[20] Rudin L, Osher S., Fatemi E., Nonlinear total variation based noise removal algorithm, *Physica D*, 1992, 60: 259-268

[21] Sethian J.A., *Level set methods*, Cambridge University Press, 1996

[22] Chan T. F., Osher S., and Shen J., The digital TV filter and nonlinear denoising, *IEEE Trans. on Image Processing*, 2001, 10(2): 231-241

[23] Catte F., et al., Image selective smoothing and edge detection by nonlinear diffusion, *SIAM J. Numerical Analysis*, 1992, 29(1): 182-193

[24] Alvarez L., Lions P.L. and Morel J.M., Image selective smoothing and edge detection by nonlinear diffusion II, *SIAM J. Numerical Analysis*, 1992, 29(3): 845-866

II. Scattering from Randomly Rough Surfaces

Wave Propagation,Scattering and Emission in Complex Media
Edited by Ya-Qiu Jin
Science Press and World Scientific,2004

75

Modeling Tools for Backscattering from Rough Surfaces

A.K.Fung[1]and K.S. Chen[2]

[1]EE Dept., Box 19016, University of Texas at Arlington, Arlington, TX 76019, U.S.A.

[2]CSRSR, National Central University, Chungli, Taiwan, China

Abstract In this study we show that backscattering models are available for a variety of surface conditions including surfaces with a narrow or a broad surface spectrum. In particular, we shall consider the application of three backscattering models *in algebraic form* to field measurements and numerical simulation results. All three models have a wider range of applicability than most existing models currently available in the literature: (1) a scattering model for surfaces whose high spatial frequency components are Gaussian correlated and may shift the relative amount of the contributions between high and low frequency components as specified by the user so that when there is a larger contribution from the higher spectral components the contribution from the lower spectral components will decline and vice versa, (2) a scattering model for nearly exponentially correlated surfaces which may contain a varying amount of spectral components smaller than or comparable to the correlation length of the surface as specified by the user and (3) a scattering model with a two-scale correlation function, an algebraic sum of two exponential-like functions, the relative weighting of which can be adjusted by the user. It is believed that most surface backscattering problems can be explained with one of the three models. A general conclusion from model applications is that the roughness components of a given surface is represented more explicitly by its spectrum than its correlation function. It is also found that the exponential correlation tends to overestimate backscattering at large angles of incidence.

1. INTRODUCTION

The roughness of ground surfaces has large variations in correlation and rms height properties. To model scattering from such surfaces different surface correlation functions or surface spectra are needed. It is desirable to have closed form representations for the Fourier and Bessel transforms of the nth power of the correlation functions so that the scattering model will have an algebraic form instead of an integral. To this aim three normalized correlation functions are considered below:

$$(1) \ \rho_1(r) = \left(1 + \frac{r^2}{L^2}\right)^{-x}, \ r \geq 0, x > 1$$

$$(2) \ \rho_2(r) = \exp\left[-\frac{r}{L}(1 - e^{-r/x})\right], \ r \geq 0, x > 0$$

$$(3) \ \rho_3(r) = \left\{\sigma_1^2 \exp\left[-\frac{r}{L_1}(1 - e^{-r/x_1})\right] + \sigma_2^2 \exp\left[-\frac{r}{L_2}(1 - e^{-r/x_2})\right]\right\} / (\sigma_1^2 + \sigma_2^2).$$

The first correlation function will be referred to as the *x-power* correlation. It allows many choices of the exponent x from unity on up. Its functional form changes with x as shown in Fig.1, where we have plotted an exponential and a Gaussian function with the same correlation parameter L to

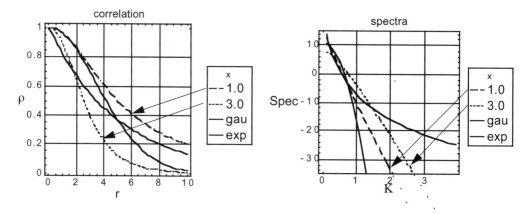

Fig.1 Normalized correlation $\rho = (1 + r^2/L^2)^{-x}$ and its corresponding surface spectrum based on $L = 5$ cm.

serve as references. It is Gaussian-like over small correlation distances near the origin and does not show any agreement with the exponential function. However, its spectral form can be close to an exponential spectrum over small wave numbers, K, i.e., over the low spatial frequency region, when $x = 1$. When $x = 3$, its spectrum has the Gaussian appearance but contains many more high spatial frequency components than the corresponding Gaussian. The rms slope of the surface represented by this correlation is $s = (\sigma\sqrt{2x})/L$, where σ is the standard deviation of the surface height. Clearly, we can choose the value of x to obtain different spectral forms but this correlation function cannot approach an exponential regardless of the choice of x.

The second correlation function is an *exponential-like* function. It can approximate an exponential correlation, when $x \ll L$. It has an rms slope given by $s = \sigma\sqrt{2/(xL)}$. When $x \to 0$, this correlation function becomes exponential and its rms slope tends to infinity. For $x > 0$ this correlation function acts like a Gaussian over a small region near the origin thus affecting backscattering at large angles of incidence. An illustration of this correlation and its spectrum is given in Fig.2 along with an exponential and a Gaussian with the same L to provide comparisons. The exponential-like correlation always approaches the exponential function over large lag distances.

From Fig.2 we see that as x increases, it can cause a substantial drop in the high frequency components of the surface spectrum. *Note that there is not much difference between the surface correlation with x =1 and the exponential function over all lag distances but the difference between the high frequency portions of the corresponding spectra are significant.* It shows (1) a small difference in the correlation properties near the origin can be important to scattering at large angles of incidence, and (2) the surface spectrum is a better indicator of scattering properties because it is more sensitive to changes in the high frequency spectral components.

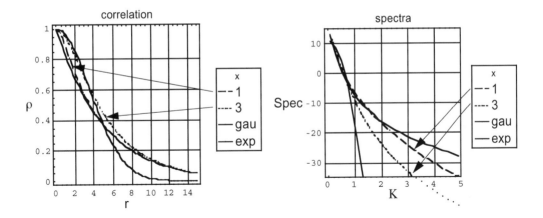

Fig.2 Normalized correlation $\rho = \exp[-(r/L)(1 - e^{-r/x})]$ and its corresponding surface spectrum based on $L = 5$ cm. Its rms slope is $\sigma\sqrt{2/(xL)} = 0.189$ when $x_1 = 1$ and equals 0.1095 when $x_1 = 3$.

For surfaces with moderate rms slopes Fig.2 shows that when such a surface possesses a *single correlation parameter it should* contain significantly less high frequency components than the corresponding exponential spectrum. Thus, the exponential correlation function cannot be used even though the scattering properties over small angles of incidence appears to be exponential. If such a surface does produce a high level of scattering at large angles of incidence, this can be due to the presence of another roughness scale as represented by the third correlation function which is a *two-scale correlation*. Its rms slope is given by $\sqrt{2}\sqrt{\sigma_1^2/(x_1L_1) + \sigma_2^2/(x_2L_2)}$. An illustration using the same surface parameters as in Fig.2 for the first term plus choosing $\sigma_2 = 0.1, L_2 = 2, x_2 = 0.5$ all in cm for the second term is shown in Fig. 3.

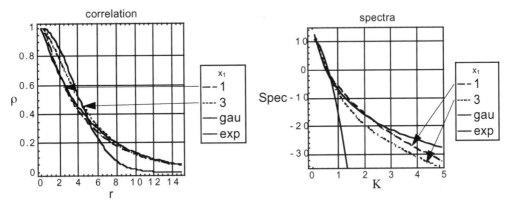

Fig.3 A two-scale correlation $\rho = \{\sigma_1^2\exp[-(r/L_1)(1 - e^{-r/x_1})] + \sigma_2^2\exp[-(r/L_2)(1 - e^{-r/x_2})]\}/(\sigma_1^2 + \sigma_2^2)$ and its corresponding surface spectrum based on $\sigma_1 = 0.3, L_1 = 5, \sigma_2 = 0.1, L_2 = 2, x_2 = 0.5$ all in cm. Two values of x_1 are shown. Its rms slope is 0.179 when $x = 3$.

Relative to Fig.2 we do not see much difference between the corresponding correlation functions but the spectra in the high spectral value region are quite different. The additional term in the correlation function is intended to increase the spectral components in the high frequency region. This will enhance scattering at large angles of incidence. *Note that the added term in the surface correlation enables us to better approximate an experimentally measured surface correlation or surface spectrum*, while maintaining a finite value for the rms slope of the surface.

2. The Backscattering Model

The backscattering model used here is a special case of the bistatic surface scattering model given by Fung et al. [2002]. The general forms of the backscattering coefficients for vertical, σ_{vv}^0 and horizontal, σ_{hh}^0, polarizations are given by (1)

$$\sigma_{pp}^0 = \frac{k^2}{4\pi} \exp[-4k_z^2\sigma^2] \left\{ \left| (2k_z\sigma)f_{pp} + \frac{\sigma}{4}(F_{pp1} + F_{pp2}) \right|^2 w(2k\sin\theta, 0) \right.$$

$$\left. + \sum_{n=2}^{\infty} \left| (2k_z\sigma)^n f_{pp} + \frac{\sigma}{4} F_{pp1}(2k_z\sigma)^{n-1} \right|^2 \frac{w^{(n)}(2k\sin\theta, 0)}{n!} \right\} \qquad (1)$$

where $p = v, h, f_{vv} = 2R_v/\cos\theta, f_{hh} = -2R_h/\cos\theta$ and

$$F_{vv1} = \frac{4k}{\sqrt{\varepsilon_r - \sin^2\theta}} \left\{ (1 - R_v)^2 \varepsilon_r \cos\theta + (1 - R_v)(1 + R_v)\sin^2\theta(\sqrt{\varepsilon_r - \sin^2\theta} - \cos\theta) \right.$$

$$\left. -(1 + R_v)^2 \left[\cos\theta + \frac{\sin^2\theta}{2\varepsilon_r}(\sqrt{\varepsilon_r - \sin^2\theta} - \cos\theta) \right] \right\}$$

$$F_{vv2} = 4k\sin^2\theta \left[(1 - R_v)^2 \left(1 + \frac{\varepsilon_r\cos\theta}{\sqrt{\varepsilon_r - \sin^2\theta}} \right) - (1 - R_v)(1 + R_v) \left(3 + \frac{\cos\theta}{\sqrt{\varepsilon_r - \sin^2\theta}} \right) \right.$$

$$\left. + (1 + R_v)^2 \left(1 + \frac{1}{2\varepsilon_r} + \frac{\varepsilon_r\cos\theta}{2\sqrt{\varepsilon_r - \sin^2\theta}} \right) \right]$$

$$F_{hh1} = \frac{-4k}{\sqrt{\varepsilon_r - \sin^2\theta}} \left\{ (1 - R_h)^2 \cos\theta + (1 - R_h)(1 + R_h)\sin^2\theta(\sqrt{\varepsilon_r - \sin^2\theta} - \cos\theta) \right.$$

$$\left. -(1 + R_h)^2 \left[\varepsilon_r\cos\theta + \frac{\sin^2\theta}{2}(\sqrt{\varepsilon_r - \sin^2\theta} - \cos\theta) \right] \right.$$

$$F_{hh2} = -4k\sin^2\theta\left[(1-R_h)^2\left(1+\frac{\cos\theta}{\sqrt{\varepsilon_r-\sin^2\theta}}\right) - (1-R_h)(1+R_h)\left(3+\frac{\cos\theta}{\sqrt{\varepsilon_r-\sin^2\theta}}\right)\right.$$

$$\left. + (1+R_h)^2\left(1+\frac{1}{2}+\frac{\cos\theta}{2\sqrt{\varepsilon_r-\sin^2\theta}}\right)\right]$$

The quantities $w, w^{(n)}$ are the surface spectra corresponding to the two-dimensional Fourier transforms of the surface correlation coefficient $\rho(x, y)$ and its nth power, $\rho^n(x, y)$, defined as follows in polar forms:

$$w(\kappa, \varphi) = \int_0^{2\pi}\int_0^\infty \rho(r, \phi)e^{-j\kappa r\cos(\varphi-\phi)}r\,dr\,d\phi \text{ and}$$

$$w^{(n)}(\kappa, \varphi) = \int_0^{2\pi}\int_0^\infty \rho^n(r, \phi)e^{-j\kappa r\cos(\varphi-\phi)}r\,dr\,d\phi \tag{2}$$

If the surface roughness is independent of the view direction, the correlation coefficient is isotropic depending only on r. In this case (2) becomes

$$w(\kappa) = 2\pi\int_0^\infty \rho(r)J_0(\kappa r)r\,dr \text{, and } w^{(n)}(\kappa) = 2\pi\int_0^\infty \rho^n(r)J_0(\kappa r)r\,dr \tag{3}$$

where $J_0(\kappa r)$ is the zeroth order Bessel function. It is worth noting that the first term in (1) reduces to the first-order perturbation model when $k\sigma$ is small.

In (1) R_v, R_h are the Fresnel reflection coefficients which can be generalized by replacing them with a reflection transition function [Wu et al., 2001] that allows the argument of the Fresnel reflection coefficients to change from the incident angle to the specular angle as the operating frequency changes from low to high or roughness from small to large. They are defined as follows:

$$R_{vt} = R_v(\theta) + [R_{v0} - R_v(\theta)](1 - S_t/S_{t0})$$
$$R_{ht} = R_h(\theta) + [R_{h0} - R_h(\theta)](1 - S_t/S_{t0}) \tag{4}$$

where R_{v0}, R_{h0} are the Fresnel reflection coefficients evaluated at the specular angle which means normal incidence for backscattering.

$$S_t = \frac{|F_t|^2 \displaystyle\sum_{n=1}^\infty \frac{(k\sigma\cos\theta)^{2n}}{n!}w^{(n)}(2k\sin\theta)}{\displaystyle\sum_{n=1}^\infty \frac{(k\sigma\cos\theta)^{2n}}{n!}\left|F_t + \frac{2^{n+2}R_{v0}}{e^{(k\sigma\cos\theta)^2}\cos\theta}\right|^2 w^{(n)}(2k\sin\theta)},$$

$$F_t = 8R_{v0}^2 \sin^2\theta \left(\frac{\cos\theta + \sqrt{\varepsilon_r - \sin^2\theta}}{\cos\theta\sqrt{\varepsilon_r - \sin^2\theta}} \right)$$

$$S_{t0} = \left| 1 + \frac{8R_{v0}}{F_t\cos\theta} \right|^{-2} \text{ is the limit of } S_t \text{ as } k\sigma \to 0.$$

The functional form of this transition function indicates that $1 - S_t/S_{t0} \to 1$ as frequency or surface rms height becomes large and goes to zero as frequency or surface rms height is small. For very large dielectric values the magnitudes of R_{vt}, R_{ht} approach unity. Thus, the reflection coefficients with transitional properties, R_{vt}, R_{ht}, provide a desired change when either frequency or roughness changes. Note that this is an estimate which may not correctly handle all roughness, frequency and angular changes correctly especially at large angles of incidence. In the low frequency region backscattering is expected to follow the $R_h(\theta)$-based curve. After transition, it should tend towards the $R_h(0)$-based curve in the high frequency region.

3. APPLICATION OF THE *X*-POWER CORRELATION

The spectrum corresponding to the nth power of the *x*-power correlation function, $[1 + (r/L)^2]^{-x}$, $x > 1$ takes the following form in the scattering model depicted in (1)

$$w^{(n)}(2k\sin\theta) = \begin{cases} \dfrac{2\pi L^2}{3n - 2}, & \sin\theta = 0 \\[2ex] \dfrac{2\pi L^2 (kL\sin\theta)^{xn-1} BesselK[1-xn, 2kL\sin\theta]}{Gamma[xn]} \end{cases} \tag{5}$$

When we change the surface correlation function, the scattering models defined by equations (1) predict a different angular trend. The use of the *x*-power correlation function leads to a nearly linear variation over small incident angles and a fast drop-off over large incident angles which resembles the results based on the Gaussian correlation.

With the above representations for surface spectra, we illustrate the scattering model behaviors for the like polarizations and show comparisons with measured and simulated data in the following subsections.

3.1 Theoretical trends for like polarization with *x*-power correlation

Effects of changing x

We shall first illustrate how the backscattering coefficients change with *x*. Then, we do more detailed illustration for $x = 1.5$. This choice is due partly because we have some numerically simulated data at $x = 1.5$ with which we shall make comparisons to see how well the model works. When we carry out comparisons with field measurements, other choices of *x* will be considered. In Fig.4 we show the behaviors of vertically and horizontally polarized backscattering versus the incident angle. There is a wide range of angular curves generated by changing x of the x-power correlation function. However, there is no angular shape that resembles the exponential which is often observed in field measurements over an angular region near the vertical. The fast rise between normal and 5 degree incidence, when x is equal to one, is due to the tail of this correlation function being higher than the exponential function with the same L. It is not clear whether such a change is physically possible in non-coherent scattering. For this reason we recommend x *values larger than one in practical applications.*

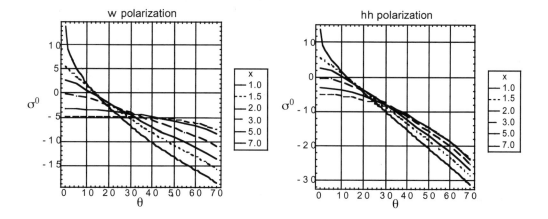

Fig.4 Changes in the form of the angular form of the backscattering coefficients due to changes in x with f = 5 GHz, σ = 0.3 cm, L = 3 cm, and relative dielectric constant, ε = 80-j50.

Effects of rms height

For further illustration we select x = 1.5 and show the backscattering behavior according to (1) by selecting rms heights of 0.2, 0.3, 0.5, 0.7, 0.9 cm at a correlation lengths of 3.0 cm and a frequency of 5 GHz for a surface with dielectric constant, $\varepsilon_r = 80 - j50$ in Fig.5. This large dielectric constant is selected so that we can see better the changes in the spacing between vertical (vv) and horizontal polarization (hh). As the rms height increases, the backscattering coefficients for both vv and hh increase until the rms height reaches about 0.5 cm. Further increase in the rms height causes a decrease in backscattering over the small incident angle region for both polarizations. This is expected since the angular curve should be more isotropic for rougher surfaces. Note that over all incident angles the vv polarized backscattering is always larger than or equal to hh. Fig.5 also shows that the rise in hh polarization with the rms height is significantly faster than vv. It is close to a factor of two at 70 degrees. Intuitively, one would expect the spacing between vv and hh polarizations to narrow as roughness increases, because for very rough surfaces there should be negligible polarization dependence. This is, indeed, the case as shown in Fig.6 where we see that as the surface rms height increases, the spacing between vertical and horizontal polarization curves narrows.

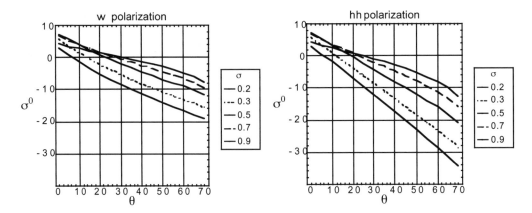

Fig.5 Polarized backscattering coefficients at 5 GHz for $L = 3$ cm, $\varepsilon_r = 80 - j50$ and 1.5-power correlation showing an increase with increasing rms height until $\sigma = 0.5$ cm for (a) vv and (b) hh polarization. Then, it drops at near nadir angles. vv is always higher than or equal to hh due to the use of a large dielectric constant.

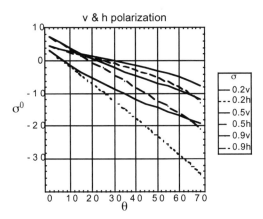

Fig.6 The spacing between vv and hh decreases as the surface rms height increases showing a decrease in polarization dependence for rougher surfaces. f = 5 GHz, $L = 3$, $\varepsilon_r = 80 - j50$ with 1.5-power correlation.

Effects of correlation length

When we increase the correlation length, L, it gives a faster rate of decrease of the backscattering coefficient with the incident angle. This observed effect is common to all other forms of the surface correlation function but the angular shape and the amount of change will vary. What is typical of a 1.5-power correlation is that it maintains a nearly linear trend over small incident angles for both vv and hh polarizations, whereas the exponential and Gaussian correlations would generate an angular backscattering curve with a positive and a negative curvature respectively. Another major property of the 1.5-power correlation is that it may cause a backscattering curve to have a positive curvature first at a smaller incident angle and then a negative curvature at a larger incident angle as shown in Fig.7 for $L = 2$ cm in vv polarization. In most cases shown in Fig.7 the linear trends in backscattering are followed by a drop with a negative curvature.

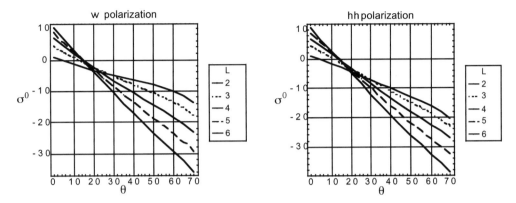

Fig.7 Faster decrease of vv and hh backscattering coefficients at 5 GHz with increasing correlation length L for $\varepsilon_r = 13 - j0.2$, $\sigma = 0.5$ cm and 1.5-power correlation.

An increase in the correlation length also closes the gap between vv and hh polarizations with 1.5-power correlation. This point is illustrated in Fig.8.

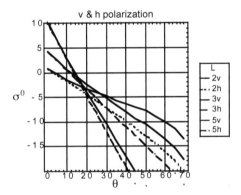

Fig.8 An increase in the correlation length of a 1.5-power correlation function causes the spacing between vv and hh polarization to narrow very quickly. f = 5 GHz, $\sigma = 0.5$ cm and $\varepsilon_r = 13 - j0.2$.

Effects of frequency

A change in frequency amounts to changing both rms height and the correlation length simultaneously by the same proportion. Hence, as frequency increases, we anticipate both a faster drop-off at large angles of incidence same as when the correlation length increases (Fig.9) and a narrowing of the spacing between vv and hh polarizations similar to an increase in rms height (Fig.5). This is illustrated in Fig.9. These properties are similar to those with the Gaussian correlation.

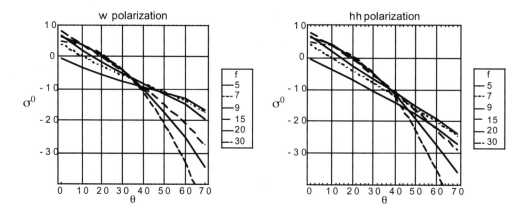

Fig.9 Frequency variations in GHz with $\sigma = 0.3$, $L = 2$ cm, and $\varepsilon_r = 16 - j0.2$ showing a faster drop-off in backscatter coefficient with the incident angle for (a) vv polarization and (b) hh polarization.

Unlike an increase in rms height, a higher frequency does not cause isotropic scattering because the surface roughness properties remain unchanged. The narrowing of spacing between vv and hh backscattering curves is faster with frequency change, because an increase in either rms height or the correlation length is a cause for narrowing. This point is shown in Fig.10.

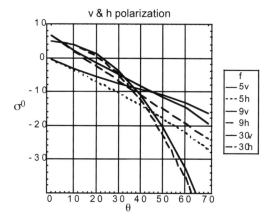

Fig.10 The spacing between vv and hh polarization narrows quickly as frequency increases. The selected parameters are $\sigma = 0.3$, $L = 2$ cm, and $\varepsilon_r = 16 - j0.2$. The vv and hh scattering coefficients act as upper and lower bounds of the Kirchhoff scattering coefficient in the high frequency region.

For a surface with a spectrum wider than the corresponding Gaussian, as frequency increases, the backscattering curve will approach the geometric optics in the small incident angle region but not necessarily in the large incident angle region. This is because away from the specular direction the roughness spectral component responsible for scattering decreases in size and can become comparable or smaller than the exploring wavelength.

3.2 Comparison with Measurements and Simulations

Comparisons with field measurements

As an illustration of this model application, we show comparisons with backscattering measurements from an unknown asphalt rough surface in Fig.11 and Fig.12 at 8.6 GHz.

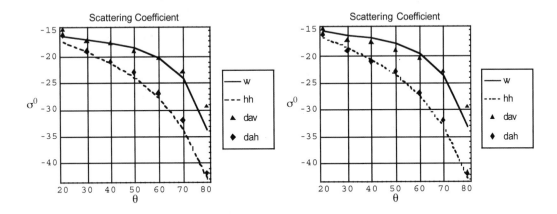

Fig.11 Comparisons of vv and hh polarization with radar measurements taken at 8.6 GHz over an asphalt surface.
Data are taken from Ulaby et al [1986] reproduced in Fung [1994, p.95].$x = 1.5$, $\sigma = 0.13cm$, and $L = 0.47cm$, (a)
$\varepsilon = 5$ and (b) $\varepsilon = 4 - j3$.

In Fig.11 (a) and (b) we consider the effect of changing the relative values of the real and imaginary parts of the dielectric constant. We see that a decrease in the real part and an increase in the
imaginary part of the dielectric constant causes the backscattering curve to increase more near 20
degrees incidence than at large angles thus showing a more peaked backscattering curves and a
better fit to the data. This action also improves the matching at 70 and 80 degrees, although the
improvements are small.

To see the impact of the *x*-power correlation we show in Fig.12 (a) a similar fit as in Fig.11 (b)
but with x reduced to 1.2 from 1.5 while keeping the dielectric value unchanged at $\varepsilon = 5$. Here,
we see that there is a similar effect on backscattering due to a reduced x to a reduction in the real
part of the dielectric constant, while keeping its magnitude the same. By keeping the real part of
the dielectric constant unchanged as in Fig.11 (a) we can use a smaller x value to raise the
backscattering curve at small incident angles. Unlike changing the dielectric constant the
reduction in x influences mainly the small angles leaving the agreement at large angles relatively
unaffected. In Fig.12 (b) we show a fit mainly to the data beyond 20 degrees incidence, because
leaving out the 20 degree points the data show a different angular trend. The purpose here is to
demonstrate the change in the angular shape of the backscattering curves when we increase x to
2.8. It is seen that a larger x causes a larger downward bending in the smaller angular region and
simultaneously raises the level of vv at large angles of incidence. This action causes a larger
separation between vv and hh at the large angular region allowing a better fit for vv at 80 degree
incidence. However, it requires a much larger dielectric value ($\varepsilon = 8 - j3$) to bring the scattering

level over small incident angles back to where it should be. The overall effect favors both the fits at small and large angular regions, but the widening between vv and hh polarizations in the 50 to 60 degree range leads to a poorer fit at those angles.

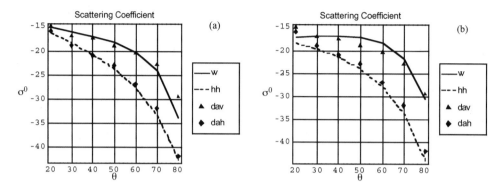

Fig.12 Comparisons of vv and hh polarization with radar measurements taken at 8.6 GHz over an asphalt surface. Data are taken from Ulaby et al [1986] reproduced in Fung [1994, p.95]., $\sigma = 0.13cm$, and $L = 0.47cm$, (a) $x = 1.2$, $\varepsilon = 5$ and (b) $x = 2.8$, $\varepsilon = 8 - j3$.

Now consider a multi-frequency data set (1.5, 4.75 and 9.5 GHz) reported by Oh et al. [1992] which is also available in Qin et al. [2002]. The following information has been reported about the surface: $\sigma = 1.12cm$ and $L = 8.4cm$ with dielectric values of $\varepsilon_r = 15.3 - j3.7$ at 1.5 GHz, $\varepsilon_r = 15.2 - j2.1$ at 4.75 GHz and $\varepsilon_r = 13.14 - j3.8$ at 9.5 GHz. The comparisons between this data set and the model predictions are given in Fig.13 and Fig.14.

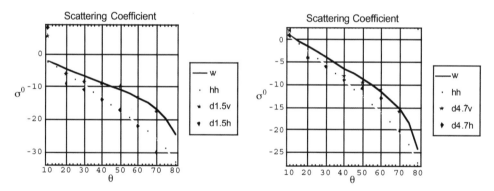

Fig.13 (a) Backscattering of vv and hh polarizations based on x-power correlation from a surface (a) with $\sigma = 1.12$ cm, $L = 8.4cm$ and $x = 1.8$ at 1.5 GHz, and (b) with $\sigma = 0.6$ cm, $L = 4cm$ and $x = 2.5$ at 4.75 GHz.

In these figures x has been selected to be 1.8 at 1.5 GHz and 2.5 at higher frequencies. It is believed that most real ground surfaces have more than one scale of roughness. Furthermore, when the surface correlation length is larger than a wavelength, roughness scales smaller than a wavelength becomes important and scales longer than a correlation length is becoming the reference plane for scattering. As a result, we have used different rms heights and correlation lengths to fit data at higher frequencies. While general agreements are obtained between the model and measurements, there is a clear difference in angular trends. We shall address this issue when we apply the multi-scale correlation to the same data set in Section 6.

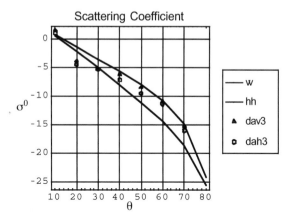

Fig.14 Backscattering of vv and hh polarizations based on the x-power correlation $\sigma = 0.35$ cm, $L = 2$ cm and $x = 2.5$ at 9.5 GHz. The reported dielectric constant is $\varepsilon_r = 13.14 - j3.8$.

Comparisons with simulations

Next, we show a comparison with two-dimensional (2D) moment method simulations over a range of frequencies from 1 to 15 GHz at an incident angle of 10 degrees in vv and hh polarizations. The simulated backscattering points were taken from Wu et al. [2001]. As shown in Fig.15 at 10 degrees there is no appreciable difference between the backscattering curves using the incident or the specular angle. Hence, this comparison provides an example showing that the new IEM model gives the correct frequency dependence.

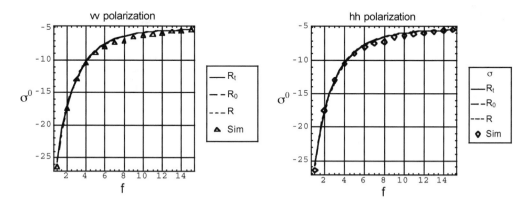

Fig.15 At 30 degree incidence and the parameters, $\sigma = 0.429 cm$, $L = 3 cm$, $\varepsilon_r = 3 - j0.1$ with Gaussian correlation. Simulation points are from Wu et al. [2001]. (a) hh polarization (b) vv polarization.

In order to show that the model defined in (1) does provide the correct transition in frequency from low to high, we have included two additional backscattering curves one using only the incident angle in the Fresnel reflection coefficients and the other only the specular angle in Fig.16 and Fig.17.

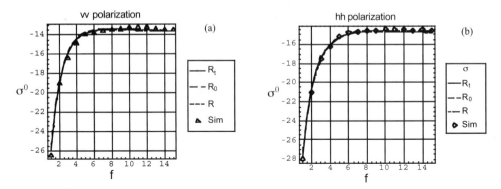

Fig.16 At 30 degree incidence and the parameters, $\sigma = 0.429cm$, $L = 3cm$, $\varepsilon_r = 3 - j0.1$ with Gaussian correlation. Simulation points are from Wu et al. [2001]. (a) hh polarization (b) vv polarization.

Fig.16 is for an incident angle of 30 degrees and Fig.17 is for 50 degrees. These larger incident angles are chosen because there is a clear difference in the Fresnel reflection coefficients at 0 and 30 or 50 degrees. How they affect the backscattering curves depend on our formulation of the surface current. In Fig.17 we do see different backscattering curves based on $R(0)$ and $R(\theta)$, but the curves are so close together that they do not require an additional transition function.

At 50 degree incidence, we see a clear separation between the two backscattering curves denoted by dash lines. Thus, a transition from one to the other is needed. In Fig.17 we see that the transition takes place from about 6 GHz to 12 GHz. It also shows that without the transition function we would only know the upper and lower bounds but not an estimate of the backscattering coefficient at a given frequency.

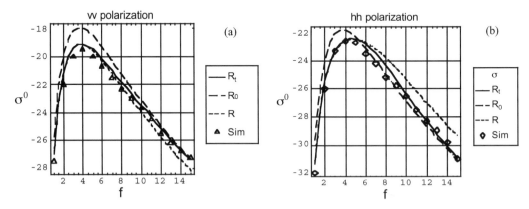

Fig.17 At 50 degree incidence and the parameters, $\sigma = 0.429cm$, $L = 3cm$, $\varepsilon_r = 3 - j0.1$ with Gaussian correlation. Simulation points are from Wu et al. [2001]. (a) hh polarization (b) vv polarization.

Next, we show additional two 2D simulations under different roughness conditions for vv and hh polarizations versus the incident angle. We would expect that under small roughness condition the simulated backscattering points will be closer to the backscattering curve calculated based on reflection coefficients evaluated at the incident angle. For large roughness we expect the reflection coefficient evaluated at the specular angle to give a closer agreement. These points are illustrated in the figures below. Fig.18 shows a case where roughness scales are somewhat larger than what is permitted by the perturbation model. We expect the simulation to follow mostly the backscattering curve with $R(\theta)$ which is, indeed, the case for both polarizations.

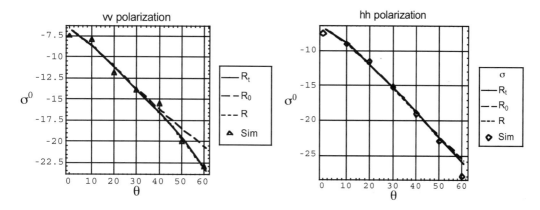

Fig.18 Comparisons of vv and hh polarization with moment method simulations by Wu et al. [2001]. $k\sigma = 0.449$, $kL = 3.142$ with 1.5 power correlation

When the roughness scales are increased to $k\sigma = 0.719$ - cm, scattering is far away from the perturbation condition. The theoretical backscattering curve has moved away from the $R(\theta)$ curve when θ is larger than 40 degrees. The deviation, however, is within the fluctuation of the simulation points. Thus, Fig.19 can serve as an indication of change but not a proof that a transition is under way.

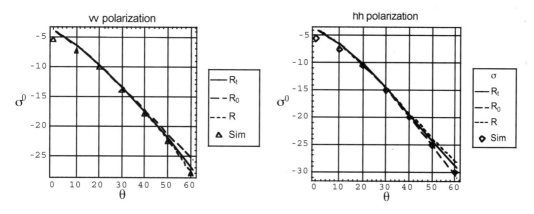

Fig.19 Comparisons of vv and hh polarization with moment method simulations by Wu et al. [2001]. $k\sigma = 0.719$, $kL = 5.027$ with 1.5 power correlation

A further increase to $k\sigma = 0.988$ cm meets the condition for geometric optics. The backscattering curve at large angles should follow the curve with $R(0)$ for both polarizations. Although the amount of change in Fig.20 remains too small to provide a definite proof, the trends are in agreement with what we expect.

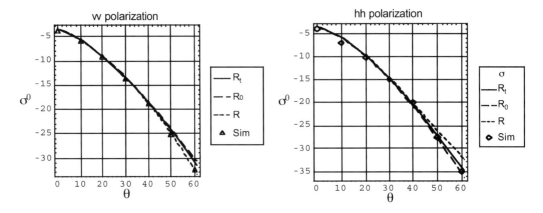

Fig.20 Comparisons of vv and hh polarization with moment method simulations by Wu et al. [2001]. $k\sigma = 0.988$, $kL = 6.912$ with 1.5 power correlation

4. EXPONENTIAL-LIKE CORRELATION

In the previous section we considered a correlation function that is like a Gaussian over small lag distance and generates a nearly linear response over small angles of incidence but it can never approach an exponential. For this reason we now consider a correlation function, $R(r)$, that can approach an exponential and has an rms slope. We refer to this function as the exponential-like correlation,

$$R(r) = \sigma^2 \exp\left[-\frac{r}{L}(1 - e^{-r/x})\right] \tag{6}$$

where $L \geq x$. The rms slope of this correlation function is $\sigma\sqrt{2/(xL)}$.

An analytic expression for the Bessel transform of the nth power of this correlation function is

$$w^{(n)}(2k\sin\theta) = \sum_{m=0}^{\infty} \left(\frac{n}{L}\right)^m \frac{L_e^{m+2}}{m!}\Gamma(m+2)$$

$$Hypergeometric2F1\left[\frac{2+m}{2}, \frac{3+m}{2}, 1, -(2kL_e\sin\theta)^2\right] \tag{7}$$

where $L_e = (xL)/(nx + mL)$. The above expression is efficient for numerical calculation because even though the upper limit for the series is infinity, generally, only ten terms are needed for convergence.

4.1 Theoretical trends for exponential-like correlation

Effects of changing x

In this section we show the backscattering angular trends based on $R(r)$ in (6) for vv and hh polarizations. The surface parameters are taken to be $\sigma = 0.15$ cm and $L = 3$ cm with a relative dielectric constant of 25 . Calculations are carried out at 10 GHz for x values of 0.1, 1.0, 1.5, 2.0, 2.5, 3 as shown in Fig.21. At 10 GHz the effective roughness parameters are $k\sigma = 0.314$ and $kL = 6.28$. For large x there is a significant drop over large angles of incidence for both polarizations. This is because a larger x means a smaller rms slope and a smoother surface (Fig.2) thus reducing backscattering at large angles of incidence. As x decreases, the tail part (large-angle region) of the backscattering curve begins to increase and shows a flatter response. This is because a smaller x yields a larger rms slope representing a rougher surface (Fig.2) which tends to raise backscattering in the large angular region.

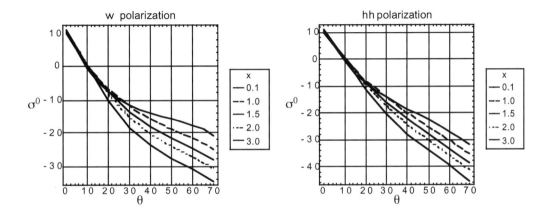

Fig.21 Polarized backscattering based on $R = \sigma^2 \exp[-(r/L)(1 - e^{-r/x})]$, $\sigma = 0.15$ cm and $L = 3$ cm at f = 10 GHz. The backscattering level over large angles of incidence decreases with an increase in x.

The surface spectrum with a large x will be higher than that with a smaller x over small lag distances (Fig.2). Then, there is a crossover point and the levels of the two spectra interchanged. At a sufficiently low frequency, only the first portion of the spectrum are effective causing the backscattering level of the surface with larger x to be higher, thus reversing the backscattering levels of the two surfaces discussed in the previous paragraph. As an example, consider $\sigma = 1.1$ cm and $L = 8$ cm. A variation of x from 0.05 to 0.5 does not cause much change over small angles of incidence. Over large angles increasing values of x causes an increase in the backscattering level instead of a decrease, *exactly the opposite to the trend observed earlier* for surfaces with small scale roughness. This is shown in Fig.22. Further increase of x from 0.5 to 3 yields the same trend as before, i.e., scattering decreases with increasing x over large angles of incidence, because a spectrum with a larger x has a shorter lag distance before its level drops off below the one with a smaller x value. In addition over small angles of incidence the backscattering level is increasing with x instead of remaining unchanged. This additional change in trend behavior is shown in Fig.23. The backscattering curves in this figure are in direct proportion to their corresponding surface spectra with larger incident angles corresponding to higher surface wave numbers.

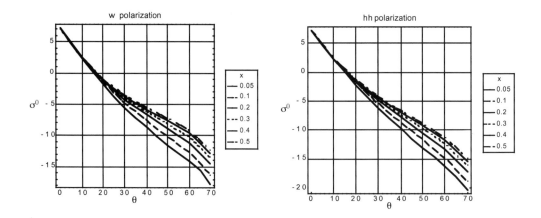

Fig.22 Polarized backscattering based on $R = \sigma^2 \exp[-(r/L)(1 - e^{-r/x})]$, $\sigma = 1.10$ cm and $L = 8$ cm at f = 5 GHz. As x increases from 0.05 cm to 0.5 cm, the backscattering level over large angles of incidence increases also.

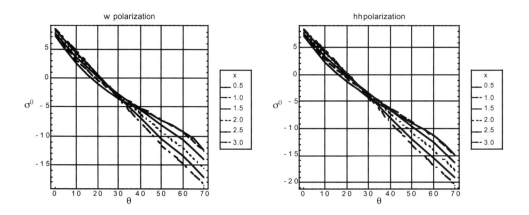

Fig.23 Polarized backscattering based on $R = \sigma^2 \exp[-(r/L)(1 - e^{-r/x})]$, $\sigma = 1.10$ cm and $L = 8$ cm at f = 5 GHz. As x increases from 0.5 cm to 3 cm, backscattering level over large angles of incidence decreases but it increases over small angles of incidence.

Effects of frequency

Next, we want to show how the backscattering curves vary with frequency, while holding x constant at $x = 1.5$ (Fig.24) keeping other surface parameters unchanged at $\sigma = 0.15$ cm, $L = 3$ cm, and $\varepsilon_r = 25$. From Fig.24 we see that as frequency increases from 1 GHz the scattering curves change from a more isotropic scattering into one that shows a peak near vertical incidence. The isotropic scattering behavior is typical of small scale roughness. As frequency increases, the roughness appears larger to the incident wavelength resulting in a gradual increase in overall scattering strength and peaking near vertical incidence. A given incident frequency can only sense *a range of roughness spectral components responsible for scattering.* Spectral components too large or too small compared to the exploring wavelength are not in this *effective range.* At low frequencies all spectral components appear small, i.e., the high wave number portion of the effective range is not filled up. As frequency increases, the low wave number components of the surface spectrum begin to act as large scale contributors to scattering. There is peaking and raising of the overall scattering level, as the number of large scale roughness increases. Simultaneously, there is an appearance of saturation spreading towards the small incident angle region. This saturation appearance is due to many more low wave number spectral components appearing as large scale contributors. It *does not mean* that backscattering has approached geometric optics condition as is evident from the angular shape of the backscattering curve. In fact, the very presence of roughness scales small or comparable to the incident wavelength implies that the geometric optics condition is not satisfied. Their presence is evidenced by the separation between vv and hh as shown in Fig.25.

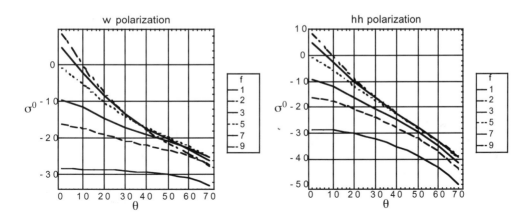

Fig.24 Polarized backscattering based on exponential-like correlation at $x = 1.5$ with $\sigma = 0.15$ cm, $L = 3$ cm and $\varepsilon_r = 25$. Starting at about 5 GHz, there is an appearance of saturation in backscattering. Actually, this is due to the specific shape of the surface spectrum instead of approaching the geometric optics.

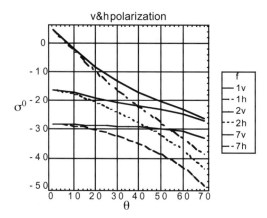

Fig.25 Backscattering of vv and hh based on exponential-like correlation at $x = 1.5$ with $\sigma = 0.15$ cm, $L = 3$ cm and $\varepsilon_r = 25$. Separation between vv and hh is significant, although saturation appearance has occurred at f = 7 GHz.

In the previous paragraph we showed a case where there is an appearance of saturation but the result does not indicate geometric optics because there is a separation between vv and hh polarizations. There is another situation where vv tends to hh *without* approaching the geometric optics. This is when the surface rms slope is large. An illustration of this point is shown in Fig.26, where the size of rms height has been increased from 0.15 to 0.75 cm with $x = 1$ cm and the correlation length is 4 cm. By plotting vv and hh together we see that for large rms slope, vv is approximately equal to hh as shown in Fig.26.

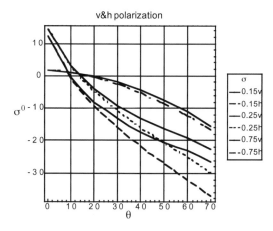

Fig.26 Polarized backscattering based on exponential-like correlation at $f = 10$ GHz with $x = 1$ cm, $L = 4$ cm and $\varepsilon_r = 25$.

4.2 Comparison with measurements

The correlation discussed in this section has an rms slope and can also generate a backscattering curve just like an exponential function. This explains why an exponential correlation can be used in an analytic surface scattering model as an approximation in the low frequency region. In the moderate and high frequency regions the use of an exponential correlation may overestimate scattering at large angles of incidence.

The first set of multi-frequency data to compare is the set over dry asphalt reported in Ulaby et al. [1986] at 8.6, 17, and 35.6 GHz as shown in Fig.27 and Fig.28.

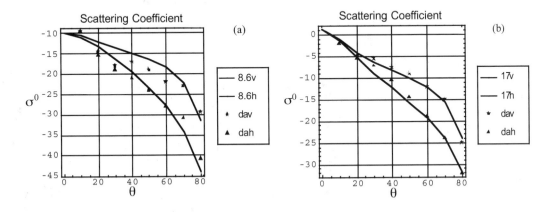

Fig.27 Comparisons with measurements from an asphalt surface reported by Ulaby et al. [1986, p. 1809] at (a) 8.6 GHz and (b) 17 GHz. Chosen model parameters are $\sigma = 0.12$ cm, $L = 0.6$ cm and a dielectric constant of 5.

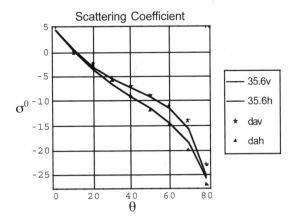

Fig.28 Comparisons with measurements from an asphalt surface reported by Ulaby et al. [1986, p. 1809] at 35.6 GHz. Chosen model parameters are $\sigma = 0.12$ cm and $L = 0.6$ cm with a dielectric constant of 12.

This is a surface consists mainly of small scale roughness. The selected surface parameters are given in the figure legends as σ = 0.12 cm and L = 0.47 cm. The chosen x value is 0.03 for both 8.6 GHz and 35.6 GHz. A value of x equal to 0.2 is used for 17 GHz. Generally, the model is able to predict frequency changes in dealing with surface of small rms height. The dielectric constant used for all three frequencies is 16 which is too high for a dry asphalt surface, unless there is moisture immediately below the surface. Thus, while we have good agreement in angular trends with measurements, it is not clear whether there is agreement in level. In the next data set acquired by Oh et al [1992] we shall be dealing with a surface with a large correlation length. Its rms height is again smaller than the exploring wavelengths. Hence, we still expect to obtain reasonable agreements without adjusting the surface parameters.

This data set was acquired by Oh et al. [1992]. Here, the surface parameters have been given as σ = 0.4 cm, L = 8.4 cm and the dielectric constants are $15.6 - j3.7$, $15.4 - j2.15$, and $12.3 - j3.55$ at frequencies of 1.5, 4.75 and 9.5 GHz respectively. For this data set x has been chosen to be 0.1 based on data matching at 1.5 GHz. It turns out that fairly good matching can be realized over all three frequencies without any adjustment of the surface parameters as shown in Fig.29.

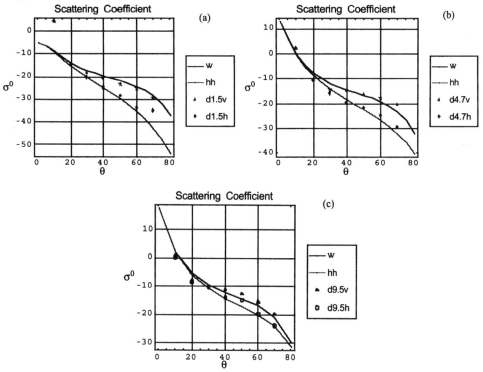

Fig.29 Comparison with data using surface parameters provided by Oh et al.[1992] at (a) 1.5 GHz, (b) 4.75 GHz and (c) 9.5 GHz. The value of x has been chosen to be 0.1 and σ = 0.4 cm, L = 8.4 cm.

5. A TWO-SCALE CORRELATION FUNCTION

The two-scale correlation function is useful for surfaces that have a wide spectral content. It can make up a portion of the spectral content missed by the use of a single-scale correlation. The two-scale correlation considered here is the sum of two exponential-like functions whose spectrum can be written as

$$
w(K) = \sum_{m=0}^{n} \frac{n! a^m b^{n-m}}{(n-m)! m!} \left\{ \frac{L_{e1}^2}{[1+(KL_{e1})^2]^{3/2}} \right.
$$

$$
+ \sum_{p=1}^{\infty} \left(\frac{m}{L_1}\right)^p \frac{L_{p1}^{p+2}}{p!} \Gamma(p+2) Hypergeometric2F1\left[\frac{p+2}{2}, \frac{p+3}{2}, 1, -(KL_{p1})^2\right]
$$

$$
+ \sum_{p=1}^{\infty} \left(\frac{m}{L_1}\right)^p \frac{1}{p!} \sum_{q=0}^{\infty} \left(\frac{n-m}{L_2}\right)^q \frac{L_e^{q+2+p}}{q!} \Gamma(q+2+p) Hypergeometric2F1\left[\frac{q+2+p}{2}, \frac{q+3+p}{2}, 1, -(KL_e)^2\right]
$$

$$
\left. + \sum_{q=0}^{\infty} \left(\frac{n-m}{L_2}\right)^q \frac{L_{q1}^{q+2}}{q!} \Gamma(q+2) Hypergeometric2F1\left[\frac{q+2}{2}, \frac{q+3}{2}, 1, -(KL_{q1})^2\right] \right\} \qquad (8)
$$

where $L_{e1} = \dfrac{L_1 L_2}{(n-m)L_1 + mL_2}$; $L_{p1} = \dfrac{x_1 L_{e1}}{x_1 + pL_{e1}}$; $L_{q1} = \dfrac{x_2 L_{e1}}{x_2 + qL_{e1}}$; $L_x = \dfrac{x_1 x_2}{qx_1 + px_2}$

$$
L_e = \frac{L_x L_{e1}}{L_x + L_{e1}}
$$

Although the theoretical sums in (8) are to infinity, usually ten terms are sufficient due to the quick convergence of these hyper geometric functions. It is understood that $L_1 > L_2 > x_1 > x_2$.

The surface measurement by Oh et al.[1992] for a fairly rough surface was considered in Section 3. It is the one with rms height of 1.12 cm and a correlation length of 8.4 cm with all dielectric values provided as $\varepsilon_r = 15.6 - j3.7$, $\varepsilon_r = 15.4 - j2.15$ and $\varepsilon_r = 13 - j3.8$ respectively at 1.5, 4.75 and 9.5 GHz. By adding a small scale roughness, $\sigma_2 = 0.35$ cm and $L_2 = 1.8$ cm, and decreasing the rms height σ_1 from 1.12 cm to 0.5 cm and 0.2 cm at 4.75 GHz and 9.5 GHz, while leaving L_1 unchanged at 8.4 cm, we can realize a much better fit to this set of data than was given in Section 3. This is, in part, because the correlation function used in Section 3 was not the proper one. The results and selected scales are shown in Fig.30 and Fig.31. There is no change in the dielectric values provided by Oh et al. [1992].

The rationale here is that for this surface there is actually a smaller scale roughness sitting on top of the one measured by Oh et al. [1992]. As frequency increases to 9.5 GHz, the correlation length of 8.4 cm is more than twice the wavelength. Thus, there is the possibility that coherency cannot be maintained over the full length of 8.4 cm due to the disturbance of the smaller scale. Furthermore, when the smaller scale roughness is effective in scattering, the larger scale roughness is acting as its reference plane of scattering. For this reason, the rms height of the large scale roughness is not detected to the same proportion as when there is no small scale roughness present. It is for this reason that we reduce the rms height, σ_1 as frequency increases. All other surface parameters remain fixed over the range of frequency considered.

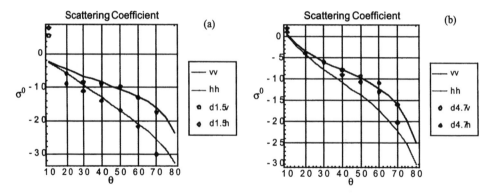

Fig.30 Comparisons with data by Oh et al.[1992] for the surface, $\sigma = 1.12$ and $L = 8.4$. The two-scale correlation function is used with (a) $\sigma_{1,2} = 1.12, 0.35$, $L_{1,2} = 8.4, 1.8$, $x_{1,2} = 1, 0.5$ and $\varepsilon_r = 15.6 - j3.7$ at 1.5 GHz and (b) $\sigma_{1,2} = 0.5, 0.35$, $L_{1,2} = 8.4, 1.8$, $x_{1,2} = 1, 0.5$ and $15.4 - j2.15$ at 4.75 GHz.

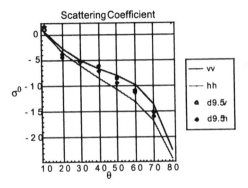

Fig.31 A comparison between model and data acquired by Oh et al. [1992] for the surface, $\sigma = 1.12$ and $L = 8.4$ at 9.5 GHz. The two-scale correlation function is used with $\sigma_{1,2} = 0.2, 0.35$, $L_{1,2} = 8.4, 1.8$, $x_{1,2} = 1, 0.5$ and $\varepsilon_r = 13 - j3.8$. The small scale is the same as in Fig 30.

6. DISCUSSIONS

The above study shows that for idealized surfaces as provided by numerical simulation it is possible to obtain very good match between theoretical model predictions and simulated data without adjusting surface parameters at all frequencies. For real soil surface agreement between model and measurements is possible only over a narrow range of frequency. This is because soil surfaces usually have more than one scale of roughness and it is not a truly continuous surface as assumed in theoretical models. At a wavelength comparable to or longer than the correlation length of the surface, there is no problem because surface irregularities have been smooth out by the wavelength. At higher frequencies small irregularities will appear as additional roughness scale not yet accounted for.

Another problem with multi-scale surfaces is that when there is a small scale riding on a much larger one (a factor of two or more in correlation length), the larger scale will not be seen by an exploring wavelength comparable in size to the small scale, *because the coherency needed to see the large scale over two or more wavelength distance is destroyed by the presence of the small scale.* For this reason, the full size of the rms height of the large scale roughness should not be used as input to the model because it is not acting as a roughness scale in scattering. Instead, it is serving as a reference plane for the scattering of the small scale roughness.

In this study the x-power correlation may be viewed as an extension of the Gaussian correlation into higher spectral region, while the exponential-like correlation is an extension of the exponential correlation into the lower spectral region. Both of these correlation functions are able to cover a wider range of spectral regions than what is available in an exponential or a Gaussian function. Further extension of the range of correlation function is shown to be possible by adding two exponential-like functions together. Thus, the models presented have a wide coverage and are believed to be able to deal with most of the practical cases that one may encounter.

7. REFERENCES

Fung, A.K. Microwave Scattering and Emission Models and Their Applications, Artech House, 1994

Fung, A.K., An improved IEM model for bistatic scattering from rough surfaces, J. of Electromagnetic Applications, Vol. 16, No. 5, pp. 689-702, 2002

Oh, Y. Sarabandi, K., and F.T.Ulaby, An empirical model and an inversion technique for radar scattering from bare soil surface, IEEE GRS Vol. 30, no. 2, pp. 370-381, 1992

Ulaby, F.T., Moore, R.K. and A. K. Fung, Microwave Remote Sensing, Volume 3, Artech House, 1986

Wu, T.D. Chen, K.S., Shi, J. and A.K. Fung, A transition model for the reflection coefficient in surface scattering, IEEE GRS Vol. 39, no. 9, pp. 2040-2050, 2001

Pseudo-Nondiffracting Beams from Rough Surface Scattering

E. R. Méndez

División de Física Aplicada, Centra de Investigación

Científica y de Educación Superior de Ensenada,

Apdo. Postal 2732, Ensenada, B.C. 22800, México

T. A. Leskova, A. A. Maradudin

Department of Physics and Astronomy,

and Institute for Surface and Interface Science,

University of California, Irvine, CA 92697, U.S.A.

In this paper we present an approach to the design of a two-dimensional circularly symmetric randomly rough Dirichlet surface that, when illuminated at normal incidence by a scalar plane wave, scatters it with a specified dependence of its intensity along the normal to the surface in the absence of the roughness. This approach is illustrated by applying it to the design of a surface that produces a constant scattered intensity within a finite region of this normal, and no scattered intensity outside this region. It is validated by the results of rigorous computer simulation calculations of scattering from the surface designed in this way. It is then shown that a random surface of this type can be used to produce a pseudo-nondiffracting beam in reflection, namely a scattered beam with a finite beam aperture, a finite propagation range, a variation in transverse beam profiles, and an intensity peak in the direction of propagation.

There is a great deal of interest at the present time in nondiffracting beams. They are important for applications including optical interconnection, high precision autocollimation, optical alignment, laser machining, and laser surgery [1–6]. The nondiffracting beam introduced by Durnin[1, 2] is a solution of the free-space wave equation of the form $E(\rho, z) = J_0(\alpha\rho)\exp(i\beta z)$, in which $\alpha^2 + \beta^2 = k^2$, where k is the wave number, $J_0(x)$ is the zero-order Bessel function, and (ρ, θ, z) are the cylindrical coordinates. This beam has an infinite extent in the transverse plane, and is capable of propagating to infinity in the z-direction without spreading. Such an ideal nondiffracting beam contains an infinite amount of energy, and is impossible to realize in practice. Consequently, most recent studies of nondiffracting beams have focused on pseudo-nondiffracting beams, which have a finite beam aperture [2–14]. Such beams have a finite propagation range, have variation in transverse beam profiles, and intensity peaks in the direction of propagation. Nevertheless, the propagation length can extend to several tens of centimeters, long enough for many applications.

Several approaches have been developed to produce three-dimensional [2, 6, 8] and even two-dimensional [15] pseudo-nondiffracting beams. The latter are characterized by a constant intensity along the direction of propagation z and a beam-like shape in one of its transverse directions, say x.

In this paper we present an approach to the production of a three-dimensional pseudo-nondiffracting beam that is based on designing a two-dimensional circularly symmetric random Dirichlet surface that, when illuminated at normal incidence by a scalar plane wave, produces a scattered field that is a three-dimensional pseudo-nondiffracting beam. This surface is defined by the equation $x_3 = H(r)$, where $r = (x_1^2 + x_2^2)^{\frac{1}{2}}$ is the radial coordinate in the plane $x_3 = 0$. The region $x_3 > H(r)$ is vacuum, while the region $x_3 < H(r)$ is the scattering medium.

We approach the problem of designing such a surface by first solving the problem of designing a two-dimensional circularly symmetric random Dirichlet surface that, when illuminated at normal incidence by a scalar plane wave, produces a scattered field with a prescribed distribution of intensity along the axis normal to the surface in the absence of the roughness (the positive x_3-axis). This approach is used to design a surface that produces a constant scattered intensity along a finite segment of the positive x_3-axis, and produces zero scattered intensity along the remainder of that axis. That the surface generated in this way produces the desired distribution of scattered intensity is confirmed by the results of rigorous computer simulation calculations. It is then shown that the scattered intensity is a rapidly decreasing function of the coordinate r transverse to the x_3-axis for each value of x_3. This suggests that if a surface is designed to produce a constant scattered intensity along the x_3-axis in the interval $0 < x_3 < z_0$, where z_0 is some large distance, then the intensity of the resulting scattered beam will be confined to the neighborhood of the x_3-axis, that is it will be a pseudo-nondiffracting beam. This suggestion is also confirmed by the results of rigorous computer simulation calculations.

The physical system we consider initially in this paper consists of vacuum in the region

$x_3 > \zeta(\mathbf{x}_\parallel)$, where $\mathbf{x}_\parallel = (x_1, x_2, 0)$ is a position vector in the plane $x_3 = 0$, and the scattering medium in the region $x_3 < \zeta(\mathbf{x}_\parallel)$. The surface profile function $\zeta(\mathbf{x}_\parallel)$ is assumed to be a single-valued function of $\mathbf{x}_\parallel$ that is differentiable with respect to x_1 and x_2, and to constitute a random process, but not necessarily a stationary one. It is further assumed that the Dirichlet boundary condition is satisfied on this surface.

The surface $x_3 = \zeta(\mathbf{x}_\parallel)$ is illuminated from the vacuum side by a scalar plane wave of frequency ω,

$$\psi(\mathbf{x}|\omega)_{inc} = \exp[i\mathbf{k}_\parallel \cdot \mathbf{x}_\parallel - i\alpha_0(k_\parallel)x_3], \tag{1}$$

where

$$\alpha_0(k_\parallel) = [(\omega/c)^2 - k_\parallel^2]^{\frac{1}{2}} \qquad k_\parallel < \omega/c \tag{2a}$$

$$= i[k_\parallel^2 - (\omega/c)^2]^{\frac{1}{2}} \qquad k_\parallel > \omega/c. \tag{2b}$$

We have assumed a time dependence of the field of the form $\exp(-i\omega t)$, but have suppressed explicit reference to it.

An application of Green's second integral identity to the region $x_3 > \zeta(\mathbf{x}_\parallel)$ yields the result that the total field in this region, $\psi(\mathbf{x}|\omega)$ can be written in the form [16],

$$\theta(x_3 - H(r))\psi(\mathbf{x}|\omega) = \psi(\mathbf{x}|\omega)_{inc} - \frac{1}{4\pi} \int d^2x'_\parallel [g_0(\mathbf{x}|\mathbf{x}')]_{x'_3=\zeta(\mathbf{x}'_\parallel)} L(\mathbf{x}_\parallel|\omega), \tag{3}$$

where $\theta(z)$ is the Heaviside unit step function, $g_0(\mathbf{x}_\parallel|\mathbf{x}'_\parallel)$ is the scalar free-space Green's function,

$$g_0(\mathbf{x}|\mathbf{x}') = \frac{e^{i\frac{\omega}{c}|\mathbf{x}-\mathbf{x}'|}}{|\mathbf{x} - \mathbf{x}'|}, \tag{4}$$

and the source function $L(\mathbf{x}_\parallel|\omega)$ is given by

$$L(\mathbf{x}_\parallel|\omega) = \left(-\frac{\partial\zeta(\mathbf{x}_\parallel)}{\partial x_1}\frac{\partial}{\partial x_1} - \frac{\partial\zeta(\mathbf{x}_\parallel)}{\partial x_2}\frac{\partial}{\partial x_2} + \frac{\partial}{\partial x_3}\right)\psi(\mathbf{x}|\omega)\Bigg|_{x_3=\zeta(\mathbf{x}_\parallel)}. \tag{5}$$

The fact that the total field in the vacuum region $\psi(\mathbf{x}|\omega)$ satisfies the Dirichlet boundary condition

$$\psi(\mathbf{x}|\omega)\Bigg|_{x_3=\zeta(\mathbf{x}_\parallel)} = 0 \tag{6}$$

on the surface $x_3 = \zeta(\mathbf{x}_\parallel)$ has been used in the derivation of Eq. (3). The scattered field is therefore given by

$$\psi(\mathbf{x}|\omega)_{sc} = -\frac{1}{4\pi} \int d^2x'_\parallel [g_0(\mathbf{x}|\mathbf{x}')]_{x_3=\zeta(\mathbf{x}'_\parallel)} L(\mathbf{x}'_\parallel|\omega). \tag{7}$$

In the Kirchhoff approximation, which we adopt here for simplicity, $L(\mathbf{x}_\parallel|\omega)$ is given by

$$L(\mathbf{x}_\parallel|\omega) = 2 \left(-\frac{\partial\zeta(\mathbf{x}_\parallel)}{\partial x_1}\frac{\partial}{\partial x_1} - \frac{\partial\zeta(\mathbf{x}_\parallel)}{\partial x_2}\frac{\partial}{\partial x_2} + \frac{\partial}{\partial x_3} \right) \psi(\mathbf{x}|\omega)_{inc} \Bigg|_{x_3=\zeta(\mathbf{x}_\parallel)} \tag{8a}$$

$$= -2i \left(\frac{\partial\zeta(\mathbf{x}_\parallel)}{\partial x_1}k_1 + \frac{\partial\zeta(\mathbf{x}_\parallel)}{\partial x_2}k_2 + \alpha_0(k_\parallel) \right)$$
$$\times \exp[i\mathbf{k}_\parallel \cdot \mathbf{x}_\parallel - i\alpha_0(k_\parallel)\zeta(\mathbf{x}_\parallel)]. \tag{8b}$$

When $x'_3 = \zeta(\mathbf{x}'_\parallel)$, and $x_1, x'_1 \ll x_3$, $x_2, x'_2 \ll x_3$, we have that

$$|\mathbf{x} - \mathbf{x}'| \doteq [(x_1 - x'_1)^2 + (x_2 - x'_2)^2 + (x_3 - \zeta(\mathbf{x}'_\parallel))^2]^{\frac{1}{2}}$$
$$= x_3 \left[1 - \frac{2\zeta(\mathbf{x}'_\parallel)}{x_3} + \frac{\zeta^2(\mathbf{x}'_\parallel)}{x_3^2} + \frac{(x_1 - x'_1)^2 + (x_2 - x'_2)^2}{x_3^2} \right]^{\frac{1}{2}}. \tag{9}$$

In particular, on the x_3-axis, $x_1 = x_2 = 0$, we see that

$$|\mathbf{x} - \mathbf{x}'| \cong x_3 - \zeta(\mathbf{x}'_\parallel) + \frac{x'^2_1 + x'^2_2}{2x_3} + \cdots \tag{10}$$

Consequently, if we now restrict our attention to the case of normal incidence $\mathbf{k}_\parallel = 0$, then by the use of Eqs. (7), (4), (9) and (10) we can write the scattered field along the x_3-axis in the form

$$\psi(0,0,x_3|\omega)_{sc} \cong \frac{i}{2\pi}\frac{\omega}{c}\frac{e^{i\frac{\omega}{c}x_3}}{x_3} \int d^2x'_\parallel e^{-2i\frac{\omega}{c}\zeta(\mathbf{x}'_\parallel)+i\frac{\omega}{c}\frac{x'^2_\parallel}{2x_3}}. \tag{11}$$

We now make the assumption that the surface profile function $\zeta(\mathbf{x}_\parallel)$ is a function of $\mathbf{x}_\parallel$ only through its magnitude, $|\mathbf{x}_\parallel| = r$, and write

$$\zeta(\mathbf{x}_\parallel) = H(r). \tag{12}$$

This assumption, together with the assumption of normal incidence, ensures that the intensity of the scattered field is also circularly symmetric. With this assumption Eq. (9) becomes

$$\psi(0,0,x_3|\omega)_{sc} = i\left(\frac{\omega}{c}\right)\frac{e^{i\frac{\omega}{c}x_3}}{x_3} \int_0^\infty dr\, r\, e^{-2i\frac{\omega}{c}H(r)+i\frac{\omega}{c}\frac{r^2}{2x_3}}. \tag{13}$$

We now make the change of variable $r^2 = t$, and introduce the definition

$$H(\sqrt{t}) = h(t). \tag{14}$$

As a result we obtain

$$\psi(0, 0, x_3|\omega)_{sc} = i \left(\frac{\omega}{2c}\right) \frac{e^{i\frac{\omega}{c}x_3}}{x_3} \int_0^\infty dt\, e^{-2i\frac{\omega}{c}h(t)+i\frac{\omega}{c}\frac{t}{2x_3}}. \tag{15}$$

The intensity of the scattered field along the x_3-axis is therefore

$$\begin{aligned}
I(x_3) &\equiv |\psi(0, 0, x_3|\omega)_{sc}|^2 \\
&= \left(\frac{\omega}{2cx_3}\right)^2 \int_0^\infty dt \int_0^\infty dt'\, e^{i\frac{\omega}{2cx_3}(t-t')} e^{-i\frac{2\omega}{c}[h(t)-h(t')]}.
\end{aligned} \tag{16}$$

Our goal, now, is to find the function $h(t)$ that produces a specified form for $I(x_3)$. As it stands, the expression for $I(x_3)$ given by Eq. (16) is too difficult to invert to obtain $h(t)$. We therefore make an approximation analogous to passing to the geometrical optics limit of the Kirchhoff approximation. Namely, we expand the difference $h(t) - h(t')$ about $t' = t$, and retain only the leading nonzero term:

$$h(t) - h(t') \cong (t - t')h'(t), \tag{17}$$

where the prime denotes differentiation with respect to the argument. With this approximation the average of $I(x_3)$ with respect to the ensemble of realizations of the surface profile function becomes

$$\langle I(x_3) \rangle = \left(\frac{\omega}{2cx_3}\right)^2 \int_0^{Nb^2} dt \int_0^{Nb^2} dt'\, e^{i\frac{\omega}{2cx_3}(t-t')} \left\langle e^{-2i\frac{\omega}{c}h'(t)(t-t')} \right\rangle, \tag{18}$$

where N is a large positive integer that tends to infinity, while b is a characteristic length that will be defined more precisely below. Equation (18) can be rewritten more conveniently as

$$\begin{aligned}
\langle I(x_3) \rangle &= \left(\frac{\omega}{2cx_3}\right)^2 \int_0^{Nb^2} dt \int_{-t}^{Nb^2-t} du\, e^{-i\frac{\omega}{2cx_3}u} \left\langle e^{2i\frac{\omega}{c}uh'(t)} \right\rangle \\
&= \left(\frac{\omega}{2cx_3}\right)^2 \sum_{n=0}^{N-1} \int_{nb^2}^{(n+1)b^2} dt \int_{-t}^{Nb^2-t} du\, e^{-i\frac{\omega}{2cx_3}u} \left\langle e^{2i\frac{\omega}{c}uh'(t)} \right\rangle.
\end{aligned} \tag{19}$$

This is the equation that we can invert to obtain $h(t)$ in terms of $I(x_3)$.

To accomplish this we assume the following form for $h(t)$,

$$h(t) = \frac{a_n}{b}t + b_n, \qquad nb^2 \leq t \leq (n+1)b^2, \quad n = 0, 1, 2, \ldots \tag{20}$$

where $\{a_n\}$ are independent identically distributed random deviates. Consequently, the probability density function (pdf) of a_n, $f(\gamma) = \langle \delta(\gamma - a_n) \rangle$, is independent of n. The $\{b_n\}$ are determined from the condition that the surface profile function $h(t)$ be a continuous function of t, and are given by

$$b_n = b_0 + (a_0 + a_1 + \cdots + a_{n-1} - na_n)b \qquad n \geq 1. \tag{21}$$

With this form for $h(t)$ Eq. (19) becomes

$$\langle I(x_3) \rangle = \left(\frac{\omega}{2cx_3}\right)^2 \sum_{n=0}^{N-1} \int_{nb^2}^{(n+1)b^2} dt \int_{-t}^{Nb^2-t} du \, e^{-\frac{i\omega}{2cx_3}u} \left\langle e^{2i\frac{\omega}{c}\frac{u}{b}a_n} \right\rangle$$

$$= \left(\frac{\omega}{2cx_3}\right)^2 \sum_{n=0}^{N-1} \int_{nb^2}^{(n+1)b^2} dt \int_{-t}^{Nb^2-t} du \, e^{-i\frac{\omega}{2cx_3}u} \int_{-\infty}^{\infty} d\gamma f(\gamma) e^{2i\frac{\omega}{c}\frac{u}{b}\gamma}$$

$$= \left(\frac{\omega}{2cx_3}\right)^2 N^2 b^4 \int_{-\infty}^{\infty} d\gamma f(\gamma) \left(\text{sinc}\left[\frac{\omega}{c}\left(\gamma - \frac{b}{4x_3}\right)Nb\right]\right)^2. \tag{22}$$

where $\text{sinc}\,x = \sin x/x$. Now, in the limit as $N \to \infty$,

$$(\text{sinc}Nx)^2 \xrightarrow[N \to \infty]{} \frac{\pi}{N}\delta(x). \tag{23}$$

With the aid of this result Eq. (22) simplifies to

$$\langle I(x_3) \rangle = \frac{\pi}{4}\frac{\omega}{c}\frac{Nb^3}{x_3^2} f\left(\frac{b}{4x_3}\right). \tag{24}$$

Thus, the mean scattered intensity along the x_3-axis is given in terms of the pdf of a_n. With the change of variable $b/(4x_3) = \gamma$, we obtain the result that

$$f(\gamma) = \frac{1}{4\pi}\frac{c}{\omega}\frac{1}{Nb}\frac{\langle I(\frac{b}{4\gamma}) \rangle}{\gamma^2}. \tag{25}$$

Since x_3 has been assumed to be positive, we see that $f(\gamma)$ is nonzero only for positive values of γ. Because $f(\gamma)$ is nonzero only for positive γ, all the slopes $\{a_n\}$ are positive. The function $H(r)$ is therefore a monotonically increasing function of r.

For example, if we seek to design a surface that produces a constant scattered intensity within the interval $z_1 < x_3 < z_2$ of the x_3-axis, and zero scattered intensity along the rest of the x_3-axis,

$$\langle I(x_3) \rangle = A\theta(x_3 - z_1)\theta(z_2 - x_3) \qquad z_2 > z_1, \tag{26}$$

where $\theta(z)$ is the Heaviside unit step function, we obtain from Eq. (25) the result that

$$f(\gamma) = A\frac{c}{4\pi\omega}\frac{1}{Nb}\frac{1}{\gamma^2}\theta\left(\gamma - \frac{b}{4z_2}\right)\theta\left(\frac{b}{4z_1} - \gamma\right). \tag{27}$$

The constant A is determined from the normalization condition for $f(\gamma)$,

$$\int_{-\infty}^{\infty} d\gamma f(\gamma) = 1 = A\frac{c}{4\pi\omega}\frac{1}{Nb}\int_{\frac{b}{4z_2}}^{\frac{b}{4z_1}} d\gamma \frac{1}{\gamma^2}$$

$$= A\frac{c}{\pi\omega}\frac{1}{Nb^2}(z_2 - z_1), \tag{28}$$

from which we obtain

$$A = \frac{\pi\omega}{c}\frac{Nb^2}{z_2 - z_1}. \tag{29}$$

It therefore follows that

$$\langle I(x_3) \rangle = \pi\left(\frac{\omega}{c}\right)\frac{Nb^2}{z_2 - z_1}\theta(x_3 - z_1)\theta(z_2 - x_3) \tag{30}$$

and

$$f(\gamma) = \frac{b}{4(z_2 - z_1)}\frac{1}{\gamma^2}\theta\left(\gamma - \frac{b}{4z_2}\right)\theta\left(\frac{b}{4z_1} - \gamma\right). \tag{31}$$

From the result given by Eq. (31) a long sequence of $\{a_n\}$ is obtained, e.g. by the rejection method[17], and the surface profile function constructed on the basis of Eqs. (20), (21), and the fact that $t = r^2$. Thus, we can write the function $H(r)$ as

$$H(r) = \frac{a_n}{b}r^2 + b_n, \quad \sqrt{n}b \leq r \leq \sqrt{n+1}b, \qquad n = 0, 1, 2, \ldots \tag{32}$$

In Fig. 1 we present a segment of one numerically generated realization of the surface profile function $H(r)$ derived from the pdf $f(\gamma)$ given by Eq. (31). The parameters employed in generating this segment were $z_1 = b$, $z_2 = 2b$, and $b = 200\lambda$, where λ is the wavelength of

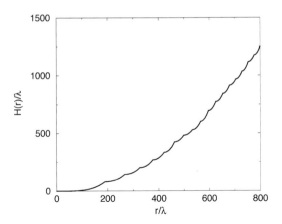

FIG. 1: A segment of one realization of a numerically generated surface profile function $H(r)$ for a two-dimensional Dirichlet surface that produces the distribution of scattered intensity along the x_3-axis given by Eq. (31). The parameters employed are $z_1 = b$, $z_2 = 2b$, and $b = 200\lambda$.

the incident light. We see that, as expected, $H(r)$ is a monotonically increasing function of r.

The procedure now is to generate a large number N_p of realizations of the random surface, and for each realization to calculate the corresponding function $|\psi(0, 0, x_3)_{sc}|^2$. The N_p results for $|\psi(0, 0, x_3)_{sc}|^2$ obtained in this way are summed and the sum then divided by N_p to yield the function $\langle I(x_3) \rangle$.

In Figs. 2, 3, and 4 we present preliminary results for the intensity $\langle I(\mathbf{x}) \rangle$ obtained on the basis of the Kirchhoff aproximation for the surface defined by the pdf given by Eq. (31). In Fig. 2 we present a plot of $\langle I(x_3) \rangle$ for the surface defined by the pdf given by Eq. (31), in the case that the parameters used in this calculation were the same as those used in generating the surface profile function plotted in Fig. 1. It is seen that the resulting $\langle I(x_3) \rangle$ indeed is a constant for x_3 in the interval $(b, 2b)$ and, on the scale of this figure, vanishes for x_3 outside this interval. In fact, the values of $\langle I(x_3) \rangle$ for x_3 outside the interval (b,2b) are $10^3 \div 10^4$ times smaller than its values for x_3 within this interval.

In Fig. 3 a we present a plot of $\langle I(x_3) \rangle$ for the surface defined by the pdf given by Eq. (31). The parameters employed are $z_1 = 50b$, $z_2 = 200b$; $b = 2000\lambda$, and $N = 20$.

In Fig. 4 we plot $\langle |\psi(\mathbf{x}|\omega)_{sc}|^2 \rangle$ in the x_1x_3−plane for the same surface used in obtaining

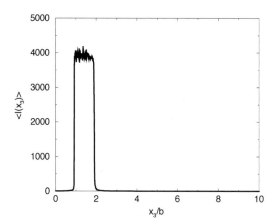

FIG. 2: A plot of $\langle I(x_3) \rangle$ as a function of x_3 estimated from $N_p = 8000$ realizations of the surface profile function for the case of scattering from a surface designed to produce the distribution of scattered intensity along the x_3-axis given by Eq. (26). The parameters employed are $z_1 = 1b$, $z_2 = 2b$; $b = 200\lambda$, and $N = 20$.

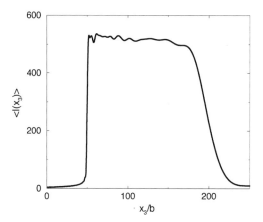

FIG. 3: A plot of $\langle I(x_3) \rangle$ as a function of x_3 estimated from $N_p = 8000$ realizations of the surface profile function for the case of scattering from a surface designed to produce the distribution of scattered intensity along the x_3-axis given by Eq. (26). The parameters employed are $z_1 = 50b$, $z_2 = 200b$; $b = 2000\lambda$, and $N = 20$.

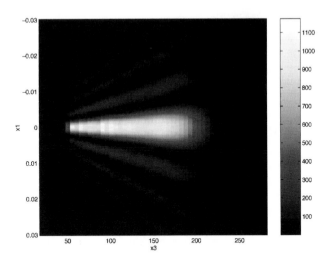

FIG. 4: A plot of $\langle |\psi(\mathbf{x})_{sc}|^2 \rangle$ as a function of x_3 and r estimated from $N_p = 8000$ realizations of the surface profile function for the case of scattering from a surface designed to produce the distribution of scattered intensity along the x_3-axis given by Eq. (26). The parameters employed are $z_1 = 50b$, $z_2 = 200b$, $b = 2000\lambda$, and $N = 20$.

the results plotted in Fig. 3. This result shows that as a function of r the scattered intensity is localized about the x_3-axis ($r = 0$) for all values of x_3, while it has the form depicted in Fig. 2 along the x_3-axis itself. An explanation of why $\langle |\psi(\mathbf{x}|\omega)_{sc}|^2 \rangle$ decreases with increasing r for each value of x_3 is presented in the Appendix.

A particularly interesting case of the surface designed on the basis of the pdf given by Eq. (31) is the one in which $z_1 = 0$ while z_2 has a large value. In Fig. 5 we have plotted $\langle |\psi(\mathbf{x}|\omega)_{sc}|^2 \rangle$, calculated in the case that $z_1 = 10b$, $z_2 = 2000b$, and $b = 2000\lambda$. It is seen that the resulting surface acts as a generator of a pseudo-nondiffracting beam, namely the scattered intensity has a constant value for $0 < x_3 < z_2$, that decays to zero with increasing r for all values of x_3.

To solve the scattering problem outside the framework of the Kirchhoff approximation for each realization of the surface profile function $H(r)$ we return to Eq. (7) and write the scattered field as

$$\psi(\mathbf{x}|\omega)_{sc} = -\frac{1}{4\pi} \int d^2 x'_{\parallel} [g_0(x|x')]_{x'_3 = H(r')} L(\mathbf{x}'_{\parallel}|\omega). \qquad (33)$$

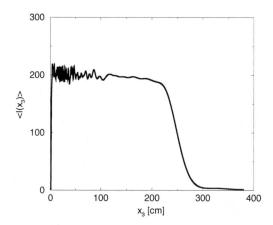

FIG. 5: A plot of $\langle|\psi(\mathbf{x})_{sc}|^2\rangle$ as a function of x_3 and r estimated from $N_p = 80000$ realizations of the surface profile function for the case of scattering from a surface designed to produce the distribution of scattered intensity along the x_3-axis given by Eq. (26) with $z_1 = 1.26$cm, $z_2 = 253$cm, $\lambda = 0.6328\mu$m, and $b = 2000\lambda$.

We next use the representation of the Green's function $g_0(\mathbf{x}|\mathbf{x}')$ given by

$$g_0(\mathbf{x}|\mathbf{x}') = \sum_{\ell=-\infty}^{\infty} \exp[-i\ell(\phi_x - \phi_{x'})][G_\ell(r, x_3|r', x_3')], \tag{34}$$

where ϕ_x and $\phi_{x'}$ are the azimuthal angles of $\mathbf{x}$ and $\mathbf{x}'$, respectively, and

$$G_\ell(r, x_3|r', x_3') = i \int_0^\infty dq \frac{q}{\alpha_0(q)} J_\ell(qr) J_\ell(qr')$$
$$\times \exp[i\alpha_0(q)|x_3 - x_3'|]. \tag{35}$$

In Eq. (35) $J_\ell(z)$ is a Bessel function of order ℓ. The use of the representation (34) for $g_0(\mathbf{x}|\mathbf{x}')$, together with the expansion

$$L(\mathbf{x}_\||\omega) = \sum_{m=-\infty}^{\infty} i^m \exp(im\phi_x)\ell_m(r|\omega) \tag{36}$$

for the source function $L(\mathbf{x}_\||\omega)$, transforms Eq. (33) into

$$\psi(\mathbf{x}|\omega)_{sc} = -\frac{1}{2} \sum_{m=-\infty}^{\infty} i^m \exp(im\phi_x) \int_0^\infty dr' G_m(r, x_3|r', H(r'))r'\ell_m(r'|\omega). \tag{37}$$

The integral equation satisfied by the coefficient function $\ell_m(r)$ in Eq. (36) is obtained by first setting $x_3 = H(r) + \eta$ in Eq. (3), where η is a positive infinitesimal, and using the boundary condition (6). The result is

$$\psi(\mathbf{x}|\omega)_{inc}|_{x_3=H(r)} = \frac{1}{4\pi} \int d^2x_{\parallel}' [g_0(\mathbf{x}|\mathbf{x}')] \begin{array}{c} \\ x_3' = H(r') \\ x_3 = H(r) + \eta \end{array} L(\mathbf{x}_{\parallel}|\omega). \tag{38}$$

Let us now restrict ourselves to the case of normal incidence, $\mathbf{k}_{\parallel} = 0$. In this case we have that

$$\psi(\mathbf{x}|\omega)_{inc}\bigg|_{x_3=H(r)} = \exp[-i(\omega/c)H(r)]. \tag{39}$$

On substituting Eqs. (36) and (39) into Eq. (38), we find that the integral equation satisfied by $\ell_m(r|\omega)$ is

$$2\exp[-i(\omega/c)H(r)] = \int\limits_0^\infty dr' G_0(r|r')r'\ell_0(r'|\omega) \qquad m = 0 \tag{40a}$$

$$0 = \int\limits_0^\infty dr' G_\ell(r|r')r'\ell_m(r'|\omega) \qquad m \neq 0, \tag{40b}$$

where

$$\begin{aligned} G_\ell(r|r') &= G_\ell(r, H(r) + \eta|r', H(r')) \\ &= G_{-\ell}(r|r'). \end{aligned} \tag{41}$$

The homogeneous nature of Eq. (40b) has the consequence that $\ell_m(r|\omega) \equiv 0$ for all $m \neq 0$.

Equation (40a) is solved numerically by converting it into a matrix equation by the use of the midpoint method [18]. In obtaining the solution it is important to take into account the logarithmic singularity of $G_0(r|r')$ as $r \to r'$. It is given by [19]

$$G_0(r|r') \sim -\frac{1}{\pi(rr')^{\frac{1}{2}}} \ell n \left[\frac{\omega}{2c}|H'(r)|^{\frac{1}{2}}|r - r'||H'(r')|^{\frac{1}{2}} \right]. \tag{42}$$

Since, in the case of normal incidence only the coefficient function $\ell_0(r|\omega)$ in the expansion (36) is nonzero, when this result is combined with Eq. (37) we find that

$$\psi(\mathbf{x}|\omega)_{sc} = -\frac{1}{2} \int\limits_0^\infty dr' G_0(r, x_3|r', H(r'))r'\ell_0(r'|\omega). \tag{43}$$

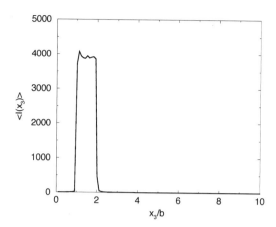

FIG. 6: A plot of $\langle I(x_3) \rangle$ as a function of x_3 estimated from $N_p = 30,000$ realizations of the surface profile function for the case of scattering from a surface designed to produce the distribution of scattered intensity along the x_3-axis given by Eq. (30). The parameters employed are $z_1 = b$, $z_2 = 2b$, $b = 200\lambda$, and $N = 4$.

In the particular case of the scattered field along the x_3-axis, for which $r = 0$, Eq. (43) becomes

$$\psi(0, 0, x_3|\omega)_{sc} = -\frac{1}{2} \int\limits_0^\infty dr' G_0(0, x_3|r', H(r'))r'\ell_0(r'|\omega). \tag{44}$$

The function $\ell_0(r|\omega)$ appearing in Eqs. (43) and (44) is the solution of Eq. (40a).

In Fig. 6 we present a plot of $\langle I(x_3) \rangle$ obtained by combining the result for $\psi(0, 0, x_3|\omega)_{sc}$ obtained by the use of Eqs. (40a) and (43) for the surface defined by the pdf given by Eq. (31). The parameters used in this calculation were the same as those used in generating the surface profile function plotted in Fig. 1 and the intensity distribution plotted in Fig. 2 except that $N = 4$.

In conclusion, we have shown how to design a two-dimensional circularly symmetric random Dirichlet surface that, when illuminated at normal incidence by a scalar plane wave, produces a specified distribution of the intensity of the scattered field along the x_3-axis. Exact scattering calculations carried out on the basis of the approach presented in this work show that the resulting surface produces the distribution of scattered intensity it was designed to produce. As a by-product of the latter calculations we showed that surfaces of this type can produce a scattered field that has the properties of a pseudo-nondiffracting beam.

ACKNOWLEDGMENTS

The research of E. R. M. was supported by CONACYT Grant 3804 P-A. The research of T. A. L. and A. A. M. was supported in part by Army Research Office Grant DAAD 19-02-1-0256.

APPENDIX. THE R-DEPENDENCE OF $\langle|\psi(\mathbf{x}|\omega)sc|^2\rangle$

Some understanding of why $\langle|\psi(\mathbf{x}|\omega)_{sc}|^2\rangle$ is localized about the x_3-axis $(r = 0)$ in the results presented in Figs. 3 and 4 can be obtained by examining the expression for $\psi(\mathbf{x}|\omega)_{sc}$ obtained in the Kirchhoff approximation in the limit that $x_3 \gg x_1, x_1'$ and $x_3 \gg x_2, x_2'$. In this limit we see from Eq. (9) that with $x_3' = \zeta(\mathbf{x}_\parallel')$

$$|\mathbf{x} - \mathbf{x}'| \cong x_3 - \zeta(\mathbf{x}_\parallel') + \frac{x_\parallel^2 + x_\parallel'^2}{2x_3} - \frac{x_\parallel x_\parallel' \cos(\phi_x - \phi_{x'})}{x_3}. \tag{A.1}$$

With the use of this result and the expression for the source function $L(\mathbf{x}_\parallel|\omega)$ given by Eq. (8b), we find that in the case of normal incidence, $\mathbf{k}_\parallel = 0$, the scattered field is given by

$$\psi(\mathbf{x}|\omega)_{sc} \cong i\left(\frac{\omega}{c}\right) \frac{e^{i\frac{\omega}{c}(x_3+\frac{r^2}{2x_3})}}{x_3} \int_0^R dr' r' J_0\left(\frac{\omega r}{cx_3}r'\right) e^{i\frac{\omega}{2cx_3}r'^2 - i\frac{2\omega}{c}H(r')}, \tag{A.2}$$

where R is a large radius that tends to infinity. We introduce the definition

$$I(r, x_3|\omega) \equiv \int_0^R dr' r' J_0\left(\frac{\omega r}{cx_3}r'\right) e^{i\frac{\omega}{2cx_3}r'^2 - i\frac{2\omega}{c}H(r')}. \tag{A.3}$$

It then follows that

$$|\psi(\mathbf{x}|\omega)_{sc}|^2 = \left(\frac{\omega}{cx_3}\right)^2 |I(r, x_3|\omega)|^2. \tag{A.4}$$

With the result for $H(r)$ given by Eq. (32), we can rewrite Eq. (A.3) in the form

$$I(r, x_3|\omega) = \sum_{n=0}^{N-1} e^{-i\frac{2\omega}{c}b_n} \int_{\sqrt{n}b}^{\sqrt{n+1}b} dr' r' J_0(\beta r') e^{-i\alpha_n r'^2}, \tag{A.5}$$

where

$$\alpha_n = \frac{\omega}{c}\left(\frac{2a_n}{b} - \frac{1}{2x_3}\right), \qquad \beta = \frac{\omega r}{cx_3}. \tag{A.6}$$

Thus the radius R in Eq. (A.2) is equal to $\sqrt{N}b$. The change of variable $r'^2 = t$ transforms Eq. (A.5) into

$$I(r, x_3|\omega) = \frac{1}{2}\sum_{n=0}^{N-1} e^{-i\frac{2\omega}{c}b_n} \int_{nb^2}^{(n+1)b^2} dt J_0(\beta\sqrt{t}) e^{-i\alpha_n t}. \tag{A.7}$$

Although the integral in Eq. (A.7) can be evaluated numerically, it is faster to evaluate it if an analytic expression for it is available. Thus, let us consider the integral

$$J = \int_a^b dt J_0(\beta\sqrt{t})e^{-i\alpha t} \tag{A.8a}$$

$$= \sum_{k=0}^{\infty} (-1)^k \frac{(\beta/2)^{2k}}{(k!)^2} \int_a^b dt\, t^k e^{-i\alpha t}, \tag{A.8b}$$

where we have used the expansion of $J_0(z)$ in powers of z. We use the result given by the indefinite integral

$$\int dt\, t^k e^{-i\alpha t} e \frac{e^{-i\alpha t}}{-i\alpha} \sum_{\ell=0}^{k} \frac{t^{k-\ell}}{(i\alpha)^{\ell}} \frac{k!}{(k-\ell)!} \tag{A.9}$$

to rewrite Eq. (A.8b) as

$$
\begin{aligned}
J &= \frac{e^{-i\alpha t}}{-i\alpha} \sum_{k=0}^{\infty} (-1)^k \frac{(\beta/2)^{2k}}{k!} \sum_{\ell=0}^{k} \frac{t^{k-\ell}}{(i\alpha)^{\ell}} \frac{1}{(k-\ell)!} \Bigg|_a^b \\
&= \frac{e^{-i\alpha t}}{-i\alpha} \sum_{k=0}^{\infty} (-1)^k \frac{(\beta/2)^{2k}}{k!} \left[\frac{t^k}{k!} + \frac{1}{i\alpha} \frac{t^{k-1}}{(k-1)!} + \frac{1}{(i\alpha)^2} \frac{t^{k-2}}{(k-2)!} + \cdots \right] \Bigg|_b^a \\
&= \frac{e^{-i\alpha t}}{-i\alpha} \left[J_0(\beta\sqrt{t}) + \frac{(\beta/2)}{i\alpha\sqrt{t}} J_{-1}(\beta\sqrt{t}) + \frac{(\beta/2)^2}{(i\alpha\sqrt{t})^2} J_{-2}(\beta\sqrt{t}) + \cdots \right]_a^b \\
&= \frac{e^{-i\alpha t}}{-i\alpha} \sum_{n=0}^{\infty} (-1)^n \frac{(\beta/2)^n}{(i\alpha\sqrt{t})^n} J_n(\beta\sqrt{t}) \Bigg|_a^b, \tag{A.10}
\end{aligned}
$$

where we have used the result that

$$J_{-\ell}(z) = \sum_{k=0}^{\infty} (-1)^k \frac{(z)^{2k-\ell}}{k!(k-\ell)!} = (-1)^{\ell} J_{\ell}(z). \tag{A.11}$$

On combining Eqs. (A.7) and (A.10) we obtain for the function $I(r, x_3|\omega)$

$$
\begin{aligned}
I(r, x_3|\omega) = &-\frac{1}{2} \sum_{n=0}^{N-1} \frac{e^{-i\frac{2\omega}{c}b n}}{i\alpha_n} \sum_{k=0}^{\infty} i^k \left(\frac{\beta}{2\alpha_n b} \right)^k \\
&\times \left\{ e^{i\alpha_n (n+1)b^2} \frac{J_k(\sqrt{n+1}\beta b)}{(n+1)^{k/2}} - e^{-i\alpha_n n b^2} \frac{J_k(\sqrt{n}\beta b)}{n^{k/2}} \right\}. \tag{A.12}
\end{aligned}
$$

This expression converges rapidly for $(\beta/(2\alpha_n b)) \leq 1$, i.e. for small values of r, $r/b < 4a_n(x_3/b) - 1$. From this result the mean intensity $\langle |\psi(\mathbf{x}|\omega)_{sc}|^2 \rangle$ is calculated from Eq. (A.4).

In the case of large r a more useful result follow from the generating function for Bessel functions

$$\sum_{n=-\infty}^{\infty} t^n J_n(z) = e^{\frac{z}{2}(t-\frac{1}{t})},\tag{A.13}$$

so that

$$\begin{aligned}
\sum_{n=0}^{\infty} t^n J_n(z) &= e^{\frac{z}{2}(t-\frac{1}{t})} - \sum_{n=-\infty}^{-1} t^n J_n(z) \\
&= e^{\frac{z}{2}(t-\frac{1}{t})} - \sum_{n=1}^{\infty} t^{-n} J_{-n}(z) \\
&= e^{\frac{z}{2}(t-\frac{1}{t})} - \sum_{n=1}^{\infty} (-1)^n \frac{J_n(z)}{t^n}.
\end{aligned}\tag{A.14}$$

On combining Eqs. (A.10) and (A.14) we find that

$$\begin{aligned}
J &= \left[\frac{e^{i\frac{\beta^2}{4\alpha}}}{-i\alpha} + \frac{e^{-i\alpha t}}{i\alpha} \sum_{n=1}^{\infty} i^n \left(\frac{2\alpha\sqrt{t}}{\beta} \right)^n J_n(\beta\sqrt{t}) \right]_a^b \\
&= \frac{1}{i\alpha} \sum_{n=1}^{\infty} i^n \left(\frac{2\alpha}{\beta} \right)^n \left[e^{-i\alpha b} b^{n/2} J_n(\beta\sqrt{b}) - e^{-i\alpha a} a^{n/2} J_n(\beta\sqrt{a}) \right].
\end{aligned}\tag{A.15}$$

When we combine Eqs. (A.7) and (A.15) we obtain finally that with the use of the Kirchhoff approximation and the approximation given by Eq. (A.1), the function $I(r, x_3|\omega)$ takes the form

$$\begin{aligned}
I(r, x_3|\omega) = \frac{1}{2} \sum_{n=0}^{N-1} &\frac{e^{-i\frac{2\omega}{c}b_n}}{i\alpha_n} \sum_{k=1}^{\infty} i^k \left(\frac{2\alpha_n b}{\beta} \right)^k \\
&\times \left\{ e^{-i\alpha_n(n+1)b^2} [(n+1)]^{k/2} J_k(\sqrt{n+1}\beta b) \right. \\
&\left. \quad - e^{i\alpha_n n b^2} [n]^{k/2} J_k(\sqrt{n}\beta b) \right\}.
\end{aligned}\tag{A.16}$$

The sum on k converges rapidly for $(2\alpha_n b/\beta) \leq 1$, i.e. for values of $r/b > 4a_n(x_3/b) - 1$, and its evaluation is faster than the direct numerical integration of the defining integral (A.3). Moreover, we see that each term in this sum is an oscillatory function of r that decreases with increasing r more and more rapidly as the summation index k increases. This is the origin of the localization of $\langle |\psi(\mathbf{x})_{sc}|^2 \rangle$ about the x_3-axis.

From the result given by Eq. (A.16) the mean intensity $\langle |\psi(\mathbf{x}|\omega)_{sc}|^2 \rangle$ is calculated from Eq. (A.4).

---

[1] J. Durnin, "Exact solutions for nondiffracting beams, I. The Scalar Theory," J. Opt. Soc. Am. A**4**, 651-654 (1987)

[2] J. Durnin, J. J. Miceli, Jr., and J. H. Eberly, "Diffraction-free beams," Phys. Rev. Lett. **58**, 1499-1501 (1987)

[3] C. Ozkul, S. Leroux, N. Anthore, M. K. Amara, and S. Rasset, "Optical amplification of diffraction-free beams by photorefractive two-wave mixing and its application to laser Doppler velocimetry," Appl. Opt. **34**, 5485-5491 (1995)

[4] J. Turunen, A. Vasara, and A. T. Friberg, "Holographic generation of diffraction-free beams," Appl. Opt. **27**, 3959-3962 (1988)

[5] R. P. McDonald, J. Chrostowski, S. A. Boothroyd, and B. A. Syrett, "Holographic formation of a diode laser nondiffracting beam, "Appl. Opt. **32**, 6470-6474 (1993)

[6] R. P. McDonald, S. A. Boothroyd, T. Okamoto, J. Chrostowski, and B. A. Syrett, "Interboard optical data distribution by Bessel beam shadowing," Opt. Commun. **122**, 169-177 (1996)

[7] F. Gori, G. Guattari, and C. Padovani, "Bessel-Gauss beams," Opt. Commun. **64**, 491-495 (1987)

[8] N. Davidson, A. A. Friesem, and E. Hasman, "Efficient formation of nondiffracting beams with uniform intensity along the propagation direction, "Opt. Commun. **88**, 326-330 (1992)

[9] J. Sochacki, A. Kolodziejcryk, Z. Jareszewicz, and S. Bará, "Nonparaxial design of generalized axicons," Appl. Opt. **31**, 5326-5330 (1992)

[10] T. Aruga, "Generation of long-range nondiffracting narrow light beams," Appl. Opt. **36**, 3762-3768 (1997)

[11] J. Rosen, "Synthesis of nondiffracting beams in free-space," Opt. Lett. 369-371 (1994)

[12] J. Rosen, B. Salik, and A. Yariv, "Pseudo-nondiffracting beams generated by radial harmonic functions," J. Opt. Soc. Am. A**12**, 2446-2457 (1995)

[13] L. Vicari, "Truncation of nondiffracting beams," Opt. Commun. **70**, 263-266 (1989)

[14] R. Liu, B. Z. Dong, G. Z. Yang, and B. Y. Gu, "Generation of pseudo-nondiffracting beams with use of diffractive phase elements designed by the conjugate-gradient method," J. Opt.

Soc. Am. A**15**, 144-151 (1998)

[15] J. Rosen, B. Aslik, A. Yariv, and H.-K. Liu, "Pseudo-nondiffracting slitlike beam and its analogy to the pseudonondispersing pulse," Opt. Lett. **20**, 423-425 (1995)

[16] P. Tran and A. A. Maradudin, "Scattering of a scalar beam from a two-dimensional randomly rough hard wall: enhanced backscattering," Phys. Rev. B**60**, 3936-3939 (1992)

[17] W. H. Press, S. A. Teukolsky, W. T. Vetterling, and B. P. Flannery, *Numerical Recipes in Fortran, 2nd Edition* (Cambridge University Press, New York, 1992), pp. 281-282

[18] Ref. [17], p. 129

[19] T. A. Leskova, A. A. Maradudin, and E. R. Méndez, "Scattering of a scalar plane wave from a two-dimensional rough circularly symmetric Dirichlet surface," (unpublished work)

Wave Propagation,Scattering and Emission in Complex Media
Edited by Ya-Qiu Jin
Science Press and World Scientific,2004

Surface Roughness Clutter Effects in GPR Modeling and Detection

Carey Rappaport

CenSSIS, Northeastern University, Boston, MA, USA

rappaport@neu.edu

Abstract Computational modeling of ground penetrating radar (GPR) wave propagation in air/soil and scattering from buried dielectric targets shows that the largest source clutter is due to the rough ground interface. Conclusions are drawn from the observed features that are somewhat unexpected. One important observation is that while target resolution increases with increasing frequency, the target features are harder to separate from the background clutter of the rough ground interface. The limitation on frequency is not the lack of penetration depth in lossy soil, but rather the greater phase effects of ground surface dips and bumps. Similarly, a greater soil moisture level increases the wave decay rate, but it is the reduction in wavelength that makes the rough surface appear electrically larger, which increases clutter and makes the target harder to resolve. Means of minimizing the rough surface clutter distortion include active ground surface subtraction in the time domain and convolution with a delayed time window in the frequency domain.

1. Introduction

The problem of detecting buried dielectric targets – such as nonmetallic antipersonnel mines – with ground penetrating radar (GPR) is important and challenging. Although there are many means of non-invasively observing below the ground surfaces, GPR is commercially available and relatively inexpensive, environmentally robust, fast, and can be used at varying distances from the target. However, GPR has been disappointing in detecting shallow buried plastic objects in the field. Because the dielectric constant and electrical conductivity of the target is often similar to that of the surrounding soil and its size is comparable to the thickness of soil layer above it, detection and discrimination of the target are difficult. In addition, the soil dielectric constant may not be well characterized, and the ground surface will usually be rough, often with surface height variations of the order of the target burial depth. While there are many sources of clutter obscuring the mine target signal—including volumetric inhomogeneities (rocks, roots, metal fragments) and surface vegetation — the largest single source of undesirable signal is the ground surface itself. Since the ground presents a larger impedance mismatch with the air above it than with the low-contrast, nonmetallic target within it, its contribution to clutter is quite significant.

To quantify the frequency dependent effects of realistic soil backgrounds and surface geometries, we have computationally investigated the scattering of buried dielectric targets in accurately modeled soil with random rough surface boundaries. It is essential that the computational models used capture the relevant wave scattering details of the random soil variation for any given frequency. Thus we chose the finite difference frequency domain (FDFD) method, which allows the arbitrary discretization of the computational space into squares in two dimensional calculations and cubes in three dimensions. The FDFD method is preferable to the FDTD time domain algorithm in the present analysis, because it computes the scattering response at a single frequency, giving the field distribution throughout the problem space. An additional advantage of FDFD is that there is no need to rely on special methods to handle frequency-dependent soil media in the time domain [1-3].

The FDFD method can be made relatively fast. For a 300 by 200 point grid, terminated on each side by a Perfectly Matched Layer (PML) absorbing boundary condition [4], the entire evaluation on a 1.6 GHz Pentium Pro running the Matlab 5.3 sparse matrix solver is about one minute. Using a preconditioned GMRES iterative solution method, the CPU time for large grids of N unknown field values grows as $N \ln N$.

One important requirement for efficient FDFD code is the PML material absorber ABC. Since the PML is composed of just layers of propagation media with particular values of electric and magnetic conductivity, the sparse, symmetric structure of the simultaneous equation matrix is unaffected by the ABC.

Care must be used to tune the PML to the air/soil interface. With the usual PML boundary, it is assumed that this ABC terminates a region of free space. The electric and corresponding magnetic conductivity of the PML sub-layers build up from zero to the maximum value at its termination. For the PML termination to a uniform conductive scattering space, the conductivity of the first sublayer must be slightly greater than that of the scattering space. Subsequent sublayer permittivity and conductivity increase according the anisotropic space mapping principle [5-7], with corresponding magnetic conductivity increasing to keep the impedance of each PML sublayer constant. Improvement in the effectiveness of the PML in 2-D Helmholtz Equation based FDFD is attained by distributing the material dependence over three sublayers according to the formula [8]:

$$\frac{E_{i+1,j}k'_{i-1/2} - E_{i,j}(k'_{i+1/2} + k'_{i-1/2}) + E_{i-1,j}k'_{i+1/2}}{k'_i \, k'_{i+1/2} \, k'_{i-1/2} \, \Delta^2} + \frac{E_{i,j+1} - 2E_{i,j} + E_{i,j-1}}{\Delta^2} + k_0^2(x,y)E_{i,j} = 0 \quad (1)$$

for a PML ABC at $x = 0$ or $x = x_{Max}$, where the relative wavenumber is $k'_i = 1 - j\sigma_i/\omega\varepsilon_0$ for the i-th sublayer with PML conductivity $\sigma_i = \sigma_0(1/N)^p$. A good choice of PML parameters is n=8 and $\sigma_0 = 0.021/\Delta$, and p=3.7 [7]. The background wavenumber $k_0(x,y) = \omega\sqrt{\mu\varepsilon}$ for background material permittivity ε and permeability μ.

To compute the scattered fields due to plane wave incidence on a target in a lossy half-space, the incident field in the nominal planar half-space is first determined analytically, using the standard transmission formulas. Then this field across the support of both the target and the non-planar interface perturbations is used as the excitation. That is, the source-free Helmholtz equation for total field:

$$[\nabla^2 + k^2(x,y)]E_z^{tot} = 0 \quad (2)$$

is rewritten in terms of material variations from the constant soil background, $k(x,y) = k_{back} + k_{scat}$, and in terms of incident (in both the air and soil backgrounds) and scattered electric field, $E_z^{tot} = E_z^{inc} + E_z^{scat}$. Here k_{back} is either k_{air} or k_{soil}, and k_{scat} is k_{targ} over the support of the target, or k_{air} or k_{soil} over the surface perturbations. Since E_z^{inc} solves the Helmholtz equation with background k_{back}, Equation (1) becomes:

$$[\nabla^2 + k^2(x,y)]E_z^{scat} = -[\nabla^2 + k^2(x,y)]E_z^{inc} = -k_{soil}^2 O(x,y)E_z^{inc} \quad (3)$$

where the object function is given for various cases as:

$$O(x,y) = \begin{cases} (k_{t\arg}/k_{soil})^2 - 1 \\ (k_{air}/k_{soil})^2 - 1 \\ 1 - (k_{air}/k_{soil})^2 \end{cases} \qquad \begin{array}{l} \text{Over the target} \\ \text{Over the surface depression} \\ \text{Over the surface protrusion} \end{array}$$

when the scatterer is the buried target, a depression down into the nominal planar ground surface, or a protrusion up above the planar ground surface, respectively.

Note that this formulation is exact, unlike the Born approximation. Approximation is unnecessary since FDFD calculates field values across the grid for all types of materials in any shape or form.

2. Target Shape Scattering Characteristics

It is well known that sharp metallic corners are diffraction sources and objects with corners scatter waves strongly. However, for low contrast non-metallic objects, corners have a much smaller effect on scattered field. The degree of scattering by corners can be studied by examining the scattered fields throughout space. For the buried object problem, it has been suggested that by observing the corner scattered features, it would be possible to characterize the shape of the object. To test this conjecture, the fields can be model in and above the surface of a lossy dispersive soil model half-space.

The first row of Figure 1 shows the geometries of five representative buried two-dimensional scattering objects and their scattered fields for two test frequencies [9]. In each case the target with material characteristics of TNT or plastic (dielectric constant, $\varepsilon' = 2.9$, loss tangent tan $\delta = 0.001$) is embedded in a half space of material with electrical characteristics of dry sand ($\varepsilon' = 2.5$, loss tangent tan $\delta = 0.01$) [10]. The targets each have the same cross-sectional area of about 100 cm^2, and are buried at a nominal depth of 5 cm. In the first case the target is circular, obviously with no corners, while the second and third have square cross-sections with right angle corners. The fourth target models a mine with plastic fins. The last target is an irregularly shaped object typical of a naturally occurring clutter item, such as a rock.

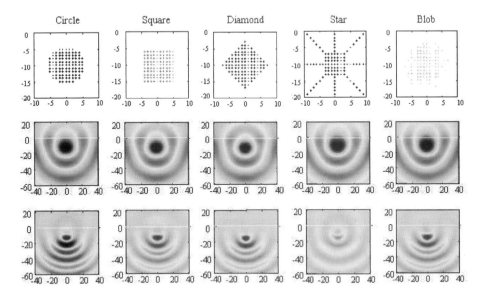

Figure 1: First row: nominal geometry of five buried target shapes each with approximately the same area of 100 cm^2, 5 cm below ground surface. Soil assumed to be dry sand, targets modeled with electrical characteristics of TNT. Second row: real parts of fields scattered from the target shapes for 500 MHz excitation. Third row: real parts of fields scattered for 1 GHz excitation.

The second row of Figure 1 shows the real part of the field scattered by each object in the sand half-space background due to a 500 MHz normally incident plane wave. Note that the field incident in the air above the ground surface (indicated by 0 on the vertical axes), the field specularly reflected from this interface, and the plane wave field transmitted through the interface, have all been suppressed to show the scattering field details. The real part of the scattered field is shown rather than its magnitude because it more clearly shows the wave behavior. With intensity represented by shades of fill, it is clear that at 500 MHz there is almost no observable difference between targets with corners, smooth sides, flat reflective surfaces, or concavities. Although the scattered field above the ground surface is fairly strong, providing detection information, it is practically impossible to distinguish the various targets from their scattered field.

The third row of Figure 1 shows the scattered field for a 1 GHz normally incident plane wave. For this frequency, there are noticeable differences in the scattered fields, but also that the intensity of the waves above the ground surface are much lower than for the 500 MHz case. This is shown quantitatively in Figure 2, which gives the magnitude of the field exactly at the planar ground surface. The differences between the scattered fields are clear only at the higher frequency, but the overall scattered intensity is overall twice to ten times smaller than for the 500 MHz excitation.

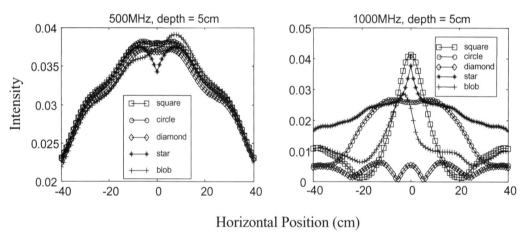

Figure 2: Magnitude of scattered field at the ground surface for the target shapes of Figure 1: 500 MHz excitation (left) and 1 GHz (right).

3. Rough Surface Modeling Results

For the idealized geometry with a perfectly planar ground surface boundary, it would indeed be possible to identify objects with corners and to determine their orientations — as well as distinguish manmade from irregular objects — with increasingly higher frequency excitation. In the field, however, the ground is never flat. Surface roughness must be incorporated into the scattering analysis.

To study the clutter effects of rough ground surface on the target-scattered signal as a function of frequency, a series of 2-D simulations were conducted. Figure 3 shows a depth cross section of a typical geometry incorporating a rectangular mine-like target nominally buried 10 cm in sand, under a surface with an average height variation of 1 cm. The surface roughness has relatively small derivatives.

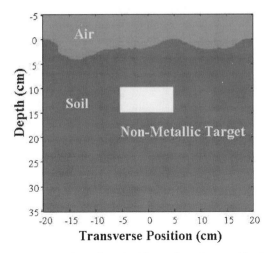

Figure 3: Geometry of cross section of sand with rough surface and buried TNT target used for 2-D FDFD plane wave simulation.

The field scattered from this rough surface scenario is compared to two other scattering geometries: first, the rectangular target buried below an ideal planar sand surface, and second just the rough surface alone, without the target. In each case the excitation is a normally incident plane from above with one of four possible frequencies: 480, 960, 1920, and 3840 MHz. Figure 4 shows the scattered wave pattern for 480 MHz excitation. The upper left plot in Figure 4 (and the three subsequent figures) is the planar boundary case, representing the scattered field with the analytically determined waves incident, reflected, and transmitted from the planar ground surface suppressed.

The upper right image in Figure 4 indicates the scattered field for the rough surface with the target. Again the analytically determined incident, reflected and transmitted fields have been removed, showing the scattering due to the perturbations of the rough surface along with that of the target. It is these depressions and protrusions of the rough surface which give rise to the confusing clutter in realistic detection applications. Since the contrast ratio between the target and the sand background is about 0.85 while the ratio between sand and air is 2.5, the random surface scattering is predominant. This effect is clearly shown by observing the lower left image in Figure 4, which shows the scattering from just the rough surface alone.

The difference between the fields of Figure 4b and 4c gives the field scattered by just the mine under the rough surface. Although this scattered field pattern is similar to the ideal planar interface case of Figure 4a, it is important to realize that it is generally unavailable, since the field scattered by the rough surface separately, without a target, cannot be measured.

Figure 5 shows the same set of scattered field calculations for 960 MHz excitation. In the upper left image, Figure 5a, the shape of the rectangular target is becoming more clearly discernable than at 480 MHz. However, the waves scattered by the same rough surface perturbations are becoming dominant. The difference field in Figure 5d still resembles the ideal field of Figure 5a, and would help to characterize the target as a rectangular anomaly, but this target scattered field is significantly less intense than the rough surface only scattered field of Figure 5c.

480 MHz

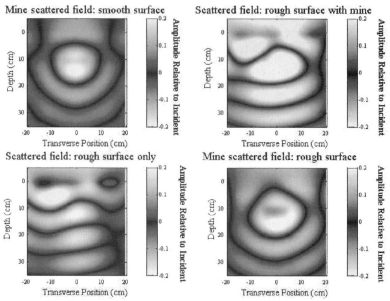

Figure 4: Scattered 480 MHz field from the rectangular TNT target in a sand half-space with a) planar boundary and b) rough boundary; c) the field scattered from just the rough surface perturbations without the mine target, and d) the numerical difference between the fields of b) and c).

960 MHz

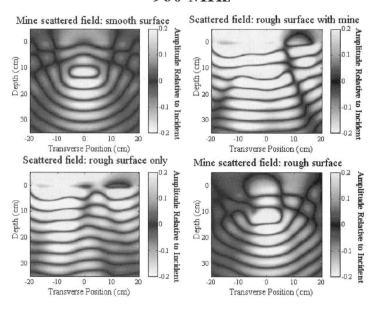

Figure 5: Scattered 960 MHz field from the rectangular TNT target in a sand half-space with a) planar boundary and b)rough boundary; c) the field scattered from just the rough surface perturbations, and d) the numerical difference between the fields of b) and c).

1920 MHz

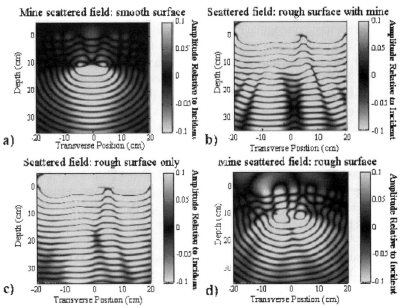

Figure 6: Scattered 1920 MHz field for the geometry of Figure 4

3840 MHz

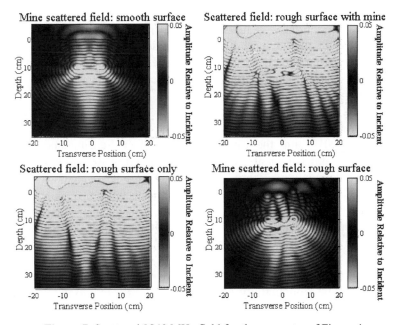

Figure 7: Scattered 3840 MHz field for the geometry of Figure 4

Doubling and redoubling the frequency to consider the possible target characterization improvements results in the images of Figures 6 and 7. Here, both the ideal planar ground surface target scattered field and the numerical difference field images, Figures 6a and 6d show and 7a and 7d show quite recognizable scattering features. The forward scattered field clearly delineates the wide rectangular shape, and the backscattered field shows the interference effects from the finite top surface of the flat target. However, in both the 1920 and 3840 MHz cases, the field scattered from the rough surface perturbations overwhelms the target-scattered field. Even though the bulk of the reflected signal—that due to reflection from the nominal planar interface—has been removed, the roughness itself adds so much clutter that it is not possible to observe any target signal in its presence. The field above the ground with and without the target: Figures 6b and 6c for 1920 MHz, and 7b and 7c for 3840 MHz, are indistinguishable from each other.

4. Discussion and Conclusions

By increasing the illuminating frequency of ground penetrating radar, the resolving capability of an ideal underground sensing and imaging system improves. However, as the incident field wavelength shortens, the random rough ground surface perturbations become more significant.

The calculated scattered fields from various two-dimensional nonmetallic mine-like target shapes indicate that although the target models are relatively simple, there is very little shape-distinguishing feature information available when the incident wave is at the typical GPR frequency of 500 MHz. At 1 GHz, shape features of 10 cm wide targets are quite evident, but a wide aperture (~50 cm) of multiple sensors across the ground surface is needed to measure the variations of scattered field. Clearly, for non-metallic antipersonnel mine detection, frequencies above 500 MHz must be employed if there is to be any hope of distinguishing mines in terms of the shape of target anomalies in soil.

However, if the probing frequency becomes too high relative to the average size of the rough ground height variations, the fields scattered from these perturbations dominate the scattered field. Since the clutter field due to the random roughness cannot be measured or determined separately from the target, nor can its specific distribution be predicted, it will always overwhelm the target signal at higher frequencies.

Acknowledgements

The authors gratefully acknowledge the support of The Army Research Office MURI grant DAAG55-97-1-0013, and the Center for Subsurface Sensing and Imaging Systems (CenSSIS) at Northeastern University, under the Engineering Research Centers Program of the National Science Foundation (Award Number EEC-9986821).

References

[1] Rappaport, C. and Weedon, W., "A General Method for FDTD Modeling of Wave Propagation in Arbitrary Frequency- Dispersive Media," *IEEE Transactions on Antennas and Propagation*, pp. 401-410, March 1997
[2] Rappaport, C., Wu, S., and Winton, S. "FDTD Wave Propagation Modeling in Dispersive Soil Using a Single Pole Conductivity Model," *IEEE Transactions on Magnetics,* vol. 35, May 1999, pp. 1542-1545

[3] Luebbers, R., Hunsberger, F., Kunz, K., Standler R., and Schneider, M., "A Frequency-Dependent Finite Difference Time Domain Formulation for Dispersive Materials," *IEEE Trans. Electromag. Compat.*, Vol. 32, No. 3, March 1990, pp. 222-227

[4] Berenger, J "A Perfectly Matched Layer for the Absorption of Electromagnetic Waves", *Jour. of Comp. Phys.*, vol. 114, pp 185-200, Oct. 1994

[5] Rappaport, C., "Interpreting and improving the PML absorbing boundary condition using anisotropic lossy mapping of space," *IEEE Transactions on Magnetics*, vol. 32, no. 3, May 1996, pp. 968-974

[6] Rappaport, C., and Winton, S., "Using the PML ABC for Air/Soil Wave Interaction Modeling in the Time and Frequency Domains," *International Journal of Subsurface Sensors and Applications,* vol. 1, no. 3, July 2000, pp. 289-304

[7] Marengo, E. and Rappaport, C. and Miller, E., "Optimum PML ABC Conductivity Profile in FDFD", *IEEE Transactions on Magnetics*, vol. 35, May 1999, pp. 1506-1509

[8] Rappaport, C., Kilmer, M., and Miller, E., "Accuracy Considerations in Using the PML ABC with FDFD Helmholtz Equation Computation," *International Journal of Numerical Modeling*, vol. 13, no. 471, September 2000, pp. 471-482

[9] Rappaport, C., Wu, S., Kilmer, M., and Miller, E., "Distinguishing shape details of buried non-metallic mine-like objects with GPR", *SPIE Aerosense Conference*, Orlando, FL, April 1999, pp. 1419-1428

[10] von Hippel, A., *Dielectric Materials and Applications,* Wiley, New York: 1953

Wave Propagation,Scattering and Emission in Complex Media
Edited by Ya-Qiu Jin
Science Press and World Scientific,2004

128

Scattering from Rough Surfaces with Small Slopes

M. Saillard, G. Soriano

Institut Fresnel, UMR CNRS 6133

Faculté St-Jérôme, case 162, 13397 Marseille Cedex 20, France

marc.saillard@fresnel.fr

1. Introduction

The boundary integral formalism, combined with fast numerical solvers, is a very efficient way to deal rigorously with the time-harmonic scattering from a rough surface separating two semi-infinite homogeneous media. However, applying a method of moments (MoM) to an integral equation leads to a linear system with full complex matrices. If the exact matrix elements are used, the number of operations for solving such a system is proportional to N^2, where N is the number of unknowns. In addition, storing all the matrix elements in memory is not possible. This is why several numerical methods have been developed to avoid full matrix storage and to make the computation time scaling as $N \log N$. This may be achieved by representing the matrix as a superposition of matrices with specific properties, in order to speed up the matrix-vector products that occur in iterative solvers. As a result, a slight approximation of the kernel of the integral equation leads to a drastic gain in terms of computation time and memory requirements. In this paper, we show that additional approximations can speed up the computation further, and we compare with existing approximate methods.

2. A boundary integral formalism

From the four Stratton-Chu equations, one can derive various couples of integral equations for the two surface currents. For bounded scatterers, only some of them avoid irregular frequencies and ensure uniqueness of the solution. Since we consider here an infinite rough surface illuminated by a finite beam, no irregular frequency occurs and we take benefit of this freedom to build a set of two weakly singular coupled integral equations [1]. Indeed, noticing that the hypersingular term of the second derivative of the free space Green's function is independent of the wavenumber k, thus is the same for integral operators associated with the upper or the lower medium, taking the difference of these two operators cancels both the $1/r^3$ and the $1/r^2$ terms, making the resulting kernels behave as $1/r$ around 0.

$$\left(\frac{1+\varepsilon_r}{2}+\mathbf{M}_0-\varepsilon_r\mathbf{M}\right)\mathbf{n}\times\mathbf{E}+\frac{i}{\omega\varepsilon_0}(\mathbf{P}_0-\mathbf{P})\mathbf{n}\times\mathbf{H}=\mathbf{n}\times\mathbf{E}^{inc} \tag{1}$$

$$(1+\mathbf{M}_0-\mathbf{M})\,\mathbf{n}\times\mathbf{H}-\frac{i}{\omega\mu_0}(\mathbf{P}_0-\mathbf{P})\mathbf{n}\times\mathbf{E}=\mathbf{n}\times\mathbf{H}^{inc} \tag{2}$$

with $\mathbf{M}\mathbf{j}_p=\mathbf{n}_p\times rot_p\int G_{pq}\mathbf{j}_q\,dS_q$ and $\mathbf{P}\mathbf{j}_p=\mathbf{n}_p\times rot_p\,rot_p\int G_{pq}\mathbf{j}_q\,dS_q$, G_{pq} denoting the free space Green's function linking points p and q.

From a numerical point of view, such a weak singularity is much easier to integrate and allows the use of piecewise constant basis functions, leading to faster computation of the matrix elements. Some comparisons with methods using more regular basis functions have shown similar accuracy for same surface sampling [2].

3. The Method of Moments

To transform the integral equations into a linear system of equations, the unknowns are written as a finite superposition of basis functions, and the integral equations themselves are projected onto a set of test functions. The resulting linear system is dense, with no symmetry property. A lot of work is currently devoted to speed up the solution of these systems. The key point consists in taking benefit from the very specific geometry of the scatterer, which is considered here to be planar in average. The basic result, called "short range interaction", claims that the unknown currents at a given point on the surface only depend on the geometry and on the incident field in the vicinity of this point.

It amounts to formally decomposing the matrix as a sum of a strong interaction matrix, describing interactions between neighboring points (distance less than r_d), and of a weak interaction one.

$$A = A^s + A^w \tag{3}$$

The first one is a sparse matrix, of which filling ratio depends on the radius of strong interaction r_d, but the second one is still a dense complex matrix without any property. Several methods have been developed to avoid the storage of the weak interaction matrix and to allow fast computation of the matrix vector products occurring in the iterative solution of the linear system.

Here we focus on the Sparse-Matrix Canonical Grid method (SMCG) [3], which uses an expansion of the weak interaction matrix as a power series of slope, defined here as the ratio of vertical to horizontal distance. If a regular sampling of the mean plane is used, such a decomposition permits one to write this matrix as superposition of block-Toeplitz matrices. Consequently, the matrix-vector products involving the weak interaction matrix can be performed very fast thanks to FFTs. A lot of memory is also saved, since only one column needs to be stored for each of these matrices. Of course, the convergence of the expansion is questionable, but this problem has been overcome thanks to the Multi-Level version of the method [4].

The SMCG method results in a two-level iterative scheme, which consists in solving a set of linear systems tending toward the true system. The main matrix is the superposition of the strong interaction matrix and of the zeroth order matrix in slope, called the Flat Surface matrix (A^{FS}), and remains the same during the whole process, while the right hand side is updated at each iteration of the outer level through a multiplication by the upper order matrices:

$$A^w = \sum_{j \geq 0} A_j^w = A^{FS} + \sum_{j > 0} A_j^w \tag{4}$$

$$(A^s + A^{FS}) x^{(n)} = b^{(n)} \tag{5}$$

$$b^{(n)} = b - \sum_{j > 0} A_j^w b^{(n-1)} \tag{6}$$

4. The Small Slope approximation

A lot of work, often motivated by remote sensing applications, has been dedicated during the last decade to the scattering from surfaces with small slopes, leading to well-known approximate methods like the Small Slope Approximation (SSA) [5], the Integral Equation Method (IEM) [6], the Operator Expansion Method (OEM) [7], etc.. These methods generally lead to analytical formulae involving one surface integral at first order, one surface integral plus one integration in the Fourier space at second order, and so on. In general the second order term is very difficult to compute accurately, and additional approximations are introduced. This term is important if the cross-polarized field has to be estimated, since the first-order does not account for multiple scattering.

At lowest order, OEM uses the Meecham-Lysanov approximation, which simply consists in replacing the distance R between two points by their horizontal distance in the mean plane. The first two terms of the expansion of kR write

$$kR = k\sqrt{\|\mathbf{x}-\mathbf{x'}\|^2 + (h(\mathbf{x})-h(\mathbf{x'}))^2} \approx k\|\mathbf{x}-\mathbf{x'}\| + k|h(\mathbf{x})-h(\mathbf{x'})|\frac{|h(\mathbf{x})-h(\mathbf{x'})|}{\|\mathbf{x}-\mathbf{x'}\|} \tag{7}$$

Neglecting the last term could be interpreted as a second-order small slope approximation. In fact, since kR also occurs in the phase, the last term has to remain much smaller than 2π. Therefore, the domain of validity of the approximation will be determined in terms of the dimensionless parameter khs where h denotes the rms height and s the rms slope. In the case of one-dimensional surfaces and s polarization, the domain of validity of the method has been estimated to $khs < \frac{1}{4}$ [8], thanks to comparisons with rigorous computations. However, it is important to remind that no depolarization occurs in this case. Hence, the accuracy of the method has to be reconsidered for two-dimensional dimensions surfaces.

Introducing this approximation in SMCG simply consists in approximating the matrix A by the flat surface matrix A^{FS}. However, since the diagonal elements have to be computed separately to integrate analytically the singular part of the kernel, calculating the exact diagonal elements has the same computational cost. This can be interpreted as reducing the strong interaction matrix A^s to its diagonal ($r_d = 0^+$). Finally, the approximate linear system writes

$$(A^s(r_d = 0^+) + A^{FS})x = b \tag{8}$$

As compared with original SMCG, the sparse matrix is restricted to one diagonal and the outer iteration level has disappeared. The approximate linear system is solved by the BiConjugate Gradient Stabilized of order 2. For perfectly conducting surfaces, each iteration requires 4 matrix-vector products, there are 12 FFTs per matrix-vector product, and $4N\log(4N)$ operations per FFT.

5. Impedance approximation

If the illuminated medium is lossy with high permittivity, the integral (impedance) operator linking the electric and magnetic surface currents can be approximated by a local one. Such an approximation is often referred to as an impedance approximation. Following [9], a second order expansion with respect to skin depth d of the impedance operator has been implemented in our code, leading to a local relationship involving the surface curvature tensor.

$$\mathbf{n}\times\mathbf{E} = Z\,\mathbf{n}\times\mathbf{H} \quad \text{with} \quad \frac{Z}{\eta_0}\mathbf{n}\times\mathbf{H} = -ik_0 d\,\mathbf{n}\times\left[1 + d\left(S - \frac{\mathrm{Tr}(S)}{2}\right)\right]\mathbf{n}\times\mathbf{H} \tag{9}$$

where η_0 and k_0 denote the impedance and wavenumber in vacuum respectively, while S denotes the extrinsic curvature tensor, and its trace $\mathrm{Tr}(S)$ equals the sum of the surface main curvatures.

As a result, the number of unknowns is divided by a factor of two. Compared to the perfectly conducting case, the unknown is the same, namely the electric surface current, but the matrix elements are three times longer to compute. This is a very low price for taking finite conductivity into account, since the perfectly conducting model is not relevant in many cases. For instance, the accuracy of the impedance approximation has been checked for metals in Optics, as well as for sea surface and wet soils in the microwave range. For the last example, the relative permittivity was set at $25 + i\,3$, and it required ten times more computation time to solve the exact integral equation, to get the same result. Some of our numerical results are described in the next section.

6. Numerical results

The numerical results exhibited here combine both impedance and small slope approximations. It allowed us to perform fast computation of the scattering from multi-scale surfaces with high permittivity. As an example, we have considered a banded power law spectrum (K^{-4}), fitting a 10 m/s wind driven ocean surface spectrum up to 6m, illuminated under 20° incidence at λ = 24 cm (*L* band). The complex permittivity of the sea has been set to ε = 73.5 + i 63.

In Fig. 1, we have compared the rigorous solution of the integral equation (SMFSIA) with that of the small slope integral equation described in the previous section (SSIE) and with the first-order Small-Slope Approximation (SSA) of Voronovich. The three methods have been applied to the same 100 surface samples, illuminated by the same Gaussian incident beam. Plots represent the co-polarized (higher levels) and cross-polarized (lower levels) cross-sections in the plane of incidence. Let us recall that SSA is unable to predict cross-polarization in the plane of incidence at first order.

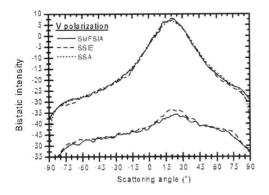 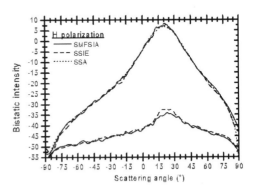

Fig. 1 : Comparison between rigorous (SMSFIA) and small slope approximations (SSA, SSIE), for band limited (up to 6 m) power law surface spectrum (K^{-4}) fitting lower scales of the ocean surface. λ = 24 cm. Each plot presents both co- and cross-polarization.

This surface has a rms height h = 0.25λ and a rms slope s = 0.112, leading khs = 0.176, well suited for this approximation, but out of the range of validity of the small perturbation method and of Kirchhoff approximation. Except around the specular direction for cross-polarization, there is an excellent agreement between the various methods. Here, for each surface, the computational costs of SSIE are 16 Mo for matrices and 6 minutes for solution (82944 unknowns).

The second example concerns a Gaussian surface with rms height h =0.3λ and correlation length l_c = λ, leading to a rms slope s = 0.212 and khs = 0.8, which by far exceeds the expected domain of validity. The dielectric constant is that of a wet soil, ε = 25 + i 3. When the small slope integral equation is used, the extinction theorem [10] may reach up to 20%, indicating an inaccurate solution of the linear system, as expected. In fact the error in estimating the cross-polarized component is much more than 20%, since it is always more than 3 dB (factor of 2) here. However, it is amazing to notice in Fig. 2 that the co-polarized cross-section remains accurate, except at large scattering angles (if one admits that the oscillations result from a poor averaging process). The same has been observed for Gaussian surfaces with khs up to 1.2, but we have no explanation yet.

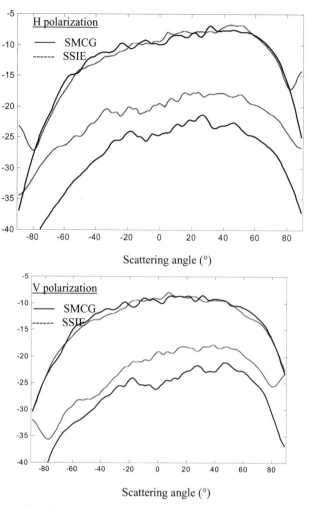

Fig. 2 : Comparison between rigorous (multi-level SMCG) and small slope approximation (SSIE), for a Gaussian surface with rms height $h = 0.3\lambda$ and correlation length $l_c = . \lambda$. Each plot presents both co- and cross-polarization.

7. Conclusion

The Small Slope Integral Equation method (SSIE) predicts multiple scattering at moderate numerical costs and competes with statistical approximate methods like the second-order small slope approximation or the operator expansion method, of which multiple integrals are also difficult to compute accurately, especially for power-law spectra. In addition, these approximate methods require such computations for each scattering angle, which is not suited for bistatic scattering. With SSIE, once the surface current is estimated, computing the bistatic scattering cross-section only involves a surface integral. Let us also recall that the statistical formula of SSA or OEM are obtained under the assumption of Gaussian height distribution; otherwise, a Monte-Carlo process is also required. Finally, it must be emphasized that an indicator of the error may be provided with SSIE. Here, we have used the extinction theorem to detect when the solution is not accurate. It has been noticed that, as claimed in [9] for one-dimensional perfectly conducting surfaces, when the parameter khs remains smaller than ¼, the extinction theorem is fulfilled with about 1% accuracy, and SSIE compares well with the rigorous solution. Hence, the extinction theorem is a good test to check the accuracy of SSIE, and it is not time consuming (one surface integral). When the parameter *khs* is increased, the extinction theorem is also regularly increasing, but it cannot be easily interpreted in terms of error. Indeed, through our computations, it has been noticed that SSIE predicts accurately the co-polarized bistatic cross-section for high values of *khs*, and that the error on the cross-polarized part was bigger than the residue of extinction theorem. The latter must be considered as a lower bound in this case. Some other means to estimate the error, probably more accurate, exist but require the computation of exact matrix elements.

References

[1] G. Soriano, M. Saillard, *Scattering of Electromagnetric Waves from Two-Dimensional Rough Surfaces with Impedance Approximation*, J. Opt. Soc. Am. A **18** (2001) 124-133

[2] M.Y. Xia, C.H. Chan, S.Q. Li, B. Zhang, L. Tsang, *An Efficient Algorithm for Electromagnetic Scattering from Rough Surfaces Using a Single Integral Equation and Multilevel Sparse-Matrix Canonical-Grid Method*, IEEE Trans. Ant. Propagat., to appear

[3] K. S. Pak, L. Tsang, C. H. Chan, J. Johnson, *Backscattering enhancement of electromagnetic waves from two-dimensional perfectly conducting random rough surfaces based on Monte-Carlo simulations*, J. Opt. Soc. Am. A **12** (1995) 2491-2499

[4] S.-Q. Li, C. H. Chan, M.-Y. Xia, B. Zhang, L. Tsang, *Multilevel expansion of the sparse-matrix canonical grid method for two-dimensional random rough surfaces*, IEEE Trans. Ant. Propagat.**49** (2001) 1579-1589

[5] A. A. Voronovich, *Wave scattering from rough surfaces*, Springer-Verlag, Berlin (1994)

[6] A. K. Fung, *Microwave scattering and emission models and their applications*, Artech House, Boston (1994)

[7] D. M. Milder, *An improved formalism for electromagnetic scattering from a perfectly conducting rough surface*, Radio Science **31** (1996), 1369-1371

[8] E. Thorsos and D. R. Jackson, *Application of the operator expansion method to scattering from one-dimensional moderately rough Dirichlet random rough surfaces*, J. Acoust. Soc. Am. **86** (1989) 261-277

[9] A. M. Marvin, V. Celli, *Relation Between the Surface Impedance and the Extinction Theorem on a Rough Surface*, Phys. Rev. B **50** (1994) 14546-14553

[10] M. Nieto-Vesperinas, *Scattering and Diffraction in Physical Optics*, Wiley & Sons, New-York (1991)

Polarization and Spectral Characteristics of Radar Signals Reflected by Sea-Surface

V.A.Butko, V.A.Khlusov, L.I.Sharygina

Tomsk State University of Control Systems and Radioelectronics

40 Lenin av., Tomsk, 634050, Russia. E-mail: gssh@mail.tomsknet.ru

Abstract   In the paper there are described results of experimental investigation of polarization and spectral characteristics of 3 cm radar signals scattered by sea surface at low grazing angles. Peculiarities of the sea surface scattering co-variation matrix and Doppler spectra of co- and cross-polarized back-scattered signals in linear polarization basis have been analyzed.

Equipment and philosophy of measurement

The measuring complex includes coherent-pulse radar with fast polarization variation of the illuminated signal from horizontal to vertical and simultaneous reception and registration of the signals of both polarizations. Frequency of polarization re-switching is equal the half of the pulse repetition frequency 1 kHz. Pulse duration is 0.8 µs. The equipment allows to measure all four back scattering matrix (BSM) elements simultaneously for each element of resolution by range during two periods of pulse repetition. Measurements were done for paths 3-9 km and grazing angles 0.25-0.6 degrees.

A pulse quasi-coherent radar put at the height of 40 m over the sea level was used for measurements. The radar sight allowed to sound the sea surface in azimuth sector of 90°. Transmitting-receiving antenna had symmetrical in both plates pattern with the ray width at the half power level 3°. Measurements were carried out in autumn. Conditions of wave refraction were about standard, sea roughness was from 2 to 5 numbers.

In the process of measurements there were registered co- and cross-polarization quadrature components of back-scattered signals, duration of each measurement session 1-2 s long. 276 measuring sessions were performed.

Doppler spectra and coefficients of BSM elements mutual correlation were calculated while processing each resolution element by range. Calculation of Doppler spectra was done by the method of periodograms with Hamming weighting function. Spectrum resolution varied from 2.7 to 5.4 Hz depending on the séance duration and number of averaged periodograms.

Spectral characteristics of co- and cross-polarized components of back-scattering

Fig.1 shows Doppler spectra of sea-surface back-scatterings in case of sea-way 2 numbers and different combinations of transmitted and received signals polarization. In this figure, as in the following ones, the type of polarization is pointed out by two letters. The first letter shows the polarization of the received signal, the second – the transmitted polarization. H means horizontal polarization, V – vertical. Vertical axis shows the spectral power density of the illuminated part of the sea-surface BSM element.

Fig.1 shows that co-and cross-components of the signals back-scattered by quiet sea have bell-like shape, spectra of cross-components coincide, which is a result of BSM symmetry for the same antenna used to transmit and receive signals. Spectrum of horizontally polarized co-polarization components is wider than in case of vertical polarization. Spectral shifts of co- and cross-polarized back-scattered vertical and horizontal signals are about the same.

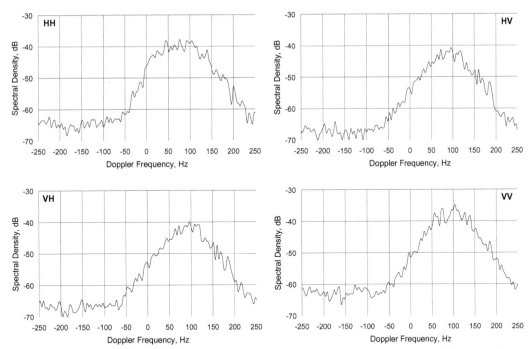

Fig.1 Spectra of co-polarized and cross-polarized components of the signals scattered by sea surface.
Distance 4.5 km, 2 number sea-way

When the sea-way grows spectrum form and width change. Doppler spectra in Fig.2 show that measurements were done at sea-way of 5 numbers. Spectra were received for vertical illuminated polarization only. Doppler frequencies characterizing this regime are twice higher than in case of pulse by pulse re-switching and was ± 500 ??. Spectra in Fig. 2a and 2b correspond signals received from distances 7.5 and 7.86 km. The form of spectra of co- and cross-polarized components in case of horizontal polarization of the instant wave looks like shown in Fig.2.

In case of rough sea (wind velocity more than 4 m/s) there are observed "bell-like' (Fig.2a) and "square" (Fig.2b) spectra. The names of the spectra are conventional and are used to emphasize that spectra are different in case of rough sea. When the sea is quiet spectra of signals from different distances and azimuth directions look about the same and have a bell-like form. So, spatial-temporary homogeneity of back-scattered Doppler spectra is destroyed when the sea becomes rough.

Fig.3 shows spectra of back-scatterings by the 3-number sea (wind velocity 4 m/s) with the curves approximating behavior of spectral power density (dotted curves). At the level above -20 dB of spectral power density Gaussian law is good for approximation:

$$S(F) = \left[1 + \left(|F| / \Delta F \right)^n \right]^{-1}, \tag{1}$$

where F is frequency shift from maximum, ΔF is the width of approximating spectrum at -3 dB level. When $n = 2$ (1) well describes Lorenz spectral line. It corresponds the case when a scatterer moving with constant velocity has final "time of life". Results of spectra of different sessions processing show that n is in interval from 2 to 3.

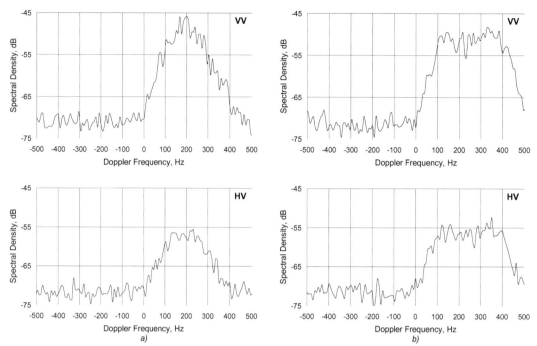

Fig.2. "Bell-shaped" and "square" spectra of vertically polarized back-scattered signals.
5 number sea-way

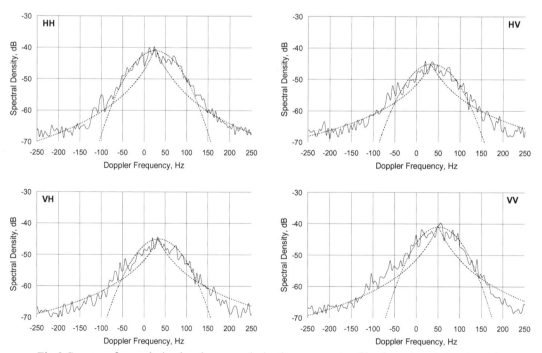

Fig.3 Spectra of co-polarized and cross-polarized components of back scattered signals with
approximating curves. Distance 4.75 km, 3 number sea-way

So, Doppler spectra form of back-scattered signals, when the sea is quiet (2-3 numbers) differs from the Gaussian one, which is predicted by the Bragg's model of scattering in case of normal distribution of velocities. In case of high frequency fluctuations the main deposit into the scattered signal make parts of the sea surface moving with constant velocity and having final "time of life".

Quantitative characteristics of back-scattered spectra are their width and shift. The width of spectra depends on the sea-way and was equal to 80-200 Hz in case of 2-3 number sea. It corresponds 45-110 Hz of Gaussian approximation at half power level. There were observed variations in relation between spectrum widths of horizontal and vertical polarizations but on a whole the spectrum width in case of horizontal polarization is 1.2-1.3 times wider than for the vertical one. The spectrum width of cross-polarized components is about the same as of co-polarized one for horizontally polarized instant wave. There was not observed visible dependence of the spectrum width on the distance and direction of sounding.

At stronger wind (>4-5 m/s, sea-way >3 numbers) the form of spectral curve is not bell-like and even can be "square" (see fig. 2.6). Spectrum width in such a case at -10 dB level is 280…360 Hz.

Spectrum shift was -100…250 Hz. Negative values were observed while sounding along the wind, positive – against the wind. The shift is zero while sounding across the wind. Analysis showed that the shift is maximal when the radar ray is oriented along or against the wind. As a rule, sifts of co-polarized waves is 1.2-1.4 times larger in case of vertical polarization compared with horizontal one.

Fig.4 shows dependence of the spectrum middle frequency of co-polarized bask-scattered signals on the direction of sounding. Shifts about zero were observed at azimuth 318 and 324 degrees. According to meteorologists the wind direction was 225 degrees, which corresponds zero shifts of scattered spectra.

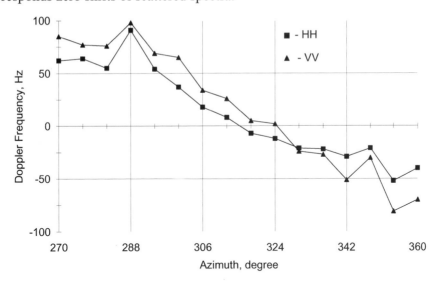

Fig.4. Dependence of Doppler spectrum shift (the middle frequency) of co-polarized components of the signals scattered back in the direction of sounding. Distance 5 km, 3 number sea-way

Polarization characteristics of back-scattering

Polarization properties of a fluctuating radar object at reciprocal conditions can be described by three real $\sigma_{hh}, \sigma_{vv}, \sigma_{hv}$ and three complex $\dot{\rho}_{hhvv}, \dot{\rho}_{hhhv}, \dot{\rho}_{vvhv}$ values. $\sigma_{hh}, \sigma_{vv}, \sigma_{hv}$ are equal RCS of a target in case of co-polarization $(\sigma_{hh}, \sigma_{vv})$ and cross-polarization $(\sigma_{hv} = \sigma_{vh})$ illuminated by horizontally and vertically polarized waves. They define average power of scattered signals. Complex values $\dot{\rho}_{hhvv}, \dot{\rho}_{hhhv}, \dot{\rho}_{vvhv}$ are normalized coefficients of mutual correlation of co- $(\dot{\rho}_{hhvv})$ and cross-polarized $(\dot{\rho}_{hhhv}, \dot{\rho}_{vvhv})$ back-scattered components. First two letters and last two letters in indexes point to those BSM elements between which correlation coefficient calculated.

Coefficients of depolarization $D_h = \sigma_{hv}/\sigma_{hh}$ and $D_v = \sigma_{hv}/\sigma_{vv}$ for horizontally and vertically polarized instant waves as well as modules of mutual correlation were calculated while processing the experimental data. Coefficients of depolarization of horizontally polarized instant waves lie within interval -3..-9 dB with average -5.7 dB. In case of vertically polarized waves these figures are -6.5..-12 dB with average -7.5 dB. The value of σ_{vv} as a rule is 2-4 dB bigger than σ_{hh}. When the sea roughness grows coefficients of horizontally polarized wave depolarization also grows by 1.5 dB. Depolarization of vertically polarized waves does not change.

Estimates of normalized modules of co-and cross-polarized components coefficients of mutual correlation are about zero. Estimates of co-and cross-polarized components coefficients of mutual correlation for different types of polarization lies on the interval 0.25...0.4 with average 0.35. It means that different scatterers takes part in forming signals back-scattered by sea surface of horizontal and vertical polarization.

Conclusion

There have been experimentally received estimates of characteristics of sea surface BSMs in the linear polarization basis at the frequency 10 GHz and different conditions of the sea. For quiet sea (2-3 numbers) homogeneity of back-scattered spectra is typical, which would destroy in a case of a rough sea. The form of the spectra differs greatly from Gaussian when the spectrum density is lower than −20 dB. This fact as well as the difference of spectral parameters in case of different polarization proves that there are other than Bragg's mechanisms of wave scattering by the sea surface.

The average power of back-scattered by sea co-polarized component in case of vertical polarization is 2-4 dB higher than in case of horizontally polarized instant wave. Cross-polarized component is 6-12 dB lower for vertically polarized instant wave and 3-9 dB for horizontal polarization. Co- and cross-components fluctuations of back-scattered signals practically are not correlated in case of illumination of vertical and horizontal polarization. Coefficient of mutual correlation of elements on the matrix main diagonal is less than 0.4. It means that different scatters form the signals of vertical and horizontal polarization.

Wave Propagation,Scattering and Emission in Complex Media
Edited by Ya-Qiu Jin
Science Press and World Scientific,2004

Simulation of Microwave Scattering from Wind-Driven Ocean Surfaces

M. Y. Xia[1], C. H. Chan[2], G. Soriano[3], M. Saillard[3]

[1]Department of Electronics, Peking University, Beijing 100871, China
[2]Department of Electronic Engineering, City University of Hong Kong
Hong Kong SAR, China
[3]Institut Fresnel, UMR CNRS 6133, Faculté Saint Jérôme, case 162
13397 Marseille Cedex 20, France

Abstract   An efficient approach for fast analysis of vector wave scattering from random rough surfaces is described in this paper. Numerical examples are provided to demonstrate the merits of the scheme. The underlying methodology lies on using a single integral equation formulation combined with a multilevel sparse-matrix canonical-grid method. Triangular patches are employed to tightly model the surfaces. A localized approximation method is employed for the highly conductive seawater medium. A beam decomposition technique is incorporated to cope with an extremely large surface. Monte-Carlo simulation results are presented as bistatic scattering coefficients for wind-driven ocean surfaces illuminated by microwave beams at L-band. A special attention is paid to low grazing incidences due to its practical importance.

1. INTRODUCTION

Electromagnetic scattering from random rough surfaces has drawn an increasing attention in the past decades because of its broad applications. These include microwave remote sensing of wind field over ocean, microwave remote sensing of soil moisture and ice layer, and planet surface investigation using radio waves in astronomy, etc. A state-of-the-art review may be found in [1]. With the advent of high performance computers, a remarkable advancement on numerical approaches has been made in recently years and a review may be found in [2]. Analytical models have limited domains of validity and for situations that they are not applicable, rigorous numerical methods are desirable. In general, for practical applications and comparisons with experiment results, a simulated surface size area must be sufficiently large, often in the order of tens of thousand square wavelengths. Discretization of the governing equations results in a large system of linear equations that its matrix solution is almost intractable. The purpose of this paper is to report our recent progress toward an efficient procedure for scattering modeling of random rough surfaces and to validate the codes for simulations of microwave beams impinging on wind-driven ocean surfaces.

For a vector wave scattering from a dielectric surface of several thousand square wavelengths, a Method of Moments (MoM) equation up to millions of unknowns has to be solved. Thus crucial to an efficient algorithm is to achieve a considerable reduction both in memory storage requirement and CPU solution time. One of the most successful approaches used so far is based on the Sparse-Matrix Canonical-Grid (SMCG) strategy [3]-[4], by which the interaction matrices are written as two parts, the near-strong parts and the far-weak ones. The strong parts are very sparse and stored. Upon a Taylor series expansion of the Green's functions about a canonical-grid, the weak parts are cast into Toeplitz structures so that the fast Fourier transform (FFT) technique applies. A multilevel version of the method extended to suit surfaces of larger roughness was presented in [5]. Because the Green's function in a highly lossy medium decays rapidly, a two-grid SMCG method was suggested to maintain the accuracy of a fine discretization in the dielectric region while achieve the efficiency using a coarse grid in the air region [6]. Recently, the multilevel SMCG method has been accomplished adopting a single integral equation formulation with the Rao-Wilton-Glisson (RWG) triangular basis functions [7]. This allows the number of unknowns to be reduced by a factor of two or three and the rough surfaces to be modeled more accurately using the triangular patches. The number of unknowns representing the surface equivalent currents solved by this method exceeds 12 millions on a supercomputer platform [8].

In addition to the various versions based on the SMCG strategy, other approaches have also been reported in the literature, including the Fast Multipole Methods (FMM) [9]-[10] and the Finite-Difference Time-Domain (FDTD) scheme [11]. The impedance approximation method [12] seems to be particularly suitable when the lower medium is highly conductive such as in the case of seawater.

In this paper, based on the scheme developed in [7] with increased efficiency and flexibility, some approximate methods are integrated into the codes, which may be activated to speed up the computing. A Localized Approximation Method (LAM) is introduced for a highly conductive medium by ignoring the far-weak interactions and using a coarse discretization in the air region. This method is not exactly the same as the impedance approximation method employed in [12]. A Beam Decomposition Technique (BDT) is also incorporated, to tackle an extremely large surface. The BDT, once called beam simulation method proposed in [13] and used in [14], is a viable means when the surface is too big to be analyzed as a whole, by splitting a prohibitively large problem into a set of smaller problems that are tractable. These smaller surfaces are necessarily overlapped leading to extensive solution time.

In Section 2, we will give an outline of the formulation and various techniques that are integrated into the numerical solver. Numerical examples are presented in Section 3. Since a great challenge in passive and active remote sensing is to develop accurate models for bistatic scattering from sea surfaces, the results are expected to serve this goal. Grazing incidence is of great importance for costal and ship-board radar and deserves a special attention. Some concluding remarks are given in the last Section.

2. THEORY

2.1 Single Integral Equation (SIE) Formulation

The geometry of a random rough surface is shown in Fig.1. To adopt the single magnetic field integral equation formulation [15], it is assumed that there is an *effective electric* current distribution $\bar{J}^{eff}$ on the lower side of the interface so that the fields in the lower region can be expressed by

$$\bar{E}_2 = i\omega\mu_0 \left\langle \bar{\bar{G}}_2, \bar{J}^{eff} \right\rangle_{S^-} \tag{1}$$

$$\bar{H}_2 = \left\langle \nabla \times \bar{\bar{G}}_2, \bar{J}^{eff} \right\rangle_{S^-} \tag{2}$$

where the inner product is evaluated underneath the surface, and $\bar{\bar{G}}_2$ is known as the electric dyadic Green's function for a homogeneous space filled with the lower medium, i.e.,

$$\bar{\bar{G}}_2 = \left(\bar{\bar{I}} + \frac{1}{k_2^2} \nabla\nabla \right) \frac{e^{ikR'}}{4\pi R'}, \quad R' = \left| \bar{r}' - \bar{r}'' \right|$$

As the field point P' approaches the interface from the lower region, a pair of *equivalent* surface electric and magnetic currents can be defined as

$$\bar{J}^e = \bar{J} = \hat{n} \times \bar{H}_2 = \left\langle \hat{n} \times \nabla \times \bar{\bar{G}}_2, \bar{J}^{eff} \right\rangle_{S^-} \tag{3}$$

$$-\bar{J}^m = -\bar{M} = \hat{n} \times \bar{E}_2 = i\omega\mu_0 \left\langle \hat{n} \times \bar{\bar{G}}_2, \bar{J}^{eff} \right\rangle_{S^-} \tag{4}$$

Then the scattering field in the air region may be found through

$$\bar{H}^s = i\omega\varepsilon_0 \left\langle \bar{\bar{G}}_1, \bar{J}^m \right\rangle_{S^+} + \left\langle \nabla \times \bar{\bar{G}}_1, \bar{J}^e \right\rangle_{S^+} \tag{5}$$

where the inner product is evaluated on the upper side of the interface. The total field in the air region is $\bar{H}_1 = \bar{H}^{inc} + \bar{H}^s$ with $\bar{H}^{inc}$ the incident magnetic field. Continuity condition of the tangential magnetic fields demands that $\hat{n} \times \bar{H}_2 = \hat{n} \times \bar{H}_1$, i.e.,

$$\bar{J}^e = \bar{J}^{inc} + i\omega\varepsilon_0 \left\langle \hat{n} \times \bar{\bar{G}}_1, \bar{J}^m \right\rangle_{S^+} + \left\langle \hat{n} \times \nabla \times \bar{\bar{G}}_1, \bar{J}^e \right\rangle_{S^+} \tag{6}$$

where $\bar{J}^{inc} = \hat{n} \times \bar{H}^{inc}$.

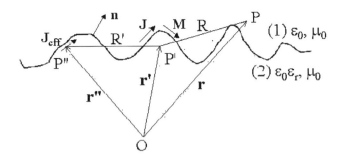

Fig.1 Geometry of rough surface scattering using SIE approach.

Using the MoM with the RWG basis functions [16], we convert (3)-(4) and (6) into sets of equations:

$$\left. \begin{array}{l} \bar{I}^e = (\bar{\bar{P}}_2 - \bar{\bar{D}}^+) \cdot \bar{I}^{eff} \\ -\bar{I}^m = \eta_0 \bar{\bar{Q}}_2 \cdot \bar{I}^{eff} \end{array} \right\} \tag{7}$$

$$\bar{I}^e = \bar{I}^{inc} + \frac{1}{\eta_0} \bar{\bar{Q}}_1 \cdot \bar{I}^m + (\bar{\bar{P}}_1 + \bar{\bar{D}}^-) \cdot \bar{I}^e \tag{8}$$

where $\bar{D}^+$ and $\bar{D}^-$ are diagonal matrices with diagonal elements $D_{ii}^+ = \Omega_i^+ / 2\pi$ and $D_{ii}^- = \Omega_i^- / 2\pi$, here Ω_i^+ and Ω_i^- satisfying $\Omega_i^+ + \Omega_i^- = 2\pi$ are respectively the exterior and interior angles subtended by the triangle pair $T_n = T_n^+ + T_n^-$ connected to the i-th edge. The column vectors $\bar{I}^e, \bar{I}^m, \bar{I}^{eff}, \bar{I}^{inc}$ are the expansion coefficients of their corresponding vectors, which are interpreted as the average values of the currents flowing across the edge, for instance,

$$\bar{J}^{inc} = \sum_{i=1}^{N} I_i^{inc} \vec{f}_i(\vec{r}), \qquad I_i^{inc} = \frac{1}{l_i} \int_{l_i} (\hat{n} \times \hat{l}_i) \cdot \bar{J}^{inc} \, dl \tag{9}$$

where $\vec{f}_i$ ($i = 1, 2, \cdots, N$) are the RWG basis functions associated with the N interior edges of the triangulated surface; and $\hat{l}_i$ labels the direction of the i-th edge, which is so defined that $(\hat{n} \times \hat{l}_i)$ coincides with the current direction across the edge from T_n^+ to T_n^-. The elements of the matrices $\bar{\bar{P}}$ and $\bar{\bar{Q}}$ are

$$P_{1,2}(i,j) = \frac{1}{l_i} \int_{l_i} \int_{T_j} (\hat{l}_i \cdot \nabla G_{1,2} \times \vec{f}_j) dS' dl \tag{10}$$

$$Q_{1,2}(i,j) = \frac{ik_1}{l_i} \int_{l_i} \int_{T_j} \left[(\hat{l}_i \cdot \vec{f}_j) G_{1,2} + \frac{1}{k_{1,2}^2} \frac{\partial G_{1,2}}{\partial l_i} (\nabla_s' \cdot \vec{f}_j) \right] dS' dl \tag{11}$$

where G_1 and G_2 are the scalar Green's functions of the upper and lower media. Substituting (7) into (8), one obtains

$$\left[\overline{\overline{Q}}_1 \cdot \overline{\overline{Q}}_2 - (\overline{\overline{D}}^+ - \overline{\overline{P}}_1)(\overline{\overline{D}}^+ - \overline{\overline{P}}_2) \right] \cdot \overline{I}^{\it eff} = \overline{I}^{\it inc} \tag{12}$$

This is the MoM equation based on a single magnetic field integral equation using an *effective electric* current on the lower side of the interface. Other SIE formulations may be derived in the similar way using either an effective electric or an effective magnetic current on *either* side of the interface.

Once the effective electric current $I^{\it eff}$ is solved, the equivalent currents $\overline{I}^e$ and $\overline{I}^m$ can be retrieved by (7), then the scattering fields may be found by (5). Instead of calculating the scattering fields, we can compute the normalized bistatic scattering coefficients using the equivalent currents, which is defined by

$$\gamma_{ab}(\theta_s, \phi_s) = \lim_{r \to \infty} \frac{r^2 |E_a^s|^2}{2\eta_0 P_b^{\it inc}} = \frac{|S_a(\hat{k}_s)|^2}{2\eta_0 P_b^{\it inc}} \tag{13}$$

where a and b indicate the polarization states of the scattering and incident waves, respectively, and $P_b^{\it in}$ is the power impinged on the surface.

$$S_v(\hat{k}_s) = \frac{ik_1}{4\pi} \int_S \left(\hat{v}_s \cdot \eta_0 \overline{J}^e + \hat{h}_s \cdot \overline{J}^m \right) e^{-ik_1 \hat{k}_s \cdot \vec{r}'} dS' \tag{14}$$

$$S_h(\hat{k}_s) = \frac{ik_1}{4\pi} \int_S \left(\hat{h}_s \cdot \eta_0 \overline{J}^e - \hat{v}_s \cdot \overline{J}^m \right) e^{-ik_1 \hat{k}_s \cdot \vec{r}'} dS' \tag{15}$$

where

$$\hat{v}_s = \cos\theta_s \cos\phi_s \hat{x} + \cos\theta_s \sin\phi_s \hat{y} - \sin\theta_s \hat{z}$$

$$\hat{h}_s = -\sin\phi_s \hat{x} + \cos\phi_s \hat{y}$$

$$\hat{k}_s = \sin\theta_s \cos\phi_s \hat{x} + \sin\theta_s \sin\phi_s \hat{y} + \cos\theta_s \hat{z}$$

with (θ_s, ϕ_s) being the common spherical coordinates.

2.2 Multilevel Sparse-Matrix Canonical-Grid (MSMCG) Method

To use the MSMCG method solving (12), each interaction matrix $(\overline{\overline{Q}}_1, \overline{\overline{Q}}_2, \overline{\overline{P}}_1, \overline{\overline{P}}_2)$ is written as two parts, e.g., $\overline{\overline{Q}}_2 = \overline{\overline{Q}}_2^s + \overline{\overline{Q}}_2^w$, where $\overline{\overline{Q}}_2^s$ denotes the *near-strong* interactions within a given distance between source points and field points, and $\overline{\overline{Q}}_2^w$ the *far-weak* interactions beyond the separation. The elements of the strong parts are calculated by exact numerical integration and stored in computer memory. However, the elements of the weak parts are not calculated and stored in common ways. We expand the Green's function and its gradient as Taylor series about a three-dimensional cubic grid as shown in Fig.2,

$$G = \sum_{k=0}^{K} \frac{1}{k!} \frac{\partial^k G}{\partial \xi^k} (\xi - \xi')^k = \sum_{k=0}^{K} C_k (\xi - \xi')^k$$

$$= \sum_{k=0}^{K} C_k \sum_{l=0}^{k} \binom{l}{k} \xi^l (-\xi')^{k-l}$$

(16a)

$$\nabla G = \sum_{k=0}^{K} \vec{D}_k (\xi - \xi')^k = \sum_{k=0}^{K} \vec{D}_k \sum_{l=0}^{k} \binom{l}{k} \xi^l (-\xi')^{k-l}$$

(16b)

where K truncates the number of terms of the Taylor series; C_k and $\vec{D}_k$ are functions of the canonical-grid coordinates. For example, the expressions of the zero-order and first-order terms are:

$$C_0 = G(R_c)$$
$$C_1 = G'(R_c) z_c / R_c$$
$$\vec{D}_0 = G'(R_c) \hat{R}_c$$
$$\vec{D}_1 = (z_c / R_c)[G''(R_c) - G'(R_c)/R_c]\hat{R}_c + (1/R_c)G'(R_c)\hat{z}$$

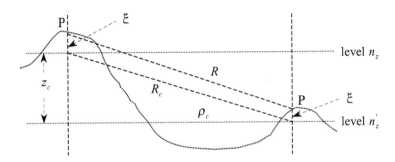

Fig.2 Expansion of the Green's function and its gradient about a canonical-grid coordinate system.

Inserting (16a)-(16b) into (10)-(11) and approximating the linear integration by one-point rule (the midpoint) and the surface integration over a triangle by three-points rule (the midpoints of the edges), we can show that the weak elements, say Q^w, have the following form:

$$Q^w(n,n') = \sum_{k=0}^{K} \sum_{\ell=0}^{k} f_{k\ell}(n) U_k(n-n') g_{k\ell}(n')$$

(17)

where $f_{k\ell}(n)$ is associated with an observation point while $g_{k\ell}$ associated with a source location. The matrix $\overline{\overline{U}}_k$ has the Toeplitz structure that may be stored as a vector with a length of $2N-1$. For a three-dimensional canonical grid system, the grid index in (17) should be understood as $n = (n_x, n_y, n_z)$ and $n - n' = (n_x - n'_x, \cdots)$. Note that (n, n') in (17) refers to the grid indices, while (i, j) in (10)-(11) refers to the edge indices. There exists a corresponding relation between them using an index transform, that is, the midpoint of the i-th edge corresponds to a closest grid coordinate (n_x, n_y, n_z).

The reduction of memory storage requirement is achieved as follows. The storage of the matrix $\overline{\overline{Q}}$ is on the order of $O(N^2)$ by common means. However, the storage of $\overline{\overline{Q}}^s$ is proportional to $O(N)$ because at each observation point only a few near source elements within a given distance are taken into account; the storage of $\overline{\overline{U}}_k$ is $O(2N\ )$, and so is $\bar{f}_{k\ell}$ and $\bar{g}_{k\ell}$ with N zeros padded to each of them to make use of the FFT. Thus, the total storage of $(\overline{\overline{Q}}^s + \overline{\overline{Q}}^w)$ is scaled down to $O(\alpha N)$ where α is a constant typically less than a few tens. The memory storage saving is huge to think that N can be several millions.

The equation (12) is solved using the Conjugate Gradient Method (CGM), by which no matrix inversion but matrix-vector multiplication is involved. The multiplication of the strong parts with a column vector takes little time, as the matrices are very sparse. Due to the approximation (17), the multiplication of the weak parts with a column vector can be significantly accelerated using the FFT technique, i.e.,

$$\overline{\overline{Q}}^w \cdot \bar{x} = \sum_{k=0}^{K} \sum_{\ell=0}^{k} \bar{f}_{k\ell} \, \text{FFT}^{-1}\{\text{FFT}[\overline{U}_k] \, \text{FFT}[\bar{g}_{k\ell}\bar{x}]\} \tag{18}$$

where the products between two column vectors are effected between their corresponding elements. Direct computation of the left-hand side of (18) is on the order of $O(N^2)$. Because the computing complexity of the three FFTs is $3\times(2N\log 2N)$ and that of the three vector products is $3\times(2N)$, the total computing burden of the right-hand side is only $O(\beta N)$ where $\beta = \frac{1}{2}(K+1)(K+2)(6\log 2N + 6)$ which may be seen as a constant as $\log 2N$ varies very slowly with N. Thus, the saving of CPU time for an iterative solution is substantial as β is typically several hundreds while N on the order of million.

2.3 Alternate Approximation Methods

The procedure described above is rigorous and the results should be exact. However, in some special cases, simplifications may be made to speed up the analysis with a tolerable loss of accuracy. It is noted that for a highly lossy medium, the Green's function decays rapidly and may be ignored beyond an appropriate distance. At microwave frequencies, both the real and imaginary parts of the permittivity of the seawater are very high. Thus we may discard $\overline{\overline{Q}}_2^w$ and $\overline{\overline{P}}_2^w$, the *far-weak* interaction matrices in the seawater region, and call this simplification as Localized Approximation Method (LAM). Doing so will save about half the CPU time in solving the equation (12). The *near-strong* matrices $\overline{\overline{Q}}_2^s$ and $\overline{\overline{P}}_2^s$ are retained and their banded-widths depend on the distance beyond which the *far-weak* interactions have been ignored. As an extreme case, if only the self-interaction elements are kept for the strong matrices so that they become diagonal, we call this extremity the Impedance Approximation Simplification (IAS), which is not the same as the impedance approximation method proposed in [12]. We may go further to ignore even the self-interactions (the self-interaction for $\overline{\overline{P}}_2$ is actually already extracted), i.e., totally discard the matrices $\overline{\overline{P}}_2$ and $\overline{\overline{Q}}_2$, then (12) will be reduced to the Perfectly Electric Conducting (PEC) case. Thus (12) is a general form valid for any kind of medium in the lower region, from conductor to lossy and dielectric bodies.

In general, numerical analysis using MoM requires that the discretization of the surface be about one tenth wavelength of the denser medium or on the scale of the skin-depth for a highly lossy medium. However, for a perfectly conducting surface, sampling at one tenths free-space wavelength is sufficient. For seawater at microwave bands, the ocean behaves like a metallic body, which may allow us to use much coarser meshing. In this case, the strong elements of $\overline{\overline{Q}}_2^s$ and $\overline{\overline{P}}_2^s$ must be carefully evaluated. If this conjecture applies, the number of unknowns will be reduced substantially and a great deal of memory storage and CPU solution time can be saved. We will provide examples to check the accuracy of this approach.

2.4 Beam Decomposition Technique (BDT)

The above algorithm is rather robust for a range of surface sizes and roughness. On a low-cost cluster platform that consists of 32 PC and uses the Message Passing Interface (MPI) for communicator, up to six million unknowns has been solved. Should a surface is so large that a computer may run out of core memory, an incorporated BDT may be activated. This happens if the number of processors is too few or the memory for each processor is too small, or the surface exceeds many thousand square wavelengths such as the case of low grazing incidences. By BDT, the incident wave is decomposed as many narrower beams, so that the total scattering field may be synthesized coherently from the partial scattering fields of the individual beams. The incident wave used in our codes is

$$\overline{E}_B^{inc} = \int_{-\infty}^{\infty}\int_{-\infty}^{\infty} \overline{S}(k_x,k_y)\Phi_B(k_x-k_{ix},k_y-k_{iy})e^{i(k_xx+k_yy-k_zz)}dk_xdk_y \tag{19}$$

where $k_z = \sqrt{k^2-k_x^2-k_y^2}$, and $\overline{S}(k_x,k_y)=\hat{p}_i\cdot(\hat{v}\hat{v}+\hat{h}\hat{h})$, with $\hat{p}_i$ being $\hat{v}_i$ for vertically polarized (TM waves) or $\hat{h}_i$ for horizontally polarized (TE waves) incidences. The definitions of $\hat{h}$ and $\hat{v}$ are $\hat{h}=(\hat{z}\times\hat{k})/|\hat{z}\times\hat{k}|$ and $\hat{v}=\hat{h}\times\hat{k}$, so that ($\hat{v},\hat{h},\hat{k}$) forms an orthonormal system. The incident magnetic field is obtained by replacing $\overline{S}$ with $\hat{k}\times\overline{S}/\eta=\hat{p}_i\cdot(\hat{v}\hat{h}-\hat{h}\hat{v})/\eta$, where η is the intrinsic impedance of free-space. The function Φ_B is a Gaussian tapered spectrum with center incident direction in (k_{ix},k_{iy},k_{iz}) , where $k_{iz}=\sqrt{k^2-k_{ix}^2-k_{iy}^2}$, i.e.,

$$\Phi_B(k_x-k_{ix},k_y-k_{iy})=\frac{B_xB_y}{4\pi}\exp\left\{-\tfrac{1}{4}[(k_x-k_{ix})^2B_x^2+(k_y-k_{iy})^2B_y^2]\right\} \tag{20}$$

where B_x and B_y control the tapering or the beam-widths and are typically chosen to be $B_x = L_x/5$ and $B_y = L_y/5$, where L_x and L_y are the dimensions of the simulated area.

A smaller incident beam is written in the same form as (19) but replacing Φ_B with Φ_b , where Φ_b is the same as (20) by replacing B_x and B_y with b_x and b_y , while b_x and b_y control the tapering of the narrower beams and are typically chosen to be one half or one fourth of B_x and B_y , respectively.

Thus the large incident beam (19) may be synthesized by a set of smaller beams as

$$\overline{E}_B^{inc}(x,y,z) = \int_{-\infty}^{\infty}\int_{-\infty}^{\infty} U(X,Y)\overline{E}_b^{inc}(x-X,y-Y,z)\,dX\,dY$$

$$\approx \sum_{m=-M_x}^{M_x}\sum_{n=-M_y}^{M_y} W_{mn}\,\overline{E}_b^{inc}(x-m\Delta X,y-n\Delta Y,z) \tag{21}$$

where the number of small beams is supposed to be $(2M_x+1)*(2M_y+1)$ and the moving steps of the small beams are ΔX and ΔY in the x - and y - directions. For instance, if $\Delta X = l_x/4$ and $l_x = L_x/2$, then $M_x = (L_x-l_x)/2\Delta X = 2$. The weighting factor is $W_{mn} = U(m\Delta X,n\Delta Y)\Delta X\,\Delta Y$ and $U(X,Y)$ is the inverse Fourier transform of Φ_B/Φ_b , specifically,

$$U(X,Y) = \frac{1}{4\pi^2} \int_{-\infty}^{\infty} \int_{-\infty}^{\infty} \frac{B_x B_y}{b_x b_y} e^{-\left[(k_x-k_{tx})^2(B_x^2-b_x^2)+(k_y-k_{ty})^2(B_y^2-b_y^2)\right]/4} e^{i(k_x X+k_y Y)} dk_x dk_y$$

$$= \frac{1}{\pi} \frac{B_x B_y}{b_x b_y} \frac{1}{\sqrt{B_x^2-b_x^2}} \frac{1}{\sqrt{B_y^2-b_y^2}} e^{-\left[X^2/(B_x^2-b_x^2)+Y^2/(B_y^2-b_y^2)\right]} e^{i(k_{tx}X+k_{ty}Y)} \tag{22}$$

Now that the large incident beam is decomposed as many smaller beams, each of which illuminates a portion of a large surface, the original big problem may be translated into a set of smaller problems. By virtue of the linearity of the Maxwell equations, the equivalent surface electric and magnetic currents due to the large beam will be synthesized following the same way as (21), and so are the scattered fields.

2.5 Parallel Computing

The above algorithms are particularly suitable for parallel computing. Once a random rough surface satisfying some given statistical properties is generated, each processor starts to compute a part of the strong matrices and stores the elements in its own memory using a sparse matrix storage strategy. When the strong matrices are multiplied with a column vector, each processor yields a local vector and the resultant vector is obtained by summing the local ones. For the multiplication of the weak matrices with a column vector by the FFT algorithm described in Section 2.2, the MPI version of the FFTW software [17] has been resorted, by which each processor only stores a portion of the transformed vector. Thus, the storage requirement for each processor is a portion of the strong matrices and a portion of an array involved in the FFT calculations. Essential immediate storage is carefully distributed to individual computers subject to least communications for later uses.

The endpoint of the numerical analyses in this paper is to calculate the bistatic scattering coefficient by (13). Each processor computes the scattering integration of (14)-(15) only from a portion of the surface, and the whole integration is retrieved by coherently adding the partial values. If the BDT is employed, the above procedure applies to each portion of the surface. Then the resultant integration is recovered following the weighting summation (21). That is, replace on the left-hand side $\overline{E}_B^{inc}$ with $\overline{E}_B^s$, the total scattering field from the whole surface illuminated by a single big beam, and on the right-hand side $\overline{E}_b^{inc}$ with $\overline{E}_b^{s'}$, the scattering field from a portion of the surface illuminated by a smaller beam.

3. NUMERICAL EXAMPLES

The ocean spectrum used in this paper is due to Durden and Vesecky [18] and restated by Yueh [19], which reads

$$W(k,\phi) = \frac{1}{2\pi k} S(k)\Phi(k,\phi) \tag{23}$$

See [19] for a detailed interpretation. The spectral range used in this paper is taken to be $1 \le k \le 120 \, m^{-1}$, and the surface dimension used for our simulations is chosen to be about 7.31 m by 7.31 m to sufficiently resolve the lower limit of the selected spectral range. At wind speed 10 m/s, the root-mean-square (rms) height and slope is about 5 cm and 0.12, respectively. The relative permittivity of the seawater is taken to be $74 + 61i$ at $15^o C$ and 35 psu salinity [20] at L-band. Thus at 1.5 GHz, the simulated area is about 36.6 by 36.6 square wavelengths.

The correctness and accuracy of the SIE formulation has been demonstrated in [7]. Thus the first example here is to check the accuracy of the Localized Approximation Method (LAM) and the Impedance Approximation Simplification (IAS) mentioned in Section 2.3. A comparison of the co-polarized components obtained by LAM and IAS with that of the numerically exact results is shown in Fig.3. The sampling density for the exact results is 14 points per linear wavelength, while that for the approximate results is only 7 points per linear wavelength. Thus the number of unknowns for the exact results is 785,408 using the RWG basis functions, while that for the approximate results is only 196,096, about one fourth of the former. It can be seen that both the LAM and IAS can well capture the features of the scattering coefficients, though the IAS is less accurate when compared with the LAM. More numerical experiments have been conducted, and it concludes that the IAS tends to overestimate the scattering coefficient in all cases. In view of this, the LAM will be adopted in the following simulations, and the separation for the weak interaction is set to about half free-space wavelength, beyond which the weak interactions in the seawater medium are ignored.

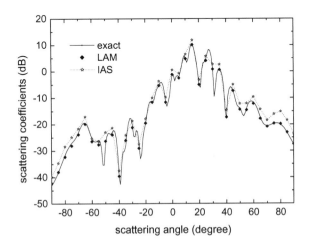

Fig.3 Comparison of the localized approximation method (LAM) and impedance approximation simplification (IAS) with the exact results for a vertically polarized wave incident at -20 degree.

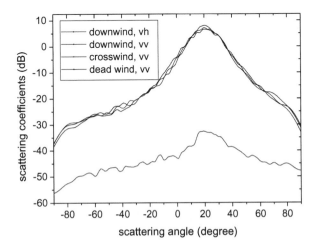

Fig.4 Simulation results of 50 realizations for a TM waves incident in different direction relative to the wind direction.

Some Monte-Carlo simulations are shown in Fig.4 for 50 realizations. It is observed that the peak in the specular scattering direction for a wave incident in the crosswind direction is slightly stronger than that in the downwind and dead wind direction. For the crosswind case, the peak is 8.1 dB, while for the downwind and dead wind cases they are 6.8 dB and 6.6 dB, respectively. The 50 sample surfaces are the same for the three azimuth angles of incidences relative to the wind direction. On the other hand, it shows that for the three incident azimuths, downwind, crosswind or dead wind, the half-power scattering angular widths are around 15 degrees.

As mentioned in Section 2.4, if the surface is too large to be simulated as a whole, we can activate the BDT to coherently analyze it portion by portion. To this end, we first give an example to check the accuracy of the method. Fig.5 is a comparison of the synthesized results with the exact results, which shows in reasonable agreement. The original large incident beam is split into 25 smaller beams, each of which illuminates on an area of one fourth the large surface. The BDT translates a big problem into a set of small problems, so that the stringent memory storage is lifted but more CPU time is required. For this example, the CPU time to solve a small problem is about one fourth of that of the big problem, but there are 25 small ones to handle.

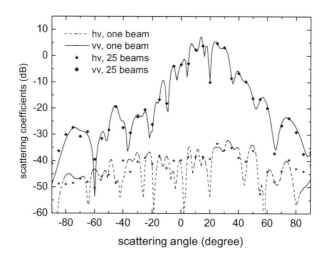

Fig.5 Comparison of the results obtained using 25 small beams with that using a single large beam.

Finally we give an example for low grazing incidence. The size of the surface simulated is 292.58 by 36.57 square wavelengths (58.52 by 7.31 square meters at 1.5 GHz). As the length/width ratio of the surface is 0.125, this allows us to use a gazing incident angle slightly greater than 82.4 degrees, and here we choose 85 degrees. To test the convergence of the scattering coefficients, we calculate the ensemble average of 20 realizations, which is shown in Fig. 6. The surfaces are modeled by 1,048,576 triangular patches, which result in 1,570,560 unknowns using the single integral equation formulation and the RWG basis functions. Using the present simulation codes, however, this is not a "big" problem for a cluster of 16 PCs, each of which has one 677 MHz CPU and possesses 512 MB RAM. The analysis time for each realization is about 20 hours for both states of polarization combined.

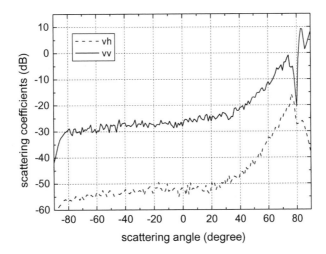

Fig.6 Scattering coefficients for low grazing incidence at -85 degree.

4. CONCLUDING REMARKS

In this paper, an efficient procedure is presented for vector wave scattering from random rough surfaces, and numerical results at the microwave frequency are provided for wind-driven ocean surfaces. The governing equation adopted is the single magnetic integral equation, while the key technique is based on the multilevel sparse-matrix canonical-grid scheme, which enables us to achieve a substantial reduction both in memory storage requirement and CPU solution time. To speedup the analysis for scattering from the highly lossy seawater, a challenging topic in microwave remote sensing of oceans, the Localized Approximation Method (LAM) and the Beam Decomposition Technique (BDT) are incorporated. The anisotropy of the wind-driven ocean surfaces may be identified from the simulation results, that is, the scattering in crosswind direction is stronger than that in downwind and dead-wind direction. The scattering angular widths, however, are quite similar in all the cases considered. Grazing incidences are attempted particularly, and some preliminary results are given. Unlike Gaussian surfaces, ocean surfaces are multi-scale or composite in nature and thus the simulation results are both spectral-dependent and frequency-dependent, i.e., the scattering coefficients depend on the spectral range selected and the operating frequency used. Currently we are optimizing the codes and making comparisons with other methods, toward an efficient and reliable tool for modeling of microwave scattering from ocean-like surfaces.

Acknowledgement: Project 60171001 supported by NSFC and a Joint Hong Kong/France Research Grant No. 9050161.

References

1. M. Saillard and A. Sentenac, "Rigorous solutions for electromagnetic scattering from rough surfaces," *Waves in Random Media,* Vol.11, pp.R103-R137, 2001

2. K.F. Warnick and W.C. Chew, "Numerical simulation methods for rough surface scattering," *Waves in Random Media,* Vol.11, pp.R1-R30, 2001

3. L. Tsang, C. H. Chan and K. Pak, "Backscattering enhancement of a two-dimensional random rough surface (three-dimensional scattering) based on Monte-Carlo simulation," *J. Opt. Soc. Am. A*, vol.11, pp.711-715, 1994

4. K. Pak, L. Tsang and J.T. Johnson, "Numerical simulations and backscattering enhancement of electromagnetic waves from two-dimensional dielectric random rough surface with the sparse-matrix canonical-grid method," *J. Opt. Soc. Am. A*, vol.14, pp.1515-1529, 1997

5. S.Q. Li, C.H. Chan, M.Y. Xia, B. Zhang and L. Tsang, "Multilevel expansion of the sparse-matrix canonical-grid method for two-dimensional random rough surfaces," *IEEE Trans. Antennas & Propagat.,* vol.49, pp.1579-1589, 2001

6. Q. Li, C. H. Chan and L. Tsang, "Monte-Carlo simulation of wave scattering from lossy dielectric random rough surfaces using the physics-based two-grid method and the canonical-grid method," *IEEE Trans. Antennas & Propagat.* vol.47, pp.752-763, April 1999

7. M.Y. Xia, C.H. Chan, S.Q. Li, B. Zhang and L. Tsang, "An efficient algorithm for electromagnetic scattering from rough surfaces using a single integral equation and multilevel sparse-matrix canonical-grid method," to appear in *IEEE Trans. Antennas & Propagat*

8. M. Y. Xia and C. H. Chan, "Parallel analysis of electromagnetic scattering from rough surfaces," *Electronics Letters.* vol. 39, no.9, pp.710-712, May 2003

9. R.L. Wagner, J.M. Song and W.C. Chew, "Monte-Carlo simulation of electromagnetic scattering from two-dimensional random rough surfaces," *IEEE Trans. Antennas & Propagat.*, vol.45, pp.235-245, February 1997

10. V. Jandhyala, B. Shanker, E. Michielssen, and W.C. Chew, "Fast algorithm for the analysis of scattering by dielectric rough surfaces," *J. Opt. Soc. Am. A*, vol.15, pp.1877-1885, 1998

11. A.K. Fung, M.R. Shah and S. Tjuatja, "Numerical simulation of scattering from two-dimensional randomly rough surfaces," *IEEE Trans. Geoscience & Remote Sensing,* vol.32, no.5, pp.986-994, 1994

12. G. Soriano and M. Saillard, "Scattering of electromagnetic waves from two-dimensional rough surfaces with an impedance approximation," *J. Opt. Soc. Am. A*, vol.18, pp.124-133, January 2001

13. M. Saillard and D. Maystre, "Scattering from random rough surfaces: a beam simulation method," *J. Optics (Paris)*, vol.19, no.4, pp.173-176, 1988

14. H.D. Ngo and C.L. Rino, "Application of beam simulation to scattering at low grazing angles, 1. Methodology and validation," *Radio Science*, vol.29, no.6, pp.1365-1379, 1994

15. M.S. Yeung, "Single integral equation for electromagnetic scattering by three-dimensional homogeneous dielectric objects," *IEEE Trans. Antennas Propagat.*, vol.47, no.10, pp.1615-1622, October 1999

16. S.M. Rao, D.R. Wilton and A.W. Glisson, "Electromagnetic scattering by surfaces of arbitrary shape," *IEEE Trans. Antennas & Propagat.*, vol.30, pp.409-418, May 1982

17. M. Frigo and S.G. Johnson, "The fastest Fourier transform in the west," Tech. Rep., MIT-LCS-TR-728, 1997. (also see http://www.fftw.org)

18. S.L. Durden and J.F. Vesecky, "A physical radar cross-section model for a wind-driven sea with swell," *IEEE Journal of Oceanic Engineering,* Vol. GE-10, pp.445-451, October 1985

19. S.H. Yueh, "Modeling of wind direction signals in polarimetric sea surface brightness temperature," *IEEE Trans. Geoscience & Remote Sensing,* vol.35, pp.1400-1418, November 1997

20. L.A. Klein, C.T. Swift, "Improved model for dielectric-constant of sea-water at microwave-frequencies," *IEEE Trans. Antennas & Propagat.*, vol.25, no.1, pp.104-111, 1977

Wave Propagation,Scattering and Emission in Complex Media
Edited by Ya-Qiu Jin
Science Press and World Scientific,2004

HF Surface Wave Radar Tests at the Eastern China Sea[*]

Xiong Bin Wu, Feng Cheng, Shi Cai Wu, Zi Jie Yang

BiYang Wen, Zhen Hua Shi, JianSheng Tian, HengYu Ke, HuoTao Gao

School of Electronic Information, Wuhan Univeristy, Wuhan 430079, China

Abstract   The HF surface wave radar system OSMAR2000 adopts Frequency Modulated Interrupted Continuous Waveform (FMICW) and its 120m-antenna array is transmitting/ receiving co-used. MUSIC and MVM are applied to obtain sea echo's direction of arrival (DOA) when extracting currents information. Verification tests of OSMAR2000 ocean surface dynamics detection against in-situ measurements had been accomplished on Oct. 23~29, 2000. Ship detection test was carried out on Dec.24, 2001. It shows that OSMAR2000 is capable of detecting 1000 tons ships with a wide beam out to 70 km. This paper introduces the radar system and the applied DOA estimation methods in the first, and then presents ship detection results and some sea state measurement results of surface currents and waves. The results indicate the validity of the developed radar system and the effectiveness of the applied signal processing methods.

Key Words   HF surface wave radar, Ship Detection, Doppler spectrum, MUSIC, MVM, Phase Stability

1.   Introduction

Coast-based High Frequency Surface Wave Radar (HFSWR) can provide large coverage ocean surface dynamics surveillance in all weather conditions[1]. Vertically polarized HF radiowaves possesses the feature of sea surface propagating over-the-horizon, and the roughness of the surface will cause backscatter of the incident radiowaves. In 1955, Crombie proposed that the first order interaction between radiowave and ocean waves can be explained using the model of Bragg resonance scattering[2]. Wait confirmed Crombie's proposal in 1966 and pointed out that the magnitude of the first order echo has relationship with the height of the ocean wave with certain wavelength[3]. In 1972, Barrick derived the radar cross-section equations for the first and the second order sea echo[4],[5]. The equations indicate abundant ocean surface dynamics information, i.e. that of the currents, waveheight spectrum and wind field, is contained in the ocean echo, and they are the theoretical foundation for ocean surface dynamics surveillance by HFSWR. Many experimental studies on HFSWR ocean surface detection had been carried out in the Unit States, England, Canada, Germany, Australia and Japan, etc [6].

The Radiowave Propagation Laboratory (RPL) of Wuhan University began the HFSWR study since 1987, and in 1993 developed the first such radar system specialized in ocean surface surveillance in China. The so-called OSMAR (OSMAR-Ocean State Measuring & Analyzing Radar) had measured the radial sea currents out to 30 km at the coast of China Guangxi Province[7]. From 1997 to 1999, the RPL of Wuhan University had developed two HFSWR systems named OSMAR2000 in which the FMICW waveform[8],[9] and phased array technique were applied. The system was installed at the beach of the Zhujiajian Island of Zhejiang Province (29°53′24″N, 122°24′46″E) and put into self-test running in early December of 1999. Sea echoes were observed 6 days after the first power

* **Foundation Project:** State 863 plan project (863－818－01－02) of China，National science foundation project (60201003) of China
  About the author: Wu Xiongbin (born in 1968), E-mail: xbwu@public.wh.hb.cn

on operation. In the followed months, problems such as distance overlapping, zero-frequency interference and signal isolation on transmitting/receiving co-used antenna etc. were solved and the system was upgraded with enhanced reliability. A unique phased array operating mode, in which one of the unit in the array transmits while all the units receive sea echoes, is adopted after much experience was got from the self-test experiments. In data processing, the digital beam forming (DBF) algorithm is applied to achieve 15° angular resolution for wind and wave parameter extraction, while MUSIC(MUltiplex SIgnal Classification)[10] and MVM(Minimum Variation Method)[11] algorithms are applied in currents bearing determination which can afford 2.5°~5° resolution. The second test site was setup at Songlanshan (29°25′54″N, 121°57′42″E) of Zhejiang Province in September, 2000. Synthesized vector currents maps were got as result of synchronized observation of the two sites since October. Verification tests were carried out then against the SeaSonde[12] of the CODAR company and the traditional means of sea-state measurements. The systems are now installed at the Zhujiajian island and the Shengshan island sending regular observation results to Shanghai Oceanic Environment Prediction Center.

The paper emphasize on the OSMAR2000 detection of surface currents. Firstly the OSMAR2000 system components and features are introduced. Then the principles of DOA determination algorithms are briefly described. Detection result and verification test results are put forward. Finally analysis on the results and conclusion are given.

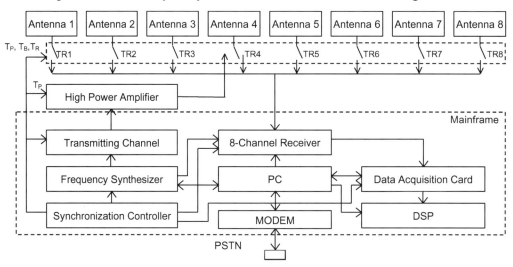

Fig. 1 Sketch map of OSMAR2000 components

TR1~TR8 represent T/R (transmitting/receiving) switch, T_P, T_B and T_R are transmitting pulse, blinding pulse and T/R control pulse, respectively.

2. About the OSMAR2000 System

OSMAR2000 operates within 6~9MHz. Its hardware includes the antenna array and its feeding system, eight-channel receiver, frequency synthesizer, synchronization controller, transmitting channel, 200W solid-state transmitter, data acquisition module and signal processing module. Figure 1 is the sketch map of OSMAR2000 components. The 120m linear phased array is composed of eight two-element Yagi units. In each unit, the fed antenna is 7m high and the reflector is 10m high and 8m in-between. The grounding net is in 2m×2m grids and extended into the seawater. One antenna unit in the middle of the array transmits power under the control of T/R switch while all of the eight units receive backscattered sea echoes. This operating manner is so called as "T/R bi-functional array with one transmitting and eight receiving" style. Figure 2 is a picture of the OSMAR2000 phased array. The duty factor of the radar waveform is 0.44 and the average transmitting power is 90W. The transmitting beamwidth is 120° and the detecting range covers 10~100km. Distance resolution can be set to 1km, 2.5km and 5km. Sweep period is 726.4ms. Echoes' distance spectrum output from the first FFT processed in the DSP is sent to PC after each sweep. The coherent accumulation of the echoes in each range bin lasts 12'40" which is the time of 1024 sweeps. The second FFT gives Doppler spectrum of each range bin. Current velocity resolution corresponding to 1024 points Doppler power spectrum is about ±1.4cm/s.

Fig.2 Linear array of OSMAR2000 on the beach of Zhujiajian Island, Zhejiang Province, China

Software package is composed of the system software and the application software. The former functions radar control and data acquisition, including setting the operating parameters, system testing, data sampling, FFT and data transferring. The later extracts sea state parameter from the echoes data and displays the results, including calculating radial currents, synthesizing vector currents, inversion of waveheight and wind field parameters. All the software is developed in VC++ and operates in the Windows98 system.

Key techniques of the system includes: (1) application of the FMICW pulse compression; (2) small phased array working in the mode of "one transmitting and eight receiving"; (3) adaptive phase and amplitude calibration of phased array utilizing the sea echoes; (4) wideband HF T/R switch; (5) DOA determination using MUSIC and MVM; (6) wind and wave parameters inversion from the echoes Doppler power spectrum.

Two radar sites needed for vector current detection locate 70km apart at Zhujiajian Island and the Songlanshan coast of Zhejiang Province. The operating frequency is chosen according to output of a HF interference spectrum monitor. Phase and amplitude calibration of the phased array is achieved utilizing the sea echoes. The sea state results are sent to Shanghai Oceanic Environment Prediction Center via public telephone network.

3. DOA Estimation Algorithms

Currents information can be extracted from the first order echo spectrum[2][4]. First echoes are intensified radio backscattering from two oppositely propagating wave trains along the incidence direction whose wavelength is half of that of the incident radiowave λ_0. In absence of currents, narrow beam radar will observe first order echoes in the spectrum as two impulses locating symmetrically at $\pm\omega_B$, where $\omega_B = \sqrt{2gk_0}$ (in which g is acceleration of gravity and k_0 is wave number of radiowave) is the Doppler shift caused by the absolute phase velocity of the wave. Ocean currents carrying the two wave trains will cause an additional Doppler shift Δ on $\pm\omega_B$, and this Δ tells that the radial component of the ocean current's velocity is $v_r = \lambda_0\Delta/2$ [4].

OSMAR2000 is wide beam radar that the array units receive echoes from various directions. Radars operating at the low part of the HF band need hundreds or kilometers of array aperture size to achieve narrow beams. OSMAR2000 cannot form narrow beam as its array length is only 120m. The radial current velocity varies when its DOA varies. Thus the two impulse-like first order peaks in case of narrow beam radar transform to be two first order bands for wide beam radar. Figure 3 is an OSMAR2000 Doppler spectrum in which the first order regions have been grayed and marked. Each point in the regions corresponding to a certain radial current velocity is due to the contribution of first order echoes from certain directions. The problem of getting currents' spatial distribution is equivalent to find the DOA for each first order spectrum point[13]. MUSIC[10] and MVM[11] algorithms have been applied in solving the problem.

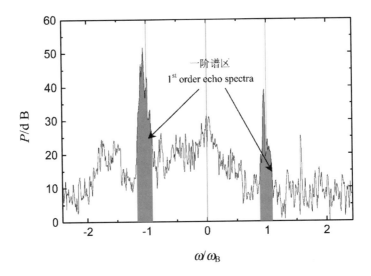

Fig.3 Doppler spectrum from 40km radar cell obtained at 7.5MHz on Oct.29, 2000 at Zhujiajian Island.

3.1 MUSIC Algorithm

MUSIC is based on the geometric explanation of the signal parameter estimation problem, in which the observed data space is classified into two orthogonal subspaces: the signal subspace and the noise subspace. The signal subspace is spanned by those eigenvectors of the data covariance matrix $\mathbf{R}$ that are corresponding to the signals, while the noise subspace is spanned by the remained eigenvectors with least eigenvalues. DOA problem is equivalent to finding out vector(s) in the array manifold vector cluster $\mathbf{a}(\theta)$ (where θ scans within the radar beam scope) closest to the signal subspace. Square of the Euclidean distance of $\mathbf{a}(\theta)$ to signal subspace is

$$\|d\|^2 = \mathbf{a}^{\mathrm{H}}(\theta)\mathbf{E}_{\mathrm{N}}\mathbf{E}_{\mathrm{N}}^{\mathrm{H}}\mathbf{a}(\theta), \tag{1}$$

where the matrix $\mathbf{E}_{\mathrm{N}}$ is composed of the eigenvectors in the noise subspace, and the superscript stands for matrix conjugating transpose. In absence of noise, $\mathbf{a}(\theta)$ of the signal direction(s) will locate in the signal subspace and $d=0$. When noise presents, DOA of the signal can be estimated by finding the direction θ that minimizes the above formula whose reciprocal is defined as the MUSIC spectrum, i.e.

$$P_{\mathrm{MUSIC}}(\theta) = \frac{1}{\mathbf{a}^{\mathrm{H}}(\theta)\mathbf{E}_{\mathrm{N}}\mathbf{E}_{\mathrm{N}}^{\mathrm{H}}\mathbf{a}(\theta)}. \tag{2}$$

Peaks in the spectrum tell the estimation of signal bearing(s). In the MUSIC application of currents mapping, problems such as judging the first order scopes in the Doppler spectrum and judging the source number (i.e. the direction number) of each radial current velocity should be solved first [13]. MUSIC presents higher angular resolution than other space spectrum estimation algorithm[14].

3.2 MVM Algorithm

MVM is a mainlobe-constrained adaptive algorithm, in which a self-adaptive weight vector takes the place of the fixed weight vector in the conventional beam forming technique. In fact, MVM forms a special array response pattern that allows the receipt of signals coming from the needed direction, while rejects signals from other directions in a optimizing manner adaptive to signals. When expressed in mathematics, it is minimizing the power output of the array while ensuring the constant signal output from the needed direction, i.e.

$$\begin{cases} \min_{\mathbf{W}} \mathbf{W}^{\mathrm{T}}\mathbf{R}\mathbf{W}^* \\ \mathbf{W}^{\mathrm{T}}\mathbf{a}(\theta_{\mathrm{d}}) = 1 \end{cases} \tag{3}$$

where θ_{d} is the direction whose signal power needed to be estimated, $\mathbf{a}(\theta_{\mathrm{d}})$ is the manifold column vector at θ_{d} (standing for the array response at θ_{d}), and $\mathbf{W}$ is the weight vector on array units output signals. The superscripts T and * stand for matrix transpose and conjugate, respectively. By Lagrange method, the $\mathbf{W}$ can be expressed as

$$\mathbf{W} = \frac{(\mathbf{R}^*)^{-1}\mathbf{a}^*(\theta_{\mathrm{d}})}{\mathbf{a}^{\mathrm{H}}(\theta_{\mathrm{d}})\mathbf{R}^{-1}\mathbf{a}(\theta_{\mathrm{d}})}. \tag{4}$$

Then the array power output under the MVM restriction is

$$P_{\mathrm{MVM}}(\theta_{\mathrm{d}}) = \mathbf{W}^{\mathrm{T}}\mathbf{R}\mathbf{W}^* = \frac{1}{\mathbf{a}^{\mathrm{H}}(\theta_{\mathrm{d}})\mathbf{R}^{-1}\mathbf{a}(\theta_{\mathrm{d}})}. \tag{5}$$

By scanning θ_{d}, MVM directional spectrum estimation can be obtained.

The background of MVM spectrum is smooth and is a direct indication of noise level at the array output. Peaks in the spectrum correspond to signal power estimations at those directions. Thus MVM can be applied to measure the relative source intensity[15]. In HFSWR application, MVM can be used to restore the amplitudes of the first and the second order sea echoes from every direction in the radar beam. Radial current velocity at a certain direction can be read directly from the first order peak's shift from $\pm\omega_B$ in the Doppler spectrum corresponding to that direction. Besides, wind and wave parameters can also be inversed from the second order spectrum of that direction, and this is the predominance of MVM over MUSIC (which can be only applied in currents mapping).

4. Ship Detection Result

Ship detection test was carried out on Dec.24, 2001. The ship on test is about 1000 ton. OSMAR2000 is a wide beam radar which is not designed for target detection. The test shows that OSMAR2000 is capable of detecting 1000 ton ships out to 70 km. Figure 4 shows the ship detection result where the ship's GPS trace is also marked. Figure 5 shows Doppler power spectrum of the ship echoes in the near 70 km range bins. For wide beam radars, we proposed a detection method named phase stability checking to identify targets from sea echoes, and the false alarm rate should be set a little higher to guarantee targets to be found. Target tracking information is then adopted to restrict the false alarm rate.

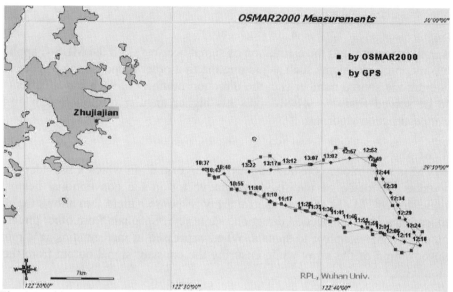

Fig. 4 Tracking of the ship in the target detection test in comparison to that recorded by GPS

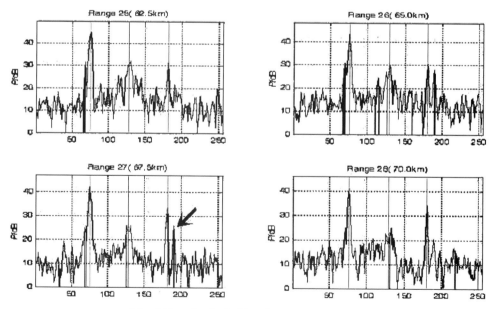

Fig. 5 Target echo in Doppler power spectrum
(spectrum 190 in range 27 as indicated by the arrowhead)

5. Sea State Detection Results

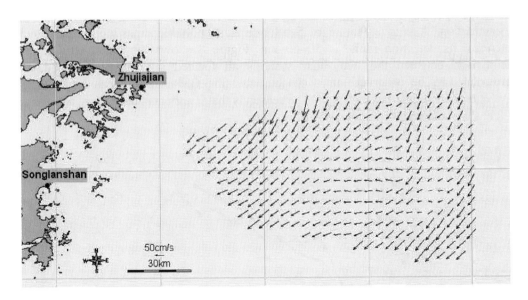

Fig. 6 Vector surface currents distribution mapped by OSMAR2000

Single site radar can only map the radial currents distribution and vector surface currents can be obtained by synthesizing radial currents maps from two or more radar sites[16]. Vector currents will be synthesized within the overlapped area of the two sectors. Currents in the test sea area are regular half-day tidal currents that are affected by both the tides and the coastwise currents of the Eastern China Sea. Figure 6 is vector surface currents distribution mapped by OSMAR2000 at 15:00 LT on Oct. 23 of 2000. Figure 7 shows a wave height distribution map near Zhujiajian during the Sinlaku typhoon at 10:00 LT on Sept.7, 2002.

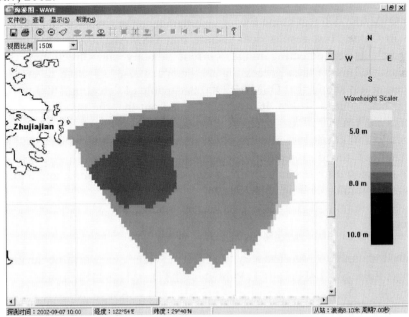

Fig. 7 Wave height distribution near Zhujiajian during the Sinlaku typhoon at 10:00 LT on Sept.7, 2002. Typical waveheight is around 8m within the radar beam illuminating area.

A comparison had been made between OSMAR2000 measured radial currents and that of the SeaSonde HFSWR produced by the CODAR Company of the United States.

SeaSonde operates at 12MHz and adopts FMICW waveform. MUSIC is applied in DOA determination. A unique feature of SeaSonde is its portable cross-loop/monopole type antennae. Its detection range is 50~60 km. Figure 8 shows the radial current results comparison between the two radar systems all located at Zhujiajian Island. Good agreement can be observed. It is obvious that the OSMAR2000 has higher angular resolution and wider viewing angle. The reason is that it adopts phased array and the array unit has more efficient radiation characteristics.

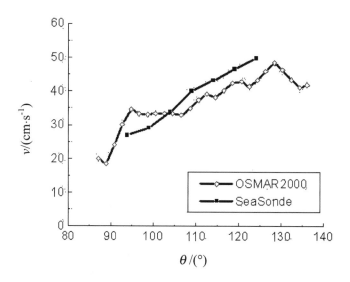

Fig.8 Comparison between OSMAR2000 detected radial currents and SeaSonde measurements of 40km radar cell at 11:30LT on Oct.23, 2000 at Zhujiajian.

6. Conclusion

The OSMAR2000 test at the Eastern China Sea is the first two-site radar detection of sea state in China. Test results demonstrate that OSMAR2000 has achieved its design requests such as the detecting range, current velocity and direction, and the system hardware as well as its data processing methods are also effective in current detection. Ship detection test shows that OSMAR2000 is capable of detecting 1000 tons ships with a wide radar beam out to 70 km. It must be pointed out that the maximum detecting range for currents can reach 250 km in daytime when noise and interference level is low, but it will reduce to 150 km or less at night when interference at the low HF band is intensified as the result of the disappearing of ionosphere D layer. OSMAR2000 provides an important equipment for oceanic environment surveillance in China, and it also provides a platform for HF radio oceanography study.

References

[1] Leigh Barnes, HF Radar-The Key to Efficient Wide Area Maritime Surveillance, *EEZ Technology*, Edition 3, London: ICG Publishing LTD. pp115~118, 1998
[2] Crombie, D D. Doppler spectrum of sea echo at 13.56Mc/s. *Nature*, 1955, 175(4459): 681~682
[3] Wait, J R. Theory of HF ground wave backscatter from sea waves. *J. Geophys., Res.*, 1966, 71:4839~4842
[4] Barrick, D E. First-order theory and analysis of MF/HF/VHF scatter from the sea. *IEEE Trans Antennas Propagation*, 1972, AP-20(1): 2~10

[5] Barrick, D E. Remote sensing of sea state by radar, In: DERR, V.E. (Ed.) Remote sensing of the troposphere. Washington DC, Government Printing Office, 1972

[6] Teague C C, Vesecky J F and Fernandez D M. HF radar instruments, past to present. *Oceanography,* 1997, 10(2): 72~75

[7] HOU Jiechang, WU Shicai, YANG Zijie, *et al.* Remote sensing of ocean surface currents by HF radar. Chinese Journal of Geophysics. 1997, **40**(1):18~26 (Ch)

[8] Khan R H, Mitchell D K. Waveform analysis for high frequency FMICW radar. IEE Proceedings-F, 1991, **138**(5):411~419

[9] YANG Zijie, KE Hengyu, WEN Biyang, *et al.* Waveform parameters design for sea state detecting HF ground wave radar. *Journal of Wuhan University* (*Natural Science Edition*), 2001, **47**(5):528~531 (Ch)

[10] Schmidt R O. Multiple emitter location and signal parameter estimation. *IEEE Trans Antenna and Propagat*, 1986, **34** (3): 276~280

[11] Capon J, High-resolution frequency-wavenumber spectrum analysis. *Proc IEEE*, 1969, **57** (10): 1408~1418

[12] Barrick D. E., EEZ Surveillance-The Compact HF Radar Alternative, *EEZ Technology*, Edition 3, ICG Publishing LTD. pp125~129，1998

[13] Barrick, D.E, B.J. Lipa, Evolution of bearing determination in HF current mapping radars, *Oceanography*, 1997, **10**(2):72~75

[14] LIU De-shu, LUO Jin-qing, ZHANG Jian-yun. *Spatial Spectrum Estimation and its Applications.* Hefei: China Science & Technology University Press, 1997(Ch)

[15] SUN Cao, LI Bin. *Theory and Application of Weighted Sub-Space Fitting Algorithms.* Xi'an: Northwest Industrial University Press, 1994(Ch)

[16] ZHOU Hao, WEN Bi-yang, LIU Hai-wen, Ocean surface vector current synthesization of HF ground wave radar. *Journal of Wuhan University* (*Natural Science Edition*), 2001, **47**(5):634~637(Ch)

[17] Wu Shi-cai, Yang Zi-jie, Wen Bi-yang, et al. Implementation Scheme of HF Ground Wave Radar Technique in Surveillance of Ocean Environment, in Marine Monitoring Subject in the State 863 Plans of China. School of Electronic Information, Wuhan University, Wuhan, China, 1998(Ch)

III. Electromagnetics of Complex Materials

Wave Propagation,Scattering and Emission in Complex Media
Edited by Ya-Qiu Jin
Science Press and World Scientific,2004

163

Wave Propagation in Plane-Parallel Metamaterial and Constitutive Relations

Akira Ishimaru, John Thomas, Seung-Woo Lee, Yasuo Kuga

Department of Electrical Engineering, University of Washington

253 EE/CSE Building, Campus Box 352500, Seattle, WA 98195-2500 USA

E-mail: ishimaru@ee.washington.edu

Abstract   This paper first discusses wave characteristics when a metamaterial with plane boundaries is illuminated by a line source. The metamaterial has arbitrary permittivity and permeability with small loss.For a half-space metamaterial,we discuss the conditions for the existence of forward or backward surface waves, forward or backward lateral waves, Zenneck waves, and relations to Brewster's angle. For a slab of metamaterial, we discuss the focusing characteristics when both permittivity and permeability are close to -1, and show that the results are so sensitive to a slight variation of parameters that perfect focusing is practically not possible. We also discuss the development of generalized constitutive relations for metamaterials based on the quasi-static Lorentz theory, including the consistency constraint and the reciprocity relations, and show a method to calculate the 6×6 constitutive relation matrix for a given three-dimensional periodic array of metallic inclusions with arbitrary shape. Thr results are applied to array of split-ring resonators and helices to obtain numerical values for the constitutive relation matrices.

1   TRANSMISSION PROPERTIES OF MATERIAL WITH NEGATIVE REFRACTIVE INDEX

We have been attracted to work on problems involved in the design of substances with specified permittivity ϵ and permeability μ. These substances have been called, variously, metameterials or photonic crystals. Smith et al. made a metamaterial [1] with simultaneously negative ϵ and μ. J.B. Pendry [2] showed, theoretically, that a plane layer of material with relative permittivity ϵ_r and relative permeability μ_r both equal to -1 will transmit a perfect image of a flat object. Recently, Ziolkowski and Heyman [3] published a thorough investigation of wave propagation and focusing properties of a slab of a double negative medium (DNG) with emphasis on a time-domain point of view with a lossy Drude model. We present limitations on focusing at a single frequency by examining details of the transmission coefficient.

In 1968 V.G. Veselago [4] studied properties of DNG materials before any such material had been created. He made arguments for the possibility of such a material and pointed out unusual effects (e.g., reversed Doppler shift) of such a material. All of these effects follow from the result that the index of refraction is negative. Following Pendry's paper [2] there has been some discussion of how a DNG material leads to negative index of refraction. The result that $n = (\epsilon_r \mu_r)^{1/2}$ is negative for ϵ_r and μ_r negative follows from the requirement for a passive substance and consideration of the limit of a slightly lossy substance, as we shall show in more detail below. A layer of $n = -1$ material bends rays back so that the angle of refraction is the negative of the angle of reflection, but it also has the unusual feature that evanescent waves are focused with enhanced amplitude to provide a perfect image. The geometric condition for imaging is that the sum of the object distance from the incident side of the layer and the image distance from the other side be equal to the thickness of the layer.

We calculate in some detail the transmission of a wave from a 2-dimensional line source through a layer with ϵ_r and μ_r close to -1. For this 2-D problem, the electromagnetic propagation can be completely separated into a p-polarization (incident E field parallel to the plane of incidence) and s-polarization (incident E field perpendicular (senkrecht) to the plane of incidence). A delta function line source images to a delta function in the specified limit. However, for slight deviations from the limit, the magnitude of the transmission coefficient for the evanescent spectrum peaks and then starts to fall off exponentially (instead of continuing its exponential rise) [7].

The 2-dimensional electromagnetic problem of reflection and transmission from a homogeneous uniform layer has been presented in many texts [5] [6] . From this solution we can write the transmission coefficient for a plane wave in the following form, which is equivalent to equation (20) of Pendry [2].

$$T = \frac{1}{\cos\left(k_{z2}d_2\right) + \frac{i}{2}\left(\frac{Z_2}{Z_1} + \frac{Z_1}{Z_2}\right)\sin\left(k_{z2}d_2\right)} \tag{1}$$

where

$\quad d_2 \quad = \quad$ thickness of layer of with permittivity ϵ_2 and permeability μ_2

$\quad k_{z2} \quad = \quad (\epsilon_{2r}\mu_{2r}k_o{}^2 - k_x{}^2)^{1/2} = -j(k_x{}^2 - n_2{}^2 k_o{}^2)^{1/2}$

$\quad Z_2 \quad = \quad$ wave impedance in medium 2 (the layer)

$\quad Z_1 \quad = \quad$ wave impedance in medium 1, thefree space on either side of the layer.

The plane of incidence is taken as the X-Z plane, with the material boundaries at $z = 0$ and $z = d_2$. The wave propagation has a positive Z component. The free-space wave number $k_o = \omega/c$, where ω is the angular frequency and c the speed of light. We have introduced the relative constants ϵ_{2r} and μ_{2r} and the index of refraction $n_2 = (\epsilon_{2r}\mu_{2r})^{1/2}$ in medium 2. We use the engineer's convention with time variation of cw components proportional to $e^{j\omega t}$. Thus, for a passive medium, ϵ_{2r} and μ_{2r} have negative imaginary parts and the branch of the square root for n_2 must also be chosen to give a negative imaginary part. The wave impedances have the following forms.

$$Z_m \quad = \quad \frac{k_{zm}}{\omega\epsilon_m}, \qquad \text{s-polarization.}$$

$$Z_m \quad = \quad \frac{\omega\mu_m}{k_{zm}}, \qquad \text{p-polarization.}$$

with $m = 1$ in free space and $m = 2$ in the layer. In this general form, since the transmission coefficient T depends on the symmetric combination $Z_2/Z_1 + Z_1/Z_2$, the p-polarization and s-polarization cases will have similar forms for T, with the change of ϵ_{2r} to μ_{2r} in key places.

To be more explicit $\epsilon_1 = \epsilon_o$, $\mu_1 = \mu_o$, and $k_{z1} = (k_o{}^2 - k_x{}^2)^{1/2}$. Also, for p polarization

$$\frac{Z_2}{Z_1} = \frac{1}{\epsilon_{2r}}\frac{k_{z2}}{k_{z1}} = \frac{1}{\epsilon_{2r}}\frac{(n_2{}^2 k_o{}^2 - k_x{}^2)^{1/2}}{(k_o{}^2 - k_x{}^2)^{1/2}},$$

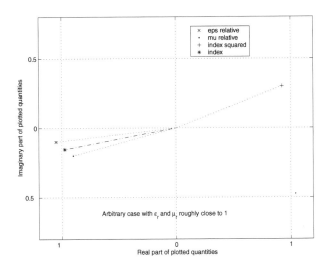

Figure 1: Illustration of permittivity, permeability and index of refraction for a slightly lossy material with dominant negative real parts of permittivity and permeability.

and for s polarization

$$\frac{Z_1}{Z_2} = \frac{1}{\mu_{2r}}\frac{k_{z2}}{k_{z1}} = \frac{1}{\mu_{2r}}\frac{(n_2^2 k_o^2 - k_x^2)^{1/2}}{(k_o^2 - k_x^2)^{1/2}} \cdot$$

We treat the p-polarization case explicitly here. The selection criterion for the correct branch of these square roots is to choose signs that give attenuation of any wave (real or evanescent) in the two media.

2   Negative Refractive Index and Behavior of k_{z1} and k_{z2}

Noting that ϵ_{2r} and μ_{2r} have negative imaginary part, and the refractive index has the negative imaginary part, we show ϵ_{2r}, μ_{2r} and n in the complex plane (Figure 1). We also show in Figure 2, the complex propagation constants k_{z1} and k_{z2} as a function of k_x. Note that in the propagating region, the k_z's are of opposite sign in the negative refractive medium whereas in the evanescent region the signs are the same.

3   Spectrum of a Line Source and Its Image

We consider a line magnetic source at a position $z = -d_1$ relative to the layer of negative refractive material between the planes $z = 0$ and $z = d_2$. The geometry and coordinates are illustrated (Layer is truncated in X direction.) in Figure 3. The following paragraphs show the δ-function spectrum of the E-field of the source and derive the spectrum of the E-field in the image plane. For the perfect lens material of $\epsilon_{2r} = \mu_{2r} = -1$, these spectra are identical.

A magnetic current source of $I_m \delta(x)\delta(z+d_1)$ in the Y direction (out of the paper in Figure 3) is chosen because it drives a p-polarized wave. The source magnetic field satisfies the 2-D

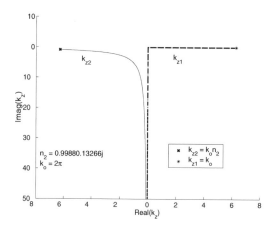

Figure 2: Complex plane paths of longitudinal wavenumbers k_{z1} and k_{z2} as a function of k_x for a slightly lossy DNG medium.

wave equation outside this line source.

$$\left(\frac{\partial^2}{\partial x^2} + \frac{\partial^2}{\partial z^2} + k_o^{\,2}\right)H_y = j\omega\epsilon_o J_m = -j\omega\epsilon_o I_m\big(-\delta(x)\delta(z+d_1)\big).$$

The source magnetic field is, thus, a scalar multiple $-j\omega\epsilon_o I_m$ of the Green's function.

$$H_y(x,z) = -j\omega\epsilon_o I_m\left(\frac{-j}{4}H_o^{(2)}(k_o|\bar{\rho} + d_1\hat{z}|)\right)$$

where $\bar{\rho} = x\hat{x} + z\hat{z}$, $\hat{x}$ and $\hat{z}$ are unit coordinate vectors, and $H_o^{(2)}$ is the zero-order Hankel function of the second kind. This solution can also be expressed in a Fourier transform as

$$H_y(x,z) = -\frac{\omega\epsilon_o I_m}{4\pi}\int_{-\infty}^{\infty}\frac{\exp\left(-jk_x x - jk_{z1}|z+d_1|\right)}{k_{z1}}\mathrm{d}k_x.$$

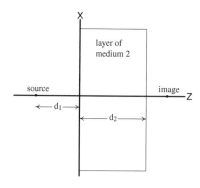

Figure 3: Coordinate system and location of the source and image relative to the DNG layer.

The electric field of this wave is obtained from Ampere's law, and in particular the transverse incident E field in the region $z > -d_1$ is

$$E_x^{inc} = -\frac{1}{j\omega\epsilon_o}\frac{\partial H_y}{\partial z} = -\frac{I_m}{4\pi}\int_{-\infty}^{\infty}\exp\left(-jk_xx - jk_{z1}(z+d_1)\right)\mathrm{d}k_x. \qquad (2)$$

In the plane $z = -d_1$, this gives $E_x^{inc} = (-I_m/2)\delta(x)$. Furthermore, in the region $z > d_2$, the fields have only a transmitted component with

$$E_x^{tr} = \frac{-I_m}{4\pi}\int_{-\infty}^{\infty}T(k_x)\exp\left(-jk_xx - jk_{z1}(z-d_2+d_1)\right)\mathrm{d}k_x$$

where $T(k_x)$ is given by (1). In the limit that $\epsilon_{2r} = \mu_{2r} = -1$, $T(k_x) = exp(jk_{z1}d_2)$ (where $k_x^2 + k_{z1}^2 = k_o^2$), and thus

$$E_x^{tr} = \frac{-I_m}{4\pi}\int_{-\infty}^{\infty}\exp\left(-jk_{z1}(z-2d_2+d_1) - jk_xx\right)\mathrm{d}k_x. \qquad (3)$$

As a restatement of Pendry's conclusion [2], in the plane $z = 2d_2 - d_1$ (and in the region where $z > d_2$, so for a focal plane to exist, $d_2 > d_1$), we have a delta-function image of the delta-function source and so by superposition a perfect image of any flat source.

We examine the lack of perfection for the cases where ϵ_{2r} and μ_{2r} deviate slightly from -1 by examining in further detail the spectrum of E_x^{tr}. If we let $I_m = -2$ and consider the factor $1/(2\pi)$ to be associated with the inverse transform definition, then the spectrum in the region $z > d_2$ is

$$\widetilde{E}(k_x) = T(k_x)\exp\left(-jk_{z1}(z-d_2+d_1)\right). \qquad (4)$$

We generated calculations of $\widetilde{E}(k_x)$ in the image plane. Plots of the results are shown in Figure 4. A perfectly focusing layer would produce a flat line of magnitude 1 (logarithm = 0) in the graphs of Figure 4. We notice that the focusing loses accuracy for the evanescent waves at about the same values of k_x for comparable absolute values of the deviations from -1. When the deviations are real, or quite close, there is a spike in the amplitude just before it starts to decrease.

For large k_x, $\ln|T\exp(-jk_{z1}d_2)|$ is asymptotic to a straight line with slope $-2d_2$. The asymptotic form is

$$|T(\exp(-jk_{z1}d_2))| \sim \frac{4}{|\delta_e|^2}\exp\left(-2\sqrt{k_x^2 - k_o^2}d_2\right). \qquad (5)$$

As the plots of Figure 4 illustrate, $\ln|Te^{-jk_{z1}d_2}| \to 0$ for $k_x < k_o$. The shoulder (half-width) of this spectrum may then be regarded as the point where $|T| = 1$. Let Δk denote the full width and $M_1 = \ln(2/|\delta_e|)$. Then

$$\Delta k/2 = \sqrt{(M_1/d_2)^2 + k_o^2}\ .$$

The focal spot width ΔW (full width at half height) is then related inversely to Δk roughly by $\Delta W\Delta k \approx 2\pi$. Calculations based on the asymptotic spectrum verify this formula is accurate within 20% for $d_2 = 1$.

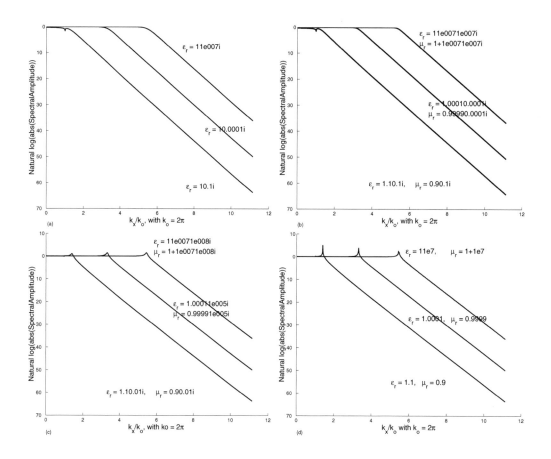

Figure 4: Log of magnitude of spectrum of line source image through a half-wavelength layer, p polarization. In 5(a), each case has $\mu_r = \epsilon_r$ and deviations from -1 are only in the imaginary part. In 5(b), the deviation in ϵ_r is of the form $\delta(-1 - j)$, and the deviation in mu_r is of the form $\delta(1 - j)$. In 5(c) the deviation in ϵ_r is of the form $\delta(-1 - 0.1j)$, and the deviation in mu_r is of the form $\delta(1 - 0.1j)$. In 5(d) the deviations are only in the real part, and of the form $-\delta$ for ϵ_r and δ for mu_r. The small parameter δ is positive.

4 LIMITATION ON PERFECT LENS

Materials designed to provide negative index of refraction will have unusual and surprising reflection, refraction, and transmission properties. One of these is the possibility to create an image with a planar layer. But this image will have many practical limits, which will be related to the fact that such materials will have a loss component, will have to be carefully designed to get ϵ_{2r} and μ_{2r} close to -1, and will have strong frequency dispersion. From the above, we note that the perfect lens occurs only when ϵ_{2r} and μ_{2r} are exactly equal to -1, and the image is so sensitive that a slight deviation of ϵ_{2r} and μ_{2r} from -1 causes a significant departure from the perfect image.

In conclusion we agree with Pendry's calculation [2] of the limit of the transmission coefficient, which yields amplification of the evanescent waves to provide a perfect image. We examined the limiting behavior of a passive medium as it approaches this limit of $\epsilon_{2r} = \mu_{2r} = -1$. For a given ϵ_{2r} and μ_{2r} (both very close to -1), then for $|k_x|$ less than some "turn-around" value the transmission coefficient is close to the ideal amplifying factor. However, for larger $|k_x|$, T rapidly diverges from that ideal perfect-focus amplification.

5 REFLECTION, TRANSMISSION AND EXCITATION OF A SLAB OF ARBITRARY PERMITTIVITY AND PERMEABILITY

We have examined reflection and transmission at the vacuum interface of a slab of material or metamaterial with arbitrary permittivity ϵ and permeability μ. Our objective is to understand the range of properties that could be provided by such materials. The material is assumed uniform and isotropic. Excitation sources include plane waves and the radiation from an idealized line source. Since the p-polarization and s-polarization cases can be treated as equivalent by interchanging permittivity and permeability (with allowance for a change in sign of reflection coefficient), we concentrate on the p-polarization case. For a line source, this is an idealized line of magnetic current. Then, we have categorized the small loss cases by studying reflection properties in the 2-dimensional plane of Real(ϵ) versus Real(μ).

We present a systematic characterization of the poles, zeros, and branch points involved in the reflection coefficient from a semi-infinite half space of metamaterial. We divide the Real(ϵ) – Real(μ) plane into a sequence of regions based on the magnitude of S^2 and the index of refraction. The complex quantity S is defined following Wait as k_x/k_o at the pole of the reflection coefficient, where k_x is the transverse propagation constant along the slab surface. S may also be regarded as a generalized sine of the angle of incidence. In each different region we find the asymptotic behavior from the poles and branch cuts. We classify these various contributions in the classical forms of primary wave, reflected wave, surface wave, Zenneck wave, and lateral wave. We have found unusual combinations such as a region with forward surface wave and backward lateral wave or with backward surface wave and backward lateral wave. We discuss the range of the parameters ϵ and μ to give the conditions for the existence of forward or backward surface waves, forward or backward lateral waves, Zenneck waves, relations to Brewster's angle and the locations of the poles in the proper and improper Riemann surfaces.

6   Generalized Constitutive Relations for Metamaterials Based on the Quasi-static Lorentz Theory

In recent years, there has been an increasing interest in the development of new materials with characteristics which may not be found in nature. Examples are metamaterials [1],[9], left-handed media [2]-[4],[10],[11], composite media [12],[13], and chiral media [14]-[20]. They have a broad range of applications including artificial dielectrics, lens, absorbers, antenna structures, optical and microwave components, frequency selective surfaces, and composite materials. In addition, left-handed material has a negative refractive index, its permeability and permittivity are both negative, and was conjectured to be capable of producing a perfect lens [2], although its limitations have been pointed out [3],[21].

In these applications, it is important to describe the material characteristics in terms of the physical properties of the inclusions. This paper presents a generalized matrix representation of the macroscopic constitutive relations for a three-dimensional periodic array of non-magnetic inclusions based on the quasi-static Lorentz theory. The macroscopic constitutive relations are given by

$$\begin{bmatrix} \bar{D} \\ \bar{B} \end{bmatrix} = \begin{bmatrix} \bar{\bar{\epsilon}} & \bar{\bar{\xi}} \\ \bar{\bar{\zeta}} & \bar{\bar{\mu}} \end{bmatrix} \begin{bmatrix} \bar{E} \\ \bar{H} \end{bmatrix} \tag{6}$$

where $\bar{\bar{\epsilon}}$, $\bar{\bar{\xi}}$, $\bar{\bar{\zeta}}$, and $\bar{\bar{\mu}}$ are 3×3 matrices. Eq. (6) is applicable to linear medium, and is often called the E-H (or Tellegen) representation. It is convenient as the Maxwell equations appear symmetric and boundary conditions are generally given in terms of $\bar{E}$ and $\bar{H}$ [22],[23]. Note, however, that physically $\bar{E}$ and $\bar{B}$ are the fundamental fields and $\bar{D}$ and $\bar{H}$ are the derived fields related to $\bar{E}$ and $\bar{B}$ through constitutive relations [23]. Therefore, in the E-B (or Boys-Post) representation, we have

$$\begin{bmatrix} \bar{D} \\ \bar{H} \end{bmatrix} = \begin{bmatrix} \bar{\bar{\epsilon}}_p & \bar{\bar{\alpha}}_p \\ \bar{\bar{\beta}}_p & \bar{\bar{\mu}}_p^{-1} \end{bmatrix} \begin{bmatrix} \bar{E} \\ \bar{B} \end{bmatrix}. \tag{7}$$

(6) and (7) are equivalent for linear medium and related through the following:

$$\begin{aligned} \bar{\bar{\epsilon}} &= \bar{\bar{\epsilon}}_p - \bar{\bar{\alpha}}_p \bar{\bar{\mu}}_p \bar{\bar{\beta}}_p, & \bar{\bar{\mu}} &= \bar{\bar{\mu}}_p \\ \bar{\bar{\xi}} &= \bar{\bar{\alpha}}_p \bar{\bar{\mu}}_p, & \bar{\bar{\zeta}} &= -\bar{\bar{\mu}}_p \bar{\bar{\beta}}_p. \end{aligned} \tag{8}$$

The medium with the constitutive relation (6) is called the "bi-anisotropic medium" [5],[15],[24]. If $\bar{\bar{\epsilon}}$, $\bar{\bar{\xi}}$, $\bar{\bar{\zeta}}$, and $\bar{\bar{\mu}}$ are scalars, this is called the "bi-isotropic" or "chiral" medium.

It has been shown [22],[25] that for the constitutive relations of the bi-anisotropic medium to be compatible with Maxwell's equations, the following consistency constraint holds:

$$\text{Trace}([\bar{\bar{\xi}}][\bar{\bar{\mu}}]^{-1} + [\bar{\bar{\mu}}]^{-1}[\bar{\bar{\zeta}}]) = 0 \tag{9}$$

in the E-H formulation.

It has also been shown [23] that the constitutive relations for the bi-anisotropic non-gyrotropic medium satisfy the following reciprocity relations.

$$[\bar{\bar{\epsilon}}]^t = [\bar{\bar{\epsilon}}], \qquad [\bar{\bar{\mu}}]^t = [\bar{\bar{\mu}}], \qquad [\bar{\bar{\xi}}]^t = -[\bar{\bar{\zeta}}] \tag{10}$$

where t denotes transpose.

These relations (9) and (10) are useful to verify the accuracy of the calculations as explained in Section 11.

There are two important questions. First is how to obtain these parameters $\bar{\bar{\epsilon}}$, $\bar{\bar{\xi}}$, $\bar{\bar{\zeta}}$, and $\bar{\bar{\mu}}$ for a given material. The second is how to describe the wave characteristics in such a medium.

In this paper, we address the first question. We derive the explicit expressions of these matrix parameters for a given configuration of the inclusions in a host material. The inclusions are arranged in a three-dimensional array and consist of non-magnetic materials with complex dielectric constants. The derivation is based on the quasi-static Lorentz theory and, therefore, applicable to inclusions whose sizes and spacings are small compared with a wavelength. For our calculations, the spacings near the resonance frequency in Figure 9 are 0.1λ, which is generally accepted as the limit for quasi-static Lorentz theory [26]. Beyond the resonance frequency, the Lorentz theory becomes increasingly less accurate. It should be noted that for the Lorentz theory, the inclusion size should also be much smaller than the spacing.

The 6×6 constitutive relation matrix is expressed in terms of the interaction matrix and the polarizability matrix which can be numerically calculated by using the existing electromagnetic codes. Examples of split ring resonators and an array of helices are shown to illustrate a negative refractive index medium and a chiral medium. Numerical examples are also used to verify the consistency constraint and the reciprocity relations for a bi-anisotropic medium. This paper shows a method of calculating those constitutive relations for a given metamaterial and composite medium. This is obviously a first step towards designing and producing metamaterials with desired characteristics. In this paper, we use the time dependence of $\exp(j\omega t)$. Generalized constitutive relations discussed here are taken from our previous work [8].

7 FORMULATION OF THE PROBLEM

Let us consider a medium consisting of a three-dimensional periodic array of inclusions in a host material (Figure 5). Each inclusion may be a wire, a ring, a helix, or a split ring proposed by Pendry et al. [9]. The spacings along x, y and z directions are a, b and c, respectively. Under the influence of an applied electromagnetic field, the inclusion produces electric and magnetic multipoles. In this paper, to be consistent with the Lorentz theory [26], we limit ourselves to the electric and magnetic dipoles expressed in a quasi-static approximation.

Let us first note that in the E-B representation, we have in general

$$
\begin{aligned}
\bar{D} &= \epsilon_0 \bar{E} + \bar{P}(\bar{E}, \bar{B}) \\
\bar{H} &= \frac{1}{\mu_0} \bar{B} - \bar{M}(\bar{E}, \bar{B})
\end{aligned}
\tag{11}
$$

where the electric polarization $\bar{P}$ and the magnetic polarization $\bar{M}$ are functions of $\bar{E}$ and $\bar{B}$. In the Lorentz theory, we only consider the dipole term representing each inclusion immersed in the uniform effective field $(\bar{E}_\ell, \bar{B}_\ell)$. The effective field $(\bar{E}_\ell, \bar{B}_\ell)$ acts on the inclusion to produce the electric and magnetic dipoles. The effective field consists of the external applied field $(\bar{E}, \bar{B})$ and the interaction field $(\bar{E}_i, \bar{B}_i)$ which are produced by all the inclusions except

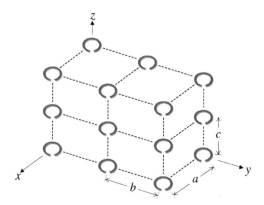

Figure 5: Three-dimensional array of inclusions whose dielectric constant is ϵ_r and conductivity is σ. Host material has a dielectric constant ϵ_b.

the particular inclusion under consideration in the infinite array.
$\bar{P}$ and $\bar{M}$ can then be expressed as:

$$\begin{aligned} \bar{P} &= N\,\bar{p} \\ \bar{M} &= N\,\bar{m} \end{aligned} \tag{12}$$

where $N = (abc)^{-1}$ is the number of inclusions per unit volume, and $\bar{p}$ and $\bar{m}$ are the electric and magnetic dipole moments of each inclusion produced by the effective field $(\bar{E}_\ell, \bar{B}_\ell)$. The dipole moments are then given by the generalized polarizability matrix $[\bar{\bar{\alpha}}]$ [15].

$$\begin{bmatrix} \bar{p} \\ \bar{m} \end{bmatrix} = [\bar{\bar{\alpha}}] \begin{bmatrix} \bar{E}_\ell \\ \bar{B}_\ell \end{bmatrix} \tag{13}$$

where $[\bar{\bar{\alpha}}] = \begin{bmatrix} \bar{\bar{\alpha}}_{ee} & \bar{\bar{\alpha}}_{em} \\ \bar{\bar{\alpha}}_{me} & \bar{\bar{\alpha}}_{mm} \end{bmatrix}$.

Note that $\bar{p}$, $\bar{m}$, $\bar{E}_\ell$, and $\bar{B}_\ell$ are all 3×1 vectors and $[\bar{\bar{\alpha}}]$ is a 6×6 matrix.

Let us examine the polarizability matrix $[\bar{\bar{\alpha}}]$. The electric dipole moment $\bar{p}$ and the magnetic dipole moment $\bar{m}$ are produced by the effective field $(\bar{E}_\ell, \bar{B}_\ell)$. The field $\bar{E}_\ell$ produces the current $\bar{J}_e$ and the field $\bar{B}_\ell$ produces the current $\bar{J}_m$ on the inclusion (Figure 6). In matrix form, we write

$$[\bar{p}] = \frac{1}{j\omega} \int dv \left([\bar{J}_e][\bar{E}_\ell] + [\bar{J}_m][\bar{B}_\ell] \right), \tag{14}$$

$$[\bar{m}] = \frac{1}{2} \int dv\, \bar{r} \times \left([\bar{J}_e][\bar{E}_\ell] + [\bar{J}_m][\bar{B}_\ell] \right). \tag{15}$$

Now we have the final expression for the polarizability matrix $[\bar{\bar{\alpha}}]$.

$$[\bar{\bar{\alpha}}] = \begin{bmatrix} \bar{\bar{\alpha}}_{ee} & \bar{\bar{\alpha}}_{em} \\ \bar{\bar{\alpha}}_{me} & \bar{\bar{\alpha}}_{mm} \end{bmatrix},$$

$$\bar{\bar{\alpha}}_{ee} = \frac{1}{j\omega} \int dv \, [\bar{J}_e] \qquad \bar{\bar{\alpha}}_{em} = \frac{1}{j\omega} \int dv \, [\bar{J}_m]$$

$$\bar{\bar{\alpha}}_{me} = \frac{1}{2} \int dv \, \bar{r} \times [\bar{J}_e] \qquad \bar{\bar{\alpha}}_{mm} = \frac{1}{2} \int dv \, \bar{r} \times [\bar{J}_m]. \tag{16}$$

Note that $\bar{J}_e$ is the current density produced by the electric field $E_{\ell x} = E_{\ell y} = E_{\ell z} = 1(V/m)$ and thus the unit is $(A/m^2)/(V/m)$. Similarly, $\bar{J}_m$ has the unit of $(A/m^2)/T$(tesla).

Let us next consider the effective field $(\bar{E}_\ell, \bar{B}_\ell)$. As discussed above, the effective field acts on the inclusion and produces the electric and the magnetic dipoles. The effective field consists of the external applied field $(\bar{E}, \bar{B})$ and the interaction field $(\bar{E}_i, \bar{B}_i)$.

$$\begin{bmatrix} \bar{E}_\ell \\ \bar{B}_\ell \end{bmatrix} = \begin{bmatrix} \bar{E} \\ \bar{B} \end{bmatrix} + \begin{bmatrix} \bar{E}_i \\ \bar{B}_i \end{bmatrix} \tag{17}$$

The interaction field is produced by all the dipoles except the particular inclusion under consideration. For a three-dimensional array of inclusions consisting of the electric and magnetic dipoles, the interaction field $(\bar{E}_i, \bar{B}_i)$ has been obtained and is given by [26]

$$\begin{bmatrix} \bar{E}_i \\ \bar{B}_i \end{bmatrix} = N[\bar{\bar{C}}] \begin{bmatrix} \bar{p} \\ \bar{m} \end{bmatrix}. \tag{18}$$

The interaction constant matrix $[\bar{\bar{C}}]$ is given by

$$[\bar{\bar{C}}] = \begin{bmatrix} \dfrac{1}{\epsilon_0 \epsilon_b} \bar{C} & \bar{0} \\ \bar{0} & \mu_0 \bar{C} \end{bmatrix} \tag{19}$$

where $\bar{C} = 3 \times 3$ diagonal matrix $= \begin{bmatrix} C_x & 0 & 0 \\ 0 & C_y & 0 \\ 0 & 0 & C_z \end{bmatrix}$, and $\bar{0} = 3 \times 3$ null matrix, and ϵ_b is the relative dielectric constant of the host material.

Substituting (13) and (18) into (17), we obtain

$$\begin{bmatrix} \bar{E}_\ell \\ \bar{B}_\ell \end{bmatrix} = \begin{bmatrix} \bar{E} \\ \bar{B} \end{bmatrix} + N[\bar{\bar{C}}][\bar{\bar{\alpha}}] \begin{bmatrix} \bar{E}_\ell \\ \bar{B}_\ell \end{bmatrix}. \tag{20}$$

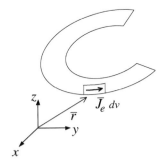

Figure 6: Current $\bar{J}_e$ on the inclusion produced by the effective field $\bar{E}_\ell$. Current $\bar{J}_m$ is also produced by the effective field $\bar{B}_\ell$.

We then obtain

$$
\begin{bmatrix} \bar{E}_\ell \\ \bar{B}_\ell \end{bmatrix} = \left[\bar{\bar{U}} - N[\bar{\bar{C}}][\bar{\alpha}] \right]^{-1} \begin{bmatrix} \bar{E} \\ \bar{B} \end{bmatrix} \tag{21}
$$

where $[\bar{\bar{U}}]$ is a 6×6 unit matrix. Finally, $\bar{D}$ and $\bar{H}$ are given by

$$
\begin{aligned}
\bar{D} &= \epsilon_0 \epsilon_b \bar{E} + N\bar{p} \\
\bar{H} &= \frac{1}{\mu_0} \bar{B} - N\bar{m}
\end{aligned} \tag{22}
$$

Substituting (13) and (21) into (22), we get

$$
\begin{bmatrix} \bar{D} \\ \bar{H} \end{bmatrix} = \begin{bmatrix} \bar{\bar{\epsilon}}_p & \bar{\bar{\alpha}}_p \\ \bar{\bar{\beta}}_p & \bar{\bar{\mu}}_p^{-1} \end{bmatrix} \begin{bmatrix} \bar{E} \\ \bar{B} \end{bmatrix}
$$

where

$$
\begin{bmatrix} \bar{\bar{\epsilon}}_p & \bar{\bar{\alpha}}_p \\ \bar{\bar{\beta}}_p & \bar{\bar{\mu}}_p^{-1} \end{bmatrix} = \begin{bmatrix} \epsilon_0 \epsilon_b \bar{U} & \bar{0} \\ \bar{0} & \frac{1}{\mu_0} \bar{U} \end{bmatrix} + N \begin{bmatrix} \bar{U} & \bar{0} \\ \bar{0} & -\bar{U} \end{bmatrix} [\bar{\alpha}] \left[\bar{U} - N[\bar{\bar{C}}][\bar{\alpha}] \right]^{-1}, \tag{23}
$$

and $[\bar{U}]$ is a 3×3 unit matrix.

The constitutive relation (6) in $E\text{-}H$ representation is then given by (8). This is the final expression for the generalized constitutive relations for a three-dimensional array of inclusions under quasi-static approximation. This is the generalization of the Lorentz-Lorenz formula and leads to the Maxwell-Garnett formula for randomly distributed spherical inclusions [5],[27].

8   INTERACTION MATRIX

The interaction matrix $[\bar{\bar{C}}]$ for a three-dimensional array of dipoles with the spacing a, b, and c in the x, y, and z directions respectively (Figure 5), has been obtained [26].

$$
C_x = f\left(\frac{b}{a}, \frac{c}{a}\right) = \left(\frac{b}{a}\right)\left(\frac{c}{a}\right)\left[\frac{\zeta(3)}{\pi} - S\left(\frac{b}{a}, \frac{c}{a}\right)\right], \quad C_y = f\left(\frac{c}{b}, \frac{a}{b}\right), \quad C_z = f\left(\frac{a}{c}, \frac{b}{c}\right) \tag{24}
$$

where

$$
\zeta(z) = \sum_{k=1}^{\infty} k^{-z}, \ \operatorname{Re}\{z\} > 1 \ (\text{Riemann Zeta function}),
$$

$$
S\left(\frac{b}{a}, \frac{c}{a}\right) = \frac{1}{\pi} \sum_{n=-\infty}^{\infty} \sum_{s=-\infty}^{\infty} \sum_{m=1}^{\infty} (2m\pi)^2 \, K_0 \left(2m\pi \left[\left(\frac{nb}{a}\right)^2 + \left(\frac{sc}{a}\right)^2\right]^{1/2}\right).
$$

K_0 is the modified Bessel function, and the term with $n = s = 0$ is excluded. For a cubic lattice, $a = b = c$ and we get $C_x = C_y = C_z = 1/3$. This is identical to the interaction constant used to derive the Clausius-Mosotti equation by calculating the internal field inside a spherical cavity surrounded by the dielectric with uniform polarization [5].

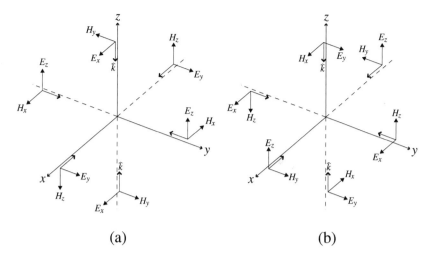

(a)   (b)

Figure 7: (a) A pair of two plane waves along $+z$ and $-z$ directions produces uniform E_x when summed, and uniform H_y when differences are taken. Similarly, $E_y \times H_z$ along $+x$ and $-x$ directions give uniform E_y and H_z, and $E_z \times H_x$ along $+y$ and $-y$ directions give uniform E_z and H_x. (b) Similarly, the waves shown here give additional plane waves which, together with (a), produce nearly uniform effective fields.

9   QUASI-STATIC CALCULATION OF $\bar{J}_e$ AND $\bar{J}_m$

As can be seen from (14) and (15), the current $\bar{J}_e$ on the inclusion is produced by $\bar{E}_\ell$ and the current $\bar{J}_m$ is produced by $\bar{B}_\ell$, and these two currents are independently produced under quasi-static approximation.

In general, in order to calculate all 6×6 elements of the polarizability matrix $[\bar{\bar{\alpha}}]$ in (16), we need to have E_x, E_y, E_z, H_x, H_y, and H_z which are uniform throughout the space $a \times b \times c$. One scheme is to have three incident plane waves along the x, y, and z axes as shown in Figure 7. By taking the sum and the difference, we can generate uniform E_x, E_y, E_z, H_x, H_y, and H_z. Note that by using the scheme in Figure 7(a) and 7(b), four plane waves are combined to give a nearly uniform field for each of six components. For example, to get uniform E_x, we take the sum of two plane waves along the z axis propagating in opposite directions in Figure 7(a) and take the sum of two plane waves along the y axis propagating in opposite directions in Figure 7(b). Note that for plane wave, $\bar{B}$ and $\bar{H}$ are simply related by $\bar{B} = \mu_0 \bar{H}$.

10   SPLIT RING RESONATOR

As an example, we consider a three-dimensional array of split ring resonators (SRRs) (Figure 8). Using the scheme shown in Figure 7, the uniform effective magnetic field is excited and μ' and μ'' are calculated at different frequencies with the formulation described in Section 7, which are shown in Figure 9(a). Similarly, taking the difference between two opposing plane waves, we calculate ϵ' and ϵ'', which are shown in Figure 9(b).

In this example, $\bar{B}_\ell = B_{\ell z}\hat{z}$ and $\bar{E}_\ell = E_{\ell x}\hat{x} + E_{\ell y}\hat{y}$. Therefore, writing $\bar{\bar{\epsilon}} = [\epsilon_{ij}]$, $\bar{\bar{\xi}} = [\xi_{ij}]$,

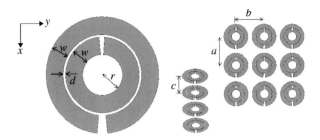

Figure 8: Split ring resonator (SRR) and definitions of distances [1].

$\bar{\bar{\zeta}} = [\zeta_{ij}]$, and $\bar{\bar{\mu}} = [\mu_{ij}]$ with i and $j = x,y,z$, we calculated:

$$
\begin{aligned}
\epsilon_{xx} &= \epsilon'_{xx} - j\epsilon''_{xx} \\
\epsilon_{yy} &= \epsilon'_{yy} - j\epsilon''_{yy} \\
\mu_{zz} &= \mu'_{zz} - j\mu''_{zz}
\end{aligned}
\tag{25}
$$

It is interesting to note the resonance behaviors discussed by Pendry, Smith and others. Also noted are the negative permeability and permittivity near resonance frequency, which have strong dispersive characteristics as already discussed by several workers [1],[4],[9],[10]. The SRR used here has the same dimensions as that in [1], and its resonance frequency in Figure 9 is close to that of [1] with selected spacings. Note that the resonance frequency is dependent on the spacings as well as the size [9], and according to our calculation, if c is increased by 10%, then the resonance frequency is increased by about 3%. In Figure 9(b), it can be found that ϵ_{yy} shows an analogous resonance curve to that of μ_{zz} whereas ϵ_{xx} is almost constant, which agrees with the theoretical analysis by Marqués et al.[11]

If a plane wave with E_y and H_z is propagating in the x direction in this medium, the refractive index is given by

$$
n = (\epsilon_{yy}\mu_{zz})^{\frac{1}{2}} = n' - jn''.
\tag{26}
$$

This is shown in Figure 9(c). Note that n' becomes negative in the frequency range above the resonance, and the phase velocity $v_p = c_0/n$ is negative, where c_0 is the velocity of light. However, the group velocity given by

$$
\frac{v_g}{c_0} = \left[\frac{\partial(\omega n')}{\partial \omega}\right]^{-1}
\tag{27}
$$

is positive representing the signal velocity, except in the region very close to the resonance, where the group velocity becomes negative. This is the anomalous dispersion region, and the group velocity does not represent the signal velocity. The phase velocity, the group velocity (27) and the signal velocity in the anomalous dispersion region have been extensively discussed by Brillouin [29].

11 CHIRAL MEDIUM CONSISTING OF AN ARRAY OF HELICES

We consider a three-dimensional array of helices [15],[16], [18]-[20] (Figure 10). We calculated all 6×6 matrix elements. Note that the dimensions of $\bar{\bar{\xi}}$ and $\bar{\bar{\zeta}}$ are both $\gamma\mu_0 = \sqrt{\mu_0\epsilon_0} =$

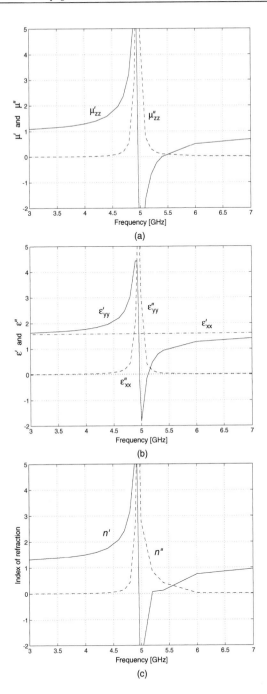

Figure 9: Plot of μ, ϵ, and n of a SRR medium. $r = 1.5$mm, $w = 0.8$mm, $d = 0.2$mm, $a = b$ = 8mm, and $c = 3.9$mm. The conductivity of the ring is 5.8×10^7[S/m]. The thickness of the ring is much greater than the skin depth.

c_0^{-1}, and therefore we normalized all elements of ξ and ζ by c_0^{-1}. $\bar{\bar{\epsilon}}$ and $\bar{\bar{\mu}}$ are normalized with ϵ_0 and μ_0, respectively.

Our calculations show at 8GHz,

$$
\begin{aligned}
\frac{[\bar{\bar{\epsilon}}]}{\epsilon_0} &= \\
\frac{[\bar{\bar{\xi}}]}{\gamma\mu_0} &= \\
\frac{[\bar{\bar{\zeta}}]}{\gamma\mu_0} &= \\
\frac{[\bar{\bar{\mu}}]}{\mu_0} &=
\end{aligned}
\left[
\begin{array}{ccc}
1.20 - j0.01 & 0 & 0 \\
0 & 0.77 - j0.08 & 0.47 + j0.10 \\
0 & 0.47 + j0.10 & 0.48 - j0.15 \\
0 & 0 & 0 \\
0 & -0.01 + j0.05 & 0.10 - j0.37 \\
0 & 0.02 - j0.05 & -0.13 + j0.54 \\
0 & 0 & 0 \\
0 & 0.01 - j0.05 & -0.02 + j0.06 \\
0 & -0.10 + j0.40 & 0.14 - j0.56 \\
1.00 - j0.00 & 0 & 0 \\
0 & 0.99 - j0.00 & 0.06 + j0.01 \\
0 & 0.06 + j0.01 & 0.50 - j0.13
\end{array}
\right] .
\tag{28}
$$

From that, we get

$$
\begin{aligned}
\text{Trace}([\bar{\bar{\xi}}][\bar{\bar{\mu}}]^{-1}) &= (-0.57 + j1.04) \times \gamma \\
\text{Trace}([\bar{\bar{\mu}}]^{-1}[\bar{\bar{\zeta}}]) &= (0.60 - j1.08) \times \gamma
\end{aligned}
\tag{29}
$$

which satisfy (9) with less than 5% difference. The elements designated by 0 are negligibly small. Since the polarizability matrices $\bar{\bar{\alpha}}_{ee}$, $\bar{\bar{\alpha}}_{em}$, $\bar{\bar{\alpha}}_{me}$, and $\bar{\bar{\alpha}}_{mm}$ are independently calculated from six components of $(\bar{E}_\ell, \bar{B}_\ell)$, (29) provides an independent check on the accuracy of the numerical calculation.

The elements (28) are for right-handed helices shown in Figure 10. We also calculated all 36 elements for left-handed helices shown below:

$$
\begin{aligned}
\frac{[\bar{\bar{\epsilon}}]}{\epsilon_0} &= \\
\frac{[\bar{\bar{\xi}}]}{\gamma\mu_0} &= \\
\frac{[\bar{\bar{\zeta}}]}{\gamma\mu_0} &= \\
\frac{[\bar{\bar{\mu}}]}{\mu_0} &=
\end{aligned}
\left[
\begin{array}{ccc}
1.20 - j0.01 & 0 & 0 \\
0 & 0.78 - j0.08 & -0.46 - j0.10 \\
0 & -0.48 - j0.10 & 0.48 - j0.15 \\
0 & 0 & 0 \\
0 & 0.01 - j0.05 & 0.10 - j0.38 \\
0 & 0.01 - j0.05 & 0.14 - j0.54 \\
0 & 0 & 0 \\
0 & -0.01 + j0.05 & -0.02 + j0.06 \\
0 & -0.10 + j0.40 & -0.14 + j0.56 \\
1.00 - j0.00 & 0 & 0 \\
0 & 0.99 - j0.00 & -0.06 - j0.01 \\
0 & -0.06 - j0.01 & 0.50 - j0.13
\end{array}
\right]
\tag{30}
$$

and

$$
\begin{aligned}
\text{Trace}([\bar{\bar{\xi}}][\bar{\bar{\mu}}]^{-1}) &= (0.57 - j1.03) \times \gamma \\
\text{Trace}([\bar{\bar{\mu}}]^{-1}[\bar{\bar{\zeta}}]) &= (-0.60 + j1.08) \times \gamma
\end{aligned}
\tag{31}
$$

We note that the reciprocity relations (10) are satisfied for (28) and (30). Also we note that for the right-handed and the left-handed medium, the diagonal elements of $\bar{\bar{\epsilon}}$ and $\bar{\bar{\mu}}$ are the same, while the off-diagonal elements have opposite sign, and the diagonal elements of $\bar{\bar{\xi}}$ and

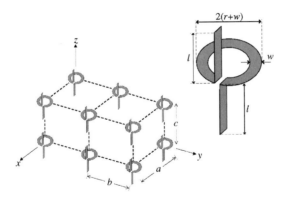

Figure 10: Three-dimensional array of helices. $r = 1.5$mm, $w = 0.8$mm, $l = 3$mm. $a = b = c = 8$mm.

$\overline{\overline{\zeta}}$ have opposite sign, while the off-diagonal elements are the same. These observations are useful to check the accuracy of the calculations.

In the paper, we derived the generalized constitutive relations (23) for metamaterials consisting of a three-dimensional array of inclusions of arbitrary shape. This formula is derived under quasi-static approximation based on the Lorentz theory and is applicable to the spacing between the inclusions, which are small compared with a wavelength. Calculations of all elements of the matrix (23) for metamaterials can be made using the interaction matrix (24) for given spacings and the polarizability matrix (16) which is calculated using three incident plane waves along the x, y, and z axes as shown in the scheme in Figure 7. Some numerical examples using an array of split ring resonators and a chiral array of helices are shown to illustrate the usefulness of the formulation and to verify the consistency constraint (9) and reciprocity relations (10). It should be noted that if the inclusion size and spacing increase, the multipole moments and the propagating constant need to be included, requiring a complete full wave analysis [26].

ACKNOWLEDGMENTS

We thankfully acknowledge support by the National Science Foundation grant ECS-9908849.

REFERENCES

[1] D. R. Smith, W. J. Padilla, D. C. Vier, S. C. Nemat-Nasser, and S. Schultz, "Composite medium with simultaneously negative permeability and permittivity," *Phys. Rev. Lett.*, vol. 84, no. 18, pp. 4184–4187, 2000

[2] J. B. Pendry, "Negative refraction makes a perfect lens," *Phys. Rev. Lett.*, vol. 85, pp. 3966–3969, 2000

[3] R. W. Ziolkowski and E. Heyman, "Wave propagation in media having negative permittivity and permeability," *Phys. Rev. E*, vol. 64, 056625-1–056625-15, 2001

[4] V. G. Vesalago, "The electrodynamics of substances with simultaneously negative values of ϵ and μ," *Sov. Phys. Usp.*, vol. 10, no. 4, pp. 509–514, 1968

[5] A. Ishimaru, *Electromagnetic Wave Propagation, Radiation, and Scattering*, pp. 36–45, Prentice Hall, New Jersey, 1991

[6] W.-C. Chew, *Waves and Fields in Inhomogeneous Media*, pp. 48–53, IEEE Press, New York, 1995

[7] J. R. Thomas and A. Ishimaru, "Transmission properties of material with relative permittivity and permeability close to -1," *Proc. SPIE*, vol. 4806, pp. 167–175, 2002

[8] A. Ishimaru, S.-W. Lee, Y. Kuga, and V. Jandhyala, "Generalized constitutive relations for metamaterials based on the quasi-static Lorentz theory," *IEEE Trans. Antennas Propagat.*, accepted for publication in special issue on metamaterials

[9] J. B. Pendry, A. J. Holden, D. J. Robbins, and W. J. Stewart, "Magnetism from conductors and enhanced nonlinear phenomena," *IEEE Trans. Microwave Theory Tech.*, vol. 47, pp. 2075–2084, 1999

[10] R. A. Shelby, D. R. Smith, and S. Schultz, "Experimental verification of a negative index of refraction," *Science*, vol. 292, pp. 77–79, 2001

[11] R. Marqués, F. Medina, and R. Rafii-El-Idrissi, "Role of bianisotropy in negative permeability and left-handed metamaterials," *Phys. Rev. B*, vol. 65, pp. 144440-1–144440-6, 2002

[12] F. Wu and K. W. Whites, "Quasi-static effective permittivity of periodic composites containing complex shaped dielectric particles," *IEEE Trans. Antennas Propagat.*, vol. 49, pp. 1174–1182, 2001

[13] B. Sareni, L. Krahenbuhl, A. Beroual, and A. Nicolas, "A boundary integral equation method for the calculation of the effective permittivity of periodic composites," *IEEE Trans. Magn.*, vol. 33, pp. 1580–1583, 1997

[14] A. Lakhtakia, V. K. Varadan, and V. V. Varadan, *Time-Harmonic Electromagnetic Fields in Chiral Media*, Springer Verlag, New York, 1989

[15] I. V. Lindell, A. H. Sihvola, S. A. Tretyakon, and A. J. Viitanen, *Electromagnetic Waves in Chiral and Bi-isotropic media*, Artech House, MA, 1994.

[16] V. V. Varadan, A. Lakhtakia, and V. K. Varadan, "Equivalent dipole moments of helical arrangements of small, isotropic, point-polarizable scatters: Application to chiral polymer design," *J. Appl. Phys.*, vol. 63, pp. 280–284, 1988

[17] N. Engheta, D. L. Jaggard, and M. W. Kowarz, "Electromagnetic waves in Faraday chiral media," *IEEE Trans. Antennas Propagat.*, vol. 40, pp. 367–374, 1992

[18] A. J. Bahr, K. R. Clausing, "An approximate model for artificial chiral material ," *IEEE Trans. Antennas Propagat.*, vol. 42, pp. 1592–1599, 1994

[19] F. Mariotte, S. A. Tretyakov, and B. Sauviac, "Isotropic chiral composite modeling: comparison between analytical, numerical, and experimental results," *Microwave Opt. Tech. Lett.*, vol 7, pp. 861–864, 1994

[20] S. A. Tretyakov, F. Mariotte, C. R. Simovski, T. G. Kharina, and J.-P. Heliot, "Analytical antenna model for chiral scatterers: comparison with numerical and experimental data," *IEEE Trans. Antennas Propagat.*, vol. 44, pp. 1006–1014, 1996

[21] A. Lakhtakia "On perfect lenses and nihility," *Int. J. Infrared Millim. Waves*, vol. 23, pp. 339–343, 2002

[22] W. S. Weiglhofer, "Constitutive relations," *Proc. SPIE*, vol. 4806, pp. 67–80, 2002

[23] J. Kong, *Electromagnetic Wave Theory*, John Wiley, New York, 1986

[24] J. A. Kong, "Theorems of bianisotropic media," *Proc. IEEE*, vol. 60, no. 9, pp. 1036-1046, 1972

[25] W. S. Weiglhofer and A. Lakhtakia, "A brief review of a new development for constitutive relations of linear bi-anisotropic media," *IEEE Antennas Propagat. Mag.*, vol. 37, no. 3, pp. 32–35, 1995

[26] R. E. Collin, *Field Theory of Guided Waves*, Chapter 12, IEEE Press, New York, 1991

[27] B. Michel, A. Lakhtakia, W. S. Weiglhofer, and T. G. Mackay, "Incremental and differential Maxwell Garnett formalisms for bi-anisotropic composites," *Composites Sci. Tech.*, vol. 61, pp. 13–18, 2001

[28] H. Braunisch, C. O. Ao, K. O'Neill, and J. A. Kong, "Magnetoquasistatic response of conducting and permeable prolate spheroid under axial excitation," *IEEE Trans. Geosci. Remote Sensing*, vol. 39, pp. 2689–2701, 2001

[29] L. Brillouin, *Wave Propagaion and Group Velocity*, Academic Press, New York, 1960

Wave Propagation,Scattering and Emission in Complex Media
Edited by Ya-Qiu Jin
Science Press and World Scientific,2004

182

Two Dimensional Periodic Approach for the Study of Left-Handed Metamaterials

T. M. Grzegorczyk[†], L. Ran[‡], X. Zhang[‡], K. Chen[‡], X. Chen[†],
J. A. Kong[†],

[†]*Research Laboratory of Electronics, Massachusetts Institute of Technology, Cambridge, MA 02139, USA.*
[‡]*Department of Information and Electronic Engineering, Zhejiang University, Hangzhou 310027, China.*

Abstract   Metamaterials exhibiting left-handed properties are nowadays composed of a succession of rods and split-ring resonators, locally organized in a periodic fashion. At a more macroscopic scale, these metamaterials are built in shapes of prisms, slabs, and lenses, each of these geometry allowing for the measurement of different properties: negative refraction index, transmission, and focusing, respectively. The slab, unlike the other geometries, can be studied using a periodic approach, since its cross-section in the longitudinal direction is invariant. This leads to the possibility of studying transmission through slabs using periodic numerical approaches, yielding efficient and accurate numerical codes. This is in particular important when performing parametric studies of split-rings and rods, to identify those frequency bands where a left-handed behavior can be observed.

In this paper, we present an integral formulation based on a 2D periodic Green's function, applied to the study of split-rings and rods. Transmission measurements are performed on a structure we have studied previously with a finite-difference time-domain (FDTD) technique, and the prediction of the frequencies were left-handed behavior occurs is shown to be in agreement with our previous results. We use this approach as a first step in the study of split-rings and rods. Ultimately, the geometry is introduced into our FDTD code, which can simulate the dielectric support of the split rings and rods. Experimental results are finally shown for a prism geometry, and clearly show frequency bands where the metamaterials exhibit negative index of refraction. Finally, we verify that the metamaterials thus realized present some isotropy for the magnetic field by performing an addition prism experiment on a rotated structure.

1   Introduction

Transmission measurements carried out experimentally or numerically are commonly used to identify frequency bands where a metamaterial can potentially exhibit left-handed properties, *i.e.* simultaneously negative values of permittivity and permeability. For this measurement, a plane wave is launched onto a slab of a certain thickness in the propagation direction, and of some extend in the two longitudinal directions. In theory, these dimensions are often infinite, while numerically or experimentally, they are made large enough compared to the electromagnetic wavelength of the source. However, for a better matching with theoretical predictions, these dimensions can also be made infinite in a numerical calculation, if a 2D periodic approach is considered.

Such approach has already been used in the past to compute the radiation from grating surfaces, arrays, and other structures exhibiting a 2D periodicity. Upon using an integral equation approach, the method requires the proper evaluation of the periodic Green's functions, which is expressed in terms of series expansions. Various methods have been proposed to efficiently calculate these series, such as Kummer's transformation in the space or spectral domain [1; 2; 3; 4], or other integral transformation as proposed in [5] and [6].

In this paper, we present a 2D periodic approach to study metamaterials composed of a periodic arrangement of rods and split-ring resonators [7; 8], as reported in [9; 10]. Due to the complex interactions between the split-rings and rods constituents, these metamaterials are still best studied today using numerical methods. We have reported in the past results obtained using the finite-difference time-domain (FDTD) technique [11], with which we have reproduced numerically the experimental measurements of a prism [9]. Despite the complexity of the medium, numerical predictions were in good agreement with measurements. However, the advantage of being able to deal with such a complex medium was counterbalanced by a computation time that did not allow for fast computation, for example needed in parametric studies of the split-rings and rods geometries. In addition, obtaining a successful prism experiment requires first to identify the frequency band where the material can exhibit negative index of refraction, which is a step typically done by transmission measurements

In view of corroborating the FDTD results presented in [11] and performing fast transmission measurements for the parametric study of split-rings and rods, we have implemented an Method of Moments (MoM) code based on a 2D periodic Green's function using the method presented in [6] and extensively described in [12; 13]. The evaluation of the Green's functions is done using Ewald's method, which we show to be both accurate and fast for all points of interest (in our case, all points lying in the first elementary cell). Transmission curves for a typical geometry are given for the sake of illustration.

In the second part of this paper, we show some experimental results obtained from measurements of metamaterials based on two designs of split-ring resonators: the first one is directly inspired from [10] while the second one is reported in [11] and studied here with our 2D periodic approach. Measurements are performed on prism structures, and it is shown that in both cases there exist a frequency band where the power is bent toward negative angles, indicating a negative index of refraction. In addition, the sensitivity of these measurements is investigated in terms of distance between the bottom and top plates of the parallel plate waveguide. Finally, the isotropy of the metamaterials is verified by performing power measurements refracted from rotated prisms. In both cases we show that negative refraction still occur.

2 Formulation of the problem

We shall first briefly present the calculation of the 2D periodic Green's functions used in our approach. Without loss of generality, we consider the medium to be periodic along $\hat{x}$ and $\hat{y}$, *i.e.* in the (xy) plane, as shown in Fig. 1. The unit cell has dimensions a_x in the $\hat{x}$ direction and a_y in the $\hat{y}$ direction. Although we shall deal with normally incident electromagnetic waves (in the

$\hat{z}$ direction), the formulation is presented without restriction and, therefore, also applicable to oblique incidences. Finally, an $\exp(-i\omega t)$ time dependence is assumed and omitted throughout this paper.

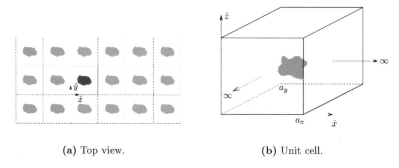

(a) Top view.   (b) Unit cell.

Figure 1: Geometry of the problem: sources are periodic in $\hat{x}$ and $\hat{y}$.

3   Evaluation of the periodic Green's function

Many ways of calculating the Green's function associated with the geometry of Fig. 1 have been presented in the literature. The very first one is to use the image theory and write a spatial summation of the form:

$$G(\bar{r}) = \sum_{\bar{R}} \exp(i\bar{k}_i \cdot \bar{R}) \frac{e^{ik_0|\bar{r}-\bar{R}|}}{4\pi|\bar{r}-\bar{R}|} , \tag{1}$$

where $\bar{R} = n_1\hat{a}_x + n_2\hat{a}_y$, $\bar{k}_i$ is the incident wavevector, and n_1 and n_2 are two integers. In the case of a rectangular lattice ($\hat{a}_x = a_x\hat{x}$ and $\hat{a}_y = a_y\hat{y}$) as in our case, we have:

$$G(X,Y,Z) = \sum_{n_1=-N_s}^{N_s} \sum_{n_2=-N_s}^{N_s} e^{i(k_{ix}n_1a_x+k_{iy}n_2a_y)} \frac{e^{ik_0R_{n_1n_2}}}{4\pi R_{n_1n_2}} , \tag{2}$$

where $X = x - x'$, $Y = y - y'$, $Z = z - z'$, and

$$R_{n_1n_2} = \sqrt{(X - n_1a_x)^2 + (Y - n_2a_y)^2 + Z^2} . \tag{3}$$

It is known, however, that the slow rate of convergence of this expansion makes it unusable for practical applications. An immediate improvement of Eq. (2) is to transform it to the spectral domain to yield:

$$G(X,Y,Z) = \frac{i}{2\Omega} \sum_{l_1=-N_s}^{N_s} \sum_{l_2=-N_s}^{N_s} \frac{1}{k_{zl_1l_2}} e^{i\left[\left(k_{ix}+\frac{2\pi l_1}{a_x}\right)X+\left(k_{iy}+\frac{2\pi l_2}{a_y}\right)Y\right]} e^{ik_{zl_1l_2}|Z|} , \tag{4}$$

where

$$k_{zl_1l_2} = \sqrt{k_0^2 - \left(k_{ix}+\frac{2\pi l_1}{a_x}\right)^2 - \left(k_{iy}+\frac{2\pi l_2}{a_y}\right)^2} \tag{5}$$

with $\text{Im}(k_{zl_1l_2}) > 0$ and $\Omega = a_x a_y$. This expression is rapidly convergent for large values of Z, and can therefore be used for practical purposed, as mentioned in [4]. Yet, even in this case, it is suitable to apply some acceleration techniques such as Shanks technique, to further improve the efficiency of the method.

For a better efficiency, we have used Ewald's method [6; 14], which we will show to be both accurate and very rapidly convergent in all necessary situations. Ewald's method has been described in [14; 12], and its implementation within an integral equation method has been exposed in details in [13]. It is therefore already known that the Green's function can be split into two components G_1 and G_2 defined by:

$$G(X, Y, Z) = G_1(X, Y, Z) + G_2(X, Y, Z),$$ (6)

where

$$G_1(X, Y, Z) = \frac{i}{4\Omega} \sum_{l_1=-N_1}^{N_1} \sum_{l_2=-N_1}^{N_1} \frac{1}{k_{zl_1l_2}} e^{i\left[\left(k_{ix}+\frac{2\pi l_1}{a_x}\right)X + \left(k_{iy}+\frac{2\pi l_2}{a_y}\right)Y\right]}$$
$$\left\{ e^{ik_{zl_1l_2}Z}\mathrm{erfc}\left(-\frac{ik_{zl_1l_2}}{2E} - EZ\right) + e^{-ik_{zl_1l_2}Z}\mathrm{erfc}\left(-\frac{ik_{zl_1l_2}}{2E} + EZ\right) \right\},$$ (7a)

$$G_2(X, Y, Z) = \sum_{n_1=-N_2}^{N_2} \sum_{n_2=-N_2}^{N_2} \frac{1}{8\pi R_{n_1n_2}} e^{i(k_{ix}n_1a_x + k_{iy}n_2a_y)},$$
$$\left\{ e^{ik_0R_{n_1n_2}}\mathrm{erfc}\left(R_{n_1n_2}E + \frac{ik_0}{2E}\right) + e^{-ik_0R_{n_1n_2}}\mathrm{erfc}\left(R_{n_1n_2}E - \frac{ik_0}{2E}\right) \right\}.$$ (7b)

and E is Ewald's parameter chosen in a way to make the asymptotic expansions of G_1 and G_2 similar [14].

Note that the disadvantage of Eq. (7) is that is requires the evaluation of complementary error functions of complex argument. Although this does not present any theoretical difficulty, the numerical evaluation is more time consuming that the evaluation of an exponential function for example, as in Eq. (4). Therefore, evaluating numerically N terms in Eq. (7) is more time consuming that evaluating N terms in Eq. (4) such that only a much faster convergence can justify the use of Ewald's method. We shall show in the next section that the number of terms needed in Eq. (6) is indeed very small for all ranges of distances between source and observation lying within the first elementary cell, yielding a very reliable method for the computation of the 2D periodic Green's function.

The convergence of Ewald's method is illustrated for some particular cases in Fig. 2-4, where it is also compared to the convergence rate of the spectral expression of Eq. (4). Various conclusions can be drawn from these figures. First, it can be seen that the spectral series converges faster for large Z, which is a known phenomenon (Z larger than 0.1 mm are not shown since convergent issues are less stringent). Second, as it can be expected, the series expansions of the Green's functions converge to larger values when the distances between source and observation are small. A difference with the free-space case, however, is that this distance is here repeated periodically: if, for example, the source is located at ($x' = a_x/2, y' = a_y/2$), the Green's functions have a divergence at ($x = (2n + 1)a_x/2, y = (2m + 1)a_y/2$), where n and m are two integers. Fig. 2 illustrates this behavior for $a_x = a_y = 10$ mm and $X = Y = 5$ mm and 15 mm: the convergence curves obtained with the spectral expansion of Eq. (4) for these two cases are indistinguishable.

Note also that this behavior seems to hold for points that are lying on, or very close to, a periodic boundary: $X = na_x$, $Y = ma_y$, as can be seen in Fig. 4 for $X = Y = 0$ mm, $X = Y = 20$mm, and $X = Y = 30$ mm. Yet, the convergence obtained from Ewald's method, although much faster than that with the spectral series, seems to be slightly affected by the distance to the source of the first elementary cell: Fig. 4(b) shows that convergence is reached after 0, 2, and 3 terms for the cases of $X = Y = 0$ mm, $X = Y = 20$ mm, $X = Y = 30$ mm, respectively. As a consequence, in Ewald's method, although only one term is sufficient for most of the cases, we need to calculate more terms (up to five in this example) to properly compute the fields outside the first elementary cell, this number increasing with the distance at which the fields need to be evaluated.

In most practical applications, however, such as the computation of the near field or the reflection and transmission coefficients, the observation is bound to the first elementary cell, and Ewald's method can be used with one and up to three terms in the series only, as can be verified by a direct inspection of Figs. 2-4. Finally, it is worth noticing that this behavior holds for both large and small values of Z, as well as for the critical points $X/a_x = Y/a_y$ and $Z \rightarrow 0$, which were identified in [4] as being delicate to evaluate.

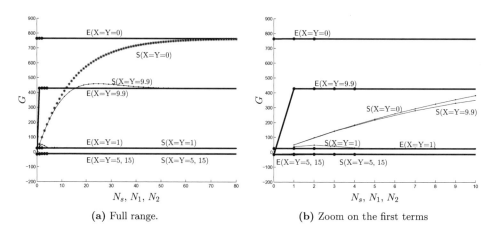

(a) Full range.   (b) Zoom on the first terms

Figure 2: Convergence of the series defined by Ewald's method (Eq. (6)) denoted by E, and the direct spectral summation (Eq. (4)) denotes by S, for different values of $X = Y$ given in millimeters and for $Z = 0.1$ mm (periods are $a_x = a_y = 10$ mm).

4   Numerical results for rods and SRRs

The Green's function described previously has been input into a MoM code and used for the study of split-ring resonators and rods, which are the building blocks of the metamaterials we are studying. For the sake of comparison with a case we have previously studied with an FDTD approach, we shall give results here for the geometry given in [11] and depicted in Fig. 5.

One important parameter that we can determine quickly for example is the frequency band at which the metamaterial may exhibit left-handed properties. This is done by performing a

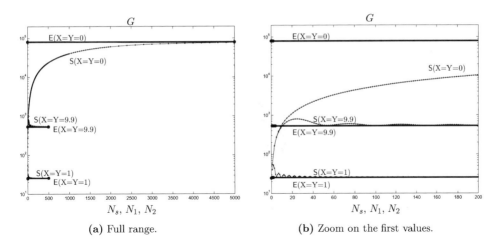

Figure 3: Convergence of the series defined by Ewald's method (Eq. (6)) denoted by E, and the direct spectral summation (Eq. (4)) denotes by S, for different values of $X = Y$ given in millimeters and for $Z = 0.001$ mm (periods are $a_x = a_y = 10$ mm).

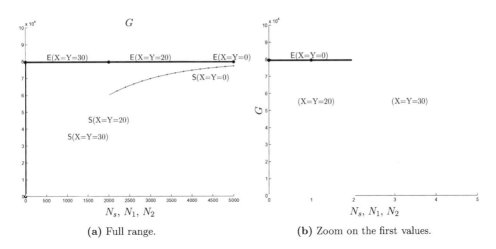

Figure 4: Convergence of the series defined by Ewald's method (Eq. (6)) denoted by E, and the direct spectral summation (Eq. (4)) denotes by S, for different values of $X = Y$ given in millimeters and for $Z = 0.001$ mm (periods are $a_x = a_y = 10$ mm).

transmission/reflection measurement on three structures: (1) a rod only structure, (2) a split-ring only structure, and (3) a structure that contains both split-rings and rods. The first and second experiments identify frequency regions were there is no transmission, corresponding to negative values of the permittivity ($\epsilon(\omega)$ where $\omega = 2\pi f$, f being the frequency) and permeability ($\mu(\omega)$), respectively. If a transmission band appears in the third experiment at those frequencies at which $\epsilon(\omega) < 0$ and $\mu(\omega) < 0$, it may imply that the metamaterial exhibit a negative refraction index (in order to be sure of this property, we need for example to simulate the prism configuration and measure the angle at which the maximum of the power leaves the structure).

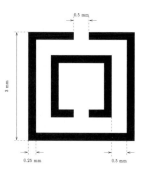

Figure 5: Geometry of the split-ring resonator we study. Rods is 1 mm away from the split-ring and has a thickness of 0.5 mm. Lattice constant is 5 mm×5 mm.

Transmission results for the structure of Fig. 5 are shown in Fig. 6 for two layers (computed using our code) and for five layers (computed using the Transmission Matrix method from one layer data). For the sake of comparison, we also show the transmission for the "split-ring only" structure, and the "rod-only" structure (in the inset). As expected, the "rod-only"

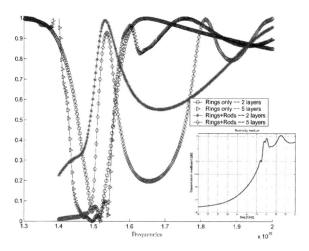

Figure 6: Transmission coefficient for 2 and 5 layers, for "rings only" and "split-rings+rods" structures. Inset: transmission coefficient for "rods only" in dB from 20 GHz to 30 GHz.

structure exhibits a plasma frequency at about 27 GHz, and yields a very low transmission below. The "split-ring" structure, however, behaves like a stop band centered at about 15 GHz. The frequency range around 15 GHz corresponds therefore to a region where the permittivity of the "rod-only" structure is negative, and the permeability of the "ring-only" structure is also negative. When both split-rings and rods are present, a passband appears close to this region, centered at about 15.2 GHz, in agreement with the predictions of [11]. This transmission is attributed to the fact that the entire structure retains the property of negative permittivity and negative permeability, therefore yielding a real wavevector, allowing propagation in the medium (as mentioned before, this conclusion can be drawn with certitude only after another experiment, such as the prism, shows a negative index of refraction).

Note that one difference with [11] is that the split-rings are not located inside a waveguide and therefore, are all in the same direction by periodicity (for example the external gaps of all the split-rings are pointing upward). A way to overcome this difference would be to use a unit cell composed of twice three split-rings in our case, the first set having their external gap pointing upward and the second set having their external gap pointing downward.

However, the 2D periodic MoM presented here can still be used to quickly determine whether an usual transmission band appear at frequencies where the separate media ("rings only" and "rods only") exhibit negative permittivity and permeability. In addition, this method is more oriented toward parametric studies and optimization of split-rings than toward the simulation of real metamaterials (with dielectric support and arbitrary macroscopic shapes) and therefore, should be used *in conjunction* with the FDTD method reported in [11].

5 Experimental measurements of metamaterials

In order to verify and corroborate the numerical predictions, we have also carried out experimental measurements on metamaterials composed of split-rings and rods, and exhibiting left-handed properties. Measurements were carried out on a prism structure [9], which building blocks were in one case those reported in [10] and in the other case those of Fig. 5.

The first measurement setup was largely inspired by [9], which we wanted to reproduce with both the design reported in the paper and the one shown in Fig. 5. The measurement environment was composed of a 1 m long parallel-plate waveguide with absorbers on the side. The quality of the absorbers was quantified, and absorption was found to be between -20 dB and -30 dB between 8 GHz and 19 GHz, as shown in Fig. 7. The waveguide was terminated

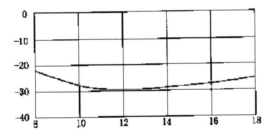

Figure 7: Measured performance of the absorbers used on the side of the parallel-plate waveguide (data from the Beijing Institute of Aeronautical Materials). Vertical axis is in dB, horizontal axes is in GHz.

by a circular parallel-plate chamber in which were located the samples of metamaterials to be measured. The source was an HP8350B, outputting a power of 15 dBm within a frequency range between 20 MHz and 20 GHz, and the power was fed to the waveguide through a 3 cm rectangular waveguide operating in its TE_{01} mode. The output power was scanned at the exit of the circular chamber with a step of 0.9 degrees.

The first set of measurements were performed on a structure identical to the one reported in [10], where split-rings and rods where printed on a FR4 substrate. The prism was formed by cutting a rectangular slab at an angle of 18.43 degrees, following a $1*3$ indentations of the material. The distance between the plates of the waveguide was set to 10 mm. The output power measured experimentally versus angle and frequency is shown in Fig. 8(a): we can see that around 9.6 GHz, the power recorded at negative angles is larger than the one recorded at positive angles, indicating that the energy bends with a negative refraction index. Fig. 8(b) is extracted from Fig. 8(a) at 9.68 GHz and clearly shows the higher power detected at negative angles. The range where negative refraction occurs was determined to be of 0.6 GHz (about

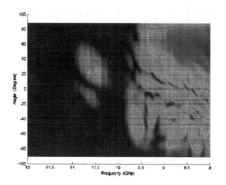
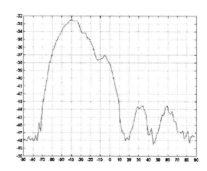

(a) Output power as function of frequency (horizontal) and angle (vertical).

(b) Output power versus angle extracted at 9.68 GHz.

Figure 8: Output power as function of frequency and angle for the geometry of split-rings and rods reported in [10].

5.7%), with a minimum angle measured at -40 degrees, corresponding to an index of refraction of about -1.5. The difference between the level of power refracted at positive and negative angles was of 11 dB.

The sensitivity of this setup was also tested, and it has been found that the negative refraction index effect starts to deteriorate if we lift the top plate, yet not immediately disappearing. At 11 mm of separation, the minimum angle dropped to -45 degrees at 9.55 GHz, with a level difference of 11.7 dB. At 12 mm, the difference in levels between power at positive and negative angles continued to drop, and the negative refraction effect was not completely clear.

The next experiment we have performed consisted in verifying the isotropy of the material. As a matter of fact, split-rings are placed in two planes in order to achieve a better isotropy for the magnetic field: although the electric field can only be in one direction (parallel to the rods), the magnetic field can theoretically have any orientation in the plane perpendicular to the rods. This property can easily be checked by reversing the prism and having the input wave impinging on the hypotenuse of the triangle, as depicted in Fig. 9. Experimental results are shown in Fig. 9(b), and indicate that the frequency band around 9.6 GHz where negative refraction occurs can still be detected. However, the experiment was more delicate to adjust in this case, and more sensitive to the measurement conditions (such as the top plate contact discussed above).

Finally, we have realized and measured the structure shown in Fig. 5. The only difference with the numerical results presented previously was that split-rings and rods were now printed on a dielectric substrate and not suspended in free-space. Before realizing and measuring this structure, however, we have simulated the real metamaterial using an FDTD approach [11].

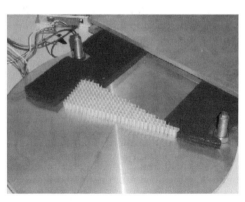
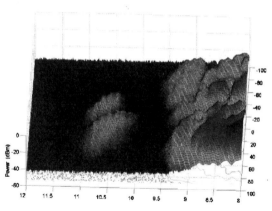

(a) Configuration of the measurement for test-ing the isotropy of the material.

(b) Output power versus angle and frequency.

Figure 9: Output power as function of frequency and angle for the geometry of split-rings and rods reported in [10].

Experimental results are shown in Fig. 10: A frequency band around 7.8 GHz where negative refraction occurs is clearly visible, although the frequency predicted by our numerical analysis were slightly higher, and losses were slightly lower. In addition, the prism was found to be

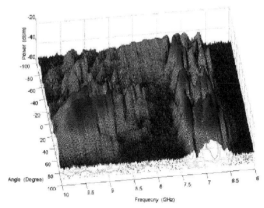

Figure 10: Output power as function of frequency and angle for the geometry of split-rings and rods described in Section 5.

very isotropic, *i.e.* waves impinging on the flat side or on the hypotenuse gave similar negative refractions at similar angles.

6   Conclusion

In this paper, we have studied the 2D periodic Green's function applied to the study of meta-materials composed of a periodic arrangement of split-ring resonators and rods. The numerical evaluation of the Green's function has been done using Ewald's method, which convergence has been shown to be very fast for all the points of interest within the first elementary cell.

Transmission results have been presented for a geometry we have already studied in a previous publication, in order to identify the frequency band in which the metamaterial may exhibit left-handed properties. Results obtained with the method presented here are in good agreement with previous predictions. Finally, experimental results are shown for two types of split-rings and rods, and the sensitivity of the measurements to the experimental conditions as well as the isotropy of the metamaterial thus created has been addressed.

This work has been supported in part by the MIT Lincoln Laboratory under Contract No. BX-8823, the Office of Naval Research under Contract No. N00014-03-1-0716, and by the Chinese National Science Foundation under Contracts 60201001 and 60271010.

References

[1] R. E. Jorgenson and R. Mittra, "Efficient calculation of the free-space periodic green's function," *IEEE Trans. Antennas Propagat.*, vol. 38, pp. 633–642, May 1990

[2] S. Singh, W. F. Richards, J. R. Zinecker, and D. R. Wilton, "Accelerating the convergence of series representing the free space periodic green's function," *IEEE Trans. Antennas Propagat.*, vol. 38, pp. 1958–1962, 1990

[3] G. S. Wallinga, E. J. Rothwell, K. M. Chen, and D. P. Nyquist, "Efficient computation of the two-dimensional periodic green's function," *IEEE Trans. Antennas Propagat.*, vol. 47, pp. 895–897, May 1999

[4] N. Guérin, S. Enoch, and G. Tayeb, "Combined method for the computation of the doubly periodic green's function," *J. Electromagn. Waves Applicat.*, vol. 15, no. 2, pp. 205–221, 2001

[5] M. E. Veysoglu, *Polarimetric Passive Remote Sensing of Periodic Surfaces and Anisotropic Media.* PhD thesis, Massachusetts Institute of Technology, 1989. master thesis

[6] P. P. Ewald, "Die berechnung optischer und elektrostatischen gitterpotentiale," *Ann. Phys.*, vol. 64, pp. 253–258, 1921

[7] J. Pendry, A. Holden, W. Stewart, and I. Youngs, "Extremely low frequency plasmons in metallic mesostructures," *Phys. Rev. Lett.*, vol. 76, pp. 4773–4776, 17 June 1996

[8] J. Pendry, A. Holten, and W. Stewart, "Magnetism from conductors and enhanced nonlinear phenomena," *IEEE Trans. Microwave Theory Tech.*, vol. 47, pp. 2075–2084, November 1999

[9] R. Schelby, D. Smith, and S. Schultz, "Experimental verification of a negative index of refraction," *Science*, vol. 292, pp. 77–79, April 2001

[10] R. Schelby, D. Smith, S. Nemat-Nasser, and S. Schultz, "Microwave transmission through a two-dimensional, isotropic, left-handed material," *Appl. Phys. Lett.*, vol. 78, pp. 489–491, January 2001

[11] C. Moss, T. M. Grzegorczyk, Y. Zhang, and J. A. Kong, "Numerical studies of left-handed metamaterials," *Progress in Electromagn. Research Book Series*, vol. 35, pp. 315–334, 2002

[12] A. W. Mathis and A. F. Peterson, "A comparison of acceleration procedures for the two-dimensional periodic green's function," *IEEE Trans. Antennas Propagat.*, vol. 44, pp. 567–571, April 1996

[13] A. W. Mathis and A. F. Peterson, "Efficient electromagnetic analysis of a doubly infinite array of rectangular apertures," *IEEE Trans. Microwave Theory Tech.*, vol. 46, pp. 46–54, January 1998

[14] L. Tsang, J. Kong, K. Ding, and C. Ao, *Scattering of Electromagnetic Waves: Numerical Simulations*. Wiley, New York (in press), 2000

Numerical Analysis of the Effective Constitutive Parameters of a Random Medium Containing Small Chiral Spheres

Yukihisa NANBU

Department of Electrical Engineering

Sasebo National College of Technology

1-1 Okishin-chou, Sasebo 857-1193, Japan

nanbu@post.cc.sasebo.ac.jp

Tsuyoshi MATSUOKA, **Mitsuo TATEIBA**

Graduate School of Information Science and Electrical Engineering

Kyushu University

6-10-1 Hakozaki, Higashi-ku, Fukuoka 812-8581, Japan

tateiba@csce.kyushu-u.ac.jp

1. INTRODUCTION

Chiral medium has been known in optics since the early part of the nineteenth century as materials with natural optical activity and has attracted a renewed attention owing to novel applications in the microwave, millimeter-wave and optical regions[1]. In recent years, to estimate the characteristics of the artificial chiral materials, studies on the analysis of the effective properties of chiral mixtures containing chiral particles and host medium have become of great interest in electromagnetic(EM) theory, material science and development of high frequency devices because the wave interaction in the mixture is complicated by the effects on multiple scattering between the particles[1–5].

In general, when a EM coherent wave propagates through a chiral mixture containing randomly distributed chiral particles in a background medium, it is attenuated by multiple scattering even if particles in the mixture are lossless ones. Therefore the effective propagation constant($K_{\text{eff}}^{\pm}$) and constitutive parameters ($\epsilon_{\text{eff}}\epsilon_0$, $\mu_{\text{eff}}\mu_0$, $\xi_{\text{eff}}\sqrt{\epsilon_0\mu_0}$) become complex value, where "+" and "−" denote the right-handed and left-handed circularly polarized(RCP and LCP) waves, respectively, and $\xi_{\text{eff}}\sqrt{\epsilon_0\mu_0}$ is the chiral admittance. In low frequency region where EM waves inside and near the chiral mixture can be approximated by the static EM fields, the multiple scattering of EM waves due to particles can be negligible. Therefore $K_{\text{eff}}^{\pm}$ becomes real value, and have been analyzed by the quasi-static methods such as Maxwell-Garnett method and Bruggeman effective medium approximation[3–4]. However, these methods are invalid for high frequency region where the multiple scattering due to particles can not be neglected, and cannot estimate the imaginary part of $K_{\text{eff}}^{\pm}$. In the high frequency region $K_{\text{eff}}^{\pm}$ should be estimated by the multiple scattering methods such as Foldy's approximation[6–8], the quasi-crystalline approximation(QCA)[7–8], QCA with coherent potential(QCA-CP)[7–10]. Numerical examples of $K_{\text{eff}}^{\pm}$ have been obtained by using these multiple scattering methods under the condition that the size of chiral particles is much smaller than the wavelength of EM wave in the medium[11–15].

In this paper, we have analyzed by using our approach the effective constitutive parameters of a random medium containing many chiral spheres. The approach produces the same K_{eff} as QCA-CP does for low dielectric particles but is superior to QCA-CP for high dielectric particles[16]. It is based on an unconventional multiple scattering method by which wave scattering can be systematically treated in a medium whose dielectric particles are randomly displaced from a uniformly ordered spatial distribution[17–18]. The effective constitutive parameters are estimated by changing the volume fraction, dielectric constant and chirality of the chiral spheres. The numerical results are compared with those of a random medium with dielectric spheres in order to make clear the characteristics of the effective constitutive parameters of the chiral random medium.

2. EFFECTIVE PROPERTIES OF CHIRAL MIXTURE

Let us consider the problem of EM wave scattering by N chiral spheres located at $\boldsymbol{r}_j$, $j = 1, 2, \cdots, N$, where the j-th sphere is specified by the dielectric constant $\epsilon_{sj}(\boldsymbol{r})\epsilon_0 = [1 + \epsilon_{dj}(\boldsymbol{r})]\epsilon_0$, the permeability $\mu_{sj}(\boldsymbol{r})\mu_0 = [1 + \mu_{dj}(\boldsymbol{r})]\mu_0$, and the chirality $\xi_{sj}(\boldsymbol{r})$ ($|\xi_{sj}|^2 < \epsilon_{sj}\mu_{sj}$). Figure 1 shows the geometry of the chiral mixture containing randomly distributed chiral spheres. The $j-$th sphere has the following constitutive relations.

$$\left.\begin{array}{rcll}
\boldsymbol{D} &=& \epsilon_{sj}(\boldsymbol{r})\epsilon_0\boldsymbol{E} &+& i\xi_{sj}(\boldsymbol{r})\sqrt{\epsilon_0\mu_0}\boldsymbol{H} \\
\boldsymbol{B} &=& -i\xi_{sj}(\boldsymbol{r})\sqrt{\epsilon_0\mu_0}\boldsymbol{E} &+& \mu_{sj}(\boldsymbol{r})\mu_0\boldsymbol{H}
\end{array}\right\} \qquad (1)$$

For simplicity, we assume that the background medium is free space. When we designate an incident EM wave by $\overline{U}_{in}(\boldsymbol{r}) = [\boldsymbol{E}_{in}, \boldsymbol{H}_{in}]^T$, the scattered EM wave by $\overline{U}_s(\boldsymbol{r}) = [\boldsymbol{E}_s, \boldsymbol{H}_s]^T$, and the total EM wave by $\overline{U}(\boldsymbol{r}) = \overline{U}_{in}(\boldsymbol{r}) + \overline{U}_s(\boldsymbol{r}) = [\boldsymbol{E}, \boldsymbol{H}]^T$, then $\overline{\overline{U}}_{in}(\boldsymbol{r})$ and $\overline{\overline{U}}(\boldsymbol{r})$ satisfy Maxwell's equations in overall region, respectively.

$$\left.\begin{array}{rcl}
\left[\overline{\overline{L}} + \overline{\overline{M}}_b\right]\overline{\overline{U}}_{in}(\boldsymbol{r}) &=& 0 \\
\left[\overline{\overline{L}} + \overline{\overline{M}}_b + \overline{\overline{M}}(\boldsymbol{r})\right]\overline{\overline{U}}(\boldsymbol{r}) &=& 0
\end{array}\right\} \; ; \qquad (2)$$

$$\overline{\overline{L}} = \begin{bmatrix} 0 & \nabla \times \overline{\overline{1}} \\ -\nabla \times \overline{\overline{1}} & 0 \end{bmatrix}, \quad \overline{\overline{M}}_b = ik_0 \begin{bmatrix} \zeta^{-1}\overline{\overline{1}} & 0 \\ 0 & \zeta\overline{\overline{1}} \end{bmatrix}, \qquad (3)$$

$$\overline{\overline{M}}(\boldsymbol{r}) = ik_0 \begin{bmatrix} \zeta^{-1}\sum_{j=1}^{N}\epsilon_{dj}(\boldsymbol{r})\overline{\overline{1}} & i\sum_{j=1}^{N}\xi_{sj}(\boldsymbol{r})\overline{\overline{1}} \\ -i\sum_{j=1}^{N}\xi_{sj}(\boldsymbol{r})\overline{\overline{1}} & \zeta\sum_{j=1}^{N}\mu_{dj}(\boldsymbol{r})\overline{\overline{1}} \end{bmatrix} \qquad (4)$$

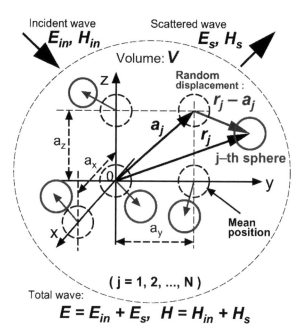

Fig.1 Geometry of the chiral mixture containing randomly distributed chiral spheres.

where $\overline{\overline{1}}$ is the 3×3 unit dyadic, $k_0 = \omega\sqrt{\epsilon_0\mu_0}$ is the wavenumber in free space, and $\zeta = \sqrt{\mu_0/\epsilon_0}$ is the intrinsic impedance of free space and the time factor $\exp(-i\omega t)$ is suppressed. In addition, we assume that $\overline{\overline{U}}_\mathrm{s}(\boldsymbol{r})$ satisfies the radiation condition, then $\overline{\overline{U}}(\boldsymbol{r})$ can be obtained as the solution of the integral equation

$$\overline{\overline{U}}(\boldsymbol{r}) = \overline{\overline{U}}_\mathrm{in}(\boldsymbol{r}) + \int_\mathrm{V} \overline{\overline{G}}(\boldsymbol{r} - \boldsymbol{r}')\overline{\overline{M}}(\boldsymbol{r}')\overline{\overline{U}}(\boldsymbol{r}')\mathrm{d}\boldsymbol{r}' \tag{5}$$

Here, $\overline{\overline{G}}(\boldsymbol{r} - \boldsymbol{r}')$ is the dyadic Green's function defined as the solution of these equations

$$\left(L + M_\mathrm{b}\right) G(\boldsymbol{r}, \boldsymbol{r}') = -\delta(\boldsymbol{r} - \boldsymbol{r}')I \;\; ; \quad G = \begin{bmatrix} G_\mathrm{ee} & G_\mathrm{em} \\ \overline{\overline{G}}_\mathrm{me} & \overline{\overline{G}}_\mathrm{mm} \end{bmatrix}, \quad I = \begin{bmatrix} 1 & 0 \\ 0 & \overline{\overline{1}} \end{bmatrix} \tag{6}$$

where $\delta(\boldsymbol{r} - \boldsymbol{r}')$ is the Dirac delta function

To make clear the characteristics of the method presented here, we assume that the chiral spheres are identical in material, shape and size and that the spheres displace randomly from an ordered distribution independently of each other.

$$\epsilon_{\mathrm{d}j}(\boldsymbol{r}) = \epsilon_\mathrm{d}(\boldsymbol{r}_j + \boldsymbol{r}_0), \;\; \mu_{\mathrm{d}j}(\boldsymbol{r}) = \mu_\mathrm{d}(\boldsymbol{r}_j + \boldsymbol{r}_0), \;\; \xi_{\mathrm{s}j}(\boldsymbol{r}) = \xi_\mathrm{s}(\boldsymbol{r}_j + \boldsymbol{r}_0) \tag{7}$$

where $\boldsymbol{r}_0$ originates from the center position of each sphere $\boldsymbol{r}_j$, which $\boldsymbol{r}_j$ is a Gaussian random vector with the mean $\langle\boldsymbol{r}_j\rangle = \boldsymbol{a}_j = (l_x a_x, l_y a_y, l_z a_z)$ and the variance $\sigma_j^2 = \langle(\boldsymbol{r}_j - \boldsymbol{a}_j)^2\rangle$ in which l_x, l_y, l_z are any integer. Therefore $\sigma_j^2 = 0$ for any j means a uniformly ordered distribution of spheres in the sense of probability. For simplicity, we assume that

$a_x = a_y = a_z = a$ and $\sigma_j^2 = \sigma^2$. In this paper, the radius of spheres is denoted by b_s and the volume fraction f is expressed by $f = n_0 v_s$ where $n_0 \equiv a^{-3}$ is the number of spheres per unit volume and $v_s = 4\pi b_s^3/3$ is the volume of one sphere.

According to [16–18], we can derive the mean dyadic Green's function $\langle\overline{\overline{G}}(\boldsymbol{r})\rangle$ in the random medium from Eq.(5). Then $\langle\overline{\overline{G}}(\boldsymbol{r})\rangle$ satisfies the following equation.

$$\left(\overline{\overline{L}} + \overline{\overline{M}}_{\text{eff}}\right) \langle\overline{\overline{G}}(\boldsymbol{r})\rangle = -\delta(\boldsymbol{r})\overline{\overline{I}} \; ; \quad \overline{\overline{M}}_{\text{eff}} = ik_0 \begin{bmatrix} \zeta^{-1}\epsilon_{\text{eff}}\overline{\overline{1}} & i\xi_{\text{eff}}\overline{\overline{1}} \\ -i\xi_{\text{eff}}\overline{\overline{1}} & \zeta\mu_{\text{eff}}\overline{\overline{1}} \end{bmatrix} \tag{8}$$

Under the condition that $|k_0 b_s \sqrt{\epsilon_s \mu_s}| \ll 1$, we can obtain the effective constitutive parameters

$$\left.\begin{aligned} \epsilon_{\text{eff}}\epsilon_0 &= \epsilon_{\text{av}}\epsilon_0 + n_0(\alpha_{\text{ee1}} - \alpha_{\text{ee2}}) + i\frac{k_0^3 n_0}{6\pi\epsilon_0}\alpha_{\text{ee1}}^2 D(f) \\ \epsilon_{\text{av}}\epsilon_0 &= \epsilon_0(1 + f\epsilon_d), \quad \epsilon_d = \epsilon_s - 1 \end{aligned}\right\} \tag{9}$$

$$\left.\begin{aligned} \mu_{\text{eff}}\mu_0 &= \mu_{\text{av}}\mu_0 + n_0(\alpha_{\text{mm1}} - \alpha_{\text{mm2}}) + i\frac{k_0^3 n_0}{6\pi\mu_0}\alpha_{\text{mm1}}^2 D(f) \\ \mu_{\text{av}}\mu_0 &= \mu_0(1 + f\mu_d), \quad \mu_d = \mu_s - 1 \end{aligned}\right\} \tag{10}$$

$$\left.\begin{aligned} \xi_{\text{eff}}\sqrt{\mu_0\epsilon_0} &= \xi_{\text{av}}\sqrt{\mu_0\epsilon_0} + n_0(\alpha_{\text{em1}} - \alpha_{\text{em2}}) + i\frac{k_0^3 n_0}{6\pi\sqrt{\epsilon_0\mu_0}}\alpha_{\text{em1}}^2 D(f) \\ \xi_{\text{av}}\sqrt{\epsilon_0\mu_0} &= f\xi_s\sqrt{\epsilon_0\mu_0} \end{aligned}\right\} \tag{11}$$

where $D(f)$ is a monotonically decreasing function of f with the characteristics[16]

$$\lim_{f\to 0} D(f) = 1 \quad \text{and} \quad \lim_{f\to 1} D(f) = 0. \tag{12}$$

We can choose $D(f)$ free under the above condition because our method is quite different from the conventional methods and does not need explicitly the pair distribution function of spheres[16]. In this paper, $D(f) = (1 - f)^4/(1 + 2f)^2$ is assumed because the Imaginary part of the effective dielectric constant was almost equal to that of QCA-CP for a medium containing randomly distributed lossless spheres of a fairly low dielectric constant[16]. Here, α_{ee1}, α_{mm1}, and α_{em1} are each component of the polarizability of one chiral sphere with radius b_s, dielectric constant $(\epsilon_{\text{av}} + \epsilon_d)\epsilon_0$, permeability $(\mu_{\text{av}} + \mu_d)\mu_0$ and chirality $(\xi_{\text{av}} + \xi_s)\sqrt{\epsilon_0\mu_0}$, located in an unbounded chiral medium $(\epsilon_{\text{av}}\epsilon_0, \mu_{\text{av}}\mu_0, \xi_{\text{av}}\sqrt{\epsilon_0\mu_0})$ [16–18]. Similarly, α_{ee2}, α_{mm2}, and α_{em2} are each component of the polarizability of an inhomogeneous chiral sphere located in an unbounded chiral medium $(\epsilon_{\text{av}}\epsilon_0, \mu_{\text{av}}\mu_0, \xi_{\text{av}}\sqrt{\epsilon_0\mu_0})$ [16–18]. For the case of $|k_0 b_s \sqrt{\epsilon_s \mu_s}| \ll 1$ used in this paper, the inhomogeneous chiral sphere can be approximated by the homogeneous one with effective radius $b_e = b_s/[W(\sigma)]^{1/3}$, dielectric constant $[\epsilon_{\text{av}} + \epsilon_d W(\sigma)]\epsilon_0$, permeability $[\mu_{\text{av}} + \mu_d W(\sigma)]\mu_0$

and chirality $[\xi_{av} + \xi_s W(\sigma)]\sqrt{\epsilon_0\mu_0}$. Here $W(\sigma)$ is the distribution function of chiral spheres and is expressed by the following equation for a Gaussian random displacement[16],

$$W(\sigma) = \sqrt{\frac{2}{\pi}}\frac{1}{\sigma^3}\int_0^{b_s}\exp\left(-\frac{r_1^2}{2\sigma^2}\right)r_1^2 dr_1 \tag{13}$$

where $W(\sigma) = 1$ for $\sigma/a \to 0$ and $W(\sigma) \simeq 0$ for $\sigma/a \to 1$. In this paper, we assume a simple relation $\sigma/a = 1 - f$, because the effective dielectric constant did not alter appreciably with changing this relation and was almost equal to that of QCA-CP for a medium containing randomly distributed lossless spheres of a fairly low dielectric constant[16]. Therefore $\sigma/a \simeq 1$ for $f \ll 1$, which means that the random displacement occurs in the region of radius a. With these specified parameters, both α_{ee1}, α_{mm1}, α_{em1} and α_{ee2}, α_{mm2}, α_{em2} can be calculated from the general formula given in [3–4].

Suppose a chiral sphere (radius: b) with parameters $(\epsilon_2\epsilon_0, \mu_2\mu_0, \xi_2\sqrt{\epsilon_0\mu_0})$ is located in an unbounded background $(\epsilon_1\epsilon_0, \mu_1\mu_0, \xi_1\sqrt{\epsilon_0\mu_0})$. Then each component of the polarizability reads:

$$\alpha_{ee} = 3\epsilon_0\frac{\epsilon_1(\epsilon_2 - \epsilon_1)(\mu_2 + 2\mu_1) - \epsilon_1(\xi_2 - \xi_1)^2 - 3(\epsilon_2 - \epsilon_1)\xi_1^2}{(\mu_2 + 2\mu_1)(\epsilon_2 + 2\epsilon_1) - (\xi_2 + 2\xi_1)^2}\cdot\frac{4}{3}\pi b^3 \tag{14}$$

$$\alpha_{mm} = 3\mu_0\frac{\mu_1(\mu_2 - \mu_1)(\epsilon_2 + 2\epsilon_1) - \mu_1(\xi_2 - \xi_1)^2 - 3(\mu_2 - \mu_1)\xi_1^2}{(\mu_2 + 2\mu_1)(\epsilon_2 + 2\epsilon_1) - (\xi_2 + 2\xi_1)^2}\cdot\frac{4}{3}\pi b^3 \tag{15}$$

$$\alpha_{em} = 3\sqrt{\epsilon_0\mu_0}\frac{3\mu_1\epsilon_1(\xi_2 - \xi_1) + \xi_1[(\mu_2 - \mu_1)(\epsilon_2 - \epsilon_1) - (\xi_2 - \xi_1)(\xi_2 + 2\xi_1)]}{(\mu_2 + 2\mu_1)(\epsilon_2 + 2\epsilon_1) - (\xi_2 + 2\xi_1)^2}\cdot\frac{4}{3}\pi b^3 \tag{16}$$

The α_{ee1}, α_{mm1}, α_{em1}, α_{ee2}, α_{mm2}, and α_{em2} can be calculated from Eqs.(14)–(16) using Table 1.

Table 1. Each component of the polarizability of one chiral sphere

α_{ee1}	Eq.(14)	$\epsilon_1 \to \epsilon_{av}$,	$\mu_1 \to \mu_{av}$,	$\xi_1 \to \xi_{av}$,
α_{mm1}	Eq.(15)	$\epsilon_2 \to \epsilon_{av} + \epsilon_d$,	$\mu_2 \to \mu_{av} + \mu_d$,	$\xi_2 \to \xi_{av} + \xi_s$,
α_{em1}	Eq.(16)	$b \to b_s$		
α_{ee2}	Eq.(14)	$\epsilon_1 \to \epsilon_{av}$,	$\mu_1 \to \mu_{av}$,	$\xi_1 \to \xi_{av}$,
α_{mm2}	Eq.(15)	$\epsilon_2 \to \epsilon_{av} + \epsilon_d W(\sigma)$,	$\mu_2 \to \mu_{av} + \mu_d W(\sigma)$,	$\xi_2 \to \xi_{av} + \xi_s W(\sigma)$,
α_{em2}	Eq.(16)	$b \to b_e = b_s/[W(\sigma)]^{1/3}$		

3. NUMERICAL RESULTS

Figures 2 and 3 show ϵ_{eff}, μ_{eff} and ξ_{eff} as a function of f for $\epsilon_s = 2, 3, 5$ and $\xi_s = 0$(dielectric), 0.5, 1.0, where $k_0 b_s = 0.1$ and $\mu_s = 1$ are assumed. In these figures, when ϵ_s is small, then the effective constitutive parameters are dependent on the chiral admittance, while are not for large ϵ_s. On the other hand, $\mu_{eff} \simeq 1$ for $0 \le f \le 1$ and $2 \le \epsilon_s \le 5$, because $\mu_s = 1$ is assumed. When $\xi_s = 0$, then $\xi_{eff} = 0$. In this case, the medium becomes an achiral medium.

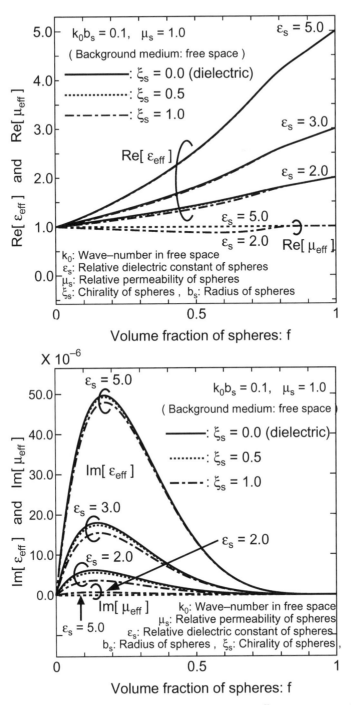

Fig.2   The effective dielectric constant ϵ_{eff}, and the effective permeability μ_{eff} as a function of the volume fraction of chiral spheres f for $\epsilon_{\mathrm{s}} = 2$, 3, and 5, where ϵ_{s}, μ_{s}, ξ_{s} are the relative dielectric constant, the relative permeability and the chirality of the chiral spheres, respectively.

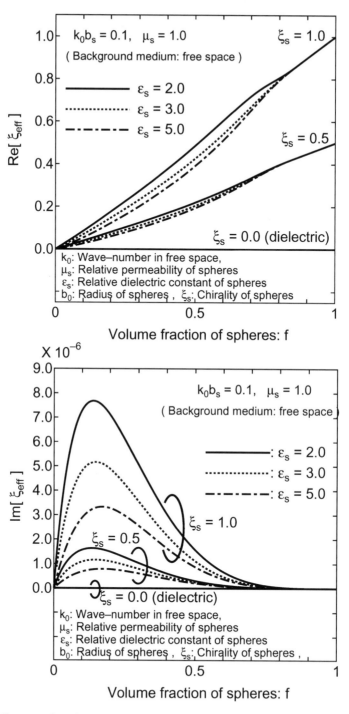

Fig.3 The effective chirality ξ_{eff} as a function of the volume fraction of chiral spheres f for $\epsilon_s = 2, 3,$ and 5, where ϵ_s, μ_s, ξ_s are the relative dielectric constant, the relative permeability and the chirality of the chiral spheres, respectively.

All the numerical results are valid for $f \leq 0.785$, because chiral spheres are completely packed at about $f = 0.785$ and deformed from the spherical form for $f > 0.785$.

4. CONCLUSION

Using our method, we have analyzed the effective constitutive parameters of a medium containing randomly distributed many chiral spheres embedded in an achiral background medium and shown the numerical data on the effective constitutive parameters of a chiral mixture. The effective parameters become increasingly different between chiral and dielectric mixture for low dielectric constant case.

REFERENCES

[1] A.Lakhtakia(ed.), *Selected Papers on Natural Optical Activity*, SPIE Opt. Eng. Press, Bellingham, WA, 1990

[2] N.Engheta,"Chiral electrodynamics and complex materials: a review of recent developments", *Proc.1992 URSI International Sympsium on EM theory*, pp.140–142, 1992

[3] I.V.Lindell, A.Sihvola, S.A.Tretyakov and A.J.Viitanen, *Electromagnetic Waves in Chiral and Bi-Isotropic Media*, Artech House, Boston·London, 1994

[4] A. Sihvola, *Electromagnetic mixing formulas and application*, The Institution of Electrical Engineering, 1999

[5] O. N. Singh and A. Lakhtakia (ed), *Electromagnetic fields in unconventional materials and structures*, WileyInterscience, (2000)

[6] A. Ishimaru, *Wave Propagation and Scattering in Random Media*, vol.2, Academic Press, 1978

[7] L. Tsang, J. A. Kong, R. T. Shin, *Theory of Microwave Remote Sensing*, Wiley Interscience, 1985

[8] L. Tsang, J. A. Kong, *Scattering of Electromagnetic Waves ·Advanced Topics*, Wiley Interscience, 2001

[9] D. Cheng, "Homogenisation of densely - distributed chiral spheres", *Electronics Letters*, vol.32, pp.2326 –2327, 1994

[10] D. Cheng and Y. M. M. Antar, "Coherent scattering in chiral mixture", *Journal of Electromagnetic Waves and Applications*, vol.13, pp.1523–1528, 1999

[11] Y.Nanbu, W.Ren, M.Hiramoto, T.Matsuoka, and M.Tateiba, "Analysis of the effective medium parameters of a medium containing randomly distributed chiral particles", *Proc. ISAP2000*, vol.2, pp.859–862, 2000

[12] Y. Nanbu, W. Ren, T. Matsuoka, M. Hiramoto, and M. Tateiba, "The effective properties of a medium containing randomly distributed chiral spheres", *Proc. 2001 URSI International Symposium on EM Theory*, pp.178–180, 2001

[13] Y. Nanbu, W. Ren, T. Matsuoka, and M. Tateiba, "Effective constitutive parameters of a sparse medium containing randomly distributed chiral spheres", *Proc. AP-S'01*,

vol.3, pp.322–325, 2001

[**14**] Y. Nanbu, W. Ren, T. Matsuoka, and M. Tateiba, "Effective constitutive parameters of a sparse medium containing high dielectric chiral spheres", *Proc. KJJC-AP/EMC/EMT 2001*, pp.121–124, 2001

[**15**] Y. Nanbu, T. Matsuoka, D. Ochi and M. Tateiba, "Numerical analysis of the effective constitutive parameters of a dense random medium containing chiral spheres", *Proc. URSI XXVIIth General Assembly*, CD-ROM, 2002

[**16**] Y.Nanbu and M.Tateiba, "A comparative study of the effective dielectric constant of a medium containing randomly distributed dielectric spheres embedded in a homogeneous background medium", *Waves in Random Media*, vol.6, pp.347-360, 1996

[**17**] M.Tateiba, "A new approach to the problem of wave scattering by many particles", *Radio Science*, vol.22, pp.881-884, 1987

[**18**] M.Tateiba, "Electromagnetic wave scattering in media whose particles are randomly displaced from a uniformly ordered spatial distribution", *IEICE Trans. Electron.*, vol.E78-C, pp.1357-1365, 1995

Wave Propagation,Scattering and Emission in Complex Media
Edited by Ya-Qiu Jin
Science Press and World Scientific,2004

Wave Propagation in Inhomogeneous Media: From the Helmholtz to the Ginzburg-Landau Equation

M.Gitterman

Department of Physics, Bar-Ilan University, Ramat-Gan 52900, Israel

Abstract The linear stability analysis of the inhomogeneous Ginzburg-Landau Equation with multiplicative noise shows that the shift of the symmetry breaking due to multiplicative noise increases with noise intensity, particularly for a small diffusion coefficient. Since the Ginzburg-Landau Equation is a time-dependent generalization of the Helmholtz Equation this approach can be also used for analysis of the propagation of electromagnetic waves in random media.

Propagation and scattering of waves in random media have attached considerable interest in different fields of science. These phenomena are described, in particular, by the Helmholtz and by the Ginzburg-Landau equation with random coefficients. The Ginzburg-Landau equation which is widely used for the study of phase transitions, is a time-dependent generalization of the Helmholtz equations describing the propagation of electromagnetic waves. Therefore, the approaches used in analysis of these two equations are similar. In fact, in this note we use the ideas of multiplicative noise in the Ginzburg-Landau equation on the appearance of ordered symmetry-breaking states for different space dimension. In contrast to additive noise which induces disorder, multiplicative noise may influence a physical system in a orderly fashion.

As was shown in the end of seventies by Horsthemke and Lefever [1], multiplicative noise in certain nonlinear differential equations which are of first-order in time and homogeneous in space, may cause the appearance of new states. On the other hand, the propagation of waves in a medium with the fluctuating dielectric constant was the subject of intensive study in sixties-seventies. This process is described by the Helmholtz equation which is inhomogeneous in space and independent of time.

Recently, in a series of very interesting articles, Van der Broeck and collaborators [2], [4] described a new manifestation of the ordered influence of multiplicative noise in a spatially extended system. Not only may new noise-induced transitions occur, but the transition found to be reentrant: the ordered state appears for intermediate values of noise intensity and disappears both for small and large noise. These interesting results were obtained within a mean-field approximation, an extended mean-field approximation (including pair correlation functions), as well as by intensive numerical simulations for two-dimensional systems. The latter was necessary for a full analysis of the properties of ordered states.

However, our goals are less ambitious; if one is interested only in the appearance of ordered symmetry-breaking states, it is enough to perform a linear stability analysis. Such an analysis is performed here using as a particular example, the non-homogeneous Ginzburg-Landau Equation of the form

$$\frac{\partial \Psi}{\partial t} = D\nabla^2 \Psi - a\Psi + b\Psi^3 + \xi(r,t)\Psi \qquad (1)$$

For the homogeneous case (D=0), this equation has been analyzed both analytically and numerically [3]. Furthermore, the numerical solution of two-dimensional version of (1) that also includes additive noise was obtained recently [4].

In order to perform the stability analysis of a disorder state $\overline{\Psi} = 0$, it is enough to consider a linearized version of (1), namely,

$$\hat{L}_0\Psi = \hat{L}_1\Psi ; \quad \hat{L}_0 \equiv \frac{\partial}{\partial t} + a - D\nabla^2 , \qquad (2)$$

where $\hat{L}_1$ is an operator (in fact, $\hat{L}_1(\xi) \equiv \xi$) defined with the help of the random function $\xi(r,t)$. For weak noise, one can use a well-known perturbation series. However, a better approximation is that widely used in the analysis of the Helmholtz equation, suggested by I. Lifshitz and described as a renormalized perturbation scheme in [5]. Let us first separate the average, Ψ_0, of the order parameter, and the random part, Ψ_1, resulting from the presence of noise. Substituting $\Psi = \Psi_0 + \Psi_1$ into (2), averaging this equation and subtracting the result from the initial equation, one gets

$$\hat{L}_0\Psi_1 = \overline{\hat{L}_1\Psi_0} ; \quad \hat{L}_0\Psi_1 \equiv \hat{L}_1\Psi_0 + (\hat{L}_1\Psi_1 - \overline{\hat{L}_1\Psi_1}) \qquad (3)$$

Restricting ourselves to the first approximation, one neglects the term $(\hat{L}_1\Psi_1 - \overline{\hat{L}_1\Psi_1})$ in (3). From the first of Eqs. (3) one finds $\Psi_1 = \hat{L}_0^{-1}\hat{L}_1\Psi_0$, where $\hat{L}_0^{-1}$ is the Green operator inverse to $\hat{L}_0$. Inserting the latter equation into the second of equation (3), one gets

$$(\hat{L}_0 - \hat{L}_1\hat{L}_0\hat{L}_1)\Psi_0 = 0 \qquad (4)$$

The Fourier component of (4) determines the dispersion relation. If one defines the Fourier series as

$$R(k,\omega) = \int R(r,t)\exp[i(kr+\omega t)]d^d rdt ; \qquad (5)$$

$$R(r,t) = \frac{1}{(2\pi)^d}\int R(k,\omega)\exp[-i(kr+\omega t)]d^d kd\omega \qquad (6)$$

then the Fourier component of (4) can be written as

$$\{-i\omega + D\kappa^2 + a -$$
$$-\frac{1}{(2\pi)^d}\int B(\rho,\tau)\exp[-\tau(a+D\kappa^2)+i\kappa\rho - i(\omega\tau + kr)]d^d\kappa d^d\rho d\tau\}\Psi(\kappa,\omega) = 0 \qquad (7)$$

where $B(\rho,\tau)$ is the space-time correlation function of the random variable ξ. For the case of statistically isotropic noise, $B(\rho,t) = B(\rho,t)$. Specifying the correlation function $B(\rho,t)$ in the form

$$B(\rho,t) = \sigma^2\exp(-\frac{\rho}{\rho_0} - \frac{\tau}{\tau_0}) \qquad (8)$$

and inserting (8) into (7) one can perform the integrations over τ, κ and ρ. Due to the different angular dependence in the integrals over κ and ρ, the results will depend on the number of space dimensions. Omitting the simple calculations, I will quote only the final results. For the one- and three-dimension cases, all the integrations can be carried out, and one finally obtains:

$$1d : y(y^2 - 1 + K^2)(y+\gamma-iK) = \frac{C}{2} \qquad (9)$$

$$3d : (y^2 - 1 + K^2)[(y+\gamma)^2 + K^2)] = C \qquad (10)$$

where

$$y^2 = 1 + a\tau_0 - i\omega\tau_0; \quad K = \kappa\sqrt{D\tau_0}; \quad \gamma = \sqrt{D\tau_0}/\rho_0; \quad C = \sigma^2\tau_0^2 \qquad (11)$$

For the two-dimensional case, the dispersion relation (7) reduces to the following form:

$$2d : y^2 - 1 + K^2 - C\int_0^\infty J_0(Kz)K_0(y^2 z)\exp(-\gamma z)zdz = 0 \qquad (12)$$

where J_0 and K_0 are the Bessel functions of zero order.

For the homogeneous case ($K=0$), $J_0(0)=1$ and one can perform integration in (12) which leads to the following dispersion relation

$$2d(K=0):\ y^2-1-\frac{4C}{3(\gamma-y^2)^2}F(2,\frac{1}{2},\frac{5}{2};\frac{\gamma-y^2}{\gamma+y^2})=0 \tag{13}$$

where $F(\alpha,\beta,\gamma;z)$ is the Gauss hypergeometric function.

Let us now turn to an analysis of the formulae obtained. In the absence of noise ($C=0$), the dispersion relation (7) has a form $-i\omega+D\kappa^2+a=0$ or

$$\mathrm{Im}(\omega)=-a-D\kappa^2 \tag{14}$$

which means that a system with $\overline{\Psi}=0$ becomes instable ($\mathrm{Im}(\omega)>0$) in a homogeneous fashion ($\kappa=0$) when a goes negative. Recall that in the Ginzburg-Landau Equation (1), $a=T-T_c$. Then, the symmetry-breaking transition occurs at $T=T_c$, and the ordered phase exists for $T\leq T_c$.

In the presence of noise the general K-dependent dispersion relations for the one- and three-dimensional cases have the forms (9) and (10), while for two dimensions, the homogeneous dispersion relation has the form (13) with y^2 defined in (11). It follows from the latter,

$$\mathrm{Im}(\omega)=-a+\frac{\mathrm{Re}(y^2)-1}{\tau_0} \tag{15}$$

It is evident from this equation that the noise-induced instability occurs when $\mathrm{Re}(y^2)>1$. This corresponds to a shift of the critical temperature $T_c\to T_c+[\mathrm{Re}(y^2)-1]/\tau_0$, i.e., multiplicative noise makes the system less stable-the symmetry breaking occurs at higher temperatures.

It now remains to solve equations (9) and (10) for different sets of parameters C and γ, which define, according to the strength of noise and its space-time correlations (11), respectively. Each of the equations (9) and (10) has four roots for $\mathrm{Re}(y^2)$, and one has to choose the maximal root. It turned out that for all sets C and γ considered, the maximal value of $\mathrm{Re}(y^2)$ corresponds to $K=0$, i.e. to spatially homogeneous states.

Table

$\gamma \backslash C$	1d			2d			3d		
	0.1	1.0	10	0.1	1.0	10	0.1	1.0	10
0.1	1.27	1.75	3.80	1.2	1.5	3.5	1.1	1.7	3.8
1.0	3.0	3.4	5.4	1.0	1.5	1.8	1.8	3.0	5.9
10	102	102	102	1.0	1.1	1.3	100	102	106

Results of numerical solutions of Eqs. (9), (13) and (10) for Max $[\mathrm{Re}\ (y)^2]$ which define, according to Eq. (15), the shift of the transition temperature due to multiplicative noise for one, two and three dimensions, respectively.

In the Table the maximal positive solutions of equations (9), (10) and (13) for $\mathrm{Re}(y^2)$ are shown for different sets of C and γ. It follows from the Table that:

    1. The shift of phase transition is increasing with the noise intensity C for a small

diffusion coefficient D ($\gamma \equiv \sqrt{D\rho_0}/\tau_0 \leq 1$), while for large D, the shift depends only slightly on the noise intensity.

2. There are no reentrant transitions similar to those found in [4] where, contrary to our analysis, both multiplicative and additive noises have been taken into account.

The above analysis supplements the intensive study [2]-[4] of the influence of multiplicative noise on phase transitions. It can be also used for the analysis of the Helmholtz equations in random media of different space dimensions.

References

[1] W. Horsthemke and R. Lefever, *Noise-induced Transitions* (Springer-Verlag, Berlin 1984)

[2] C. Van Der Broeck, J.M.R. Parrondo, and R. Toral, *Phys. Rev. Lett.* 73, 3395 (1994)

C. Van der Broeck, J.M.R. Parrondo, and A. Hernandez-Machado, *Phys. Rev. E* 49, 2639 (1994),

J.M.R. Parrando, C. Van der Broeck, J. Buceta, and F.J. dela Rubia, *Physica A* 224, 153 (1996)

[3] A. Schenze and H. Brand, *Phys. Rev. A* 20, 1628 (1979)

R. Graham, and A. Schenze, *Phys. Rev. A* 25, 1731 (1982)

L. Brenig and N. Banai. *Physica D* 5, 208 (1982)

L. Gammaitoni, F. Marchesoni, E. Menichella-Saetta, and S. Santucci, *Phys. Rev. E* 49, 4878 (1994)

[4] J. Garcia-Ojalvo, J.M.R. Parrondo, J.M. Sancho, and C. Van der Broeck, *Phys. Rev. E* 54, 6918 (1996)

[5] V.I. Tatarskii and M.E. Gerzenstein, Zh. Exp. Teor. Phys. 44, 676 (1963) [*Sov. Phys. –JETP* 17, 458 (1963)]

Wave Propagation,Scattering and Emission in Complex Media
Edited by Ya-Qiu Jin
Science Press and World Scientific,2004

207

Transformation of the Spectrum of Scattered Radiation in Randomly Inhomogeneous Absorptive Plasma Layer

G.V. Jandieri, G.D. Aburjania[*], V.G. Jandieri[+]

Physics Department, Georgian Technical University
Kostava Str., 380075 Tbilisi, Georgia
E-mail: jandieri@access.sanet.ge

[*] I. Vekua Institute of Applied Mathematics,
Tbilisi State University, University Str., 2, 380043 Tbilisi, Georgia

[+] Department of Computer Science and Communication Engineering
Kyushu University, 36 Fukuoka 812-8581
E-mail: jandieri_vacho@hotmail.com

1. Introduction

Investigations of waves propagation in strong absorptive chaotic media and the influence on statistical characteristics of multiply scattered radiation were discussed both in monographs [1,2] and in original papers [3-8]. From [1-4] it follows that small-angle incidence on the interface of homogeneous transparent and random inhomogeneous absorptive media leads to monotonic broadening of the angular power spectrum of scattered radiation with increasing the immersion depth into second medium. The width of the spectrum tends to a certain asymptotic value at large values of the depth (so-called depth regime). It was shown that a lot of essentially new phenomena arise at rather big angles of the oblique incidence of radiation on the same interface. It was found out that the maximum of the angular power distribution is monotonically displaced towards the normal direction to the interface, as with a small incidence angle. However, the spectrum width varies non-monotonically with the increase of the distance from the interface. With sufficiently strong absorption there is an interval of distances where the spectrum width substantially exceeds its depth asymptotic value. In this interval of distances from the interface, the angular power spectrum is substantially assymetrical with respect to its maximum. Those effects are a consequence of an asymmetric statement of the problem. Such asymmetry arises not only in the case of oblique illumination of the interface; it can be an internal property of a propagation medium itself owing to its anisotropy. Multiple scattering of waves in chaotically inhomogeneous absorptive anisotropic medium - turbulent collisional magnetoactive plasma has been investigated less than wave scattering in chaotic isotropic media. The evolution of statistical moment of the angular power spectrum of scattered radiation with normal incidence of electromagnetic wave on a plane boundary of semi-infinite layer of collisional turbulent plasma under the action of an inclined external magnetic field is considered in [5-7]. Such approach allows to investigate the properties of scattered radiation outside the framework of small-angle approximation and to obtain reliable results to the depth asymptotic mode. The evolution of the statistical momenta in the case of oblique incidence of radiation on a semi-infinite plane plasma layer was considered in [6].

It must be noted that the evolution of an angular power spectrum has not been considered anywhere on joint exposure to various reasons for the occurrence of asymmetry in a turbulent plasma with power phase function of single scattering. Statistical characteristics of the radiation received in plasma were calculated analytically in the complex geometrical optics approximation. That allows multiple scattering of waves at small angles to be taken into account rather easily. Strong distortions of angular power distribution caused by the asymmetric influence of absorption are investigated numerically.

Equations for the first two first moments of the angular power spectrum are derived in the geometrical optics approximation. It was found out that in a certain direction two asymmetric factors of the problem (oblique incidence and medium anisotropy) compensate each other, which interferes with the effects of gravity centre displacement anomalous broadening of the angular spectrum. By numerical calculations the evolution of the angular spectrum form versus distance growing from the boundary beyond application area of the small-angle approximation was investigated and the results obtained within the geometrical optics approximation were confirmed.

2. Formulation of the problem

Let a semi-infinite layer of a collisional magnetoactive turbulent plasma is illuminated by a plane alectromagnetic wave incident on it from vacuum. Let us choose Cartesian frame of reference so that its XY plane be the boundary between two media and the Z-axis be directed inside plasma. The coordinate plane XZ should be arranged so that it might coincide with the plane formed by a vector of an external magnetic field $\mathbf{B}_0$ and a wave vector of a refracted wave $\mathbf{k}_0$. θ_0 - is the angle between magnetic field and Z-axis, θ - is the angle between $\mathbf{B}_0$ and $\mathbf{k}_0$ vectors, θ' - is the angle of refraction with respect to the normal to the boundary, θ_i - is the angle of incidence with respect to the normal to the boundary. The electron concentration in plasma layer is the sum $p(\mathbf{r}) = p_0 + p_1(\mathbf{r})$, where p_0 - is a constant component, $p_1(\mathbf{r})$ is a random function of spatial coordinates, which describes electron concentration fluctuations. Characteristics of scattered radiation are registered at a given depth Z.

A magnetoactive plasma, in the general case, is an anisotropic medium and the components of its permittivity tensor $\hat{\varepsilon}$ are determined by a cyclic frequency of an incident wave ω, by an effective collision frequency between electrons and other particles in plasma ν_{eff}, by a plasma frequency $\omega_p = e(4\pi p / m)^{1/2}$ and by a cyclic gyrofrequency $\omega_B = e B_0 / mc$. Now we introduce those components in the frame of reference, where Z-axis is directed towards external magnetic field:

$$\varepsilon_{xx} = \varepsilon_{yy} = 1 - \frac{\omega_p^2}{2}\left(\frac{1}{\omega^2 - \omega\omega_B - i\omega\nu_{eff}} + \frac{1}{\omega^2 + \omega\omega_B - i\omega\nu_{eff}}\right) ,$$

$$\varepsilon_{xy} = -\varepsilon_{yx} = -i\frac{\omega_p^2\,\omega_B}{\omega(\omega - \omega_B - i\nu_{eff})(\omega + \omega_B - i\nu_{eff})} , \qquad (1)$$

$$\varepsilon_{zz} = 1 - \frac{\omega_p^2}{\omega^2 - i\omega\nu_{eff}} , \quad \varepsilon_{xz} = \varepsilon_{zx} = \varepsilon_{yz} = \varepsilon_{zy} = 0 ,$$

where e and m are respectively the charge and the mass of an electron, c is the velocity of light in vacuum, B_0 is the induction of an external magnetic field. The characteristic scale of inhomogeneities is assumed to exceed substantionally the wavelength. This allows utilizing the geometrical optics techniques for determining statistical characteristics of the scattered field.

3. Statistical characteristics of received radiation by using complex geometrical optics

Now let us calculate the electric field $\mathbf{E}$ in the XZ plane. An electromagnetic wave propagating in plasma satisfies a dispersion equation:

$$\mathbf{k}^2(\mathbf{r}) = \frac{\omega^2}{c^2} n^2 \, , \qquad (2)$$

where $\mathbf{k}(\mathbf{r})$ is the complex wave vector at a given point of space, n is the complex efractive index. The latter in a cold turbulent magnetoactive plasma is determined by formula [9]:

$$n^2 = 1 + \frac{2v(1-v-is)}{2(1-is)(1-v-is) - u\sin^2\theta \pm \sqrt{u^2\sin^4\theta + 4u(1-v-is)^2\cos^2\theta}} \, , \qquad (3)$$

where: $u = \omega_B^2 / \omega^2$, $v = \omega_p^2 / \omega^2$ and $s = v_{eff} / \omega$ are dimensionless plasma parameters, which account for the contribution from the magnetic field, the electron concentration and the absorption respectively; ω is a cyclic frequency.

In zero approximation without taking into account random inhomogeneities ($p_1 = 0$), the contribution into the resulting field is given only by plane initial wave incident on the layer from vacuum. The projections of its complex wave vector on the coordinate axes X and Z are purely real values. The X-projection of its wave vector k_x remains real and the Z-component k_{z2} becomes complex.

Chaotical inhomogeneities of electron concentration in a plasma give rise to fluctuations of the wave field at the observation point. Statistical characteristics of this field in the case of small-angle scattering are largely determined by fluctuations of the complex phase (φ) of an inhomogeneous plane wave [10,11]. Phase characteristics of any normal wave in the approximation of geometrical optics are given by the eikonal equation $\mathbf{k}(\mathbf{r}) = -\nabla\varphi$. Refractive index of anisotropic medium depends on the direction of the wave vector $n^2 = n^2(p(\mathbf{r}), \omega, k_x, k_y)$, therefore, acting on both sides of the dispersion equation (2) by a gradient operator, we obtain:

$$\mathbf{k} \cdot \nabla\mathbf{k} - \frac{1}{2}\frac{\omega^2}{c^2}\frac{\partial n^2}{\partial \mathbf{k}_\perp} \cdot \nabla\mathbf{k} = \frac{1}{2}\frac{\omega^2}{c^2}\frac{\partial n^2}{\partial p} \cdot \nabla p, \qquad (4)$$

where $\mathbf{k}_\perp = \{k_x, k_y\}$. Taking the electron concentration fluctuations as small enough and using perturbation method $p(\mathbf{r}) = p_0 + p_1(\mathbf{r})$ ($p_1 \ll p_0$) we expand a phase and a wave vector into series. Fluctuating components of $\mathbf{k}_1$ and φ_1 are proportional to a small dimensionless parameter p_1 / p_0. Taking into account only the first-order little summands from equation (4) to a first approximation we obtain:

$$k_z^0 \frac{\partial \varphi_1}{\partial z} + \left[k_x^0 - \frac{1}{2}\frac{\omega^2}{c^2}\frac{\partial n_0^2}{\partial k_x^0} \bigg|_{k_{0y}=0} \right] \frac{\partial \varphi_1}{\partial x} = -\frac{1}{2}\frac{\omega^2}{c^2}\frac{\partial n_0^2}{\partial p_0} p_1 \, , \qquad (5)$$

where $k_z^0 = k_{z2}$, and n_0 is unperturbed refractive index. For brevity later on index "0" will be omitted everywhere. By integrating equation (5) along its complex characteristics we obtain:

$$\varphi_1 = \frac{\alpha}{k_{z2}} \int_{-\infty}^{\infty} dk_x \int_{-\infty}^{\infty} dk_y \exp(ik_x x + ik_y y) \int_0^z d\xi \, p_1(k_x, k_y, \xi) \exp\left[-ik_x \left(\frac{\partial k_{z2}}{\partial k_x} \right)_0 (z - \xi) \right], \qquad (6)$$

where $\alpha = -\frac{1}{2}\frac{\omega^2}{c^2}\frac{\partial n^2}{\partial p}$ is the electron concentration fluctuations are expressed as their two-dimensional Fourier-transform.

Transversal correlation function of the phase is:

$$R_\varphi = <\varphi_1(x,y,z)\varphi_1^*(x+\rho_x,y+\rho_y,z)> =$$

$$= \frac{\alpha^2}{k_{z2}^2} \int_{-\infty}^{\infty} dk_x \int_{-\infty}^{\infty} dk_y \int_{-\infty}^{\infty} dk_x' \int_{-\infty}^{\infty} dk_y' \exp\left[i k_x(x+\rho_x) - i k_x'x + i k_y(y+\rho_y) - i k_y'y\right] \times$$

$$\times \int_0^z d\xi \int_0^z d\xi' <p_1(k_x,k_y,\xi) p_1^*(k_x',k_y',\xi')> \exp\left[-i k_x(\beta+i\gamma)(z-\xi) + i k_x'(\beta-i\gamma)(z-\xi')\right] \quad (7)$$

Integrating with respect to those variables in view of the correlation scale of medium parameters being substantially less than radiation registration depth z, we ultimately obtain:

$$R_\varphi(\rho_x,\rho_y,z) = 2\pi \frac{\alpha^2}{k_{z2}^2} \int_{-\infty}^{\infty} dk_x \int_{-\infty}^{\infty} dk_y \exp(i k_x\rho_x + i k_y\rho_y) \frac{\exp(i k_x\gamma z)}{2 k_x\gamma}[1 - \exp(2 k_x\gamma z)] \times$$

$$\times \Phi_p(k_x,k_y,-\beta k_x), \qquad (8)$$

where $\Phi_p(k_x,k_y,k_z)$ is the three-dimensional power spectrum of statistically homogeneous electron concentration fluctuations, and $\frac{\partial k_{z2}}{\partial k_x} = \beta + i\gamma$.

The correlation function of a complex electric field (spatial coherence function) may be written by definition:

$$R_E = <E(x,y,z)E^*(x+\rho_x,y+\rho_y,z)> = E_0^2 \exp\left[-2z\operatorname{Im}(k_{z2}) + i k_x'\rho_x\right] \times$$

$$\times <\exp\left[i\varphi_1(x,y,z) - i\varphi_1(x+\rho_x,y+\rho_y,z)\right]> . \qquad (9)$$

where E_0 is proportional to an amplitude of electric field of the initial wave.

In the nost interesting case of strong fluctuation of the phase $<\varphi_1\varphi_1^*> \gg 1$ we may suppose, as usual [10,11], that they are normally distributed. Due to fast decreasing of correlation function with growing ρ_x and ρ_y, the argument in the second exponent of the expression (9) can be expanded into an exponential series:

$$R_E(\rho_x,\rho_y) = E_0^2 \exp\left[-2z\operatorname{Im}(k_{z2}) + i k_x'\rho_x\right] \exp\left(\frac{\partial R_\varphi}{\partial\rho_x}\rho_x + \frac{1}{2}\frac{\partial^2 R_\varphi}{\partial\rho_x^2}\rho_x^2 + \frac{1}{2}\frac{\partial^2 R_\varphi}{\partial\rho_y^2}\rho_y^2\right) . \quad (10)$$

where R_φ is determined by expression (8). Derivatives in the correlation function are taken at the point $\rho_x = \rho_y = 0$.

Spatial (angular) power spectrum of a scattered field, which is of great practical importance is the Fourier-transform of a correlation function of the field [10,11]:

$$S(k_x,k_y,z) = \frac{1}{(2\pi)^2} \int_{-\infty}^{\infty} d\rho_x \int_{-\infty}^{\infty} d\rho_y \, R_E(x,y) \exp\left(-i k_x\rho_x - i k_y\rho_y\right). \qquad (11)$$

For strong fluctuations of the phase, spatial power spectrum has the Gaussian form:

$$S(k_x,k_y,z) = S_0 \exp\left\{-\frac{(k_x - k_x^0 - \Delta k_x)^2}{2<k_x^2>} - \frac{k_y^2}{2<k_y^2>}\right\}, \qquad (12)$$

where Δk_x determines the displacement of a spatial power spectrum maximum of scattered radiation caused by random inhomogeneities; $<k_x^2>$ and $<k_y^2>$ determine the width of this spectrum in the planes XZ and YZ respectively. The expressions for Δk_x, $<k_x^2>$ and $<k_y^2>$ can be obtained by differentiation of equation (8):

$$\Delta k_x = \frac{1}{i} \frac{\partial W_\varphi}{\partial \rho_x} \quad , \quad <k_x^2> = -\frac{\partial^2 W_\varphi}{\partial \rho_x^2} \quad , \quad <k_y^2> = -\frac{\partial^2 W_\varphi}{\partial \rho_y^2} \quad . \tag{13}$$

Derivatives in the correlation function are taken at the point $\rho_x = \rho_y = 0$.

As a result we obtain:

$$\Delta k_x = \frac{2\pi}{\gamma} \frac{\alpha^2}{k_{z2}^2} \int\limits_{-\infty}^{\infty} dk_x \int\limits_{-\infty}^{\infty} dk_y \, \Phi_p(k_x, k_y, -\beta k_x)\left[\exp(2k_x\gamma z)-1\right] \, , \tag{14}$$

$$<k_x^2> = \frac{2\pi}{\gamma} \frac{\alpha^2}{k_{z2}^2} \int\limits_{-\infty}^{\infty} dk_x \int\limits_{-\infty}^{\infty} dk_y \, \Phi_p(k_x, k_y, -\beta k_x)k_x\left[\exp(2k_x\gamma z)-1\right] \, , \tag{15}$$

$$<k_y^2> = \frac{2\pi}{\gamma} \frac{\alpha^2}{k_{z2}^2} \int\limits_{-\infty}^{\infty} dk_x \int\limits_{-\infty}^{\infty} dk_y \, \Phi_p(k_x, k_y, -\beta k_x) \frac{k_y^2}{k_x}\left[\exp(2k_x\gamma z)-1\right] \, , \tag{16}$$

From the formulae (14)-(16) it follows that anomalous broadening and displacement of the gravity centre take place when an imaginary component of the derivative $\partial k_{z2}/\partial k_x$ is not equal to zero. As mentioned above, both oblique incidence radiation and an anisotropy of the medium make a contribution to this imaginary part itself. Let us introduce the value of this derivative for one particular case:

Normal incidence on the layer of strongly magnetized plasma ($u \gg 1$, $v < 1$):

$$\frac{\partial k_{z2}}{\partial k_x} = -\frac{k_x\left(\cos^2\theta_0 + \varepsilon_{zz}\sin^2\theta_0\right)+k_{z2}\sin\theta_0\cos\theta_0\left(\varepsilon_{zz}-1\right)}{k_{z2}\left(\sin^2\theta_0 + \varepsilon_{zz}\cos^2\theta_0\right)+k_x\sin\theta_0\cos\theta_0\left(\varepsilon_{zz}-1\right)} \, ,$$

$$\operatorname{Im}\frac{\partial k_{z2}}{\partial k_x} = 0 \Leftrightarrow \mathbf{k} \parallel \mathbf{B_0} \tag{17}$$

The main results of these investigations were the detection of effects of anomalous broadening of the angular power spectrum and the displacement of its maximum with the appropriate statement of the problem. These effects are revealed if $k_{xm}\gamma z > 1$ (k_{xm} determines the width of the spatial spectrum of fluctuations power Φ_p). It follows from formulae (17) that the anomalous effects may take place. The investigation of the latter case shows that there is some set off direction, which is determined by roots of equation $\gamma = \operatorname{Im} \partial k_{z2}/\partial k_x = 0$. Anomalous broadening of the spectrum will take place when waves propagate at another angle.

4. Numerical analysis of the power spectrum

The peculiarities of the angular power spectrum behaviour gained with the aid of the geometrical optics method cannot claim to be all-inclusive description of multiple scattered processes of radiation in plasma. First of all, it is connected with the fact that the equation (4) has been solved by the perturbation method. Such approach to its solution means that scattering should be at a small angle. In the initial stage of wave propagation this restriction does not play a significant role, but already for region of transition to a depth asymptotic mode (i.e. in a transient state) the geometrical optic solution reflects only some qualitative aspects of the angular spectrum transformation process. There is not any universal or particular analytical method for investigating transformation of angular spectrum from a near-surface mode into a transition one and further into a depth state. Therefore the information about the behaviour of this spectrum and its dependence on the parameters of the problem can be obtained only numerically.

It is known that the influence of absorption on statistical characteristics of waves in chaotic medium substantially depends on the form of the spatial power spectrum of random inhomogeneities. Therefore the most realistic power-association model was chosen for such a spectrum (for example, for ionosphere plasma); moreover, this model for simplicity is considered to be statistically isotropic:

$$\Phi_p(k) = \begin{cases} C\left(\dfrac{\sqrt{2}}{90}\right)^{-3,5}, & k \in \left[0, \dfrac{\sqrt{2}}{90}\dfrac{\omega}{c}\right] \\ Ck^{-3,5}, & k \in \left[\dfrac{\sqrt{2}}{90}\dfrac{\omega}{c}, \sqrt{2}\dfrac{\omega}{c}\right] \\ 0, & k > \sqrt{2}\dfrac{\omega}{c} \end{cases} \tag{18}$$

where C constant is determined via a normalizing condition:

$$4\pi \int_0^{\sqrt{2}\frac{\omega}{c}} dk\, k^2 \Phi_p(k_x, k_y, k_z = 0) = <p_1^2> . \tag{19}$$

One of the most convenient approaches for the numerical solution of the problem is a statistical simulation of the radiation propagation process (Monte Carlo method). The numerical experiment was carried out with different values of plasma parameters u, v, s and the oblliquity of an external magnetic field θ_0. For each set of parameters a series of computer-based calculations of power spectra of scattered radiation has been accomplished for various initial angles of refraction of the incident wave. After completion of numerical simulations we found the angle at which the effect of compensation can be observed, and its value was compared to a root of the equation $\mathrm{Im}\,\partial k_{z2}/\partial k_x = 0$. In all the cases considered to within the error of the numerical computation, those two values match one another. The results of simulation for the magnetic field obliquity of $\theta_0 = 20$ degrees and the following plasma parameters: u = 1.25; v = 0.2; s = 0.02 are illustrated in Fig. 1. The expression $\mathrm{Im}\,\partial k_{z2}/\partial k_x\big|_{\theta_0=20°}$ for these parameters turns to zero at the angle of $\theta' = 15.9$ degrees. Fig. 1 show the plots of dependence of the first three statistical momenta of the angular power spectrum versus immersion depth into the layer. In all diagrams the depth is plotted as dimensionless units $\sigma_s(\theta_0 - \theta')\cdot z$; curve 1 corresponds to the direction of compensation, curve 2 corresponds to the refraction angle of $\theta' = 30$ degrees, curve 3 - to the case of normal incidence on the layer.

There are interesting effects of angular power spectrum transformation. The characteristic feature of magnetized plasma is a sharp decrease of both absorption and scattering of radiation towards the direction of a magnetic field. Numerical experiment shows that this fact results in "double-humped" form of the angular power spectrum in a transient mode even in the case of normal wave incidence on the plasma layer in an inclined external magnetic field (u = 10, v = 0.7, s = 0.001, $\theta_0 = 30$ degrees). In Fig. 2 the dimensionless variable k_x/k is plotted as abscissa and the value of angular power spectrum $I(k_x, k_y = 0)$ is plotted as ordinate at a depth of $\sigma_s(\theta_0 - \theta')\cdot z = 8,5$. Left maximum on that plot corresponds to waves scattering at small angles in which a larger part of the energy of the entire scattered radiation was originally concentrated. Right maximum corresponds to initially insignificant share of multiple-scattered radiation incident in the direction close to the magnetic field. With increasing the immersion depth, the first part loses its own energy due to significant absorption in the direction of its propagation. The energy loss of the second part is much less, which causes their energy equalization at a certain depth. This fact promotes the increase in dispersion of the angular spectrum of scattered radiation in the transient mode, yielding its anomalous broadening.

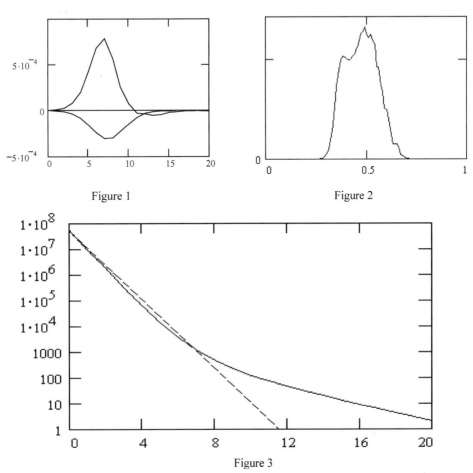

Figure 1   Figure 2

Figure 3

Another interesting result achieved by means of this numerical simulation. Calculations show that scattered radiation more slowly loses its energy when approaching to the direction of an external magnetic field (the effect of "statistical translucence" of plasma). The dependence of the total energy of scattered radiation on the immersion depth into the layer under normal incidence case is illustrated by the results of simulation in Fig. 3 (the immersion depth in terms of $\sigma_s (\theta_0 - \theta') \cdot z$ is plotted as abscissa, and the total energy of scattered radiation in conditional units is plotted as ordinate in the logarithm scale). In the absence of scattering similar dependence would appear as a straight line, denoted in Fig. 3 as a dash line.

5. Conclusion

The results obtained show that the angular power distribution of multiple scattered electromagnetic radiation in the layer of collisional turbulent magnetoactive plasma substantially depends on the factors mentioned above, responsible for the asymmetric statement of the problem (oblique wave incidence on the layer and anisotropy of the medium). Each of these factors by itself leads to distortion of the angular spectrum of scattered radiation in comparison with a completely symmetric case with normal incidence on the isotropic chaotic medium. But their mutual action may amplify or weaken the mentioned above effects of the absorption influence on the angular power spectrum. The most interesting conclusion, in our view, is the fact that the action of these factors is mutually compensated in a certain direction and the radiation propagates along this direction as with normal incidence on the isotropic medium. By applying geometrical optics method the condition of such propagation has been found and by numerical simulation the validity of this calculation has been supported.

The results obtained might be of interest for specialists developing design distant radio communication systems as well as those for remote sensing of artificial and natural plasma inhomogeneities (of the ionosphere, the solar corona, semi-conductor plasma, etc) by translucence methods.

References

[1] A.S. Monin (ed). Ocean Optics. vol. 1. 1983. Moscow: Nauka (in Russian)

[2] L.S. Dolin, I.M. Levin. Handbook of theory for underwater vision. 1991. (Leningrad: Gidrometeoizdat) (in Russian)

[3] L.S. Dolin. Dokl. Acad. Sci. USSR. 1981, vol. 260, pp. 1344-1347 (in Russian)

[4] V.G. Gavrilenko, S.S. Petrov. Radiofizika, 1985, vol. 28, pp. 1408-1412 (in Russian)

[5] A.V. Aistov, V.G. Gavrilenko, G.V. Jandieri. Radiofizika. 1999, vol. 42, pp. 1165-1171 (in Russian)

[6] V.G. Gavrilenko, A.A. Semerikov, G.V. Jandieri. Waves in Random Media. 1999, vol. 9,   pp. 427-440

[7] G.V. Jandieri, V.G. Gavrilenko, A.V. Aistov. Waves in Random Media. 2000, vol.10, pp. 435-445

[8] G.V. Jandieri, N.M. Daraselia. GEN. 2002, # 3, pp.23-28

[9] V.L. Ginzburg. Propagation of Electromagnetic Waves in Plasma. 1961 (New York: Gordon and Breach)

[10] A. Ishimaru. Wave Propagation and Scattering in Random Media. Vols. 1 and 2, Academic Press, San Diego, CA, 1978; IEEE Press-Oxford University Press Classic Reissue, 1997)

[11] S.M. Rytov, Yu.A. Kravtsov, V.I. Tatarskii. Principles of statistical Radiophysics. 1989. (Berlin: Springer)

Wave Propagation,Scattering and Emission in Complex Media
Edited by Ya-Qiu Jin
Science Press and World Scientific,2004

215

Numerical Analysis of Microwave Heating on Saponification Reaction

Kama Huang, Kun Jia

Department of Electronic Engineering, Sichuan University, Chengdu, China

Abstract   Currently, microwave is widely used in chemical industry to accelerate chemical reactions. Saponification reaction has important applications in industry; some research results have shown that microwave heating can significantly accelerate the reaction [1]. But so far, no efficient method has been reported for the analysis of the heating process and design of an efficient reactor powered by microwave. In this paper, we present a method to study the microwave heating process on saponification reaction, where the reactant in a test tube is considered as a mixture of dilute solution. According to the preliminary measurement results, the effective permittivity of the mixture is approximately the permittivity of water, but the conductivity, which could change with the reaction, is derived from the reaction equation (RE). The electromagnetic field equation and reaction equation are coupled by the conductivity. Following that, the whole heating processes, which is described by Maxwell's equations, the reaction equation and heat transport equation (HTE), is analyzed by finite difference time domain (FDTD) method. The temperature rising in the test tube are measured and compared with the computational results. Good agreement can be seen between the measured and calculated results.

Index Terms   Saponification reaction, microwave heating, and numerical analysis

1. Introduction

Most of chemical reactions are sensitive to temperature; therefore using microwave to heat reactants presents an impressive application prospect. A lot of recent reports indicated that microwave could evidently accelerate reactions, and the rate enhancement factor could reach over one thousand [1]. These enhancements were attributed to the heating of microwave in reactants, but some results are inexplicable and therefore are surmised as the hypothetical specific effects, which are called "non-thermal effects", or "athermal effect" of microwave. On the contrary, some experts resolvedly denied these specific effects because the energy of microwave photon is too low to break any chemical bonds and the electric field strength is too low to lead to induce organization according to thermal equilibrium theory. For example, the difference of the terminal product ratio between microwave heating and traditional heating is surmised as the "non-thermal effect" of microwave, but later, this result is proven to be a specific effect produced by the rapid microwave heating [1].

On the other hand, some nonlinear responses in the high-power microwave application in chemical industry, such as nonlinear reflection and absorption of microwave, require the early prediction to protect the microwave system and reactant from being burnt down. The industry application demands well-designed and efficient reactors. All of these requirements need the profound realization on the interaction between microwave and chemical reaction. But unfortunately, so far, no efficient method has been published for the analysis of the heating process.

In fact, during the microwave heating, Boltzmann distribution is not strictly satisfied. Fortunately, in most cases, the collision among the reactant molecules is faster than the change of electromagnetic fields. This means in a very short time compared with the period of microwave, Boltzmann distribution may be considered to be approximately satisfied [2]. Therefore, with the local equilibrium assumption, the effective permittivity and permeability of reactions could be derived. In this paper, the reaction system in homogenous and single phase is considered as mixtures of dilute solution. In spite of the above coarse assumption, it is very difficult to derive the effective permittivity of the reaction system due to a lack of enough information about the structure of molecules and their interactions. Meanwhile, the effective permittivity of reaction depends on the temperature, reaction rate and microwave power. The complicated processes are too difficult to get a general expression of effective permittivity and permeability for reactions.

In this paper, we present a method to study the microwave heating process on saponification reaction. Because the reaction is carried out in dilute solution, according to the preliminary measurement the effective permittivity of the solution is considered as the permittivity of H_2O, but the conductivity, which could change with the reaction, is derived from reaction equation. The change of permittivity and the conductivity with temperature rising in test tube are considered. Then, the whole heating processes, which is described by Maxwell's equation, reaction equation and heat transport equation, is analyzed by FDTD method. In order to get the enough accurate results, the time step in FDTD is set as the order of 10^{-10} second. It is well known that the reaction usually takes several minutes or hours; hence the calculation will be enormously large. Here a scaling factor in HTE was used to solve the problem [3].

In order to verify the calculated results, a type of special designed device, which is called transversal electromagnetic (TEM) cell, is employed in the experiment. It is characterized with the uniform distribution of electric and magnetic fields [4]. The temperature rising and distribution alone the axis of test tube are measured. These measured results are compared with the computed results. Good agreement can be seen between the both of results.

2. Saponification Reaction and Experimental System

1. Saponification reaction

Saponification reaction, which has important applications in industry, has been studied carefully for many years. Here we use a classical saponification reaction with $CH_3COOC_2H_5$ and $NaOH$ to verify the calculation method. The reaction equation is as follows:

$$CH_3COOC_2H_5 + NaOH \rightarrow CH_3COONa + C_2H_5OH \qquad (1)$$

reaction time = 0	C_0	C_0	0	0
reaction time = t	C_0-x	C_0-x	x	x
reaction time t ► ∞	0 →	0 ►	C_0 ►	C_0

In the above reaction, the initial concentration of the reactants is set as C_0 (mol/l) and the concentration of product is set as x (mol/l) at the reaction time t.
The rate equation can be written as:

$$\frac{dx}{dt} = k(t) \cdot (C_0 - x) \cdot (C_0 - x) \qquad (2)$$

where $k(t)$ is the rate constant $k(t) = Ae^{-\frac{E_a}{RT}}$, A is the pre-exponential factor, E_a is the activation energy, R is the gas constant, and T is the absolute temperature. Here the initial concentration are C_0=0.5 mol/l and C_0=0.7 mol/l respectively, the pre-exponential factor and activation energy are given by:

$$E_a = 61KJ, \quad A = 4.1 \times 10^9$$

By integrating the both side of equation (2), we can get the concentration of product with respect to time,

$$x(t) = C_0 - \frac{1}{[1/C_0 + \int_0^t k(t)dt]} \qquad (3)$$

Because the reaction is carried out in the dilute solution, CH_3COONa is assumed to be completely ionized. Due to the migration rate of OH^- is much higher than it of CH_3COO^-, the conductivity is mainly determined by the concentration of OH^- [5]. According to the above rate equation, the conductivity of the reaction system reducing with time can be described as

$$\sigma(t) = \sigma_0 + [x(t)/C_0] \cdot (\sigma_\infty - \sigma_0) \qquad (4)$$

where σ_0 is the initial conductivity of the solution, and σ_∞ is the terminal conductivity of the solution. When C_0 equals 0.5 mol/l, σ_0 and σ_∞ are 0.8 s/m and 0.38 s/m respectively. When C_0 equals 0.7 mol/l, σ_0 and σ_∞ are 0.94 s/m and 0.43 s/m respectively.

A complicated situation should be considered is that the above reaction is the exothermic reaction, 1 mol $CH_3COOC_2H_5$ and 1 mol $NaOH$ will produce the thermal energy of 76.09 KJ. This will be included in the heat transport equation.

Because the reaction is carried out in dilute solution, according to the preliminary experimental results, the permittivity of the solution is approximately the permittivity of H_2O, but the conductivity will change with the reaction time.

2. Experimental System

The experimental system is shown in Fig. 1

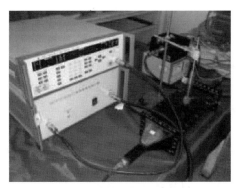

Fig. 1 Experimental system

where a special designed TEM cell is used to establish uniform electromagnetic fields. It is a kind of rectangular coaxial transmission line. In the experiment, a hole with diameter of 2 cm was made on the top of TEM cell and connected with a cut-off circular waveguide with height of 3cm to avoid the emission of microwave. The test tube with reactants is inserted into the TEM cell through the hole. The thickness of the test tube is 0.1cm.

The dimension of the cell is shown in Fig. 2

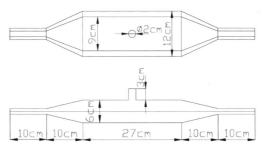

Fig. 2 The dimension of TEM cell

QF1481 microwave source and QF3860 power amplifier are used to generate 5W power, the frequency range is from 10MHz to 2.4GHz. In the experiment, the frequency is fixed at 915MHz. GH2461 load is used to absorb the output power. We use the special designed TEM cell to replace the waveguide in the experiment because the experimental system is established to test microwave effect in a wide frequency band. Since the solution is corrosive and irradiated by microwave, a traditional alcohol thermal meter with the radius of 0.2cm is used to measure the temperature inside the test tube in vivo. The disturbance of the field distribution by the thermometer has been considered in the calculation.

3. Dynamic Equations

The mode used to analysis the interaction between chemical reactions and microwave includes Maxwell's equations, heat transport equation, and reaction equation.

1. The electromagnetic fields in reactant satisfy Maxwell's equation as follows:

$$\nabla \times \vec{H} = \frac{\partial \vec{D}}{\partial t} + \sigma(t)\vec{E}$$

$$\nabla \times \vec{E} = -\frac{\partial \vec{B}}{\partial t}$$ (5)

$$\nabla \cdot \vec{D} = \rho$$

$$\nabla \cdot \vec{B} = 0$$

With the boundary conditions on the conducting surface given by

$$\hat{n} \cdot \vec{H} = 0$$

$$\hat{n} \times \vec{E} = 0$$ (6)

where $\hat{n}$ is the unit vector normal to the surface at the considered point.

2. In addition, the heat transport equation is given by [6]:

$$\rho_m C_m \frac{\partial T(\vec{r},t)}{\partial t} = k_t \nabla^2 T(\vec{r},t)$$

$$- V_s[T(\vec{r},t) - T_a] + P_d(\vec{r},t) + P_e(\vec{r},t)$$ (7)

$$P_e(\vec{r},t) = C_q \cdot A e^{-\frac{E_a}{RT}}(C_0 - x)^2$$ (8)

$$\vec{P}_d = \frac{1}{2}(\vec{E} \cdot \frac{\partial \vec{D}}{\partial t} - \vec{D} \cdot \frac{\partial \vec{E}}{\partial t}) + \vec{J} \cdot \vec{E}, \qquad \vec{J} = \sigma(t)\vec{E}$$ (9)

where $P_e(\vec{r},t)$ is the releasing power produced by the reaction ($W \cdot m^{-3}$) [2], $P_d(\vec{r},t)$ is the electromagnetic power dissipated per unit volume ($W \cdot m^{-3}$), ρ_m is the medium density ($kg \cdot m^{-3}$), C_m is the specific heat of the medium ($J \cdot K^{-1} \cdot kg^{-1}$), k_t is thermal conductivity of the medium ($W \cdot m^{-1} \cdot K^{-1}$), V_s is the product of flow and heat capacity of the cooling fluid ($W \cdot m^{-3} \cdot K^{-1}$), T_a is the temperature of the cooling fluid (K), C_q is the thermal energy produced by 1 mol reactant ($J \cdot mol^{-1}$).

3. The permittivity of H_2O with the temperature

The permittivity of H_2O can be described by the first-order equation [3]:

$$\varepsilon(\omega) = \varepsilon_1 - j\varepsilon_2 = \varepsilon_\infty + \frac{\varepsilon_s - \varepsilon_\infty}{1 + j\omega\tau} \qquad (10)$$

$$\varepsilon_s(T) = \frac{3\varepsilon_\infty T + A(\varepsilon_\infty + 2)^2}{12T} + \frac{\sqrt{(3\varepsilon_\infty T + A(\varepsilon_\infty + 2)^2)^2 + 72\varepsilon_\infty^2 T^2}}{12T} \qquad (11)$$

$$\tau(T) = \tau_0 e^{W_a/kT} \qquad (12)$$

where τ is the relaxation time, w_a is the activation energy and τ_0 is the pre-exponential factor. The values of these parameters can be found in [3].

The conductivity of the solution with respect to the time is determined by eq. (4). The above equations are used to calculate the temperature distribution and rising in the test tube.

4. Calculation Model

Generally, it is difficult to obtain an analytical result for these differential equations. Some numerical methods, such as FDTD, could be employed to solve these equations [7]. The first step in FDTD calculation is to determine the calculation domain. The calculation domain is shown in Fig. 3. In this model, we support microwave propagates along z direction, 1-D MUR absorbing boundary condition is used at z = -130mm, 130mm and x = 88mm respectively.

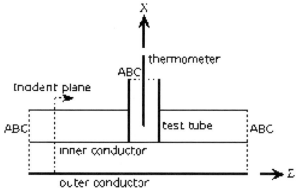

Fig.3 The calculation domain

If no cooling effects are assumed, it leads to the following equation in a three-dimension homogeneous cube grid case:

$$T^{n+1}(i,j,k) = T^n(i,j,k) + \Delta t \cdot \frac{\alpha k_t}{\rho_m C_m \Delta S^2}[T^n(i+1,j,k)$$

$$+ T^n(i-1,j,k) + T^n(i,j+1,k) + T^n(i,j-1,k) + T^n(i,j,k+1) \qquad (13)$$

$$+ T^n(i,j,k-1) - 6T^n(i,j,k)] + \Delta t \cdot \frac{\alpha P_d^n(i,j,k)}{\rho_m C_m} + \Delta t \cdot \frac{\alpha P_e^n(i,j,k)}{\rho_m C_m}$$

where the scaling factor α is inducted into HTE with the purpose to scale the actual heating duration T_{heat} to the EM computation time window T_{FDTD} [3]. The scaling factor

α is

$$\alpha = F_{heat} / T_{FDTD} \tag{14}$$

It must be chosen correctly. Too large value of α will change the assumption of equilibrium.

It's easy to get the difference case of eq. (9):

$$P_d^{n+\frac{1}{2}} = \frac{1}{2\Delta t} \sum_{i=x,y,z} (\varepsilon_1^{n+1} - \varepsilon_1^n) E_i^n E_i^{n+1}$$

$$+ \sigma \sum_{i=x,y,z} (\frac{E_i^n + E_i^{n+1}}{2})^2 \tag{15}$$

The thermal properties of the medium in the model are shown in the table 1 [8].

Table.1 The thermal properties of the medium in the calculation

	$C_m(J * kg^{-1} * K^{-1})$	$\rho_m(kg * m^{-3})$	$K_t(W * m^{-1} * K^{-1})$
Water	4180	1000	0.55
Air	3505.9	1.293	0.0261
Glass	837.4	2707.04	0.76164
Copper	385	8939	20

The whole calculating proceeding is shown in Fig. 4

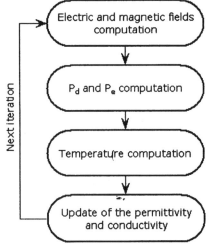

Fig. 4 Calculating proceeding of combined electromagnetic, chemical and thermal equations

5. Measured and calculated results

The relative electric field distribution in TEM cell is shown in Fig. 5. It shows the uniform distribution of fields in the test tube. The distortion of the thermal meter is not serious besides the top of the meter. The temperatures with respect to time at two points along the axis of the test tube are measured and compared with the calculated results. The two points are located at P_1 (x=45mm, z=0mm) and P_2 (x=60mm, z=0mm) respectively.

From the these figures, good agreement can be seen between the calculated and measured results in the later temperature rising. In the earlier temperature rising, the measured results are less than the calculated results; the difference is due to the slow response of the thermometer. The difference in Fig.7 is larger than it in Fig.6 because the loss of the solution is higher and the temperature rising is quicker while C_0=0.7 m/l.

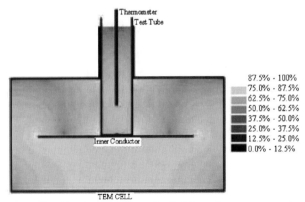

Fig.5. The relative distribution of electric field at the cross section of TEM cell

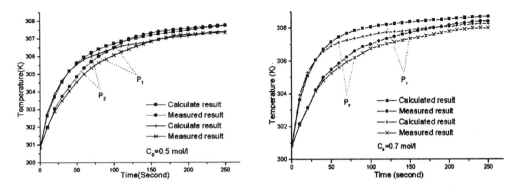

Fig.6. Measured and calculated results
at P_1, P_2 while C_0=0.5 mol/l

Fig.7. Measured and calculated results
at P_1, P_2 while C_0=0.7 mol/l

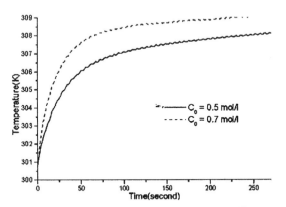

Fig.8. Temperature rising at P_1 with different C_0

From Fig. 8 we can see that the temperature rising in test tube is quite different while the initial concentration is different.

The temperature distribution on the cross section of test tube at t=250s is also calculated; the result is shown in Fig. 9. It can be seen that the temperature distribution is not uniform in the small test tube. The temperature reduces along the radius due to the heat convection on the surface. The temperature declines along the propagation direction of microwave. From these results, it is not difficult to understand that the reactor should be designed carefully to get a uniform temperature distribution.

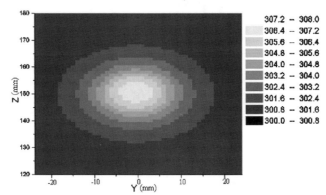

Fig.9. Temperature distribution on the cross section of test tube

6. Conclusions

The application of microwave in chemistry has presented the requirements of analysis on the phenomena, such as the "non-thermal effect" and nonlinear response. In this paper, we present a method to study the microwave heating on saponification reaction. The temperature of the reaction is obtained by solving Maxwell's equation combined with heat transport equation and reaction equation. The numerical method of FDTD is employed and the scaling factor is used to reduce the significant difference between reaction time and the period of microwave. Some details are considered in the model, such as the influence of the thermal meter on the field distribution and exothermic effect. A special designed TEM cell was used in the experiment and the temperature inside the test tube are measured. Good agreement between measured and calculated results can be seen and verifies the feasibility of the numerical method.

As we mentioned above that the calculation based on the assumption of local equilibrium because Boltzmann distribution is not strictly satisfied in the chemical reaction.

Acknowledgement

This project was supported by NSF of China.

References

[1] Jin, Q., Dai, S. and Huang, K.M., Microwave Chemistry, Sciences Press, Beijing, 1999

[2] Zhao, X.. Z. , The chemical reaction kinetics, Advance Education Press, Beijing, 1984

[3] Francois Torres and Bernard Jecko, "Complete FDTD Analysis of Microwave Heating Processes in Frequency-Dependent and Temperature-Dependent Media", *IEEE Trans. Microwave Theory Tech.*, vol. 45, pp. 108-116, January 1997

[4] Ma, Y., Huang, K., M., Wang, K. and Liu N., "Calculation of electric field distribution in TEM cell with EUT", Chinese Journal of Radio Science, vol. 15, No.1, pp75-79, March, 2000

[5] F.Daniels, R. A. Alberty, J. W. Williams, C. D. Cornwell, P. Bender, J. E. Harriman, Experimental Physical Chemistry, McGraw-Hill, Inc., New York, 1975

[6] L. Ma, D.-L. Paul, N. Pothecary, C. Railton, J. Bows, L. Barram, J.Mullin, and D. Simons, "Experimental validation of a combine electromagnetic and thermal FDTD model of a microwave heating process", *IEEE Trans. Microwave Theory Tech,* vol.43, pp.2565-2572, Nov. 1995. *IEEE Trans. Microwave Theory Tech,* vol.43, pp.299-305, Feb. 1995

[7] Allen T., Advances in Computational Electrodynamics, Artech House, INC., 1998

[8] Warren M. Rohsenow et al. (ED.), Handbook of Heat Transfer Fundamentals. (Second Edition), McGraw-Hill, 1985

IV. Scattering from Complex Targets

Wave Propagation,Scattering and Emission in Complex Media
Edited by Ya-Qiu Jin
Science Press and World Scientific,2004

225

Analysis of Electromagnetic Scattering from Layered Crossed-Gratings of Circular Cylinders Using Lattice Sums Technique

Kiyotoshi YASUMOTO, Hongting JIA

Department of Computer Science and Communication Engineering
Kyushu University, Fukuoka 812-8581, Japan

1. INTRODUCTION

Periodic dielectric or metallic structures have been a subject of continuing interest for applications to frequency selective or polarization selective devices in microwaves and optical waves. Various analytical or numerical techniques [1]-[3] have been developed to formulate electromagnetic scattering from periodic scatterers. Recently, photonic bandgap structures [4],[5] in discrete periodic systems have received a growing attention, because they have many potential applications to narrow-band filters, high-quality resonant cavities, strongly guiding devices, and substrates for antennas. A one-dimensional periodic array of infinitely long cylindrical objects is typical of discrete periodic structures. The frequency response of the array is characterized by the scattering properties of individual cylinders and the multiple scattering due to the periodic arrangement of scatterers. When the array is layered, it constitutes a two-dimensional photonic bandgap structure. In the layered system, the multiple interaction of space-harmonics scattered from each of array layers pronounces the frequency response and the photonic bandgaps are formed in which any electromagnetic wave propagation is forbidden within a fairly large frequency range.

The electromagnetic scattering from layered periodic arrays of parallel circular cylinders has been extensively investigated during the past decade. The frequency response in reflectance and the frequency range of bandgaps have been reported, using a method of cylindrical harmonic expansion [6],[7], the Fourier modal method [8],[9], the finite element method [10], the differential method [11], and time domain techniques [12]-[14]. The cylindrical harmonic expansion method [6],[7] among these approaches is a rigorous analytical one, though it requires a calculation of the periodic Green's function. This approach has been recently refined by making use of the aggregate T-matrix [15] and the lattice sums technique [16],[17] and a simpler representation of scattered fields is presented [18]-[21]. Although the layered array of parallel cylinders exhibits various interesting resonance phenomena, it is essentially a two-dimensional structure. When the array is illuminated by electromagnetic waves propagating obliquely to the cylinder axis, it is rather difficult to fully control the frequency response by the structural parameters alone. Such a deficiency in the array of infinite parallel cylinders may be resolved (improved) by introducing a three-dimensional nature through the layering process. A fully three-dimensional structure is realized [22],[23] when the axes of parallel cylinders are rotated by a constant angle in each successive layer.

In this paper, we shall discuss a rigorous analytical approach for the three-dimensional electromagnetic scattering from layered crossed-arrays of circular cylinders. The cylinders in each layer are infinitely long and parallel, while the cylinder axes are rotated by 90• in each successive layer. The layered system consists of a stacking sequence of the two orthogonal arrays. First, the reflection and transmission matrices are derived for each of two arrays by employing the electric and magnetic fields parallel to the cylinder axes as the leading fields.. The matrices are expressed in terms of the one-dimensional lattice sums [16],[17] and the three-dimensional T-matrices [24] of the isolated single cylinder for obliquely incident plane waves. Next, a coordinate transformation of the reflection and transmission matrices is introduced to one of the two arrays so that the multiple scattering process can be described using the same leading fields. Then the results for the two arrays are linked to obtain the reflection and transmission matrices of the crossed-array which constitutes a unit cell in the stacking sequence. Finally, the generalized reflection and transmission matrices for the layered crossed-arrays are calculated by concatenating those of the crossed-array successively over the repeated number of cells. Thus the three dimensional electromagnetic scattering from the layered crossed-arrays can be calculated by a simpler matrix operation on the basis of the one-dimensional lattice sums and the three-dimensional T-matrix of the isolated single cylinder.

The proposed method is used to analyze the reflection and transmission of plane waves by multilayered crossed-arrays of circular cylinders. The effects of a multiple scattering of space-harmonics between the crossed arrays are discussed. The numerical results demonstrate various interesting features of the reflectance of the layered system which depend on the frequency and incident angle of plane waves.

2. FORMULATION

The geometry considered here is shown in Fig. 1. Periodic arrays of circular cylinders are stacked in free space with a separation distance d along the y direction. The cylinders in each layer are infinitely long and parallel to each other, while the cylinder axes are rotated by 90° in each successive layer. The stacking sequence repeats every two layers. The array consisting of z-directed cylinders is referred to as the *Z-array* and the array of x-directed cylinders is as the *X-array*. The cylindrical elements and their periodic spacing may be different in *Z-array* and *X-array*. We assume an incident plane wave of unit amplitude varying as $e^{i(\xi_0 x + \beta_0 z)}$ where $\xi_0 = -k_0 \sin \theta^i \cos \phi^i$, $\beta_0 = k_0 \sin \theta^i \sin \phi^i$, and (θ^i, ϕ^i) denotes the angle of incidence as depicted in Fig. 1, $k_0 = \omega\sqrt{\varepsilon_0\mu_0} = 2\pi/\lambda_0$, and λ_0 is the wavelength in free space. The z components of the incident fields are expressed in the rectangular coordinate system as follows:

$$E_z^i = E_0 e^{i(\xi_0 x - \gamma_0 y + \beta_0 z)} \tag{1}$$

$$\tilde{H}_z^i = \tilde{H}_0 e^{i(\xi_0 x - \gamma_0 y + \beta_0 z)} \tag{2}$$

with

$$E_0 = \sin \theta^i \cos \psi^i, \quad \tilde{H}_0 = \sin \theta^i \sin \psi^i \tag{3}$$

$$\gamma_0 = \sqrt{k_0^2 - \xi_0^2 - \beta_0^2} = k_0 \cos \theta^i \tag{4}$$

where $\tilde{H}_z^i = \sqrt{\mu_0/\varepsilon_0} H_z^i$ denotes the normalized magnetic field, and ψ^i is the polarization angle of the incident electric field.

Since the structure is periodic both in x and z directions, the scattered fields consist of a set of space-harmonics varying as $e^{i(\xi_\ell x + \beta_m z)}$ where $\xi_\ell = \xi_0 + 2\ell\pi/h_z$, $\beta_m = \beta_0 + 2m\pi/h_x$, h_z and h_x are the periodic spacing in the *X-array* and *Z-array*, respectively, and ℓ and m are integers denoting the order of space harmonics. We formulate first the reflection and transmission matrices for the *Z-array* and *X-array* separately. Then the results are used to derive the reflection and transmission matrices for the *crossed-array* which doubles the *Z-array* and *X-array* and constitutes a unit cell in the stacking sequence.

The generalized reflection and transmission matrices for the layered system in Fig. 1 are obtained by concatenating those of the *crossed-array* over the repeating number of cells in the y direction.

2.1 Z-Array

Assume a Z-*array* of circular cylinders with radius a_1, permittivity ε_1, and permeability μ_1 situated in free space with ε_0 and μ_0. The z-directed cylinders are periodically spaced with a distance h_x in the x-direction as shown in Fig. 2. The array is illuminated by three-dimensional plane waves varying as $e^{i(\beta_m z + \xi_\ell x)}$. The scattering problem is formulated by employing E_z and $\tilde{H}_z = \sqrt{\mu_0/\varepsilon_0} H_z$ fields as the leading fields. The fields can be characterized in terms of the amplitudes $\{e_{z,\ell,m}\}$ and $\{\tilde{h}_{z,\ell,m}\}$ for each of space-harmonics. Let $\boldsymbol{e}_z$ and $\tilde{\boldsymbol{h}}_z$ be the column vectors of $(2L+1) \times (2M+1)$ dimensions defined as

$$\boldsymbol{e}_z = [\boldsymbol{e}_z(-M) \cdots \boldsymbol{e}_z(0) \cdots \boldsymbol{e}_z(M)]^T, \quad \boldsymbol{e}_z(m) = [e_{z,-L,m} \cdots e_{z,0,m} \cdots e_{z,L,m}]^T \quad (5)$$

$$\tilde{\boldsymbol{h}}_z = [\tilde{\boldsymbol{h}}_z(-M) \cdots \tilde{\boldsymbol{h}}_z(0) \cdots \tilde{\boldsymbol{h}}_z(M)]^T, \quad \tilde{\boldsymbol{h}}_z(m) = [\tilde{h}_{z,-L,m} \cdots \tilde{h}_{z,0,m} \cdots \tilde{h}_{z,L,m}]^T \quad (6)$$

where $m = \pm M$ and $\ell = \pm L$ denote the orders of truncation of space-harmonics. Following a similar analytical step [24] developed for analyzing the three-dimensional scattering from a periodic array of circular cylinders, the amplitudes vectors $(\boldsymbol{e}_z^r, \tilde{\boldsymbol{h}}_z^r)$ and $(\boldsymbol{e}_z^t, \tilde{\boldsymbol{h}}_z^t)$ for the reflected and transmitted waves are related to the incident ones $(\boldsymbol{e}_z^i, \tilde{\boldsymbol{h}}_z^i)$ as follows:

$$\begin{bmatrix} \boldsymbol{e}_z^r \\ \tilde{\boldsymbol{h}}_z^r \end{bmatrix} = \begin{bmatrix} \boldsymbol{R}_\| \end{bmatrix} \begin{bmatrix} \boldsymbol{e}_z^i \\ \tilde{\boldsymbol{h}}_z^i \end{bmatrix} = \begin{bmatrix} \bar{\bar{\boldsymbol{R}}}_\|^{ee} & \bar{\bar{\boldsymbol{R}}}_\|^{eh} \\ \bar{\bar{\boldsymbol{R}}}_\|^{he} & \bar{\bar{\boldsymbol{R}}}_\|^{hh} \end{bmatrix} \begin{bmatrix} \boldsymbol{e}_z^i \\ \tilde{\boldsymbol{h}}_z^i \end{bmatrix} \quad (7)$$

$$\begin{bmatrix} \boldsymbol{e}_z^t \\ \tilde{\boldsymbol{h}}_z^t \end{bmatrix} = \begin{bmatrix} \boldsymbol{F}_\| \end{bmatrix} \begin{bmatrix} \boldsymbol{e}_z^i \\ \tilde{\boldsymbol{h}}_z^i \end{bmatrix} = \begin{bmatrix} \bar{\bar{\boldsymbol{F}}}_\|^{ee} & \bar{\bar{\boldsymbol{F}}}_\|^{eh} \\ \bar{\bar{\boldsymbol{F}}}_\|^{he} & \bar{\bar{\boldsymbol{F}}}_\|^{hh} \end{bmatrix} \begin{bmatrix} \boldsymbol{e}_z^i \\ \tilde{\boldsymbol{h}}_z^i \end{bmatrix} \quad (8)$$

with

$$\bar{\bar{\boldsymbol{R}}}_\|^\mu = \begin{bmatrix} \boldsymbol{R}_\|^\mu(-M) & \cdots & 0 & \cdots & 0 \\ \vdots & \ddots & \vdots & \ddots & \vdots \\ 0 & \cdots & \boldsymbol{R}_\|^\mu(0) & \cdots & 0 \\ \vdots & \ddots & \vdots & \ddots & \vdots \\ 0 & \cdots & 0 & \cdots & \boldsymbol{R}_\|^\mu(M) \end{bmatrix} \quad (9)$$

$$\bar{\bar{\boldsymbol{F}}}_\|^\mu = \begin{bmatrix} \boldsymbol{F}_\|^\mu(-M) & \cdots & 0 & \cdots & 0 \\ \vdots & \ddots & \vdots & \ddots & \vdots \\ 0 & \cdots & \boldsymbol{F}_\|^\mu(0) & \cdots & 0 \\ \vdots & \ddots & \vdots & \ddots & \vdots \\ 0 & \cdots & 0 & \cdots & \boldsymbol{F}_\|^\mu(M) \end{bmatrix} \quad (10)$$

where $\bar{\bar{\boldsymbol{R}}}_\|^\mu$ and $\bar{\bar{\boldsymbol{F}}}_\|^\mu$ $(\mu = ee, eh, he, hh)$ represent the reflection and transmission matrices of the Z-array, and $\boldsymbol{R}_\|^\mu(m)$ and $\boldsymbol{F}_\|^\mu(m)$ $(m = -M, \cdots, 0, \cdots, M)$ denote their submatrices of $(2L+1) \times (2L+1)$ dimensions. Note that $\bar{\bar{\boldsymbol{R}}}_\|^{eh}$, $\bar{\bar{\boldsymbol{R}}}_\|^{he}$, $\bar{\bar{\boldsymbol{F}}}_\|^{eh}$, and $\bar{\bar{\boldsymbol{F}}}_\|^{he}$ describe the reflection and transmission into the cross-polarized space-harmonic components. The submatrices $\boldsymbol{R}_\parallel^\mu(m)$ and $\boldsymbol{F}_\parallel^\mu(m)$ are derived as follows [20]:

$$\boldsymbol{R}_\parallel^\mu(m) = \boldsymbol{U}_z(m)\boldsymbol{T}_z^\mu(m)\boldsymbol{P}_z(m) \qquad (\mu = ee, eh, he, hh) \tag{11}$$

$$\boldsymbol{F}_\parallel^\mu(m) = \boldsymbol{I} + \boldsymbol{V}_z(m)\boldsymbol{T}_z^\mu(m)\boldsymbol{P}_z(m) \quad (\mu = ee, hh) \tag{12}$$

$$\boldsymbol{F}_\parallel^\mu(m) = \boldsymbol{V}_z(m)\boldsymbol{T}_z^\mu(m)\boldsymbol{P}_z(m) \qquad (\mu = eh, he) \tag{13}$$

with

$$\boldsymbol{U}_z(m) = \left[\frac{2(-i)^n}{\gamma_\ell(m)h_x \sin\phi_{z,\ell}(m)}e^{in\phi_{z,\ell}(m)}\right] \tag{14}$$

$$\boldsymbol{V}_z(m) = \left[\frac{2(-i)^n}{\gamma_\ell(m)h_x \sin\phi_{z,\ell}(m)}e^{-in\phi_{z,\ell}(m)}\right] \tag{15}$$

$$\boldsymbol{P}_z(m) = \left[i^n e^{in\phi_{z,\ell}(m)}\right] \tag{16}$$

$$\sin\phi_{z,\ell}(m) = \frac{\gamma_\ell(m)}{\kappa(m)} \tag{17}$$

$$\kappa(m) = \sqrt{k_0^2 - \beta_m^2}, \quad \gamma_\ell(m) = \sqrt{\kappa^2(m) - \xi_\ell^2} \tag{18}$$

where $\boldsymbol{P}_z(m)$ is the $(2N+1)\times(2L+1)$ matrix whose (n, ℓ)-element $P_{z,n\ell}(m)$ represents the n-th coefficient in the cylindrical wave expansion of the downgoing ℓ-th space-harmonic varying as $e^{i[\xi_\ell x - \gamma_\ell(m)y]}$ for each of indicated order m, $\boldsymbol{U}_z(m)$ and $\boldsymbol{V}_z(m)$ are the $(2L+1)\times(2N+1)$ matrices whose (ℓ, n)-elements represent the transformation of the n-th cylindrical wave into the upgoing and downgoing space-harmonics varying as $e^{i[\xi_\ell x \pm \gamma_\ell(m)y]}$, respectively, $n = \pm N$ is the order of truncation in the cylindrical wave expansion, and $\boldsymbol{I}$ is the unit matrix. In Eqs.(11)-(13), the $(2N+1)\times(2N+1)$ matrices $\boldsymbol{T}_z^\mu(m)$ are the T-matrices [24] of the *Z-array* for the incidence of (ℓ, m)-th space-harmonic with the indicated order m, which are obtained as follows:

$$\boldsymbol{T}_z^{ee}(m) = [\boldsymbol{I} + \boldsymbol{\sigma}_{z,1}(m)\boldsymbol{L}_x(m)\boldsymbol{\sigma}_{z,2}(m)\boldsymbol{L}_x(m)]^{-1}[\boldsymbol{\sigma}_{z,3}(m) - \boldsymbol{\sigma}_{z,1}(m)\boldsymbol{L}_x(m)\boldsymbol{\sigma}_{z,2}(m)] \tag{19}$$

$$\boldsymbol{T}_z^{eh}(m) = [\boldsymbol{I} + \boldsymbol{\sigma}_{z,1}(m)\boldsymbol{L}_x(m)\boldsymbol{\sigma}_{z,2}(m)\boldsymbol{L}_x(m)]^{-1}\boldsymbol{\sigma}_{z,1}(m)[\boldsymbol{I} + \boldsymbol{L}_x(m)\boldsymbol{\sigma}_{z,4}(m)] \tag{20}$$

$$\boldsymbol{T}_z^{he}(m) = -[\boldsymbol{I} + \boldsymbol{\sigma}_{z,2}(m)\boldsymbol{L}_x(m)\boldsymbol{\sigma}_{z,1}(m)\boldsymbol{L}_x(m)]^{-1}\boldsymbol{\sigma}_{z,2}(m)[\boldsymbol{I} + \boldsymbol{L}_x(m)\boldsymbol{\sigma}_{z,3}(m)] \tag{21}$$

$$\boldsymbol{T}_z^{hh}(m) = [\boldsymbol{I} + \boldsymbol{\sigma}_{z,2}(m)\boldsymbol{L}_x(m)\boldsymbol{\sigma}_{z,1}(m)\boldsymbol{L}_x(m)]^{-1}[\boldsymbol{\sigma}_{z,4}(m) - \boldsymbol{\sigma}_{z,2}(m)\boldsymbol{L}_x(m)\boldsymbol{\sigma}_{z,1}(m)] \tag{22}$$

with

$$\boldsymbol{\sigma}_{z,1}(m) = [\boldsymbol{I} - \boldsymbol{\tau}_{z,1}(m)\boldsymbol{L}_x(m)]^{-1}\boldsymbol{\tau}_{z,2}(m) \tag{23}$$

$$\boldsymbol{\sigma}_{z,2}(m) = [\boldsymbol{I} - \boldsymbol{\tau}_{z,3}(m)\boldsymbol{L}_x(m)]^{-1}\boldsymbol{\tau}_{z,2}(m) \tag{24}$$

$$\boldsymbol{\sigma}_{z,3}(m) = [\boldsymbol{I} - \boldsymbol{\tau}_{z,1}(m)\boldsymbol{L}_x(m)]^{-1}\boldsymbol{\tau}_{z,1}(m) \tag{25}$$

$$\boldsymbol{\sigma}_{z,4}(m) = [\boldsymbol{I} - \boldsymbol{\tau}_{z,3}(m)\boldsymbol{L}_x(m)]^{-1}\boldsymbol{\tau}_{z,3}(m) \tag{26}$$

$$\boldsymbol{\tau}_{z,1}(m) = -\left[\frac{J_n(s_m)}{H_n^{(1)}(s_m)}\frac{\Delta_{2,n}}{\Delta_{1,n}}\delta_{n,n'}\right] \tag{27}$$

$$\boldsymbol{\tau}_{z,2}(m) = -\frac{2k_0}{\pi s_m^2 k_1}\left[\frac{K_n\delta_{n,n'}}{[H_n^{(1)}(s_m)]^2\Delta_{1,n}}\right] \tag{28}$$

$$\boldsymbol{\tau}_{z,3}(m) = -\left[\frac{J_n(s_m)}{H_n^{(1)}(s_m)}\frac{\Delta_{3,n}}{\Delta_{1,n}}\delta_{n,n'}\right] \tag{29}$$

$$\Delta_{1,n} = D_{1,n}D_{2,n} - K_n^2, \quad \Delta_{2,n} = D_{2,n}D_{3,n} - K_n^2, \quad \Delta_{3,n} = D_{4,n}D_{1,n} - K_n^2 \quad (30)$$

$$D_{1,n} = \frac{J_n'(t_m)}{t_m J_n(t_m)} - \frac{\varepsilon_0}{\varepsilon_1} \frac{H_n'^{(1)}(s_m)}{s_m H_n^{(1)}(s_m)}, \quad D_{2,n} = \frac{J_n'(t_m)}{t_m J_n(t_m)} - \frac{\mu_0}{\mu_1} \frac{H_n'^{(1)}(s_m)}{s_m H_n^{(1)}(s_m)} \quad (31)$$

$$D_{3,n} = \frac{J_n'(t_m)}{t_m J_n(t_m)} - \frac{\varepsilon_0}{\varepsilon_1} \frac{J_n'(s_m)}{s_m J_n(s_m)}, \quad D_{4,n} = \frac{J_n'(t_m)}{t_m J_n(t_m)} - \frac{\mu_0}{\mu_1} \frac{J_n'(s_m)}{s_m J_n(s_m)} \quad (32)$$

$$K_n = \frac{n\beta_m}{k_1}\left(\frac{1}{t_m^2} - \frac{1}{s_m^2}\right), \quad s_m = a_1\kappa(m), \quad t_m = a_1\sqrt{k_1^2 - \beta_m^2} \quad (33)$$

where $k_1 = \omega\sqrt{\varepsilon_1\mu_1}$, $J_n(t_m)$ and $H_n^{(1)}(s_m)$ are the n-th order Bessel function and the n-th order Hankel function of the first kind, $\delta_{n,n'}$ is the Kronecker's delta, and the matrices $\boldsymbol{\tau}_{z,1}(m)$, $\boldsymbol{\tau}_{z,2}(m)$, and $\boldsymbol{\tau}_{z,3}(m)$ are the T-matrices [24] of the z-directed circular cylinder in isolation. In Eqs.(19)-(26), $\boldsymbol{L}_x(m)$ is a square matrix whose (n, n')-element is given by the $(n-n')$-th order lattice sums [16],[17] for the Z-array defined as

$$S_{x,n-n'}(m) = \sum_{\nu=1}^{\infty} H_{n-n'}^{(1)}(\nu\kappa(m)h_x)e^{-i\nu\xi_0 h_x} + (-1)^{n-n'} \sum_{\nu=1}^{\infty} H_{n-n'}^{(1)}(\nu\kappa(m)h_x)e^{i\nu\xi_0 h_x}. \quad (34)$$

2.2 X-Array

Assume a X-array of circular cylinders with radius a_2, permittivity ε_2, and permeability μ_2 situated in free space. The x-directed cylinders are periodically spaced with a distance h_z in the z-direction as shown in Fig. 3. Employing the E_x and $\tilde{H}_x = \sqrt{\mu_0/\varepsilon_0}H_x$ fields as the leading fields, the scattering problem can be formulated in a similar way as the Z-array. The fields are characterized in terms of the amplitudes $\{e_{x,m,\ell}\}$ and $\{\tilde{h}_{x,m,\ell}\}$ of each of space-harmonics. Let $\underline{e}_x$ and $\underline{\tilde{h}}_x$ be the column vectors of $(2M + 1) \times (2L + 1)$ dimensions defined as

$$\underline{e}_x = [e_x(-L) \cdots e_x(0) \cdots e_x(L)]^T, \quad e_x(\ell) = [e_{x,-M,\ell} \cdots e_{x,0,\ell} \cdots e_{x,M,\ell}]^T \quad (35)$$

$$\underline{\tilde{h}}_x = [\tilde{h}_x(-L) \cdots \tilde{h}_x(0) \cdots \tilde{h}_x(L)]^T, \quad \tilde{h}_x(\ell) = [\tilde{h}_{x,-M,\ell} \cdots \tilde{h}_{x,0,\ell} \cdots \tilde{h}_{x,M,\ell}]^T. \quad (36)$$

Then the amplitudes vectors $(\underline{e}_x^r, \underline{\tilde{h}}_x^r)$ and $(\underline{e}_x^t, \underline{\tilde{h}}_x^t)$ for the reflected and transmitted waves are related to the incident ones $(\underline{e}_x^i, \underline{\tilde{h}}_x^i)$ as follows:

$$\begin{bmatrix} \underline{e}_x^r \\ \underline{\tilde{h}}_x^r \end{bmatrix} = [\boldsymbol{R}_x] \begin{bmatrix} \underline{e}_x^i \\ \underline{\tilde{h}}_x^i \end{bmatrix} = \begin{bmatrix} \underline{\underline{R}}_x^{ee} & \underline{\underline{R}}_x^{eh} \\ \underline{\underline{R}}_x^{he} & \underline{\underline{R}}_x^{hh} \end{bmatrix} \begin{bmatrix} \underline{e}_x^i \\ \underline{\tilde{h}}_x^i \end{bmatrix} \quad (37)$$

$$\begin{bmatrix} \underline{e}_x^t \\ \underline{\tilde{h}}_x^t \end{bmatrix} = [\boldsymbol{F}_x] \begin{bmatrix} \underline{e}_x^i \\ \underline{\tilde{h}}_x^i \end{bmatrix} = \begin{bmatrix} \underline{\underline{F}}_x^{ee} & \underline{\underline{F}}_x^{eh} \\ \underline{\underline{F}}_x^{he} & \underline{\underline{F}}_x^{hh} \end{bmatrix} \begin{bmatrix} \underline{e}_x^i \\ \underline{\tilde{h}}_x^i \end{bmatrix} \quad (38)$$

with

$$\underline{\underline{R}}_x^{\mu} = \begin{bmatrix} \boldsymbol{R}_x^{\mu}(-L) & \cdots & 0 & \cdots & 0 \\ \vdots & \ddots & \vdots & \ddots & \vdots \\ 0 & \cdots & \boldsymbol{R}_x^{\mu}(0) & \cdots & 0 \\ \vdots & \ddots & \vdots & \ddots & \vdots \\ 0 & \cdots & 0 & \cdots & \boldsymbol{R}_x^{\mu}(L) \end{bmatrix} \quad (39)$$

$$\underline{\underline{F}}^{\mu}_x = \begin{bmatrix} F^{\mu}_x(-L) & \cdots & 0 & \cdots & 0 \\ \vdots & \ddots & \vdots & \ddots & \vdots \\ 0 & \cdots & F^{\mu}_x(0) & \cdots & 0 \\ \vdots & \ddots & \vdots & \ddots & \vdots \\ 0 & \cdots & 0 & \cdots & F^{\mu}_x(L) \end{bmatrix} \tag{40}$$

where $\underline{\underline{R}}^{\mu}_x$ and $\underline{\underline{F}}^{\mu}_x$ $(\mu = ee, eh, he, hh)$ represent the reflection and transmission matrices of the X-array, and $R^{\mu}_x(\ell)$ and $F^{\mu}_x(\ell)$ $(\ell = -L, \cdots, 0, \cdots, L)$ denote their submatrices of $(2M+1) \times (2M+1)$ dimensions. Note that the arrangements of elements in e_x and $\tilde{h}_x$ are different from those in e_z and $\tilde{h}_z$ defined by Eqs.(5) and (6) for the Z-array. The submatrices $R^{\mu}_x(\ell)$ and $F^{\mu}_x(\ell)$ are derived as follows:

$$R^{\mu}_x(\ell) = U_x(\ell) T^{\mu}_x(\ell) P_x(\ell) \qquad (\mu = ee, eh, he, hh) \tag{41}$$

$$F^{\mu}_x(\ell) = I + V_x(\ell) T^{\mu}_x(\ell) P_x(\ell) \qquad (\mu = ee, hh) \tag{42}$$

$$F^{\mu}_x(\ell) = V_x(\ell) T^{\mu}_x(\ell) P_x(\ell) \qquad (\mu = eh, he) \tag{43}$$

with

$$U_x(\ell) = \left[\frac{2(-i)^n}{\gamma_m(\ell) h_x \sin \phi_{x,m}(\ell)} e^{in\phi_{x,m}(\ell)} \right] \tag{44}$$

$$V_x(\ell) = \left[\frac{2(-i)^n}{\gamma_m(\ell) h_x \sin \phi_{x,m}(\ell)} e^{-in\phi_{x,m}(\ell)} \right] \tag{45}$$

$$P_x(\ell) = [i^n e^{in\phi_{x,m}(\ell)}] \tag{46}$$

$$\sin \phi_{x,m}(\ell) = \frac{\gamma_m(\ell)}{\eta(\ell)} \tag{47}$$

$$\eta(\ell) = \sqrt{k_0^2 - \xi_\ell^2}, \ \gamma_m(\ell) = \sqrt{\eta^2(\ell) - \beta_m^2}. \tag{48}$$

In Eqs.(41)-(43), the $(2N+1) \times (2N+1)$ matrices $T^{\mu}_x(\ell)$ are T-matrices of the X-array for the incidence of (m, ℓ)-th space-harmonic with the indicated order ℓ. By choosing the x-axis as the reference axis for the cylindrical wave expansion, $T^{\mu}_x(\ell)$ are obtained as

$$T^{ee}_x(\ell) = [I + \sigma_{x,1}(\ell) L_z(\ell) \sigma_{x,2}(\ell) L_z(\ell)]^{-1} [\sigma_{x,3}(\ell) - \sigma_{x,1}(\ell) L_z(\ell) \sigma_{x,2}(\ell)] \tag{49}$$

$$T^{eh}_x(\ell) = [I + \sigma_{x,1}(\ell) L_z(\ell) \sigma_{x,2}(\ell) L_z(\ell)]^{-1} \sigma_{x,1}(\ell) [I + L_x(\ell) \sigma_{x,4}(\ell)] \tag{50}$$

$$T^{he}_x(\ell) = -[I + \sigma_{x,2}(\ell) L_z(\ell) \sigma_{x,1}(\ell) L_z(\ell)]^{-1} \sigma_{x,2}(\ell) [I + L_z(\ell) \sigma_{x,3}(\ell)] \tag{51}$$

$$T^{hh}_x(\ell) = [I + \sigma_{x,2}(\ell) L_z(\ell) \sigma_{x,1}(\ell) L_z(\ell)]^{-1} [\sigma_{x,4}(\ell) - \sigma_{x,2}(\ell) L_z(\ell) \sigma_{x,1}(\ell)] \tag{52}$$

with

$$\sigma_{x,1}(\ell) = [I - \tau_{x,1}(\ell) L_z(\ell)]^{-1} \tau_{x,2}(\ell) \tag{53}$$

$$\sigma_{x,2}(\ell) = [I - \tau_{x,3}(\ell) L_z(\ell)]^{-1} \tau_{x,2}(\ell) \tag{54}$$

$$\sigma_{x,3}(\ell) = [I - \tau_{x,1}(\ell) L_x(m)]^{-1} \tau_{x,1}(\ell) \tag{55}$$

$$\sigma_{x,4}(\ell) = [I - \tau_{x,3}(\ell) L_z(\ell)]^{-1} \tau_{x,3}(\ell) \tag{56}$$

$$\tau_{x,1}(\ell) = -\left[\frac{J_n(p_\ell)}{H^{(1)}_n(p_\ell)} \frac{\bar{\Delta}_{2,n}}{\bar{\Delta}_{1,n}} \delta_{n,n'} \right] \tag{57}$$

$$\boldsymbol{\tau}_{x,2}(\ell) = -\frac{2k_0}{\pi p_\ell^2 k_2}\left[\frac{\bar{K}_n \delta_{n,n'}}{[H_n^{(1)}(p_\ell)]^2 \bar{\Delta}_{1,n}}\right] \tag{58}$$

$$\boldsymbol{\tau}_{x,3}(\ell) = -\left[\frac{J_n(p_\ell)}{H_n^{(1)}(p_\ell)}\frac{\bar{\Delta}_{3,n}}{\bar{\Delta}_{1,n}}\delta_{n,n'}\right] \tag{59}$$

$$\bar{\Delta}_{1,n} = \bar{D}_{1,n}\bar{D}_{2,n} - \bar{K}_n^2, \quad \bar{\Delta}_{2,n} = \bar{D}_{2,n}\bar{D}_{3,n} - \bar{K}_n^2, \quad \bar{\Delta}_{3,n} = \bar{D}_{4,n}\bar{D}_{1,n} - K_n^2 \tag{60}$$

$$\bar{D}_{1,n} = \frac{J_n'(q_\ell)}{q_\ell J_n(q_\ell)} - \frac{\varepsilon_0}{\varepsilon_2}\frac{H_n'^{(1)}(p_\ell)}{p_\ell H_n^{(1)}(p_\ell)}, \quad \bar{D}_{2,n} = \frac{J_n'(q_\ell)}{q_\ell J_n(q_\ell)} - \frac{\mu_0}{\mu_2}\frac{H_n'^{(1)}(p_\ell)}{p_\ell H_n^{(1)}(p_\ell)} \tag{61}$$

$$\bar{D}_{3,n} = \frac{J_n'(q_\ell)}{q_\ell J_n(q_\ell)} - \frac{\varepsilon_0}{\varepsilon_2}\frac{J_n'(p_\ell)}{p_\ell J_n(p_\ell)}, \quad \bar{D}_{4,n} = \frac{J_n'(q_\ell)}{q_\ell J_n(q_\ell)} - \frac{\mu_0}{\mu_2}\frac{J_n'(p_\ell)}{p_\ell J_n(p_\ell)} \tag{62}$$

$$\bar{K}_n = \frac{n\xi_\ell}{k_2}(\frac{1}{q_\ell^2} - \frac{1}{p_\ell^2}), \quad p_\ell = a_2\eta_\ell, \quad q_\ell = a_2\sqrt{k_2^2 - \xi_\ell^2} \tag{63}$$

where $k_2 = \omega\sqrt{\varepsilon_2\mu_2}$, the matrices $\boldsymbol{\tau}_{x,1}(\ell)$, $\boldsymbol{\tau}_{x,2}(\ell)$, and $\boldsymbol{\tau}_{x,3}(\ell)$ are T-matrices [24] of the x-directed circular cylinder in isolation. In Eqs.(49)-(56), $\boldsymbol{L}_z(\ell)$ is a square matrix whose (n,n')-element is given by the $(n\text{-}n')$-th order lattice sums for the X-array defined as

$$S_{z,n-n'}(\ell) = \sum_{\nu=1}^{\infty} H_{n-n'}^{(1)}(\nu\eta(\ell)h_z)e^{-i\nu\beta_0 h_z} + (-1)^{n-n'}\sum_{\nu=1}^{\infty} H_{n-n'}^{(1)}(\nu\eta(\ell)h_z)e^{i\nu\beta_0 h_z}. \tag{64}$$

2.3 Crossed-Array

Assume a *crossed-array* in which the *Z-array* and *X-array* are doubled with a center to center separation of d in the y direction as shown in Fig. 4. The *Z-array* and *X-array* in the preceding sections have been treated separately by taking the electric and magnetic fields parallel to the cylinder axis in each configuration as the leading fields. When they are layered, however, the use of common leading fields is requested to describe the multiple scattering process between the two orthogonal arrays in unified manner. Then we choose the E_z and $\tilde{H}_z$ fields as the leading fields common to the *crossed-array* and rewrite the reflection and transmission matrices of the *X-array* in terms of these field components.

We introduce the matrix $\underline{\underline{\Gamma}}$ which transforms the order of elements of amplitude vectors defined by Eqs.(35) and (36) into those of Eqs.(5) and(6) as follows:

$$\underline{e}_x = \underline{\underline{\Gamma}}\,e_x, \quad \tilde{\underline{h}}_x = \underline{\underline{\Gamma}}\tilde{h}_x \tag{65}$$

with

$$\underline{\underline{\Gamma}} = \begin{bmatrix} \boldsymbol{\Gamma}_{-M,-L} & \cdots & \boldsymbol{\Gamma}_{0,-L} & \cdots & \boldsymbol{\Gamma}_{M,-L} \\ \vdots & \ddots & \vdots & \ddots & \vdots \\ \boldsymbol{\Gamma}_{-M,0} & \cdots & \boldsymbol{\Gamma}_{0,0} & \cdots & \boldsymbol{\Gamma}_{M,0} \\ \vdots & \ddots & \vdots & \ddots & \vdots \\ \boldsymbol{\Gamma}_{-M,L} & \cdots & \boldsymbol{\Gamma}_{0,L} & \cdots & \boldsymbol{\Gamma}_{M,L} \end{bmatrix} \tag{66}$$

$$\boldsymbol{\Gamma}_{m',\ell'} = [\delta_{m,m'}\delta_{\ell,\ell'}]. \tag{67}$$

where $\boldsymbol{e}_x$ and $\tilde{\boldsymbol{h}}_x$ are defined in the same form as Eqs.(5) and (6). We note also that the amplitudes of the (ℓ, m)-th space-harmonic in $(E_x, \tilde{H}_x)$ and $(E_z, \tilde{H}_z)$ fields are related as

$$e_{x,\ell,m} = -\frac{\xi_\ell \beta_m}{k_0^2 - \beta_m^2} e_{z,\ell,m} - \frac{k_0(\pm)\gamma_{\ell,m}}{k_0^2 - \beta_m^2} \tilde{h}_{z,\ell,m} \tag{68}$$

$$\tilde{h}_{x,\ell,m} = \frac{k_0(\pm)\gamma_{\ell,m}}{k_0^2 - \beta_m^2} e_{z,\ell,m} - \frac{\xi_\ell \beta_m}{k_0^2 - \beta_m^2} \tilde{h}_{z,\ell,m} \tag{69}$$

where the upper and lower signs on $\gamma_{\ell,m}$ correspond to the upgoing and downgoing waves in the y direction, respectively. Substituting Eqs.(65), (68), and (69) into Eqs.(37) and (38), the relations between the amplitude vectors in the *X-array* are rewritten in terms of the E_z and $\tilde{H}_z$ fields as follows:

$$\begin{bmatrix} \boldsymbol{e}_z^r \\ \tilde{\boldsymbol{h}}_z^r \end{bmatrix} = [\boldsymbol{R}_\perp] \begin{bmatrix} \boldsymbol{e}_z^i \\ \tilde{\boldsymbol{h}}_z^i \end{bmatrix} = \begin{bmatrix} \bar{\bar{\boldsymbol{R}}}_\perp^{ee} & \bar{\bar{\boldsymbol{R}}}_\perp^{eh} \\ \bar{\bar{\boldsymbol{R}}}_\perp^{he} & \bar{\bar{\boldsymbol{R}}}_\perp^{hh} \end{bmatrix} \begin{bmatrix} \boldsymbol{e}_z^i \\ \tilde{\boldsymbol{h}}_z^i \end{bmatrix} \tag{70}$$

$$\begin{bmatrix} \boldsymbol{e}_z^t \\ \tilde{\boldsymbol{h}}_z^t \end{bmatrix} = [\boldsymbol{F}_\perp] \begin{bmatrix} \boldsymbol{e}_z^i \\ \tilde{\boldsymbol{h}}_z^i \end{bmatrix} = \begin{bmatrix} \bar{\bar{\boldsymbol{F}}}_\perp^{ee} & \bar{\bar{\boldsymbol{F}}}_\perp^{eh} \\ \bar{\bar{\boldsymbol{F}}}_\perp^{he} & \bar{\bar{\boldsymbol{F}}}_\perp^{hh} \end{bmatrix} \begin{bmatrix} \boldsymbol{e}_z^i \\ \tilde{\boldsymbol{h}}_z^i \end{bmatrix} \tag{71}$$

with

$$\begin{bmatrix} \bar{\bar{\boldsymbol{R}}}_\perp^{ee} & \bar{\bar{\boldsymbol{R}}}_\perp^{eh} \\ \bar{\bar{\boldsymbol{R}}}_\perp^{he} & \bar{\bar{\boldsymbol{R}}}_\perp^{hh} \end{bmatrix} = \begin{bmatrix} \bar{\bar{\boldsymbol{C}}} & -\bar{\bar{\boldsymbol{D}}} \\ \bar{\bar{\boldsymbol{D}}} & \bar{\bar{\boldsymbol{C}}} \end{bmatrix} \begin{bmatrix} \bar{\bar{\boldsymbol{R}}}_x^{ee} & \bar{\bar{\boldsymbol{R}}}_x^{eh} \\ \bar{\bar{\boldsymbol{R}}}_x^{he} & \bar{\bar{\boldsymbol{R}}}_x^{hh} \end{bmatrix} \begin{bmatrix} \bar{\bar{\boldsymbol{A}}} & -\bar{\bar{\boldsymbol{B}}} \\ \bar{\bar{\boldsymbol{B}}} & \bar{\bar{\boldsymbol{A}}} \end{bmatrix} \tag{72}$$

$$\begin{bmatrix} \bar{\bar{\boldsymbol{F}}}_\perp^{ee} & \bar{\bar{\boldsymbol{F}}}_\perp^{eh} \\ \bar{\bar{\boldsymbol{F}}}_\perp^{he} & \bar{\bar{\boldsymbol{F}}}_\perp^{hh} \end{bmatrix} = \begin{bmatrix} \bar{\bar{\boldsymbol{C}}} & \bar{\bar{\boldsymbol{D}}} \\ -\bar{\bar{\boldsymbol{D}}} & \bar{\bar{\boldsymbol{C}}} \end{bmatrix} \begin{bmatrix} \bar{\bar{\boldsymbol{F}}}_x^{ee} & \bar{\bar{\boldsymbol{F}}}_x^{eh} \\ \bar{\bar{\boldsymbol{F}}}_x^{he} & \bar{\bar{\boldsymbol{F}}}_x^{hh} \end{bmatrix} \begin{bmatrix} \bar{\bar{\boldsymbol{A}}} & -\bar{\bar{\boldsymbol{B}}} \\ \bar{\bar{\boldsymbol{B}}} & \bar{\bar{\boldsymbol{A}}} \end{bmatrix} \tag{73}$$

$$\bar{\bar{\boldsymbol{R}}}_x^\mu = \underline{\underline{\Gamma}}^{-1} \underline{\underline{\boldsymbol{R}}}_x^\mu \underline{\underline{\Gamma}}, \quad \bar{\bar{\boldsymbol{F}}}_x^\mu = \underline{\underline{\Gamma}}^{-1} \underline{\underline{\boldsymbol{F}}}_x^\mu \underline{\underline{\Gamma}} \tag{74}$$

where $\bar{\bar{\boldsymbol{A}}}, \bar{\bar{\boldsymbol{B}}}, \bar{\bar{\boldsymbol{C}}}$, and $\bar{\bar{\boldsymbol{D}}}$ are the $(2L+1)(2M+1) \times (2L+1)(2M+1)$ matrices whose submatices are defined in the same form as Eqs.(9) and (10) and are given by

$$\boldsymbol{A}(m) = [\frac{\xi_\ell \beta_m}{\kappa^2(m)}\delta_{\ell,\ell'}], \quad \boldsymbol{B}(m) = [\frac{k_0\gamma_\ell(m)}{\kappa^2(m)}\delta_{\ell,\ell'}] \tag{75}$$

$$\boldsymbol{C}(m) = [\frac{\xi_\ell \beta_m}{\eta_\ell^2}\delta_{\ell,\ell'}], \quad \boldsymbol{D}(m) = [\frac{k_0\gamma_\ell(m)}{\eta_\ell^2}\delta_{\ell,\ell'}]. \tag{76}$$

The calculation of Eq.(74) together with Eqs.(65)-(67) is automatically carried out by changing the sequence of the array elements defined in the computer program.

Now the reflection and transmission matrices $(\boldsymbol{R}_\parallel, \boldsymbol{F}_\parallel)$ and $(\boldsymbol{R}_\perp, \boldsymbol{F}_\perp)$ of the *Z-array* and *X-array* for the incidence of downgoing space-harmonics have been obtained as Eqs.(7), (8), (70), and (71) using the common leading fields E_z and $\tilde{H}_z$. When the cylinders are isotropic and symmetric with respect to each array plane, it follows that the reflection and transmission matrices for the incidence of upgoing space-harmonics are also given by $(\boldsymbol{R}_\parallel, \boldsymbol{F}_\parallel)$ and $(\boldsymbol{R}_\perp, \boldsymbol{F}_\perp)$. Then the reflection and transmission matrices for the *crossed-array* are derived by linking $(\boldsymbol{R}_\parallel, \boldsymbol{F}_\parallel)$ and $(\boldsymbol{R}_\perp, \boldsymbol{F}_\perp)$ over the *Z-array* and *X-array* stacked with a separation distance d. For the incidence of downgoing space-harmonics from the upper region of the *Z-array*, after several manipulations [25], the reflection and transmission matrices $\boldsymbol{R}_\times^{(+)}$ and $\boldsymbol{F}_\times^{(-)}$ of the *crossed-array* are obtained as follows:

$$\boldsymbol{R}_\times^{(+)} = \boldsymbol{R}_\parallel + \boldsymbol{F}_\parallel \boldsymbol{W} \boldsymbol{R}_\perp \boldsymbol{X} \boldsymbol{W} \boldsymbol{F}_\parallel \tag{77}$$

$$\boldsymbol{F}_\times^{(-)} = \boldsymbol{F}_\perp \boldsymbol{X} \boldsymbol{W} \boldsymbol{F}_\parallel \tag{78}$$

with

$$\boldsymbol{X} = [\boldsymbol{I} - \boldsymbol{W}\boldsymbol{R}_\parallel \boldsymbol{W}\boldsymbol{R}_\perp]^{-1} \tag{79}$$

$$\boldsymbol{W} = [e^{i\gamma_\ell(m)d}\delta_{\ell,\ell'}]. \tag{80}$$

Similarly, the reflection and transmission matrices $\boldsymbol{R}_\times^{(-)}$ and $\boldsymbol{F}_\times^{(+)}$ for the incidence of upgoing space-harmonics from the lower region of the *X-array* are obtained as follows:

$$\boldsymbol{R}_\times^{(-)} = \boldsymbol{R}_\perp + \boldsymbol{F}_\perp \boldsymbol{W} \boldsymbol{R}_\parallel \boldsymbol{Y} \boldsymbol{W} \boldsymbol{F}_\perp \tag{81}$$

$$\boldsymbol{F}_\times^{(+)} = \boldsymbol{F}_\parallel \boldsymbol{Y} \boldsymbol{W} \boldsymbol{F}_\perp \tag{82}$$

with

$$\boldsymbol{Y} = [\boldsymbol{I} - \boldsymbol{W}\boldsymbol{R}_\perp \boldsymbol{W}\boldsymbol{R}_\parallel]^{-1}. \tag{83}$$

2.4 Layered Crossed-Arrays

Let us consider a 2N-layered arrays in which the *Z-array* and *X-array* are stacked one after the other in the y direction as shown in Fig. 1. This structure may be treated as a N-stacking sequence of the *crossed-array* whose reflection and transmission matrices $\boldsymbol{R}_\times^{(\pm)}$ and $\boldsymbol{F}_\times^{(\pm)}$ are defined by Eqs.(77),(78), (81), and (82). The generalized reflection and transmission matrices for the N-layered *crossed-arrays* are obtained [25] by concatenating $\boldsymbol{R}_\times^{(\pm)}$ and $\boldsymbol{F}_\times^{(\pm)}$ successively over N layers. For the downgoing incident plane wave, a recursive relation to determine the generalized reflection matrix $\mathfrak{R}_{\nu,\nu+1}^{(+)}$ is obtained as follows:

$$\mathfrak{R}_{\nu,\nu+1}^{(+)} = \boldsymbol{R}_\times^{(+)} + \boldsymbol{F}_\times^{(+)} \boldsymbol{W} \mathfrak{R}_{\nu+1,\nu+2}^{(+)} \Lambda_{\nu+1}^{(-)} \boldsymbol{W} \boldsymbol{F}_\times^{(-)} \tag{84}$$

with

$$\Lambda_{\nu+1}^{(-)} = [\boldsymbol{I} - \boldsymbol{W}\boldsymbol{R}_\times^{(-)} \boldsymbol{W} \mathfrak{R}_{\nu+1,\nu+2}^{(+)}]^{-1} \tag{85}$$

where $\nu = 1, 2, \cdots, N$ and $\mathfrak{R}_{N+1,N+2}^{(+)} = 0$. Note that $\mathfrak{R}_{\nu,\nu+1}^{(+)}$ includes the effects of reflections and transmissions of space-harmonics at all the layers of the *crossed-array* situated below the ν-th layer. In the same manner, the generalized transmission matrix $\mathfrak{S}_{1,N+1}$, which connects the space harmonics transmitted into region $N+1$ to the incident plane wave in region 1, is obtained as follows:

$$\mathfrak{S}_{1,N+1}^{(-)} = \boldsymbol{F}_\times^{(-)} \Lambda_N^{(-)} \Lambda_{N-1}^{(-)} \cdots \Lambda_3^{(-)} \Lambda_2^{(-)} \boldsymbol{W} \boldsymbol{F}_\times^{(-)}. \tag{86}$$

For the N-layered system, Eqs.(84) and (85) are recursively solved by starting with $\mathfrak{R}_{N+1,N+2} = 0$ to obtain $\mathfrak{R}_{1,2}$ in region 1. The results for $\Lambda_{\nu+1}^{(-)}$ are substituted into Eq.(86) to calculate $\mathfrak{S}_{1,N+1}$. Similarly, a recursive relation for the generalized reflection matrix $\mathfrak{R}_{\nu,\nu+1}^{(-)}$ for the upgoing incident plane wave is derived as follows:

$$\mathfrak{R}_{\nu+1,\nu}^{(-)} = \boldsymbol{R}_\times^{(-)} + \boldsymbol{F}_\times^{(-)} \boldsymbol{W} \mathfrak{R}_{\nu,\nu-1}^{(-)} \Lambda_\nu^{(+)} \boldsymbol{W} \boldsymbol{F}_\times^{(+)} \tag{87}$$

with

$$\mathbf{\Lambda}_\nu^{(+)} = [\mathbf{I} - \mathbf{W}\mathbf{R}_\times^{(+)}\mathbf{W}\mathbf{\Re}_{\nu,\nu-1}^{(-)}]^{-1} \tag{88}$$

where $\mathbf{\Re}_{1,0}^{(+)} = 0$. Note that $\mathbf{\Re}_{\nu+1,\nu}^{(-)}$ includes the effects of reflections and transmissions of space-harmonics at all the layers of the *crossed-array* situated above the ν-th layer. In the same manner, the generalized transmission matrix $\mathfrak{S}_{N+1,1}$, which connects the space harmonics transmitted into region 1 to the incident plane wave in region $N+1$, is obtained as follows:

$$\mathfrak{S}_{N+1,1}^{(-)} = \mathbf{F}_\times^{(+)}\mathbf{\Lambda}_2^{(+)}\mathbf{\Lambda}_3^{(+)}\cdots\mathbf{\Lambda}_{N-1}^{(+)}\mathbf{\Lambda}_N^{(+)}\mathbf{W}\mathbf{F}_\times^{(+)}. \tag{89}$$

Although the notations are a little intricate because of three-dimensional problems, the mathematics and analytical procedure are quite straightforward. The generalized reflection and transmission matrices of layered crossed-arrays are calculated by a simpler matrix operation using the T-matrix of isolated circular cylinder and the lattice sums for one-dimensional periodic array.

3. NUMERICAL RESULTS

The proposed approach has been used to analyze the reflection and transmission characteristics of layered crossed-arrays of circular cylinders. We discuss here the reflection characteristics in the wavelength range $h/\lambda_0 < 1$, because such a situation is essential for the use of periodic array systems. In this case, only the fundamental space-harmonic with $l = 0$ and $m = 0$ becomes a propagating wave. The parameters are assumed to be $h_x = h_z = h$, $a_1 = a_2 = 0.3h$, $d = h$, $\varepsilon_1/\varepsilon_0 = 5.0$, and $\mu_1 = \mu_2 = \mu_0$. The numerical results in what follow were obtained with the errors in the energy conservation less than 10^{-10} by taking into account the lowest seven space-harmonics and truncating the cylindrical harmonic expansion at $n = \pm10$ to calculate the T-matrix of the isolated circular cylinder.

We consider first the frequency response of the crossed arrays for the normal incidence of plane wave with $\theta^i = \phi^i = 0°$ and $\psi^i = 90°$. The reflectance $|R_{0,0}^{hh}|$ of the fundamental space-harmonic for TE-wave incidence with respect the z axis is plotted in Fig. 5 as functions of the normalized wavelength h/λ_0 for 32-layered crossed-arrays and single layer of crossed arrays. It is seen that as the number of layers is increased, very sharp stopbands with complete reflection are formed within several frequency ranges. Next examples are the dependence of the reflectance on the incident angle with respect to the array plane $z = 0$. Figure 6 shows the reflectances $|R_{0,0}^{ee}|$ and $|R_{0,0}^{hh}|$ as functions of θ^i where $h/\lambda_0 = 0.5$, $\phi^i = 0°$, and the polarization angles are (a) $\psi^i = 90°$ and (b) $\psi^i = 0°$. For 32-layered crossed arrays, there appear particular angles θ^i in which the incident wave is almost completely reflected or transmitted. This feature of crossed-arrays may be useful to select a particular wave component among those incoming from different directions and being in the same frequency band. Finally the effect of the incident angle ϕ^i on the reflectance is discussed. Figure 7 shows the plots of (a) $|R_{0,0}^{hh}|$ and (b) $|R_{0,0}^{eh}|$ as functions of θ^i, where $\phi^i = 45°$, $\psi^i = 90°$, and $h/\lambda_0 = 0.35$. In this case, the cross-polarized component $|R_{0,0}^{eh}|$ of reflection appears, because $\phi^i \neq 0$. For sake of comparison, the reflectances obtained for a single layer and 64-layeres of only Z-array are shown in Fig. 8 under the same parameters as those in Fig. 7. Figures 9 and 10 show similar plots of $|R_{0,0}^{ee}|$ and $|R_{0,0}^{he}|$ where $\psi^i = 0°$ and other parameters are the same as those in Fig. 7. From Fig. 7 to Fig. 10 we can see that due to the multiple interaction of waves between the orthogonal arrays, the reflectances of layered crossed-arrays are noticeably changed from those of layered parallel arrays. It is worth noting that almost complete reflection is attained over a wide incident angle around $\theta^i = 0$ for both TE wave and TM wave excitations.

4. CONCLUDING REMARKS

The problem of three-dimensional electromagnetic scattering from multilayered crossed-arrays of circular cylinders have been formulated using the lattice-sums technique and the T-matrix of cylindrical objects. The crossed-array consists of a doubling of two arrays whose cylinder axes are orthogonal to each other. The periodic spacing and geometrical parameters of the cylindrical elements may be different in the two arrays. It was shown that the generalized reflection and transmission matrices for the multilayered crossed-arrays are obtained by a simple matrix calculation based on the one-dimensional lattice sums and the three-dimensional T-matrix for a circular cylinder in isolation. The numerical examples demonstrated several interesting features in reflection characteristics of layered crossed-arrays. When the multilayered arrays are embedded in a dielectric slab, a slight modification of the approach is required. The reflection and transmission matrices at each array plane are defined for the background slab medium, and two additional reflection and transmission matrices are introduced in the concatenating process. These additional matrices are the Fresnel reflection and transmission matrices defined at the upper and lower boundaries of the slab.

Acknowledgment   This work was supported in part by the 2001 Research Grant of Hoso-Bunka Foundation.

REFERENCES

[1] V. Twersky, "On Scattering of Waves by the Infinite Grating of Circular Cylinders," IRE Trans. Antennas Propagat., Vol.10, pp.737-765, 1962

[2] K. Ohtaha and N. Numata, "Multiple scattering effects in photon diffraction for an array of cylindrical dielectric," Phys. Lett., Vol.73a, No.5,6, pp.411-413, 1979

[3] R. Petit, Ed., *Electromagnetic Theory of Gratings*, Springer-Verlag, 1980

[4] A. Scherer, T. Doll, E. Yablonovitch, H.O. Everitt, and J.A. Higgins, ed., "Special Section on Electromagnetic Crystal Structures, Design, Synthesis, and Applications," J. Lightwave Technol., Vol.17, No.11, pp.1928-2207, 1999

[5] ——, "Mini-Special Issue on Electromagnetic Crystal Structures, Design, Synthesis, and Applications," IEEE Trans. Microwave Theory Tech., Vol.47, No.11, pp.2057-2150, 1999

[6] G. Tayeb and D. Maystre, "Rigorous theoretical study of finite-size two-dimensional photonic crystals doped by microcavities," J. Opt. Soc. Am. A, Vol.14, No.12, pp.3323-3332, 1997

[7] K. Ohtaka, Tsuyoshi Ueta, and Katsuki Amemiya, "Calculation of photonic bands using vector cylindrical waves and reflectivity of light for an array of dielectric rods, Phys. Rev., B, Vol.57, No.4, pp.2550-2568, 1998

[8] M. Plihal and A. A. Maradudin, "Photonic band structure of two-dimensional systems: The triangular lattice," Phys. Rev., B, Vol.44, No.16, pp.8565-8571, 1991

[9] H. Benisty, "Modal analysis of optical guides with two-dimensional photonic band-gap boundaries," J. Appl. Phys., Vol.79, No.10, pp.7483-7492, 1996

[10] G. Pelosi, A. Cocchi, and A. Monorchio, "A Hybrid FEM-based procedure for the scattering from photonic crystals illuminated by a Gaussian beam," IEEE. Trans., Antennas Propagat., Vol.45, No.1, pp.185-186, 1997

[11] E. Popov and B. Bozhkov, "Differential method applied for photonic crystals," Appl. Opt., Vol.39, No.27, pp.4926-4932, 2000

[12] Y. Naka and H. Ikuno, "Guided mode analysis of two-dimensional air-hole type photonic crystal optical waveguides," IEICE Tech. Rep., EMT-00-78, pp.75-80, 2000

[13] M. Koshiba, Y. Tsuji, M. Hikari, "Time-domain beam propagation method and its application to photonic crystal circuits," J. Lightwave Technol., Vol.18, No.1, pp.102-110, 2000

[14] H. Satoh, N. Yoshida, and Y. Miyanaga, "Analysis of fundamental property of 2-D photonic crystal optical waveguide with various medium conditions by condensed node spatial network," Trans. of IEICE, Vol.J-84-C, No.10, pp.954-963, 2001

[15] C.W. Chew, *Waves and Fields in Inhomogeneous Media*, Van Nostrand Reinhold, New York, 1990

[16] N. A. Nicorovici and R. C. McPhedran, "Lattice sums for off-axis electromagnetic scattering by grating," Phys. Rev. E, Vol. 50, No. 4, pp. 3143-3160, 1994

[17] K. Yasumoto and K. Yoshitomi, "Efficient Calculation of Lattice Sums for Free-Space Periodic Green's Function," IEEE Trans. Antennas Propagat., Vol.47, No.6, pp.1050-1055, 1999

[18] H. Roussel, W. C. Chew, F. Jouvie, and W. Tabbara, "Electromagnetic Scattering from dielectric and magnetic gratings of fibers – a T-matrix solution," J. Electromag. Wave & Applications, Vol.10, No.1, pp.109-127, 1996

[19] K. Yasumoto and T. Ueno, "Rigorous Analysis of Scattering by a periodic Array of Cylindrical Objects Using Lattice Sums," Proceedings of 1998 China-Japan Joint Meeting on Microwaves, pp.247-250, 1998

[20] T. Kushta and K. Yasumoto, "Electromagnetic Scattering from Periodic Arrays of Two Circular Cylinders per Unit Cell," Progress In Electromagnetics Research, PIER 29, pp.69-85, 2000

[21] L. C. Botten, N.-A. P. Nicorovici, A. A. Asatryan, R. C. McPhedran, C. M. de Sterke, and P. A. Robinson, "Formulation for electromagnetic scattering and propagation through grating stacks of metallic and dielectric cylinders for photonic crystal calculations. Part I. Method," J. Opt. Soc. Am. A, Vol.17, No.12, pp.2165-2176, 2000

[22] E. Özbay, E. Michel, G. Tuttle, R. Biswas, M. Sigalas, and K.-M. Ho, "Micromachined millimeter-wave photonic band-gap crystals," Appl. Phys. Lett., Vol.64, No.16, pp.2059-2061, 1994

[23] R. Gonzalo, B. Martinez, C. M. Mann, H. Pellemans, P. H. Bolivar, and P. de Maagt, "A Low-Cost Fabrication Technique for Symmetrical and Asymmetrical Layer-by Layer Photonic Crystals at Submillimeter-Wave Frequencies," IEEE Trans. Microwave Theory Tech., Vol.50, No.10, pp.2384-2392, 2002

[24] K. Yasumoto and H. Jia, "Three-dimensional electromagnetic scattering from multilayered periodic arrays of circular cylinders," Proceedings of 2002 China-Japan Joint Meeting on Microwaves, pp.301-304, 2002

[25] K. Yasumoto, "Generalized method for electromagnetic scattering by two-dimensional periodic discrete composites using lattice sums," Proceedings of 2000 International Conference on Microwave and Millimeter Wave Technology, pp.P-29-P-34, 2002

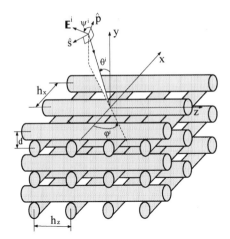

Fig.1 Geometry of multilayered *crossed-arrays* of circular cylinders.

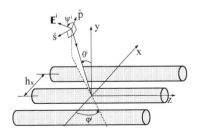

Fig.2 An array of z-directed paralell cylinders periodically spaced with a distance h_x in the x-direction.

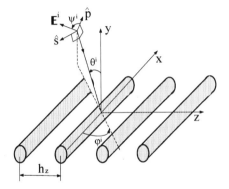

Fig.3 An array of x-directed paralell cylinders periodically spaced with a distance h_z in the z-direction.

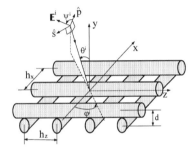

Fig.4 A *Crossed-array* in which the Z-*array* and X-*array* are doubled with a center to center separation of d in the y-direction.

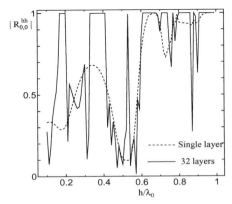

Fig.5 Reflectance $|R_{0,0}^{hh}|$ as functions of normalized wavelength h/λ_0 for $\phi^i = \theta^i = 0°$ and $\psi^i = 90°$.

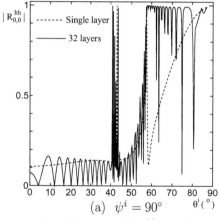
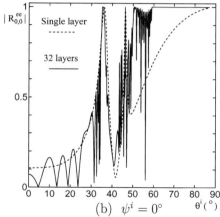

(a) $\psi^i = 90°$

(b) $\psi^i = 0°$

Fig.6 Reflectances (a) $|R_{0,0}^{hh}|$ and (b) $|R_{0,0}^{ee}|$ as functions of incident angle θ^i for $\phi^i = 0°$ and $h/\lambda_0 = 0.5$.

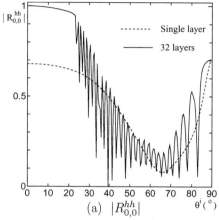
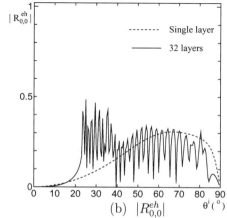

(a) $|R_{0,0}^{hh}|$

(b) $|R_{0,0}^{eh}|$

Fig.7 Reflectances (a) $|R_{0,0}^{hh}|$ and (b) $|R_{0,0}^{eh}|$ of crossed arrays as functions of incident angle θ^i for $\phi^i = 45°$, $\psi^i = 90°$, and $h/\lambda_0 = 0.35$.

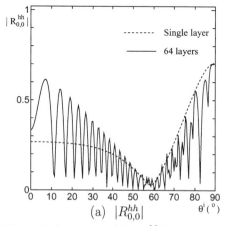
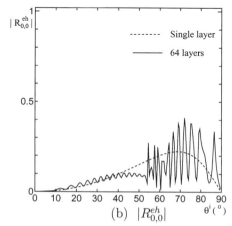

(a) $|R_{0,0}^{hh}|$

(b) $|R_{0,0}^{eh}|$

Fig.8 Reflectances (a) $|R_{0,0}^{hh}|$ and (b) $|R_{0,0}^{eh}|$ of parallel arrays as functions of incident angle θ^i for $\phi^i = 45°$, $\psi^i = 90°$, and $h/\lambda_0 = 0.35$.

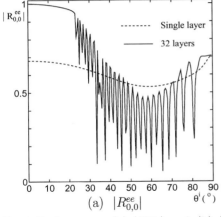
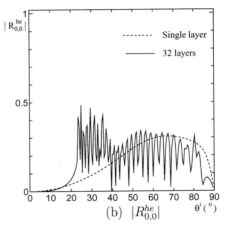

Fig.9 Reflectances (a) $|R_{0,0}^{ee}|$ and (b) $|R_{0,0}^{he}|$ of crossed arrays as functions of incident angle θ^i for $\phi^i = 45°$, $\psi^i = 0°$, and $h/\lambda_0 = 0.35$.

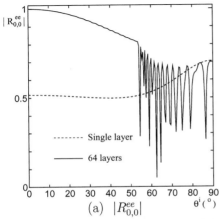
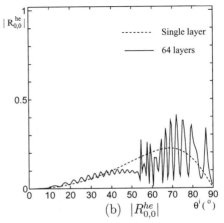

Fig.10 Reflectances (a) $|R_{0,0}^{ee}|$ and (b) $|R_{0,0}^{he}|$ of parallel arrays as functions of incident angle θ^i for $\phi^i = 45°$, $\psi^i = 0°$, and $h/\lambda_0 = 0.35$.

Scattering by a Body in a Random Medium

Mitsuo Tateiba[1], Zhi Qi Meng[2], Hosam El-Ocla[3]

[1]Dept. of Computer Science and Communication Engineering,
Kyushu University
Hakozaki 6-10-1, Higashi-ku, Fukuoka 812-8581, Japan
E-mail: tateiba@csce.kyushu-u.ac.jp

[2]Dept. of Electrical Engineering, Fukuoka University

[3]Dept. of Computer Science, Lakehead University

1 Introduction

Scattering by a body in a random medium is a key subject for developing radar and remote sensing technology. It is not easy to solve it by conventional methods. We solved it in some cases. As a result, we may minutely investigate it by considering that it consists of two issues shown below.

Consider monochromatic wave scattering from a conducting body surrounded by a random medium. Then any incident wave will be partially coherent in space in the neighborhood of the body and hence we need to deal with the scattering of spatially partially-coherent (SPC) wave. In addition, the scattered wave will be statistically coupled with the SPC incident wave through the random medium, according to the position between the transmitter and the receiver. The SPC wave scattering is expected to yield unconventional scattering characteristics compared with the perfectly coherent wave scattering (conventional scattering), and the coupling between the incident and scattered waves is also expected to produce phenomena peculiar to propagation in random medium such as backscattering enhancement.

In order to show quantitatively the above expectation, Tateiba proposed a method[1]. According to the method, the first issue: the SPC wave scattering is solved by introducing a current generator that transforms any incident wave into the surface current on the body and depends only on the body. The second issue: the coupling between the incident and scattered waves is analyzed by introducing Green's function in the random medium. Both issues require the higher order moment of Green's functions. We have numerically presented some interesting results of the first and second issues[1–6].

This paper addresses above two issues. To the first issue, numerical results of the backscattering cross section of conducting cylinder are given in detail under the condition that backscattering enhancement occurs, by changing the cross-section configuration of the cylinder and the polarization and illumination angle of the average incident wave. The spatial coherence function that characterizes the SPC incident wave in the neighborhood of the cylinder plays a central role in estimating the backscattering cross-section. To the second issue, the bistatic radar cross-section of the same cylinder is numerically analyzed by changing the random medium parameters: fluctuation intensity, scale-size and thickness. As the parameters change, the radar cross-section also changes with the angle between the transmitter and the receiver. In some case the coupling leads to a backscattering enhancement peak, two depressions on both sides of the peak and second peaks just outside the depressions. This paper shows numerical some data on both issues.

2 Formulation

Consider the two-dimensional problem of electromagnetic wave scattering from a perfectly conducting cylinder embedded in a continuous random medium, as shown in Fig.1. Here L is the thickness of the random medium surrounding the cylinder and is assumed to be larger enough than the size of the cylinder cross-section. The random medium is assumed to be described by the dielectric constant ε, the magnetic permeability μ and the electric conductivity σ, which are expressed as

$$\varepsilon = \varepsilon_0[1 + \delta\varepsilon(\boldsymbol{r})] , \quad \mu = \mu_0 , \quad \sigma = 0 , \tag{1}$$

where ε_0, μ_0 are constant and $\delta\varepsilon(\boldsymbol{r})$ is a random function with

$$\langle\delta\varepsilon(\boldsymbol{r})\rangle = 0 , \quad \langle\delta\varepsilon(\boldsymbol{r}_1)\cdot\delta\varepsilon(\boldsymbol{r}_2)\rangle = B(\boldsymbol{r}_1 - \boldsymbol{r}_2) . \tag{2}$$

Here the angular brackets denote the ensemble average and $B(\boldsymbol{r}_1 - \boldsymbol{r}_2)$ is a correlation function of the random function. For many cases, it can be approximated as

$$B(\boldsymbol{r}_1 - \boldsymbol{r}_2) = B_0 \exp\left[-\frac{(\boldsymbol{r}_1 - \boldsymbol{r}_2)^2}{l^2}\right] , \tag{3}$$

$$B_0 \ll 1 , \quad kl \gg 1 , \tag{4}$$

where B_0, l are the intensity and scale-size of the random medium fluctuation, respectively, and $k = \omega\sqrt{\varepsilon_0\mu_0}$ is the wavenumber in free space. Under the condition (4), depolarization of electromagnetic waves due to the medium fluctuation can be neglected; and the scalar approximation is valid. In addition, the small scattering-angle approximation is also valid; and the fact

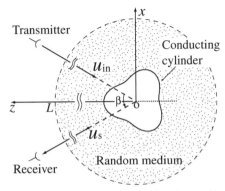

Fig.1: Geometry of the scattering problem from a conducting cylinder surrounded by a random medium.

is negligible in this case that a part of scattered waves by the random medium are incident again on the cylinder.

Suppose that the current source with the time factor $\exp(-j\omega t)$ suppressed throughout this paper is a line source, located at r_{T}, far from and parallel to the cylinder. Then the incident wave is $G(r, r_{\mathrm{T}})$ whose dimension coefficient is understood, and using the current generator Y defined on the cylinder we can give the average intensity of scattered waves u_{s} as follows [1]: for E-wave incidence,

$$
\begin{aligned}
\langle |u_{\mathrm{s}}|^2 \rangle = {} & \int_S \mathrm{d}r_1 \int_S \mathrm{d}r_2 \int_S \mathrm{d}r'_1 \int_S \mathrm{d}r'_2 \, [Y_{\mathrm{E}}(r_1; r'_1) Y_{\mathrm{E}}^*(r_2; r'_2) \\
& \langle G(r; r_1) G(r'_1; r_{\mathrm{T}}) G^*(r; r_2) G^*(r'_2; r_{\mathrm{T}}) \rangle] \,,
\end{aligned}
\tag{5}
$$

and for H-wave incidence,

$$
\begin{aligned}
\langle |u_{\mathrm{s}}|^2 \rangle = {} & \int_S \mathrm{d}r_1 \int_S \mathrm{d}r_2 \int_S \mathrm{d}r'_1 \int_S \mathrm{d}r'_2 \, \Big\{ Y_{\mathrm{H}}(r_1; r'_1) Y_{\mathrm{H}}^*(r_2; r'_2) \\
& \cdot \frac{\partial}{\partial n_1} \frac{\partial}{\partial n_2} \langle G(r; r_1) G(r'_1; r_{\mathrm{T}}) G^*(r; r_2) G^*(r'_2; r_{\mathrm{T}}) \rangle \Big\} \,,
\end{aligned}
\tag{6}
$$

where S and G denote the cylinder surface and Green's function in a random medium, respectively, $\partial/\partial n_i$ ($i = 1, 2$) denotes the outward normal derivative at r_i on S, and the asterisk denotes the complex conjugate.

The Y_{E} and Y_{H}, respectively, are expressed as

$$
Y_{\mathrm{E}}(r; r') \simeq \phi_M^*(r) A_{\mathrm{E}}^{-1} \ll \phi_M^{\mathrm{T}}(r') \,,
\tag{7}
$$

$$
Y_{\mathrm{H}}(r; r') \simeq \frac{\partial \phi_M^*(r)}{\partial n} A_{\mathrm{H}}^{-1} \ll \phi_M^{\mathrm{T}}(r') \,.
\tag{8}
$$

Here, $\phi_M = [\phi_1, \phi_2, \cdots, \phi_M]$, in which $\phi_m = H_m^{(1)}(kr) e^{jm\theta}$; $m = 0, \pm 1, \cdots, \pm N$ and $M = 2N + 1$, ϕ_M^{T} denotes the transposed vector of ϕ_M, and A_{E}, A_{H} are positive definite Hermitian matrices of $M \times M$

$$
A_{\mathrm{E}} = \begin{bmatrix} (\phi_1, \phi_1) & \cdots & (\phi_1, \phi_M) \\ \vdots & \cdots & \vdots \\ (\phi_M, \phi_1) & \cdots & (\phi_M, \phi_M) \end{bmatrix} \,,
\tag{9}
$$

$$
A_{\mathrm{H}} = \begin{bmatrix} (\partial\phi_1/\partial n, \partial\phi_1/\partial n) & \cdots & (\partial\phi_1/\partial n, \partial\phi_M/\partial n) \\ \vdots & \cdots & \vdots \\ (\partial\phi_M/\partial n, \partial\phi_1/\partial n) & \cdots & (\partial\phi_M/\partial n, \partial\phi_M/\partial n) \end{bmatrix} \,,
\tag{10}
$$

$$
(\phi_m, \phi_n) \equiv \int_S \phi_m(r) \phi_n^*(r) \mathrm{d}r \,.
\tag{11}
$$

The $\ll \phi_M^{\mathrm{T}}$, denotes the following operation of each element of ϕ_M^{T} and the incident wave u_{in} to the right of the ϕ_M^{T}:

$$
\ll \phi_m(r), u_{\mathrm{in}}(r) \gg \equiv \phi_m(r) \frac{\partial u_{\mathrm{in}}(r)}{\partial n} - \frac{\partial \phi_m(r)}{\partial n} u_{\mathrm{in}}(r) \,.
\tag{12}
$$

For a circular cylinder, (9) and (10) becomes diagonal matrices; therefore Y_{E}, Y_{H} are expressed in infinite series[7].

The calculations of (5) and (6) require the Fourth moment of Green's functions, which can be written as

$$\langle G(\boldsymbol{r};\boldsymbol{r}_1)G(\boldsymbol{r}_1';\boldsymbol{r}_\mathrm{T})G^*(\boldsymbol{r};\boldsymbol{r}_2)G^*(\boldsymbol{r}_2';\boldsymbol{r}_\mathrm{T})\rangle = G_0(\boldsymbol{r};\boldsymbol{r}_1)G_0^*(\boldsymbol{r};\boldsymbol{r}_2)G_0(\boldsymbol{r}_1';\boldsymbol{r}_{1\mathrm{T}})G_0^*(\boldsymbol{r}_2';\boldsymbol{r}_{2\mathrm{T}})\times m_\mathrm{s}\,,$$

(13)

where G_0 is Green's function in free space[11]. The m_s includes multiple-scattering effects of random medium and can be obtained by two-scale method [3, 4, 9, 10].

$$m_\mathrm{s} = \frac{k}{2\pi z}\int_{-\infty}^{\infty}\int_{-\infty}^{\infty}\mathrm{d}\eta\mathrm{d}\rho\exp\left\{-\frac{jk}{z}\eta[\rho - (x - x_\mathrm{T})]\right\}P(\rho,\eta)\,,$$

(14)

$$P(\rho,\eta) = \exp\left(-\frac{\sqrt{\pi}k^2lz}{8}\int_0^{L/z}\mathrm{d}t\,\{D[a(\sin\theta_1' - \sin\theta_2')t + \eta t]\right.$$

$$+D[a(\sin\theta_1 - \sin\theta_2)t + \eta t]$$

$$-D[a(\sin\theta_1' - \sin\theta_1)t - \rho(1 - t) + \eta t]$$

$$-D[a(\sin\theta_2' - \sin\theta_2)t - \rho(1 - t) - \eta t]$$

$$+D[a(\sin\theta_1' - \sin\theta_2)t - \rho(1 - t)]$$

$$\left.+D[a(\sin\theta_2' - \sin\theta_1)t - \rho(1 - t)]\}\right)\,,$$

(15)

$$D(x) = 2B_0\left[1 - \exp\left(-\frac{x^2}{l^2}\right)\right]\,.$$

(16)

3   Numerical Results

3.1   Bistatic RCS

We pay attention to the second issue: the statistical coupling between incident and scattered waves. This means that the spatial coherence length of incident wave around the cylinder (SCL) is much larger than the size of cylinder. The SCL will be quantitatively defined by using (18) on the next page.

As an example, parameters of the random medium are assumed to satisfy $k^2B_0Ll = 4\pi^2$. The cross section of cylinder is expressed by

$$r = a[1 - \delta\cos 3(\theta - \phi)].$$

(17)

Here we restrict the shape and size to the concavity index $\delta = 0.2$ and $ka = 3$. We calculate the bistatic RCS of cylinder σ in the cases of $\phi = 0$ and π for E-wave and H-wave incidence, respectively, and normalize them to those in free space σ_0. When we plot the normalized bistatic RCS as functions of β (the angle between incidence and observation directions), then all the results for different ϕ and polarization coincide with each other, as shown in Fig.2. There is a backscattering enhancement peak. A twin depression appears at both sides of the peak; and the depression value is less than one. The coincidence of the results means that the behavior of the normalized bistatic RCS is independent of the body configuration and the polarization of incident waves; that is, the RCS does not depend on the first issue: the SPC wave scattering but does only on the second issue. We observe that the normalized bistatic RCS tends to one as increasing β. The integration value of the normalized bistatic RCS is almost one, which fact shows that the results agree with the law of energy conservation.

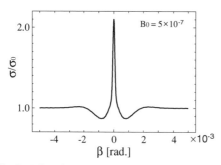

Fig.2: Bistatic RCS normalized to that in free space.

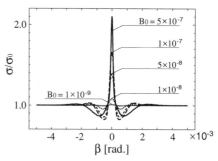

Fig.3: The effect of fluctuation intensity on the bistatic RCS.

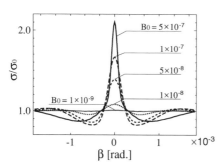

Fig.4: Enlargement of Fig.3 about the backward enhancement peak.

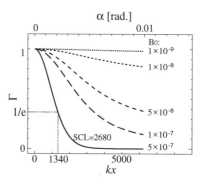

Fig.5: The degree of spatial coherence for different fluctuation intensities.

When we decrease B_0 from 5×10^{-7} to 10^{-9} gradually, the peak of the backscattering enhancement goes down and almost disappears finally, as shown in Fig.3. The size of horizontal axis in Fig.3 is tripled in Fig.4. A small twin peak clearly appears just outside the depression in the cases of $B_0 = 10^{-7}$ and 5×10^{-8}. The oscillation of bistatic RCS is considered to be caused by an interference of incident and scattered waves, i.e., the second issue. To study the interference, we analyze the degree of spatial coherence Γ:

$$\Gamma(x, z) = \frac{\langle G(\boldsymbol{r}_1; \boldsymbol{r}_t) G^*(\boldsymbol{r}_2; \boldsymbol{r}_t) \rangle}{\langle |G(\boldsymbol{r}_0; \boldsymbol{r}_t)|^2 \rangle} \,, \tag{18}$$

where $\boldsymbol{r}_1 = (x, 0)$, $\boldsymbol{r}_2 = (-x, 0)$, $\boldsymbol{r}_0 = (0, 0)$, $\boldsymbol{r}_t = (0, z)$. Figure 5 shows Γ for B_0 used in Fig.3. Then the SCL is defined as double the kx at which $|\Gamma| = \mathrm{e}^{-1} \approx 0.37$. The Γ is also described as a function of angle α at $\boldsymbol{r}_t$ between $\boldsymbol{r}_1$ and $\boldsymbol{r}_2$. Certainly, the SCL increases as B_0 decreases. For $B_0 = 5.0 \times 10^{-7}$ the SCL is about 2680 (larger enough than ka), and for $B_0 = 10^{-9}$ the SCL becomes very large. It is clear that the spatial range in which the incident and scattered waves may interfere with each other becomes wider as B_0 decreases. On the other hand, the backscattering enhancement reduces because a part of wave remains coherent for propagation of distance L in the random medium and the strength of multiple scattering due to the random medium becomes lower for small B_0.

It is concluded that the behavior of the normalized bistatic RCS is independent of the body configuration and the polarization of incident waves under the condition of SCL $\gg ka$. This condition shows that the first issue does not occur. Figures 7 and 10 described below are also depicted under the same condition. If the condition is not satisfied, then the RCS may change

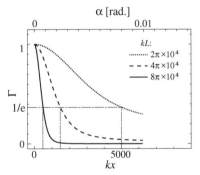

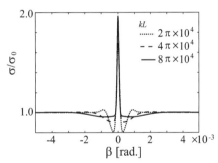

Fig.6: The degree of spatial coherence for the different thickness of random medium.

Fig.7: The effect of the random medium thickness on the bistatic RCS.

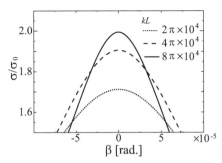

Fig.8: Enlargement of Fig.7 about the backward enhancement peak.

more complicatedly as shown in 3.2.

Figure 6 shows Γ for three cases of $kL = 8\pi \times 10^4$, $4\pi \times 10^4$, and $2\pi \times 10^4$ at $klB_0 = 2.5\pi \times 10^{-5}$. The normalized bistatic RCS for the three cases of kL is shown in Fig.7. The oscillation of bistatic RCS appears in the case of $kL = 2\pi \times 10^4$ for the same reason mentioned previously, although kL is a different parameter from B_0. Similarly, as kL decreases, the number of multiple scattering times due to the random medium decreases; as a result, the peak value at $\beta = 0$ becomes low, as shown in Fig.8.

Figure 9 shows Γ for three cases of $kl = 200\pi$, 300π, and 400π at $kLB_0 = \pi \times 10^{-2}$. The difference in SCL for the three cases of kl is not large, compared with Figs.5 and 6. Therefore the oscillation of bistatic RCS resembles each other, as shown in Fig.10. A part of Fig.10 enlarged about the backward enhancement peak is shown in Fig.11. For the three cases of kl, incident and scattered waves become almost incoherent in propagation of distance L, but the number of multiple scattering times decreases with increasing kl. Therefore the peak value of the backward enhancement also decreases with increasing kl.

3.2 Monostatic RCS

In the case of SCL $\sim ka$ and SCL $< ka$, we must consider the first issue: the SPC wave scattering and need much time to compute (5) or (6). However, if we pay attention to the monostatic RCS under the condition that the backscattering enhancement occurs; that is, we

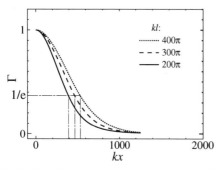

Fig.9: The degree of spatial coherence for the different scale-size of random medium.

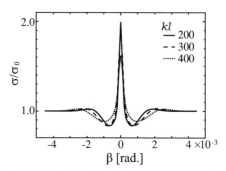

Fig.10: The effect of the random medium scale-size on the bistatic RCS.

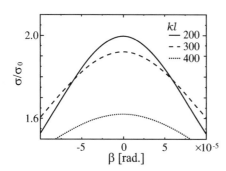

Fig.11: Enlargement of Fig.10 about the backward enhancement peak.

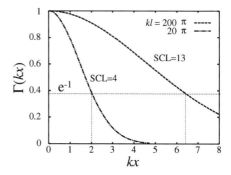

Fig.12: The degree of spatial coherence of an incident wave about the cylinder.

deal with a fixed second issue, then the forth moment of Green's function is expressed as the sum of products of the second order moment:

$$\langle G(\boldsymbol{r};\boldsymbol{r}_1)G(\boldsymbol{r}_1';\boldsymbol{r})G^*(\boldsymbol{r};\boldsymbol{r}_2)G^*(\boldsymbol{r}_2';\boldsymbol{r})\rangle$$
$$\simeq \langle G(\boldsymbol{r};\boldsymbol{r}_1)G^*(\boldsymbol{r};\boldsymbol{r}_2)\rangle\langle G(\boldsymbol{r}_1';\boldsymbol{r})G^*(\boldsymbol{r}_2';\boldsymbol{r})\rangle$$
$$+\langle G(\boldsymbol{r};\boldsymbol{r}_1)G^*(\boldsymbol{r}_2';\boldsymbol{r})\rangle\langle G(\boldsymbol{r}_1';\boldsymbol{r})G^*(\boldsymbol{r};\boldsymbol{r}_2)\rangle\ ; \tag{19}$$

therefore we can readily analyze (5) or (6) numerically and address the first issue. In calculation, we assume the parameters of random medium in order that the incident wave may be sufficiently incoherent around the cylinder and the degree of spatial coherence may behave like Fig.12.

Figures 13 and 14 show the normalized monostatic RCS of the cylinder with $\delta = 0.1, 0.2$ at two different incident angels for E-wave incidence and H-wave incidence, respectively. The twice enhancement is usually limited to SCL $\gg ka$, and an extremely big enhancement occurs at $\delta = 0.2$, $ka \sim 1.6$ and H-wave incidence. The monostatic RCS in free space becomes smaller in the neighborhood of $ka = 1.6$ as shown in Fig.15(a); also Fig.15(b) shows that the RCS in random medium becomes small but much larger than that in free space. We may roughly say that the SPC incident wave makes weak the interaction between incident and creeping waves, although the geometrical optics approximation is invalid in this case. Figure 16 also shows that the large enhancement occurs at a specific configuration of cylinder and a small SCL.

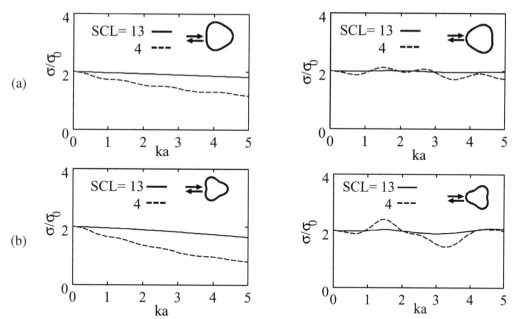

Fig.13: Normalized monostatic RCS vs. cylinder size at two different incident angles and SCLs
for E-wave incidence where (a) $\delta = 0.1$ and (b) $\delta = 0.2$

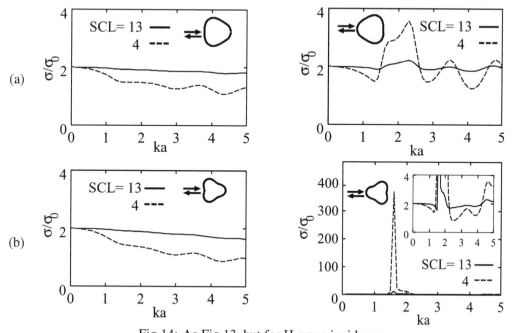

Fig.14: As Fig.13, but for H-wave incidence.

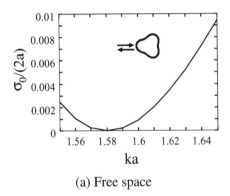

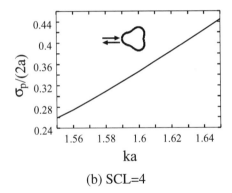

(a) Free space

(b) SCL=4

Fig.15: Monostatic RCS for $1.55 \leq ka \leq 1.65$ in case of convex part illumination of H-wave at $\delta = 0.2$.

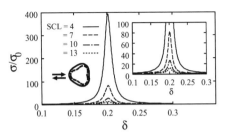

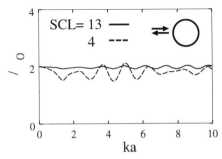

Fig.16: Normalized monostatic RCS vs. concavity index of cylinder cross-section for H-wave incidence at $ka = 1.6$.

Fig.17: Normalized monostatic RCS vs. cylinder size at $\phi = \pi$, $\delta = 0$, and different SCLs for H-wave incidence.

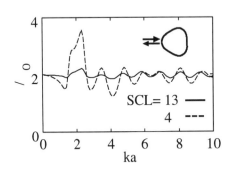

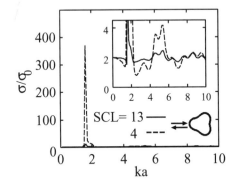

Fig.18: As Fig.17, but $\delta = 0.1$.

Fig.19: As Fig.17, but $\delta = 0.2$.

Finally Figs.17, 18 and 19 show the normalized monostatic RCS of the cylinder with $\delta = 0, 0.1, 0.2$ within $0.1 < ka < 10$ for the convex part illumination of H-wave and indicate that the random medium effect on the RCS strongly depends on the configuration of the cylinder for the case of SCL $\sim ka$ and SCL $< ka$. This fact suggests that the SPC wave scattering will become important for the development of radar and remote sensing technology in random media.

4   Conclusion

Scattering by a body in a random medium is considered to consist of two issues. The first issue is SPC wave scattering and the second is statistical coupling between incident and scattered waves. By a method using a current generator and Green's function, we have expressed scattered waves by conducting cylinders in random media and evaluated the RCS of cylinder with the concave-convex cross-section. Numerical results show clearly that the SCL of incident wave around the cylinder plays a central role in the behavior of the RCS. For the case of SCL $\gg$ the cylinder size ka, the bistatic RCS normalized to that in free space is independent of the cylinder configuration and the polarization of incident wave. This means that only the second issue characterizes the RCS. When the random medium parameters: intensity, scale-size and thickness change, we obtain interesting characteristics of the RCS. On the other hand, for the case of SCL $\sim ka$ and SCL $< ka$, the RCS is closely related to both issues. Under a fixed second issue, we show that the monostatic RCS changes complicatedly with the cylinder configuration and the polarization and incident direction of incident waves. This change is due to the first issue.

Acknowledgement

This work was supported in part by Scientific Research Grant-in-Aid (grant A:12305027) from the Japan Society for the Promotion of Science.

References

[1]   M. Tateiba, and Z. Q. Meng, "Wave scattering from conducting bodies embedded in random media — theory and numerical results," in *PIER 14: Electromagnetic Scattering by Rough Surfaces and Random Media*, ed. M. Tateiba and L. Tsang, pp. 317-361, EMW Pub., Cambridge, USA, 1996

[2]   Z. Q. Meng, and M. Tateiba, "Radar cross section of conducting elliptic cylinders embedded in strong continuous random media," *Waves in Random Media*, vol. 6, no. 4, pp. 335-345, 1996

[3]   Z. Q. Meng, N. Yamasaki, and M. Tateiba, "Numerical analysis of bistatic cross-sections of conducting circular cylinders embedded in continuous random media," *IEICE Trans. on Electronics*, vol. E83-C, No. 12, pp. 1803-1808, 2000

[4]   M. Tateiba, Z. Q. Meng, "Radar cross-sections of conducting targets surrounded by random media" (invited paper), *IEICE Trans. on Electronics*, vol. J84-C, no. 11, pp. 1031-1039, 2001 (in Japanese)

[5]   H. El Ocla and M. Tateiba, "Strong backscattering enhancement for partially convex targets in random media," *Waves in Random Media*, vol. 11, no. 1, pp. 21-32, 2001

[6]   H. El Ocla and M. Tateiba, "Analysis of backscattering enhancement for complex targets in continuous random media for H-wave incidence," *IEICE Trans. on Communications*, vol. E84-B, no. 9, pp. 2583-2588, 2001

[7]   M. Tateiba and Z. Q. Meng, "Infinite-series expressions of current generators in wave scattering from a conducting body", *Research Reports on Information Science and Electrical Engineering of Kyushu University*, vol. 4, No. 1, pp. 1-6, 1999

[8]   Yu. A. Kravtsov, and A. I. Saichev, "Effect of double passage of waves in randomly inhomogeneous media," *Sov. Phys. Usp.*, vol. 25, no. 7, pp. 494-508, 1982

[9]   R. Mazar, "High-frequency propagators for diffraction and backscattering in random media," *J. Opt. Soc. Am. A*, vol. 7, pp. 34-46, 1990

[10] R. Mazar, and A. Bronshtein, "Double passage analysis in random media using two-scale random propagators," *Waves in Random Media*, vol. 1, pp. 341-362, 1991

[11] Y. M. Lure, C. C. Yang and K. C. Yeh "Enhanced backscattering phenomenon in a random cotinuum," *Radio Science*, vol. 24, no. 2, pp. 147-159, 1989

A Rigorous Analysis of Electromagnetic Scattering from Multilayered Crossed-arrays of Metallic Cylinders

Hongting Jia, Kiyotoshi Yasumoto,

Department of Computer Science and Electrical Engineering
Kyushu University, Fukuoka 812-8581, Japan

1 Introduction

Photonic crystals are regular arrays of dielectric materials with different refractive indices or discrete metallic objects. They are classified mainly into three categories that is one-dimensional, two-dimensional, and three-dimensional crystals according to the dimensionality of the stack[1, 2]. Since they are easy to fabricate and have some special properties, they not only have widely been used as frequency selective or polarization selective devices in microwaves and optical waves, but also are expected for new application[3, 4]. A typical structure of two-dimensional photonic crystals is formed by stacking periodic arrays constituted with infinitely long cylindrical objects. In the layered structure, the multiple interaction of space-harmonics scattered from each of array layers pronounces the frequency response and the photonic bandgaps are formed in which any electromagnetic wave propagation is forbidden within a fairly large frequency range[5, 6].

The electromagnetic feature of the layered periodic arrays have been extensively investigated during the past decade, and various analytical or numerical techniques have been proposed. Some popular methods are the Fourier modal method[7, 8], the finite element method[9], the differential method[10], and time domain techniques[11]. Although these methods are very powerful, they all belong to numerical technique and cost long computer time. Recently, a method of cylindrical harmonic expansion combined with the lattice sums, T-matrix, and generalized reflection matrix[12, 13] has been proposed for calculating the reflection and transmission of photonic crystals. The method is an analytical approach, and has many advantage such as high precision, short computer time, and very fast convergence. Futuremore, the method can be easily extended to three-dimensional problems[14]. However it must rely on a numerical technique for calculating the T-matrix[15] when the cross section of scatters is not circular one. On the other hand, the two-dimensional structure is rather difficult to fully control the frequency response, even if it is illuminated by electromagnetic waves propagating obliquely relative to the cylinder axis. Hence it is of significant to study the problem of the three-dimensional photonic crystals. The three-dimensional inductive grids has been theoretically analyzed by Rayleigh hypothesis and modal expansions method[3]. Although this method is rigorous, it is not so effective for treating multilayered structures because of several reasons. The first is that the number of unknowns increases proportionally to the number of layers. The second is that a very large number of modal truncation is necessary to achieve good accuracy, due to singular behavior of fields around the corners of cylinder objects. Although this phenomenon can be largely improved by using other new set of expansion functions[16], the formula becomes much complicated.

In this paper, we shall describe a rigorous approach for the three-dimensional electromagnetic scattering from multilayered crossed-arrays of conductive rectangular cylinders using the generalized reflection and transmission matrices. The cylinders in each layer are infinitely long and parallel, while the cylinder axes are rotated by 90° in each successive layer. The layered system consists of a stacking sequence of the two orthogonal arrays. The reflection and transmission matrices are separately calculated for each of the two arrays by employing the electric and magnetic fields parallel to cylinder axes as the leading fields. This calculation is efficiently performed using a standard mode-matching technique and the reflection and transmission matrices of isolate array are obtained in closed form. Then a coordinate transformation is introduced to derive the combined matrices for the two orthogonal arrays in terms of the same leading fields. The generalized reflection and transmission matrices for the multilayered crossed-arrays are obtained using a simple recursive formula based on the two orthogonal arrays. Numerical examples have confirmed the fast convergence and high accuracy of the proposed method, and demonstrate some interesting results.

2 Formulation

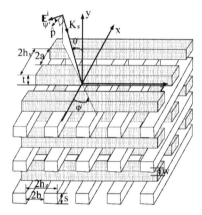

Fig.1 Geometry of multilayered *cross-arrays* of rectangular conductive cylinders.

A cross-stacked periodic structure consisting of conductive cylinders with rectangular cross-section is illustrated in Fig.1. The cylinders in each layer are infinitely long and parallel to each other, while the cylinder axes are rotated by 90° in each successive layer. The stacking repeats every two layers. The array of z-directed cylinders is referred to as the *Z-array* and the array of x-directed cylinders as the *X-array*. This structure has double period viewed from y-direction, so that it is a bigrating. When a plane wave varying as $e^{i(k_{x0}x+k_{z0}z)}$ illuminates the structure, From Floquet theorem, the scatted fields are expressed in superposition of space-harmonics varying as $e^{i(k_{xi}x+k_{zj}z)}$, where $k_{xi} = k_{x0} + i\pi/h_x$, $k_{zj} = k_{z0} + j\pi/h_z$, $k_{x0} = -k_s \sin\theta^i \cos\phi^i$, $k_{z0} = k_s \sin\theta^i \sin\phi^i$, $2h_x$ and $2h_z$ are the spacing periods in the *X-array* and *Z-array*, respectively, (θ^i, ϕ^i) denotes the angle of incidence, and i and j are integer number. Since these space-harmonics form a complete system, the properties of multilayered cross-arrays can be described by generalized reflection and transmission matrices, which can be obtained in simple matrix calculation by concatenating the reflection and transmission matrices of each layer in isolation. In what follows, we shall show the analytical steps and give the reflection and transmission matrices of each layer in closed form.

2.1   Z-Array

The z-directed cylinders are periodically spaced with a distance h_x in the x-direction as shown in Fig. 2. At a glance, the *Z-Array* is only periodic in x-direction, so that the set of space-harmonics $e^{i(k_{xi}x+k_{zj}z)}$ can be divided into many groups in each group k_{zj} is a constant, and each group forms independently a complete sub-system. We assume that the *Z-array* is illuminated by an incident plane wave to belong to j-th sub-system as

$$E_z^i(x,y) \;=\; E_0 e^{i[k_{xQ}x-k_{yQj}y]} \tag{1}$$

$$H_z^i(x,y) \;=\; \tilde{H}_0/\eta_0 e^{i[k_{xQ}x-k_{yQj}y]} \tag{2}$$

where $\eta_0 = \sqrt{\mu_0/\epsilon_0}$ and E_0 and $\tilde{H}_0$ are amplitude of incident wave. Q is integer. Then the periodic structure scatters the incident wave into a set of space harmonics as follows:

$$E_z^r(x,y) \;=\; \sum_{i=-\infty}^{\infty} C_{iQ}^e e^{i[k_{xi}x+k_{yij}y]} \tag{3}$$

$$H_z^r(x,y) \;=\; \sum_{i=-\infty}^{\infty} C_{iQ}^h/\eta_0 e^{i[k_{xi}x+k_{yij}y]} \tag{4}$$

$$E_z^t(x,y) \;=\; \sum_{i=-\infty}^{\infty} D_{iQ}^e e^{i[k_{xi}x-k_{yij}(y+t)]} \tag{5}$$

$$H_z^t(x,y) \;=\; \sum_{i=-\infty}^{\infty} D_{iQ}^h/\eta_0 e^{i[k_{xi}x-k_{yij}(y+t)]} \tag{6}$$

where $k_{yij} = \sqrt{k_s^2 - k_{xi}^2 - k_{zj}^2}$, $\mathrm{Im}(k_{yij}) \ge 0$, $i = 0 \pm 1, \pm 2 \cdots$. The fields in the region between conductors may be expanded in terms of waveguides modes as follows:

$$E_z^g(x,y) \;=\; \sum_{\nu=1}^{\infty} \left[A_\nu^e e^{-i\gamma_{a\nu}y} + B_\nu^e e^{i\gamma_{a\nu}(y+t)} \right] \sin\alpha_\nu(x+a) \tag{7}$$

$$H_z^g(x,y) \;=\; \sum_{\nu=0}^{\infty} \left[A_\nu^h e^{-i\gamma_{a\nu}y} + B_\nu^h e^{i\gamma_{a\nu}(y+t)} \right] \cos\alpha_\nu(x+a)/\eta_0 \tag{8}$$

where $\gamma_{a\nu} = \sqrt{k_1^2 - \alpha_\nu^2}$, $\alpha_\nu = \nu\pi/2a$, and $A_\nu^{e,h}$ and $B_\nu^{e,h}$ denote expanded coefficients. The tangential components of electric and magnetic fields derived from (1)-(8) should be continuous across the boundary planes $y = 0$ and $y = -t$. Taking into account the periodic condition, these boundary conditions may be expressed as follows:

$$E_{x,z}^i(x,0) + E_{x,z}^r(x,0) = \begin{cases} e^{2iih_x k_{xQ}} E_{x,z}^g(x - 2ih_x, 0) & x \in \bigcup_{i=-\infty}^{\infty} [-a + 2ih_x, a + 2ih_x] \\ 0 & \text{otherwise} \end{cases} \tag{9}$$

$$E_{x,z}^r(x,-t) = \begin{cases} e^{2iih_x k_{xQ}} E_{x,z}^g(x - 2ih_x, -t) & x \in \bigcup_{i=-\infty}^{\infty} [-a + 2ih_x, a + 2ih_x] \\ 0 & \text{otherwise} \end{cases} \tag{10}$$

$$H_{x,z}^i(x,0) + H_{x,z}^r(x,0) = e^{2iih_x k_{xQ}} H_{x,z}^g(x - 2ih_x, 0) \quad x \in \bigcup_{i=-\infty}^{\infty} [-a + 2ih_x, a + 2ih_x] \tag{11}$$

$$H_{x,z}^r(x,-t) = e^{2iih_x k_{xQ}} H_{x,z}^g(x - 2ih_x, -t) \quad x \in \bigcup_{i=-\infty}^{\infty} [-a + 2ih_x, a + 2ih_x] . \tag{12}$$

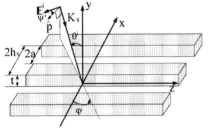

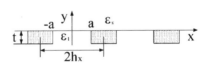

(a) Three-dimensional geometry   (b) Cross-section view at the plane $z = 0$.

Fig.2 An array of z-directed parallel cylinders periodically spaced with a distance $2h_x$ in the x-direction.

Applying the boundary condition (11) and (12) for the magnetic fields and considering the orthogonality of sets of trigonometric functions $\sin \alpha_\nu (x + a)$ and $\cos \alpha_\nu (x + a)$ on the closed interval $[-a, a]$, we have

$$A_\nu^h = \frac{-i}{a(1 + \delta_{\nu 0})[1 - e^{2i\gamma_{a\nu}t}]} \sum_{\iota = -\infty}^{\infty} k_{x\iota} G_{\nu\iota}^+ \left[\tilde{H}_0 \delta_{\iota Q} + C_{\iota Q}^h - D_{\iota Q}^h e^{i\gamma_{a\nu}t} \right] \tag{13}$$

$$B_\nu^h = \frac{i}{a(1 + \delta_{\nu 0})[1 - e^{2i\gamma_{a\nu}t}]} \sum_{\iota = -\infty}^{\infty} k_{x\iota} G_{\nu\iota}^+ \left[\left(\tilde{H}_0 \delta_{\iota Q} + C_{\iota Q}^h \right) e^{i\gamma_{a\nu}t} - D_{\iota Q}^h \right] \tag{14}$$

$$A_\nu^e = \frac{\eta_{1r}\alpha_\nu}{ak_1 \gamma_{a\nu}(1 - e^{2i\gamma_{a\nu}t})} \sum_{\iota = -\infty}^{\infty} \left\{ \left[\frac{\zeta_{1\jmath}^2}{\zeta_{s\jmath}^2} - 1 \right] \beta k_{x\iota} G_{\nu\iota}^+ \left[\tilde{H}_0 \delta_{\iota 0} + C_{\iota Q}^h - D_{\iota Q}^h e^{i\gamma_{a\nu}t} \right] \right.$$
$$\left. - \frac{\zeta_{1\jmath}^2 k_s k_{y\iota} G_{\nu\jmath}^+}{\zeta_{s\jmath}^2 \eta_{rs}} \left[E_0 \delta_{\iota Q} + C_{\iota Q}^e + D_{\iota Q}^e e^{i\gamma_{a\nu}t} \right] \right\} \tag{15}$$

$$B_\nu^e = \frac{\eta_{1r}\alpha_\nu}{ak_1 \gamma_{a\nu}(1 - e^{2i\gamma_{a\nu}t})} \sum_{\iota = -\infty}^{\infty} \left\{ \left[\frac{\zeta_{1\jmath}^2}{\zeta_{s\iota}^2} - 1 \right] \beta k_{x\iota} G_{\nu\iota}^+ \left[\left(\tilde{H}_0 \delta_{\iota 0} + C_{\iota Q}^h \right) e^{i\gamma_{a\nu}t} - D_{\iota Q}^h \right] \right.$$
$$\left. - \frac{\zeta_{1\jmath}^2 k_s k_{y\iota} G_{\nu\iota}^+}{\zeta_{s\iota}^2 \eta_{rs}} \left[\left(E_0 \delta_{\iota Q} + C_{\iota Q}^e \right) e^{i\gamma_{a\nu}t} + D_{\iota Q}^e \right] \right\} \tag{16}$$

where

$$G_{\nu\iota}^\pm = \frac{1}{k_{x\iota}^2 - \alpha_\nu^2} \left[(-1)^\nu e^{\pm i k_{x\iota} a} - e^{\mp i k_{x\iota} a} \right] \tag{17}$$

$$\zeta_{1\jmath} = \sqrt{k_1^2 - k_{z\jmath}^2}, \qquad \zeta_{s\jmath} = \sqrt{k_s^2 - k_{z\jmath}^2} \tag{18}$$

Substituting (13)-(16) into the boundary condition (9) and (10), and applying the orthogonality of the set of exponential functions $e^{i k_{x\iota} x}$ on the closed interval $[-h_x, h_x]$, a set of linear equations for the unknown coefficients of space harmonics $C_{\iota Q}^e$, $C_{\iota Q}^h$, $D_{\iota Q}^e$ and $D_{\iota Q}^h$ are derived as follows:

$$\begin{bmatrix} \boldsymbol{I} + \overline{\boldsymbol{\Gamma}}_1 & -\overline{\boldsymbol{\Gamma}}_2 & \overline{\boldsymbol{\Omega}}_1 & \overline{\boldsymbol{\Omega}}_2 \\ \overline{\boldsymbol{\Gamma}}_3 & \boldsymbol{I} + \overline{\boldsymbol{\Gamma}}_4 & \overline{\boldsymbol{\Omega}}_3 & -\overline{\boldsymbol{\Omega}}_4 \\ \overline{\boldsymbol{\Omega}}_1 & -\overline{\boldsymbol{\Omega}}_2 & \boldsymbol{I} + \overline{\boldsymbol{\Gamma}}_1 & \overline{\boldsymbol{\Gamma}}_2 \\ -\overline{\boldsymbol{\Omega}}_3 & -\overline{\boldsymbol{\Omega}}_4 & -\overline{\boldsymbol{\Gamma}}_3 & \boldsymbol{I} + \overline{\boldsymbol{\Gamma}}_4 \end{bmatrix} \begin{bmatrix} \boldsymbol{C}_Q^e \\ \boldsymbol{C}_Q^h \\ \boldsymbol{D}_Q^e \\ \boldsymbol{D}_Q^h \end{bmatrix} = \begin{bmatrix} \overline{\boldsymbol{\Gamma}}_1 - \boldsymbol{I} & \overline{\boldsymbol{\Gamma}}_2 \\ \overline{\boldsymbol{\Gamma}}_5 & \boldsymbol{I} - \overline{\boldsymbol{\Gamma}}_4 \\ \overline{\boldsymbol{\Omega}}_1 & \overline{\boldsymbol{\Omega}}_2 \\ -\overline{\boldsymbol{\Omega}}_3 & \overline{\boldsymbol{\Omega}}_4 \end{bmatrix} \begin{bmatrix} E_0 \boldsymbol{\delta}(Q) \\ \tilde{H}_0 \boldsymbol{\delta}(Q) \end{bmatrix} \tag{19}$$

$$\boldsymbol{\Gamma}_1 = \tau_1 \boldsymbol{U} \boldsymbol{K}_y, \qquad \boldsymbol{\Gamma}_2 = \tau_2 \boldsymbol{U} \boldsymbol{K}_x \tag{20}$$

$$\boldsymbol{\Gamma}_3 = \boldsymbol{K}_x \boldsymbol{K}_y^{-1}(\tau_3 \boldsymbol{I} + \tau_4 \boldsymbol{U} \boldsymbol{K}_y), \qquad \boldsymbol{\Gamma}_4 = \boldsymbol{K}_x \boldsymbol{K}_y^{-1}(\tau_5 \boldsymbol{U} + \tau_6 \boldsymbol{W}) \boldsymbol{K}_x \tag{21}$$

$$\overline{\boldsymbol{\Gamma}}_5 = \boldsymbol{K}_x \overline{\boldsymbol{K}}_y^{-1}(\tau_4 \overline{\boldsymbol{U} \boldsymbol{K}}_y - \tau_3 \boldsymbol{I}) \tag{22}$$

$$\overline{\boldsymbol{\Omega}}_1 = \tau_1 \overline{\boldsymbol{V} \boldsymbol{K}}_y, \qquad\qquad \overline{\boldsymbol{\Omega}}_2 = \tau_2 \overline{\boldsymbol{V}} \boldsymbol{K}_x \tag{23}$$

$$\boldsymbol{\Omega}_3 = \tau_4 \boldsymbol{K}_x \boldsymbol{K}_y^{-1} \boldsymbol{V} \boldsymbol{K}_y, \qquad \boldsymbol{\Omega}_4 = \boldsymbol{K}_x \boldsymbol{K}_y^{-1}(\tau_5 \boldsymbol{V} + \tau_6 \boldsymbol{X}) \boldsymbol{K}_x \tag{24}$$

$$\overline{\boldsymbol{U}} = [u_{\iota p}] = \left[\sum_{\nu=1}^{\infty} \frac{\alpha_\nu^2 \left(1 + e^{2i\gamma_{a\nu}t}\right)}{\gamma_{a\nu} \left(1 - e^{2i\gamma_{a\nu}t}\right)} G_{\nu p}^+ G_{\nu \iota}^- \right] \tag{25}$$

$$\overline{\boldsymbol{V}} = [v_{\iota p}] = \left[\sum_{\nu=1}^{\infty} \frac{\alpha_\nu^2 2 e^{i\gamma_{a\nu}t}}{\gamma_{a\nu} \left(1 - e^{2i\gamma_{a\nu}t}\right)} G_{\nu p}^+ G_{\nu \iota}^- \right] \tag{26}$$

$$\overline{\boldsymbol{W}} = [w_{\iota p}] = \left[\sum_{\nu=0}^{\infty} \frac{\gamma_{a\nu} \left(1 + e^{2i\gamma_{a\nu}t}\right)}{\left(1 + \delta_{\nu 0}\right) \left(1 - e^{2i\gamma_{a\nu}t}\right)} G_{\nu p}^+ G_{\nu \iota}^- \right] \tag{27}$$

$$\overline{\boldsymbol{X}} = [x_{\iota p}] = \left[\sum_{\nu=0}^{\infty} \frac{2\gamma_{a\nu} e^{i\gamma_{a\nu}t}}{\left(1 + \delta_{\nu 0}\right) \left(1 - e^{2i\gamma_{a\nu}t}\right)} G_{\nu p}^+ G_{\nu \iota}^- \right] \tag{28}$$

$$\boldsymbol{K}_x = [k_{x\iota}\delta_{\iota p}], \qquad\qquad \overline{\boldsymbol{K}}_y = [k_{y\iota j}\delta_{\iota p}] \tag{29}$$

$$\tau_1 = \frac{\zeta_{1j}^2 \epsilon_{rs}}{2ah_x \zeta_{sj}^2 \epsilon_{r1}}, \qquad\qquad \tau_2 = \frac{k_{zj}}{2ah_x k_0 \epsilon_{r1}} \left[\frac{\zeta_{1j}^2}{\zeta_{sj}^2} - 1 \right] \tag{30}$$

$$\tau_3 = \frac{k_{zj}}{k_0 \mu_{rs}}, \qquad\qquad \tau_4 = \frac{k_{zj} \epsilon_{rs}}{2ah_x \mu_{rs} \epsilon_{r1} k_0} \tag{31}$$

$$\tau_5 = \frac{k_{zj}^2}{2ah_x \mu_{rs} \epsilon_{r1} k_0^2} \left[\frac{\zeta_{1j}^2}{\zeta_{sj}^2} - 1 \right], \qquad\qquad \tau_6 = \frac{\zeta_{sj}^2 \mu_{r1}}{2ah_x \zeta_{1j}^2 \mu_{rs}} \tag{32}$$

$$\boldsymbol{\delta}(Q) = [\delta_{\iota Q}, \cdots, \delta_{\iota Q}, \cdots, \delta_{\iota Q}]^t \tag{33}$$

By truncating the harmonic number up to $\iota, p = \pm M$, the solution of the linear equations can be written in matrices. Since the equations (19) applies to each incident harmonic, i.e., for any integer $Q \in [-M, M]$, we have a corresponding solution $[\ \boldsymbol{C}_Q^e \quad \boldsymbol{C}_Q^h \quad \boldsymbol{D}_Q^e \quad \boldsymbol{D}_Q^h\ ]^t$. Arranging the column vectors $[\ \boldsymbol{C}_Q^e \quad \boldsymbol{C}_Q^h \quad \boldsymbol{D}_Q^e \quad \boldsymbol{D}_Q^h\ ]^t$ together in a matrix, and letting $\boldsymbol{e}_{zj} = [e_{zj,-M} \quad \cdots \quad e_{zj,0} \quad \cdots \quad e_{zj,M}]^t$ and $\tilde{\boldsymbol{h}}_{zj} = [\tilde{h}_{zj,-M} \quad \cdots \quad \tilde{h}_{zj,0} \quad \cdots \quad \tilde{h}_{zj,M}]^t$ be the column vectors of $(2M+1)$ dimensions, which denote the amplitude of space-harmonics varying as $e^{i(k_{x\iota}x + k_{zj}z)}$, it leads the reflection matrices $_j\overline{\boldsymbol{R}}^\mu$ and transmission matrices $_j\overline{\boldsymbol{T}}^\mu$ ($\mu : ee-, eh-, he-, hh-$) for downgoing incident waves. In the similar manner, the same matrices $_j\overline{\boldsymbol{R}}^\mu$ and $_j\overline{\boldsymbol{T}}^\mu$ ($\mu : ee+, eh+, he+, hh+$) corresponding to upgoing incident waves can be derived. They satisfy the following relations

$$\begin{bmatrix} \boldsymbol{e}_{zj}^r \\ \tilde{\boldsymbol{h}}_{zj}^r \\ \boldsymbol{e}_{zj}^t \\ \tilde{\boldsymbol{h}}_{zj}^t \end{bmatrix} = \begin{bmatrix} _j\overline{\boldsymbol{R}}^{ee-} & _j\overline{\boldsymbol{R}}^{eh-} & _j\overline{\boldsymbol{T}}^{ee+} & _j\overline{\boldsymbol{T}}^{eh+} \\ _j\overline{\boldsymbol{R}}^{he-} & _j\overline{\boldsymbol{R}}^{hh-} & _j\overline{\boldsymbol{T}}^{he+} & _j\overline{\boldsymbol{T}}^{hh+} \\ _j\overline{\boldsymbol{T}}^{ee-} & _j\overline{\boldsymbol{T}}^{eh-} & _j\overline{\boldsymbol{R}}^{ee+} & _j\overline{\boldsymbol{R}}^{eh+} \\ _j\overline{\boldsymbol{T}}^{he-} & _j\overline{\boldsymbol{T}}^{hh-} & _j\overline{\boldsymbol{R}}^{he+} & _j\overline{\boldsymbol{R}}^{hh+} \end{bmatrix} \begin{bmatrix} \boldsymbol{e}_{zj}^{di} \\ \tilde{\boldsymbol{h}}_{zj}^{di} \\ \boldsymbol{e}_{zj}^{ui} \\ \tilde{\boldsymbol{h}}_{zj}^{ui} \end{bmatrix}$$

$$= \begin{bmatrix} \boldsymbol{I} + \overline{\boldsymbol{\Gamma}}_1 & -\overline{\boldsymbol{\Gamma}}_2 & \overline{\boldsymbol{\Omega}}_1 & \overline{\boldsymbol{\Omega}}_2 \\ \overline{\boldsymbol{\Gamma}}_3 & \boldsymbol{I} + \overline{\boldsymbol{\Gamma}}_4 & \overline{\boldsymbol{\Omega}}_3 & -\overline{\boldsymbol{\Omega}}_4 \\ \overline{\boldsymbol{\Omega}}_1 & -\overline{\boldsymbol{\Omega}}_2 & \boldsymbol{I} + \overline{\boldsymbol{\Gamma}}_1 & \overline{\boldsymbol{\Gamma}}_2 \\ -\overline{\boldsymbol{\Omega}}_3 & -\overline{\boldsymbol{\Omega}}_4 & -\overline{\boldsymbol{\Gamma}}_3 & \boldsymbol{I} + \overline{\boldsymbol{\Gamma}}_4 \end{bmatrix}^{-1} \begin{bmatrix} \overline{\boldsymbol{\Gamma}}_1 - \boldsymbol{I} & \overline{\boldsymbol{\Gamma}}_2 & \overline{\boldsymbol{\Omega}}_1 & -\overline{\boldsymbol{\Omega}}_2 \\ \overline{\boldsymbol{\Gamma}}_5 & \boldsymbol{I} - \overline{\boldsymbol{\Gamma}}_4 & \overline{\boldsymbol{\Omega}}_3 & \overline{\boldsymbol{\Omega}}_4 \\ \overline{\boldsymbol{\Omega}}_1 & \overline{\boldsymbol{\Omega}}_2 & \overline{\boldsymbol{\Gamma}}_1 - \boldsymbol{I} & -\overline{\boldsymbol{\Gamma}}_2 \\ -\overline{\boldsymbol{\Omega}}_3 & \overline{\boldsymbol{\Omega}}_4 & -\overline{\boldsymbol{\Gamma}}_5 & \boldsymbol{I} - \overline{\boldsymbol{\Gamma}}_4 \end{bmatrix} \begin{bmatrix} \boldsymbol{e}_{zj}^{di} \\ \tilde{\boldsymbol{h}}_{zj}^{di} \\ \boldsymbol{e}_{zj}^{ui} \\ \tilde{\boldsymbol{h}}_{zj}^{ui} \end{bmatrix}$$

$$\tag{34}$$

If we apply the solution to every sub-systems, we have

$$
\begin{bmatrix} \boldsymbol{e}_z^r \\ \tilde{\boldsymbol{h}}_z^r \\ \boldsymbol{e}_z^t \\ \tilde{\boldsymbol{h}}_z^t \end{bmatrix} = \begin{bmatrix} \overset{\Leftrightarrow ee-}{\mathbf{R}_\parallel} & \overset{\Leftrightarrow eh-}{\mathbf{R}_\parallel} & \overset{\Leftrightarrow ee+}{\mathbf{T}_\parallel} & \overset{\Leftrightarrow eh+}{\mathbf{T}_\parallel} \\ \overset{\Leftrightarrow he-}{\mathbf{R}_\parallel} & \overset{\Leftrightarrow hh-}{\mathbf{R}_\parallel} & \overset{\Leftrightarrow he+}{\mathbf{T}_\parallel} & \overset{\Leftrightarrow hh+}{\mathbf{T}_\parallel} \\ \overset{\Leftrightarrow ee-}{\mathbf{T}_\parallel} & \overset{\Leftrightarrow eh-}{\mathbf{T}_\parallel} & \overset{\Leftrightarrow ee+}{\mathbf{R}_\parallel} & \overset{\Leftrightarrow eh+}{\mathbf{R}_\parallel} \\ \overset{\Leftrightarrow he-}{\mathbf{T}_\parallel} & \overset{\Leftrightarrow hh-}{\mathbf{T}_\parallel} & \overset{\Leftrightarrow he+}{\mathbf{R}_\parallel} & \overset{\Leftrightarrow hh+}{\mathbf{R}_\parallel} \end{bmatrix} \begin{bmatrix} \boldsymbol{e}_z^{di} \\ \tilde{\boldsymbol{h}}_z^{di} \\ \boldsymbol{e}_z^{ui} \\ \tilde{\boldsymbol{h}}_z^{ui} \end{bmatrix} = \begin{bmatrix} \mathfrak{R}_\parallel^- & \mathfrak{S}_\parallel^+ \\ \mathfrak{S}_\parallel^- & \mathfrak{R}_\parallel^+ \end{bmatrix} \begin{bmatrix} \boldsymbol{e}_z^{di} \\ \tilde{\boldsymbol{h}}_z^{di} \\ \boldsymbol{e}_z^{ui} \\ \tilde{\boldsymbol{h}}_z^{ui} \end{bmatrix} \quad (35)
$$

where

$$
\boldsymbol{e}_z = \begin{bmatrix} \boldsymbol{e}_{z-N} & \cdots & \boldsymbol{e}_{z0} & \cdots & \boldsymbol{e}_{zN} \end{bmatrix}^t \quad (36)
$$

$$
\tilde{\boldsymbol{h}}_z = \begin{bmatrix} \tilde{\boldsymbol{h}}_{z-N} & \cdots & \tilde{\boldsymbol{h}}_{z0} & \cdots & \tilde{\boldsymbol{h}}_{zN} \end{bmatrix}^t \quad (37)
$$

$\jmath = \pm N$ denote the orders of truncation of space-harmonics. The dimensions of sub-matrices $\mathfrak{R}_\parallel^\pm$ and $\mathfrak{S}_\parallel^\pm$ are $2(2M+1)(2N+1) \times 2(2M+1)(2N+1)$. The signs $\overset{\Leftrightarrow \mu}{\mathbf{R}_\parallel}$ and $\overset{\Leftrightarrow \mu}{\mathbf{T}_\parallel}$ $(\mu : ee\pm, eh\pm, he\pm, hh\pm)$ denotes the sub-matrices of $(2M+1)(2N+1) \times (2M+1)(2N+1)$ dimensions, whose elements $r_{\parallel mn}^\mu$ and $t_{\parallel mn}^\mu$ are defined as follows:

$$
r_{\parallel mn}^\mu = \begin{cases} {}_\jmath \bar{r}_{\imath p}^\mu & \jmath = q \\ 0 & \jmath \neq q \end{cases} \qquad t_{\parallel mn}^\mu = \begin{cases} {}_\jmath \bar{t}_{\imath p}^\mu & \jmath = q \\ 0 & \jmath \neq q \end{cases} \quad (38)
$$

where ${}_\jmath \bar{r}_{\imath p}^\mu$ and ${}_\jmath \bar{t}_{\imath p}^\mu$ are the elements of the sub-matrices ${}_\jmath \bar{\boldsymbol{R}}^\mu$ and ${}_\jmath \bar{\boldsymbol{T}}^\mu$, respectively. The integer $(\imath, \jmath)$ and (p, q) relies on m and n, respectively. Here we define these relation as follows:

$$
\jmath = \left\lceil \frac{m}{2M+1} \right\rceil - M, \qquad \imath = m - M - 1 - (2M+1)(M+\jmath) \quad (39)
$$

$$
q = \left\lceil \frac{n}{2M+1} \right\rceil - M, \qquad p = n - M - 1 - (2M+1)(M+q) \quad (40)
$$

where $\lceil \rfloor$ denotes integer operator.

2.2 X-array

The x-directed cylinders are periodically spaced with a distance $2h_z$ in the z-direction as shown in Fig. 3. Employing the E_x and $\tilde{H}_x = \eta_0 H_x$ as the leading fields. the scattering problem can be formulated in a similar way as the z-array. The field is characterized in terms of the amplitudes $e_{x,\imath,\jmath}$ and $\tilde{h}_{x,\imath,\jmath}$ of each of space-harmonics. Let $\boldsymbol{e}_x$ and $\tilde{\boldsymbol{h}}_x$ be the column vectors of $(2M+1)(2N+1)$ dimensions defined as

$$
\boldsymbol{e}_x = \begin{bmatrix} e_{x,-M,-N} & e_{x,-M+1,-N} & e_{x,-M+2,-N} & \cdots & e_{x,M,N} \end{bmatrix}^t \quad (41)
$$

$$
\tilde{\boldsymbol{h}}_x = \begin{bmatrix} \tilde{h}_{x,-M,-N} & \tilde{h}_{x,-M+1,-N} & \tilde{h}_{x,-M+2,-N} & \cdots & \tilde{h}_{x,M,N} \end{bmatrix}^t \quad (42)
$$

Then the amplitudes vectors $\boldsymbol{e}_x$ and $\tilde{\boldsymbol{h}}_x$ satisfy the following relations

$$
\begin{bmatrix} \boldsymbol{e}_x^r \\ \tilde{\boldsymbol{h}}_x^r \\ \boldsymbol{e}_x^t \\ \tilde{\boldsymbol{h}}_x^t \end{bmatrix} = \begin{bmatrix} \overset{\Leftrightarrow ee-}{\mathbf{R}_x} & \overset{\Leftrightarrow eh-}{\mathbf{R}_x} & \overset{\Leftrightarrow ee+}{\mathbf{T}_x} & \overset{\Leftrightarrow eh+}{\mathbf{T}_x} \\ \overset{\Leftrightarrow he-}{\mathbf{R}_x} & \overset{\Leftrightarrow hh-}{\mathbf{R}_x} & \overset{\Leftrightarrow he+}{\mathbf{T}_x} & \overset{\Leftrightarrow hh+}{\mathbf{T}_x} \\ \overset{\Leftrightarrow ee-}{\mathbf{T}_x} & \overset{\Leftrightarrow eh-}{\mathbf{T}_x} & \overset{\Leftrightarrow ee+}{\mathbf{R}_x} & \overset{\Leftrightarrow eh+}{\mathbf{R}_x} \\ \overset{\Leftrightarrow he-}{\mathbf{T}_x} & \overset{\Leftrightarrow hh-}{\mathbf{T}_x} & \overset{\Leftrightarrow he+}{\mathbf{R}_x} & \overset{\Leftrightarrow hh+}{\mathbf{R}_x} \end{bmatrix} \begin{bmatrix} \boldsymbol{e}_x^{di} \\ \tilde{\boldsymbol{h}}_x^{di} \\ \boldsymbol{e}_x^{ui} \\ \tilde{\boldsymbol{h}}_x^{ui} \end{bmatrix} \quad (43)
$$

where the elements $r^\mu_{x,mn}$ of matrices $\overset{\Leftrightarrow\mu}{\mathbf{R}}_x$ (μ : $ee\pm$, $eh\pm$, $he\pm$, $hh\pm$) and the elements $t^\mu_{x,mn}$ of matrices $\overset{\Leftrightarrow\mu}{\mathbf{T}}_x$ are defined as

$$r^\mu_{x,mn} = \begin{cases} {}_ir^\mu_{jq} & i = p \\ 0 & i \neq p \end{cases}, \qquad t^\mu_{x,mn} = \begin{cases} {}_it^\mu_{jq} & i = p \\ 0 & i \neq p \end{cases} \tag{44}$$

${}_ir^\mu_{jq}$ and ${}_it^\mu_{jq}$ are the elements of sub-matrices ${}_i\boldsymbol{R}^\mu$ and ${}_i\boldsymbol{T}^\mu$, which are defined as follows:

$$\begin{bmatrix} {}_i\boldsymbol{R}^{ee-} & {}_i\boldsymbol{R}^{eh-} & {}_i\boldsymbol{T}^{ee+} & {}_i\boldsymbol{T}^{eh+} \\ {}_i\boldsymbol{R}^{he-} & {}_i\boldsymbol{R}^{hh-} & {}_i\boldsymbol{T}^{he+} & {}_i\boldsymbol{T}^{hh+} \\ {}_i\boldsymbol{T}^{ee-} & {}_i\boldsymbol{T}^{eh-} & {}_i\boldsymbol{R}^{ee+} & {}_i\boldsymbol{R}^{eh+} \\ {}_i\boldsymbol{T}^{he-} & {}_i\boldsymbol{T}^{hh-} & {}_i\boldsymbol{R}^{he+} & {}_i\boldsymbol{R}^{hh+} \end{bmatrix}$$

$$= \begin{bmatrix} \boldsymbol{I}+\underline{\Gamma}_1 & -\underline{\Gamma}_2 & \underline{\Omega}_1 & \underline{\Omega}_2 \\ \underline{\Gamma}_3 & \boldsymbol{I}+\underline{\Gamma}_4 & \underline{\Omega}_3 & -\underline{\Omega}_4 \\ \underline{\Omega}_1 & -\underline{\Omega}_2 & \boldsymbol{I}+\underline{\Gamma}_1 & \underline{\Gamma}_2 \\ -\underline{\Omega}_3 & -\underline{\Omega}_4 & -\underline{\Gamma}_3 & \boldsymbol{I}+\underline{\Gamma}_4 \end{bmatrix}^{-1} \begin{bmatrix} \underline{\Gamma}_1-\boldsymbol{I} & \underline{\Gamma}_2 & \underline{\Omega}_1 & -\underline{\Omega}_2 \\ \underline{\Gamma}_5 & \boldsymbol{I}-\underline{\Gamma}_4 & \underline{\Omega}_3 & \underline{\Omega}_4 \\ \underline{\Omega}_1 & \underline{\Omega}_2 & \underline{\Gamma}_1-\boldsymbol{I} & -\underline{\Gamma}_2 \\ -\underline{\Omega}_3 & \underline{\Omega}_4 & -\underline{\Gamma}_5 & \boldsymbol{I}-\underline{\Gamma}_4 \end{bmatrix},$$

$$\tag{45}$$

where

$$\underline{\Gamma}_1 = \chi_1 \boldsymbol{U}\boldsymbol{K}_y, \qquad\qquad \underline{\Gamma}_2 = \chi_2 \boldsymbol{U}\boldsymbol{K}_z \tag{46}$$

$$\underline{\Gamma}_3 = \boldsymbol{K}_z\boldsymbol{K}_y^{-1}\left(\chi_3\boldsymbol{I} + \chi_4\boldsymbol{U}\boldsymbol{K}_y\right), \qquad \underline{\Gamma}_4 = \boldsymbol{K}_z\boldsymbol{K}_y^{-1}\left(\chi_5\boldsymbol{U} + \chi_6\boldsymbol{W}\right)\boldsymbol{K}_z \tag{47}$$

$$\underline{\Gamma}_5 = \boldsymbol{K}_z\boldsymbol{K}_y^{-1}\left(\chi_4\boldsymbol{U}\boldsymbol{K}_y - \chi_3\boldsymbol{I}\right) \tag{48}$$

$$\underline{\Omega}_1 = \chi_1\boldsymbol{V}\boldsymbol{K}_y, \qquad\qquad \underline{\Omega}_2 = \chi_2\boldsymbol{V}\boldsymbol{K}_z \tag{49}$$

$$\underline{\Omega}_3 = \chi_4\boldsymbol{K}_x\boldsymbol{K}_y^{-1}\boldsymbol{V}\boldsymbol{K}_y, \qquad \underline{\Omega}_4 = \boldsymbol{K}_z\boldsymbol{K}_y^{-1}\left(\chi_5\boldsymbol{V} + \chi_6\boldsymbol{X}\right)\boldsymbol{K}_z \tag{50}$$

$$\boldsymbol{U} = [\underline{u}_{jq}] = \left[\sum_{\nu=1}^\infty \frac{\beta_\nu^2\left(1 + e^{2i\gamma_{b\nu}s}\right)}{\gamma_{b\nu}\left(1 - e^{2i\gamma_{b\nu}s}\right)}F^+_{\nu q}F^-_{\nu j}\right] \tag{51}$$

$$\boldsymbol{V} = [\underline{v}_{jq}] = \left[\sum_{\nu=1}^\infty \frac{2\beta_\nu^2 e^{i\gamma_{b\nu}s}}{\gamma_{b\nu}\left(1 - e^{2i\gamma_{b\nu}s}\right)}F^+_{\nu q}F^-_{\nu j}\right] \tag{52}$$

$$\boldsymbol{W} = [\underline{w}_{jq}] = \left[\sum_{\nu=0}^\infty \frac{\gamma_{b\nu}\left(1 + e^{2i\gamma_{b\nu}s}\right)}{\left(1 + \delta_{\nu0}\right)\left(1 - e^{2i\gamma_{b\nu}s}\right)}F^+_{\nu q}F^-_{\nu j}\right] \tag{53}$$

$$\boldsymbol{X} = [\underline{x}_{jq}] = \left[\sum_{\nu=0}^\infty \frac{2\gamma_{b\nu}e^{i\gamma_{b\nu}s}}{\left(1 + \delta_{\nu0}\right)\left(1 - e^{2i\gamma_{b\nu}s}\right)}F^+_{\nu q}F^-_{\nu j}\right] \tag{54}$$

$$\boldsymbol{K}_z = [k_{zj}\delta_{jq}], \qquad\qquad \boldsymbol{K}_y = [k_{yj}\delta_{jq}] \tag{55}$$

$$F^\pm_{\nu j} = \frac{1}{k_{zj}^2 - \beta_\nu^2}\left[(-1)^\nu e^{\pm ik_{zj}b} - e^{\mp ik_{zj}b}\right] \tag{56}$$

$$\chi_1 = \frac{\xi_{2i}^2\epsilon_{rs}}{2bh_z\xi_{si}^2\epsilon_{r2}}, \qquad\qquad \chi_2 = \frac{k_{xi}}{2bh_zk_0\epsilon_{r2}}\left[\frac{\xi_{2i}^2}{\xi_{si}^2} - 1\right] \tag{57}$$

$$\chi_3 = \frac{k_{xi}}{k_0\mu_{rs}}, \qquad\qquad \chi_4 = \frac{k_{xi}\epsilon_{rs}}{2bh_z\mu_{rs}\epsilon_{r2}k_0} \tag{58}$$

$$\chi_5 = \frac{k_{xi}^2}{2bh_z\mu_{rs}\epsilon_{r2}k_0^2}\left[\frac{\xi_{2i}^2}{\xi_{si}^2} - 1\right] \qquad \chi_6 = \frac{\xi_{si}^2\mu_{r2}}{2bh_z\xi_{2i}^2\mu_{rs}} \tag{59}$$

$$\xi_{2i} = \sqrt{k_2^2 - k_{xi}^2}, \qquad \xi_{si} = \sqrt{k_s^2 - k_{xi}^2} \tag{60}$$

$$\gamma_{b\nu} = \sqrt{k_2^2 - \beta_\nu^2}, \qquad \beta_\nu = \frac{\nu\pi}{2b} \tag{61}$$

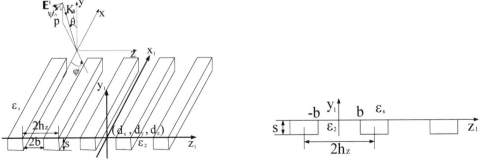

(a) Three-dimensional geometry   (b) Cross-section view at the plane $x = d_x$

Fig.3   An array of x-directed parallel cylinders periodically spaced with a distance $2h_z$ in the z-direction.

2.3   Crossed-array

Figure 4 shows a *cross-array* in which the *Z-array* and *X-array* are doubled with a center to center separation of w in the y direction. The *Z-array* and *X-array* in the preceding sections have been treated separately by taking the electric and magnetic fields parallel to the cylinder axis as the leading fields. When they are layered, the use of common leading fields is requested to describe the multiple scattering process between the two crossed arrays in unified manner. We choose the E_z and $\tilde{H}_z$ fields as common leading fields for the crossed array and express the reflection and transmission matrices of *X-array* based on these leading fields. By a simple calculation, we have

$$\begin{bmatrix} \mathfrak{R}_{l\perp}^- & \mathfrak{S}_{l\perp}^+ \\ \mathfrak{S}_{l\perp}^- & \mathfrak{R}_{l\perp}^+ \end{bmatrix} = \begin{bmatrix} \overset{\Leftrightarrow ee-}{\mathbf{R}_\perp} & \overset{\Leftrightarrow eh-}{\mathbf{R}_\perp} & \overset{\Leftrightarrow ee+}{\mathbf{T}_\perp} & \overset{\Leftrightarrow eh+}{\mathbf{T}_\perp} \\ \overset{\Leftrightarrow he-}{\mathbf{R}_\perp} & \overset{\Leftrightarrow hh-}{\mathbf{R}_\perp} & \overset{\Leftrightarrow he+}{\mathbf{T}_\perp} & \overset{\Leftrightarrow hh+}{\mathbf{T}_\perp} \\ \overset{\Leftrightarrow ee-}{\mathbf{T}_\perp} & \overset{\Leftrightarrow eh-}{\mathbf{T}_\perp} & \overset{\Leftrightarrow ee+}{\mathbf{R}_\perp} & \overset{\Leftrightarrow eh+}{\mathbf{R}_\perp} \\ \overset{\Leftrightarrow he-}{\mathbf{T}_\perp} & \overset{\Leftrightarrow hh-}{\mathbf{T}_\perp} & \overset{\Leftrightarrow he+}{\mathbf{R}_\perp} & \overset{\Leftrightarrow hh+}{\mathbf{R}_\perp} \end{bmatrix} = \begin{bmatrix} \mathbf{A}_z & -\mathbf{B}_z & 0 & 0 \\ \mathbf{C}_z & \mathbf{A}_z & 0 & 0 \\ 0 & 0 & \mathbf{A}_z & \mathbf{B}_z \\ 0 & 0 & -\mathbf{C}_z & \mathbf{A}_z \end{bmatrix} \times $$

$$\begin{bmatrix} \overset{\Leftrightarrow ee-}{\mathbf{R}_x} & \overset{\Leftrightarrow eh-}{\mathbf{R}_x} & \overset{\Leftrightarrow ee+}{\mathbf{T}_x} & \overset{\Leftrightarrow eh+}{\mathbf{T}_x} \\ \overset{\Leftrightarrow he-}{\mathbf{R}_x} & \overset{\Leftrightarrow hh-}{\mathbf{R}_x} & \overset{\Leftrightarrow he+}{\mathbf{T}_x} & \overset{\Leftrightarrow hh+}{\mathbf{T}_x} \\ \overset{\Leftrightarrow ee-}{\mathbf{T}_x} & \overset{\Leftrightarrow eh-}{\mathbf{T}_x} & \overset{\Leftrightarrow ee+}{\mathbf{R}_x} & \overset{\Leftrightarrow eh+}{\mathbf{R}_x} \\ \overset{\Leftrightarrow he-}{\mathbf{T}_x} & \overset{\Leftrightarrow hh-}{\mathbf{T}_x} & \overset{\Leftrightarrow he+}{\mathbf{R}_x} & \overset{\Leftrightarrow hh+}{\mathbf{R}_x} \end{bmatrix} \begin{bmatrix} \mathbf{A}_x & -\mathbf{B}_x & 0 & 0 \\ \mathbf{C}_x & \mathbf{A}_x & 0 & 0 \\ 0 & 0 & \mathbf{A}_x & \mathbf{B}_x \\ 0 & 0 & -\mathbf{C}_x & \mathbf{A}_x \end{bmatrix} \tag{62}$$

where

$$\mathbf{A}_z = [A_{z,mn}] = \left[\frac{k_{xi}k_{zj}}{\zeta_{sj}^2} \delta_{mn} \right], \quad \mathbf{B}_z = [B_{z,mn}] = \left[\frac{k_{yij}k_0\mu_{sr}}{\zeta_{sj}^2} \delta_{mn} \right] \tag{63}$$

$$\mathbf{C}_z = [C_{z,mn}] = \left[\frac{k_{yij}k_0\epsilon_{sr}}{\zeta_{sj}^2} \delta_{mn} \right], \quad \mathbf{A}_x = [A_{x,mn}] = \left[\frac{k_{xi}k_{zj}}{\xi_{si}^2} \delta_{mn} \right] \tag{64}$$

$$\boldsymbol{B}_x = [B_{x,mn}]\left[\frac{k_{ylj}k_0\mu_{sr}}{\xi_{si}^2}\delta_{mn}\right], \quad \boldsymbol{C}_x = [C_{x,mn}] = \left[\frac{k_{ylj}k_0\epsilon_{sr}}{\xi_{si}^2}\delta_{mn}\right] \tag{65}$$

Now the reflection and transmission matrices of the Z-array and X-array have been obtained as (35) and (62) in terms of the common leading fields E_z and $\tilde{H}_z$. Since the reflection and transmission matrices are in a local coordinate, it is necessary to remove them to the globe coordinate. We assume that the origin of the local coordinate for X-array are located at (d_x, d_y, d_z) in the globe coordinate.

$$\begin{bmatrix} \mathfrak{R}_\perp^- & \mathfrak{S}_\perp^+ \\ \mathfrak{S}_\perp^- & \mathfrak{R}_\perp^+ \end{bmatrix} = \begin{bmatrix} \Theta^{-1} & & & 0 \\ & \Theta^{-1} & & \\ & & \Theta^{-1} & \\ 0 & & & \Theta^{-1} \end{bmatrix} \begin{bmatrix} \mathfrak{R}_{l\perp}^- & \mathfrak{S}_{l\perp}^+ \\ \mathfrak{S}_{l\perp}^- & \mathfrak{R}_{l\perp}^+ \end{bmatrix} \begin{bmatrix} \Theta & & & 0 \\ & \Theta & & \\ & & \Theta & \\ 0 & & & \Theta \end{bmatrix} \tag{66}$$

where

$$\Theta = [\Theta_{mn}] = \left[e^{i[k_{xi}d_x + k_{zj}d_z]}\delta_{mn}\right]. \tag{67}$$

The electromagnetic behaviors of the space between Z-array and X-array can be described by following relation.

$$\begin{bmatrix} e_z^r \\ \tilde{h}_z^r \\ e_z^t \\ \tilde{h}_z^t \end{bmatrix} \begin{bmatrix} 0 & 0 & \Xi & 0 \\ 0 & 0 & 0 & \Xi \\ \Xi & 0 & 0 & 0 \\ 0 & \Xi & 0 & 0 \end{bmatrix} \begin{bmatrix} e_z^{di} \\ \tilde{h}_z^{di} \\ e_z^{ui} \\ \tilde{h}_z^{ui} \end{bmatrix} \tag{68}$$

where

$$\Xi = [\Xi_{mn}] = \left[e^{ik_{ylj}[w-(t+s)/2]}\delta_{mn}\right]. \tag{69}$$

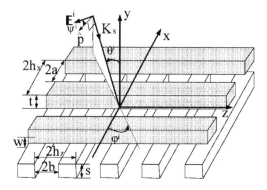

Fig.4 A *Cross-array* in which the *Z-array* and *X-array* are doubled with a center to center separation of w in the y-direction.

2.4 Recursive process for calculating generalized reflection and transmission matrices

We have derived the reflection and transmission matrices in terms of the common leading fields E_z and $\tilde{H}_z$ for characterizing the *Z-array*, the *X-array*, and the space between

Z-array and *X-array*. According to circuit theory, the reflection and transmission matrices of multilayered *cross-arrays* can be calculated by a recursive process. Assume the matrices of $(\tilde{\mathfrak{R}}_L^{\pm}, \tilde{\mathfrak{S}}_L^{\pm})$ denote the generalized reflection and transmission matrices for L-layers, who includes *Z-array* layers, *X-array* layers, and space layers, and the matrices of $(\mathfrak{R}_{L+1}^{\pm}, \mathfrak{S}_{L+1}^{\pm})$ denote alike matrices for the $(L+1)$-th layer. Then a recursive relations to determine the generalized reflection and transmission matrices of $(L+1)$ layers is expressed as follows:

$$
\begin{bmatrix} \tilde{\mathfrak{R}}_{L+1}^{-} & \tilde{\mathfrak{S}}_{L+1}^{+} \\ \tilde{\mathfrak{S}}_{L+1}^{-} & \tilde{\mathfrak{R}}_{L+1}^{+} \end{bmatrix} = \begin{bmatrix} \tilde{\mathfrak{R}}_L^{-} + \aleph \mathfrak{R}_{L+1}^{-} \tilde{\mathfrak{S}}_L^{-} & \aleph \mathfrak{S}_{L+1}^{+} \\ \wp \tilde{\mathfrak{S}}_L^{-} & \mathfrak{R}_{L+1}^{+} + \wp \tilde{\mathfrak{R}}_L^{+} \mathfrak{S}_{L+1}^{+} \end{bmatrix} \tag{70}
$$

where

$$
\aleph = \tilde{\mathfrak{S}}_L^{+} \left[\boldsymbol{I} - \mathfrak{R}_{L+1}^{-} \tilde{\mathfrak{R}}_L^{+} \right]^{-1}, \qquad \wp = \mathfrak{S}_{L+1}^{-} \left[\boldsymbol{I} - \tilde{\mathfrak{R}}_L^{+} \mathfrak{R}_{L+1}^{-} \right]^{-1} \tag{71}
$$

3   Numerical Examples

Although the notations used are a little intricate because of three-dimensional problems, the mathematics for formulation of the proposed method is straightforward. By considering the orthogonality and completeness of the space harmonics, the multilayered system can be characterized by using the independent solutions to each of layers in isolation. For the same reason, the whole system of space harmonics to crossed-array bigratings can be reduced to orthogonal mono-periodic systems in each layer. Therefore, this method can be applied to the multilayered crossed-arrays in smaller computer cost. Although the dimension of reflection and transmission matrices of the crossed-arrays is large as total, the number of the necessary sub-matrices to be calculated is very small and their dimension is also small. This is main difference to the classical methods. Hence, the increase of the number of unknowns with increase of layers is not appear. It is noted that the series sums with respect to variable ν in Eqs.(25)-(28) and Eqs.(51)-(54) converges with the rate faster than $O(\alpha_\nu^{-3})$ or $O(\beta_\nu^{-3})$.

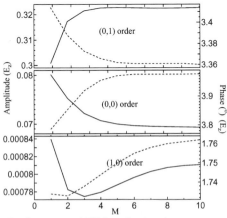

(a) Reflection of TM polarization

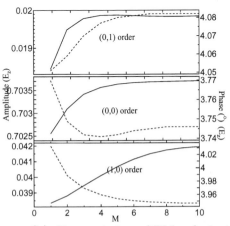

(b) Transmission of TM polarization

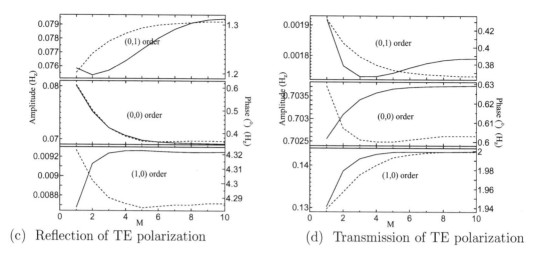

(c) Reflection of TE polarization   (d) Transmission of TE polarization

Fig.5 Convergence of the amplitudes and the phases of (0,1)-th, (0,0)-th, and (1,0)-th order harmonics as functions of the truncated number $M = N$ where $\psi^i = 45°$.

To test the convergence behavior, the amplitudes and phases of (0,1)-th, (0,0)-th, and (1,0)-th order harmonics are shown in Fig.5 as functions of the truncated number M, N for a crossed-array with $h_x = h_z = 0.45\lambda_0$, $a = b = 0.9h_x$, $t = s = 0.6h_x$, $d_x = d_z = 0$, $w = 2h_x$, $\theta^i = 0°$ $\phi^i = 180°$, $\psi^i = 45°$, and $\epsilon_{r1} = \epsilon_{r2} = \epsilon_{rs} = \mu_{r1} = \mu_{r2} = \mu_{rs} = 1$. In this figure, the solid lines denote the amplitudes and the dash lines the phases. The numerical results show a very good convergence is achieved for every space harmonics in both of TM polarization and TE polarization. In this case, all the space harmonics expect (0,0)-th order harmonic are evanescent wave.

In the next example, we shall investigate the property varying as the incident angle. Fig.6 shows computed results as function of incident angle θ^i for crossed-arrays with $h_x = h_z = 0.15\lambda_0$, $a = b = 0.9h_x$, $t = s = 0.6h_x$, $d_x = d_z = 0$, $w = 2h_x$, $\phi^i = 180°$, $\psi^i = 0°$, $\epsilon_{r1} = \epsilon_{r2} = \epsilon_{rs} = \mu_{r1} = \mu_{r2} = \mu_{rs} = 1$, and the truncation number is set $M = N = 6$. The curves show that the variation is very smaller with incident angle θ^i for one set crossed-array, whereas some interesting property appear when the crossed-arrays are stacked in multilayers. When the incident angle is less than $30°$ or more than $42.5°$, the incident wave are completely reflected into its own region, when the incident angle is set around $31.2°$, the wave direct propagates almost without any obstruct, moreover it is very sharp. The same results can obtained for $\psi^i = 90°$ due to the $x - z$ symmetry of the crossed-array.

Finally, we shall investigate the description of one set and 16 sets of crossed-arrays with $h_x = h_z$, $a = b = 0.9h_x$, $t = s = 0.6h_x$, $d_x = d_z = 0$, $w = 2.4h_x$, $\theta^i = 0°$, $\phi^i = 90°$, $\psi^i = 45°$, $\epsilon_{r1} = \epsilon_{r2} = \epsilon_{rs} = \mu_{r1} = \mu_{r2} = \mu_{rs} = 1$, and the truncation number is set $M = N = 6$. Fig.7 shows a typical spectrum of crossed arrays. Figure (a) is the basic property of one set of crossed-array. When they are stacked, we can obtain four PBG band and a very sharp complete transmittance. These characters show that the crossed-arrays can be used as PBG materials, filter, and resonance. Although we only show a few examples, optimal transmission, resonance, and PBG characteristics can be obtained with simple square-symmetric structures, made up of narrow and deep rectangular cylinders, and the relative permittivity and the relative permeability from those examples.

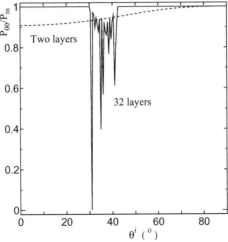

Fig.6 Reflection varying as the incident angle θ^i.

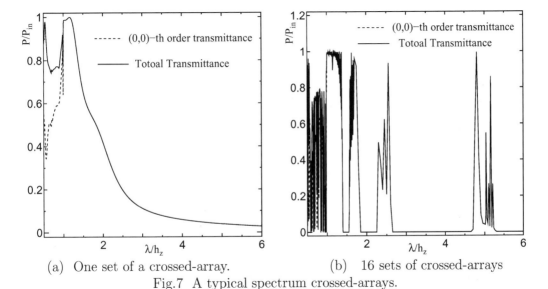

(a) One set of a crossed-array.   (b) 16 sets of crossed-arrays

Fig.7 A typical spectrum crossed-arrays.

4 Concluding Remarks

A rigorous approach for analyzing three-dimensional electromagnetic scattering from multilayered crossed-arrays of conductive rectangular cylinders using the generalized reflection and transmission matrices. The reflection and transmission matrices are separately calculated for each of the two arrays by employing the electric and magnetic fields parallel to cylinder axes as the leading fields. The numerical examples have confirmed fast convergence and high accuracy of calculating solutions with smaller truncation number. This method can be easily extended to the case of the crossed-arrays consisting of conductive cylinders with arbitrary cross section.

Acknowledgment   This work was supported in part by the 2001 Research Grant of Hoso-Bunka Foundation, Japan.

REFERENCES

[1] V. Twersky, "On scattering of waves by the infinite grating of circular cylinders," *IRE Trans. Antennas Propagat.,* Vol.10, pp.737-765, 1962

[2] K. Otaka and N. Numata, "Multiple scattering effects in photon diffraction for an array of cylindrical dielectric," *Phys. Lett.,* Vol.73a, No.5,6, pp.411-413, 1979

[3] R. Petit, Ed., *Electromagnetic Theory of Gratings,* Springer-Verlag, 1980

[4] K. Ohtaka, T. Ueta, and K. Amemiya, "Calculation of photonic bands using vector cylindrical waves and reflectivity of light for an array of dielectric rods, *Phys. Rev., B,* Vol.57, No.4, pp.2550-2568, 1998

[5] J. D. Joannopoulos, R. D. Meade and J. N. Winn, *Photonic Crystals: Molding the Flow of Light,* Princeton University Press, 1995

[6] S. Kazuaki, *Optical Properties of Photonic Crystals,* Springer-Verlag, 2001

[7] M. Plihal and A. A. Maradudin, "Photonic band structure of two-dimensional systems: The triangular lattice," *Phys. Rev., B,* Vol.44, No.16, pp.8565-8571, 1991

[8] H. Benisty, "Modal analysis of optical guides with two-dimensional photonic bandgap boundaries," *J. Appl. Phys.,* Vol.79, No.10, pp.7483-7492, 1996

[9] G. Pelosi, A. Cocchi, and A. Monorchio, "A hybrid FEM-based procedure for the scattering from photonic crystals illuminated by a Gaussian beam," *IEEE. Trans., Antennas Propagat.,* Vol.45, No.1, pp.185-186, 1997

[10] E. Popov and B. Bozhkov, "Differential method applied for photonic crystals," *Appl. Opt.,* Vol.39, No.27, pp.4926-4932, 2000

[11] Y. Naka and H. Ikuno, "Guided mode analysis of two-dimensional air-hole type photonic crystal optical waveguides," *IEICE Tech. Rep.,* EMT-00-78, pp.75-80, 2000

[12] T. Kushta and K. Yasumoto, "Electromagnetic scattering from periodic arrays of two circular cylinders per unit cell," *Progress In Electromagnetics Research,* PIER 29, pp.69-85, 2000

[13] K. Yasumoto, "Generalized method for electromagnetic scattering by two-dimensional periodic discrete composites using lattice sums," *Proceedings of 2000 International Conference on Microwave and Millimeter Wave Technology,* pp.29-34, 2000

[14] K. Yasumoto and H. Jia, "Eigenmode fields in two-dimensionally periodic arrays of circular cylinders," *Proceedings of Progress In Electromagnetics Research Symposium,* pp. 254, 2003

[15] H. Roussel, W. C. Chew, F. Jouvie, and W. Tabbara, "Electromagnetic Scattering from dielectric and magnetic gratings of fibers – a T-matrix solution," *J. Electromag. Wave & Applications,* Vol.10, No.1, pp.109-127, 1996

[16] H. Jia, K. Yasumoto and K. Yoshitomi, "Fast and efficient analysis of inset dielectric guide using Fourier transform technique with a modified perfectly matched boundary," *Progress In Electromagnetics Research,* PIER 34, pp.143-163, 2001

Wave Propagation,Scattering and Emission in Complex Media
Edited by Ya-Qiu Jin
Science Press and World Scientific,2004

Vector Models of Non-stable and Spatially-distributed Radar Objects

A.Surkov[1],V.A.Khluson[1],L.Ligthart[2],G.Sharygin[1]

[1] Tomsk State University of Control Systems and Radioelectronics
40 Lenin ave., Tomsk, 634050, Russia. E-mail: gssh@mail.tomsknet.ru
[2] International Research Center of Telecommunication-Transmission and Radar,
Delft University of Technology.
Mekelweg, 4, Delft, the Netherlands. E-mail: L.P.Ligthart@its.tudelft.nl

Abstract Complicated orthogonal signals should be used to make a correct joint estimation of the back scattering matrix (BSM) and co-ordinate parameters of a non-stable radar object [1, 2]. It is necessary to investigate the back-scattering process corrected with the object co-ordinates in case of different signal forms are considered. As the full analytic description of a compound radar target is not possible, the digital modeling of vector signals scattered by spatially distributed objects is needed.

Methods and results of computer simulation of non-stable spatially-distributed objects presented by an aggregate of independent elementary reflectors are reported in this paper. The statistical characteristics of polarization properties and motion parameters of the elementary reflectors can be chosen arbitrary. This allows to modelling real complex radar objects. The possibility to use these models in order to analyze algorithms needed for the correct BSM of real objectsis shown. The algorithm described in the paper is based on utilizating complex orthogonal LFM and PCM signals.

1. Physical model of a spatially-distributed object and its mathematical description

A spatially-distributed radar object in general case can be presented as a set of spaced elementary reflectors of electromagnetic field. As a spaced elementary reflector we'll consider a spaced heterogeneity, which back-scattered field is characterized by one "bright" point in the used frequency band. The elementary reflectors can have arbitrary polarization properties; their dimensions are much smaller than the interval of the radar space resolution. Taking into consideration all this, the algorithm of the spatially-distributed object mathematical description can be as shown in Fig. 1.

Operators S_i describe polarization properties of each of N elementary reflectors, the set of which makes spatially-distributed radar object. The delay lines with delay time $\tau_i(t)$ imitate different length of paths between different elementary reflectors and the sight of illumination and reception. Dependence of τ_i on time imitates motion of an elementary reflector with the respect to the radar antenna. In order to describe the spatially-distributed object more detailed, it is necessary to take into consideration the rotation of the i-th element own basis relative to the basis of the sounding signal $U_o(t)$. To take such rotations into consideration one should present an operator S_i in the multiplicative form:

$$\mathbf{S}_i = \tilde{\mathbf{L}}_i \cdot \mathbf{S}_{oi} \cdot \mathbf{L}_i, \tag{1}$$

where L_i is an operator of a rotation group, showing the transfer from the description in the own i-th reflector basis to the basis of the sounding signal $U_o(t)$. The symbol (~) means the operation of conjugation.

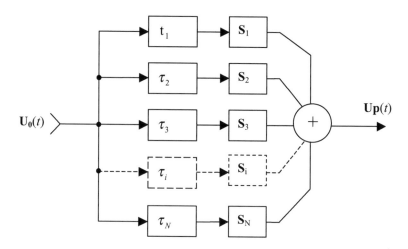

Fig. 1. A model of the spatially-distributed radar object

The operator L_i is fully defined by two parameters: ε_i – ellipticity angle of the own i-th reflector basis, θ_i – angle of this basis orientation. In general case the operator L_i looks like:

$$\mathbf{L}_i = \begin{pmatrix} \cos\varepsilon_i & j\sin\varepsilon_i \\ j\sin\varepsilon_i & \cos\varepsilon_i \end{pmatrix} \cdot \begin{pmatrix} \cos\theta_i & -\sin\theta_i \\ \sin\theta_i & \cos\theta_i \end{pmatrix} = F(\varepsilon_i) \cdot R(\theta_i) \tag{2}$$

The operator S_{oi} in (3.1) describes polarization properties of the elementary reflector in its own basis, for reciprocal medium it has diagonal (canonic) form:

$$\mathbf{S}_{0i} = \begin{pmatrix} \lambda_{1i} & 0 \\ 0 & \lambda_{2i} \end{pmatrix} \tag{3}$$

where $\lambda_{1(2)i}$ are eigen-values of the operator S_i in (1). We don't consider changes of eigen-values $\lambda_{1(2)i}$ in time – that means the reflection coefficients of waves of elementary reflector own polarization $\lambda_{1(2)i}$ don't change (with accuracy up to absolute phase on the time of sounding). The back-scattered field, described by the vector signal $\mathbf{U}_p(t)$, is a combination of fields scattered by elementary reflectors.

2. Response function of a spatially-distributed object

Due to the small size of an elementary reflector and narrow band of the most radar signals, the scattering function of a spatially-distributed object can be written as follows:

$$g_i(t) = \delta(t') \cdot \mathbf{S}_i = \delta(t') \cdot A_i \cdot \mathbf{S}_i^N \tag{4}$$

where $\delta(t')$ is delta-function at the point $t = t'$, $?_i$ – weighting coefficient, defining RCS ("electrical size") of an elementary reflector, the operator $\mathbf{S}_i = ?_i \cdot s_i^N$ describes polarization properties of an elementary reflector. We also consider the linear dimensions of an elementary reflector BSM normalized to the norm of the operator S_i. The signal reflected by an elementary reflector is in general case a convolution:

$$U_p(t) = \delta(t') \cdot S_i \cdot U_0(t) = S_i \cdot U_0(t - t')$$ (5)

Let's divide the volume of back-scattering into K zones, each of them has N elementary reflectors in the plane of the instant wave phase front at the same distance from the point of sounding. In case of single scattering the scattering function of N x K elementary reflectors, situated in the modeled volume of scattering, looks like:

$$g(t) = \sum_{j=1}^{K} \sum_{i=1}^{N} g_{ij}(t) = \sum_{j=1}^{K} \sum_{i=1}^{N} \delta(t_{ij}) \cdot S_{ij}$$ (6)

where j - the number of a zone, i - the number of an elementary reflector in the j-th zone. Geometry of the scattering volume of such a model is shown in Fig. 3.2, where V_{ij} – is a radial component of the elementary reflector Doppler velocity.

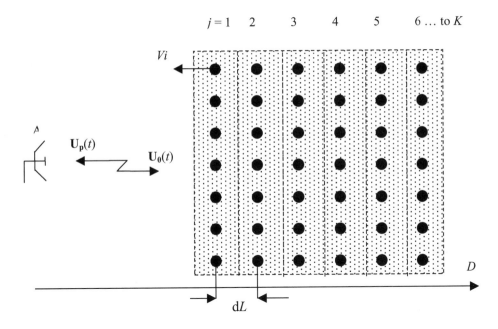

Fig.2. Geometry of a spatially-distributed radar object

When the spatial step dL (see Fig.2) is smaller than the radar resolution by range, the scattering function of an j-th zone looks like:

$$g_j(t) = \delta(t_j) \cdot \sum_{i=1}^{N} S_i \cdot \exp\{j\Omega_i t\}$$ (7)

and the scattering function of the whole modeled volume

$$g_\Sigma(t) = \sum_{j=1}^{K} g_j(t) = \sum_{j=1}^{K} \delta(t_j) \cdot \sum_{i=1}^{N} S_{ij} \cdot \exp\{j\Omega_{ij} t\}$$ (8)

3. Model of the vector signal scattered by a spatially-distributed object

So, the scattering properties of the model were defined by statistics of polarization characteristics and radial velocities of elementary reflectors, combining a radar object. There has been used the following set of parameters for each j-th strobe:

ε_i - angle of ellipticity of the own i-th elementary reflector BSM;

θ_i - angle of orientation of the own basis of the i-th elementary reflector BSM;

ϕ_i - the initial phase of the i-th elementary reflector BSM;

$\Delta\phi_i$ - phase difference of the eigen values of the i-th elementary reflector BSM;

λ_{1i} - eigen (normalized) value of the i-th elementary reflector BSM;

$?_i$ – amplitude coefficient of the i-th elementary reflector BSM;

V_i – the projection of the i-th elementary reflector velocity.

The second eigen (normalized) value λ_{2i} of the i-th elementary reflector BSM is connected with the value λ_{1i} by relation $|\lambda_{2i}|=\sqrt{1-|\lambda_{1i}|^2}$. The norm of the i-th elementary reflector BSM is assumed equal unity. Distributions of values $\varepsilon_i,\theta_i,\phi_i,\Delta\phi_i,\lambda_{1i},?_i,V_i$ were chosen as MathCAD allowed (normal, uniform, exponential, χ^2 and others). Generator of random values forming statistics of values ε_i, θ_i, ϕ_i, $\Delta\phi_i$, λ_{1i}, $?_i$, V_i, renewed at every strobe. It made the "instant" scattering parameters inside different strobes independent from each other. Number (i) of elementary reflectors in every j-th strobe is changeable in wide limits (1-500), the number of strobes is unlimited. Maximum of values ij is defined by the computer power. We used processor Athlon XP 1500+, that allowed us to make BSM calculations for $i = 500$ and $j = 50$, consuming about 10 min.

Signal scattered by a spatially-distributed object, according to Fig. 2, is defined by the convolution:

$$\mathbf{U}_p(t) = g_\Sigma(t)\cdot\mathbf{U}_0(t) = \sum_{j=1}^{K} A_{ij}\cdot\mathbf{U}_0(t-t_j)\cdot\sum_{i=1}^{N} \mathbf{S}_{ij}\cdot\exp\{j\Omega_{ij}t\} \qquad (9)$$

Distribution of the elementary reflector radial velocities V_{ij} and the conditional distribution of RCS $W(A_{ij}/V_{ij})$ are chosen according to the experimental data for different backgrounds and artificial spatially-distributed radar objects. The given model of a spatially-distributed and fluctuating radar object (Fig. 2) allows to imitate spectral characteristics of signals reflected by real backgrounds.

In Fig.3 there are shown power spectra of sea surface matrix elements received by the author of the present report. HH – spectrum of the received signal in the channel of the element S_{11} estimation, HV – spectrum of the received signal in the channel of the element S_{21} estimation, VH – spectrum of the received signal in the channel of the element S_{12} estimation and VV – spectrum of the received signal in the channel of the element S_{22} estimation. Estimation of the BSM was done by the method of temporary diversion of the orthogonal sounding signals. These signals were given as periodical pulse sequences; the illuminated signal of coherent pulse radar is in this case a wave of linear polarization, which orientation changes sequentially from pulse to pulse.

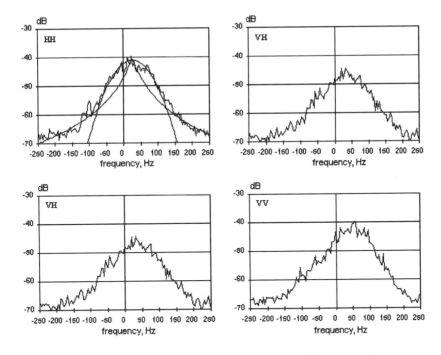

Fig. 3. Spectra of a periodical pulse signal scattered by sea surface

The level -67dB in Fig. 3 corresponds to the noise level in the channel of BSM S_{ij} estimation. In Fig. 4 there are given spectra of the periodical pulse signal, scattered by the model Fig. 1 (N=100, K=10). The sounding signal is identical to that, used in case Fig. 3.

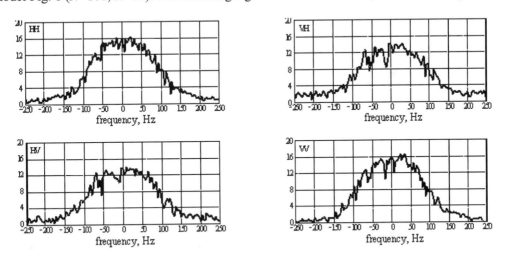

Fig. 4. Spectra produced by a model Fig. 2 illuminated by the periodical pulse signal

As follows from Fig. 4, the suggested model allows to imitate reasonably spectral characteristics of real back-scattered signals. In case of Fig. 4 the distribution of the elementary reflector velocities $W(V_{ij})$ was chosen uniform on the interval $0 \div 20$?/c, the conditional distribution $W(A_{ij}/V_{ij})$ of the amplitude $?_{ij}$ of the signal, scattered by an ij-th elementary reflector, having the velocity V_{ij}, was assumed Gaussian. The amplitude $?_{ij}$ decreases due to the velocity V_{ij} increase was -20 dB for $V = 20$m/s.

4. Conclusion

After performance of the given work and reception of results of modelling it became obvious, that the chosen techniques of have justified, and we have exact enough approximation to a reality.

There has been carried out mathematical simulation of an active radar channel in order to investigate peculiarities of the orthogonal LFM and PCM signal application in systems of correct estimation of BSM of non-stable spatially-distributed objects [3]. As a scattering volume there has been used the described above model of scattering.

References

[1] V.Klushov, L.Ligthart, G.Sharygin. "Principles of simultaneous measuring the radar objects' full scattering matrix", Proc. of MIKON'02, May 20-22, 2002, Gdansk, Poland, Vol.1, pp.155-158, 2002

[2] V.Khlusov, L.Ligthart, G.Sharygin. "Simultaneous measurement of radar objects scattering matrix all elements using complicated signals", Proc. of 8th Int. conf. "Radiolocation, Navigation, Communications", April 23-25, 2002, Voronedz, Russia, Vol. 3, pp. 1655-1667. (In Russian)

[3] D.Nosov, A.Surkov, V.Khlusov, L.Ligthart, G.Sharygin. "Simulation of algorithm of orthogonal signals forming and processing used to estimate back scattering matrix of non-stable radar objects", see this issue

Wave Propagation,Scattering and Emission in Complex Media
Edited by Ya-Qiu Jin
Science Press and World Scientific,2004

269

Simulation of Algorithm of Orthogonal Signals Forming and Processing Used to Estimate Back Scattering Matrix of Non-stable Radar Objects

D.Nosov[1],A.Surkov[1],V.A.Khlusov[1],L.Ligthart[2],G.Sharygin[1]

[1] Tomsk State University of Control Systems and Radioelectronics
40 Lenin ave., Tomsk, 634050, Russia. E-mail: gssh@mail.tomsknet.ru
[2] International Research Center of Telecommunication-Transmission and Radar,
Delft University of Technology.
Mekelweg, 4, Delft, the Netherlands. E-mail: L.P.Ligthart@its.tudelft.nl

Abstract Solving the problem of estimating the back scattering matrix (BSM) and co-ordinate parameters (range and velocity) of a non-stable radar object is based on the utilization of orthogonal signals with large bandwidth-time (BT) products [1, 2]. Results obtained from investigation allow choosing the class of orthogonal signals applicable to solve the problem. Some topics should get attention while designing devices needed for generating and processing signals with large BT: (1) the influence of the technical parameters characterizing the devices for generating and receiving orthogonal signals on errors of real radar object BSM and their co-ordinates estimation; (2) potential opportunities to decrease these errors by lowering the side-lobes of the complex signal ambiguity function.

In the paper we report simulation results of an active polarization radar, meant to provide the correct BSM estimation of a non-stable radar object, its range and velocity. The algorithm is based on utilizating noise-like signals for illumination. Furthermore, the paper describes the influence of weighting functions and limited frequency bandwidth of the transmitter and receiver on errors of BSM estimation.

1. Signals used to estimate BSM

For correct BSM estimation of non-stable spatial-distributed radar object it is necessary to measure simultaneously all BSM elements during one period of the sounding signal. It is also necessary to illuminate signals of different polarization and receive them also in different polarization channels of the receiver. Pairs of signals orthogonal in time are ideal to meet these demands. There exist many orthogonal codes. Some of them are considered in this paper.

The illuminated orthogonal signals should satisfy the following demands [3]:

(a) signals should have the same power;

(b) signals should have the same spectrum width;

(c) signals should have the same duration;

(d) signals should not be correlated at different time and frequency shifts.

They suggest utilization of complex signals with a wide base, or *BT* product, meaning that $\Delta\tau \cdot \Delta\omega \gg 1$, where $\Delta\tau$ is pulse duration, $\Delta\omega$ is spectrum width. Besides, it is important, that the polarization characteristics of a target don't change or change only slightly in the limits of the chosen "threshold of significance". It limits the choice of signals – with the explicit requirement that their duration must be shorter than the interval of time correlation of the target parameters.

Two types of signals are used below: PCM signal coded by an M-sequence or PCM signal coded by a random sequence with uniform phase distribution.

PCM-signals are coded by phase. A long pulse is divided into some shortened sub-pulses. All sub-pulses have the same duration; each is transmitted with its own phase. The phase of every sub-pulse is chosen according to the phase code.

PCM signal coded by an M-sequence are noise-like signals. Noise-like signals are sophisticated signals with continuous and practically even spectrum, which reminds the spectrum of noise within a limited frequency band. Auto-correlation function has only one peak, which width depends not on signal duration but on the signal spectrum, as in case of noise within a limited frequency band.

M-sequence is a digital one, generated by a linear generator utilizing a shift register and having maximal possible period for this method of generating. This maximal period is $L = 2^n - 1$, where L is the length of the maximal sequence (basis), n is the number of the generator stages (the length of the shift register). The codes creating M-sequences are formed nowadays by the method "try and test", though regular methods of such signal synthesis are being looked for.

M-sequences are quasi-orthogonal, that means – any two different sequences of the same length have not-zero, still rather low level of mutual correlation function outbursts. The ambiguity function of a signal coded by M-sequence is "needle" like. Such function can provide simultaneously resolution by range and by Doppler velocity without ambiguity. PCM signal coded by the random sequence with uniform phase distribution has a "needle"-like ambiguity function as an M-sequence has.

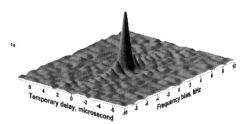

Fig. 1. Normalized ambiguity function $|\chi(\tau,\Omega)|$ of a PCM signal coded by the M-sequence. The basis is 1023.

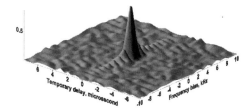

Fig.2. Normalized ambiguity function $|\chi(\tau,\Omega)|$ of a signal coded by the random sequence with uniform phase distribution. The basis is 1024.

Figures 1 and 2 show the ambiguity functions for such signals. Ambiguity function of the signal coded by M-sequence with the basis 1023 has rather low level of side-lobes: -29 dB (maximal outburst) or -42 dB (averaged level of side-lobes). Ambiguity function of the signal coded by the random sequence has level of side lobes: -25 dB (maximal outburst) or -37 dB (averaged level of side-lobes).

Mutual correlation function of two signals the basis 1023 has a low level of bursts. Isolation of such signals coded by M-sequence is 20 dB by the level of the maximal burst or 36 dB by the level of the burst average level. If the signals coded by the random sequence its isolation is 22 dB by the level of the maximal burst or 36 dB by the level of the burst average level (see fig. 3 and 4).

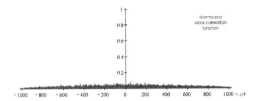

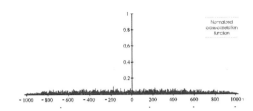

Fig.3. Normalized function of mutual correlation of the PCM signals, coded by the M-sequence

Fig.4. Normalized function of mutual correlation of the signal coded by the random sequence with uniform phase distribution

When a signal becomes longer, isolation between two noise-like signals increases. The signal with the basis 8192 has the isolation 31 dB by the maximal outburst and 45 dB by the averaged level of outbursts. Level of side lobes of the auto-correlation function is –31 dB by the maximal outburst and –45 dB by the averaged level of side lobes.

2. Algorithms of BSM elements estimation

Modeling of the radar channel consists of matrix convolution of the back-scattered vector signal $U_p(t)$ and the illuminated vector signal $U_0(t)$. There was used a model of the radar target described in [4].

The matrix convolution of the illuminated and scattered signals J (generalized response) is a function of τ (time of arrival) and Ω (Doppler frequency). It is the BSM estimate in absence of noise and interference.

$$J = \overline{U_p(t) \otimes U_0(t-\tau)\exp(j\Omega t)} = \begin{Vmatrix} J_{11} & J_{12} \\ J_{21} & J_{22} \end{Vmatrix} \Rightarrow S, \tag{1}$$

$$U_0(t) = \begin{bmatrix} \Phi_1(t) \\ \Phi_2(t) \end{bmatrix},$$

where $\Phi_1(t)$, $\Phi_2(t)$ are orthogonal signals.

3. Sources of BSM estimation errors

There are sources of errors when BSM elements are estimated simultaneously. Errors in case of complex signals such as (a) LFM and (b) PCM modulated with M-sequences, are compared in this paper. Difference of out-of-diagonal elements of BSM $J_{12} - J_{21}$ is used as a criterion of error and estimation of its value (for many real targets this difference should be zero [5]).

Further in the paper there is investigated suppression of the mutual correlation function (1) side lobes in the process of the received signal temporal weighting and frequency filtering. The degree of this suppression allows to estimate advantages of PCM-signals as well as the maximal possible dynamic range of the receivers not providing false targets.

Temporal weighting of the received signals can take place in case of scanning antennas, when the signal is modulated by the antenna pattern. If the same antenna is used for transmission and reception, the signals $U_0(t)$ and $U_p(t)$ should have the same weighting function. If the transmitting antenna does not rotate or has much wide lobe than the receiving one, only the received signal is multiplied by the weighting function. Results of modeling allow to estimate the allowable scanning speed in case of different complex signals.

Frequency response of the band filters, used in receivers, is not even. So the level of the mutual correlation function (1) side lobes changes differently in case of different signals. Modeling allows choosing the better signals and formulating the demands to receiver filer characteristics.

4. Simulation results in case of different complex (complicated) signals

There have been investigated six different cases resulting with errors when different complex signals are used.

In the <u>first and the second cases</u> the object was presented by a set of 2000 spatially diverse elementary reflectors-dipoles with uniform (in the limit of 2π) distribution of angle orientation and uniform distribution of the initial phase of these reflectors. Distribution of radial velocities V of these elementary reflectors is also chosen uniform within the interval ± 20 m/s (1st variant) and ± 100 *m/s* (2nd variant). Dependence of the back-scattered signal amplitude is Gaussian with maximum at $V = 0$. Variation of the amplitude was -20 dB when the velocity grows up to $\pm V_{max}$. In the <u>third case</u> there were three targets in one element of resolution. Each target was presented by a vibrator with orientation $\theta = 45^0$ and velocities $V_1 = -20$, $V_2 = 0$, $V_3 = 20$ m/c, correspondingly. It was assumed that amplitudes of signals scattered by each of the two moving targets were 3 dB less than the amplitude of the signal scattered by the stable target.

The basis of LFM and PCM signals in all three cases was the same, equal to 1023.

The <u>fourth variant</u> was similar to the third one, but the basis was 8192.

The <u>fifth variant</u> differs from the fourth one by the assumption that the stable vibrator had zero orientation angle while the vibrators moving with velocities -20 m/s and +20 m/s were oriented by 45^0 with the respect to the measuring (reference) basis.

In the <u>sixth variant</u> there were two targets in one resolution element of the radar with signal basis 8192. Parameters of these targets are follow:

1. $V_1 = 1$м/s, $\varepsilon_1 = -9,934^0$, $\theta_1 = 304,23^0$, $\Delta\varphi_1 = 196,193^0$, $\lambda_1 = 1$, $\phi_1 = 307,107^0$;
2. $V_2 = -1$м/s, $\varepsilon_2 = -9,934^0$, $\theta_2 = 304,23^0$, $\Delta\varphi_2 = 196,193^0$, $\lambda_2 = 1$, $\phi_2 = 307,107^0$

where V – radial target velocity; ε, θ - angle of ellipticity and orientation angle of the own target basis; ϕ - initial BSM phase; $\Delta\varphi$ - phase difference of BSM eigen values; λ - normalized BSM eigen value.

In the last three variants noise-like signal was used instead of PCM: $S(t) = A \cdot \exp\{j\omega_0 t + \varphi(t)\}$, where A – amplitude, ω_0 – carrier frequency, $\varphi(t)$ – random process uniformly distributed within the interval ($+\pi \div \pi$).

Figures 5-18 shows the results of all six variants modeling. Figures show modules of BSM elements (1) and modules of out-of-diagonal elements difference $|J_{12} - J_{21}|$. This difference should be zero for all reciprocal elements (all elementary reflectors were assumed reciprocal). If it is not zero, it can be a criterion of BSM estimation error.

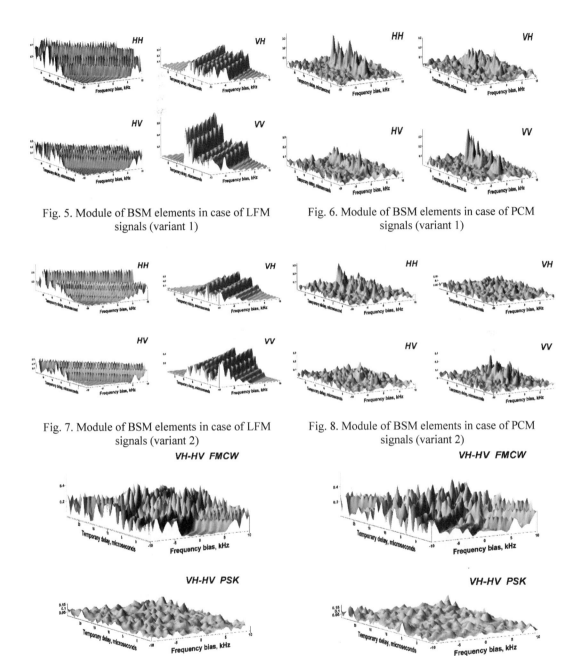

Fig. 5. Module of BSM elements in case of LFM signals (variant 1)

Fig. 6. Module of BSM elements in case of PCM signals (variant 1)

Fig. 7. Module of BSM elements in case of LFM signals (variant 2)

Fig. 8. Module of BSM elements in case of PCM signals (variant 2)

Fig. 9. Module of BSM out-of-diagonal elements difference in case of LFM and PCM signals (variant 1)

Fig. 10. Module of BSM out-of-diagonal elements difference in case of LFM and PCM signals (variant 2)

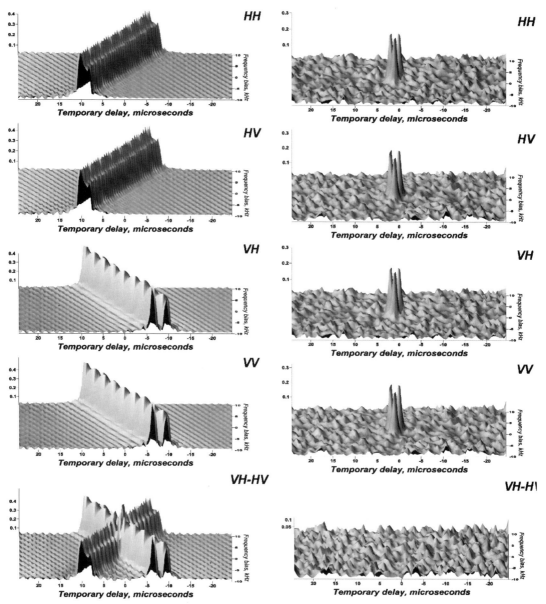

Fig. 11. Modules of BSM elements and of out-of-diagonal elements difference in case of LFM signal (variant 3)

Fig. 12. Modules of BSM elements and of out-of-diagonal elements difference in case of PCM signal (variant 3)

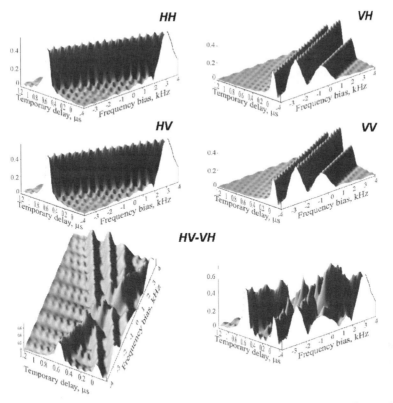

Fig. 13. Modules of BSM elements and of out-of-diagonal elements difference in case of LFM signal
(variant 4)

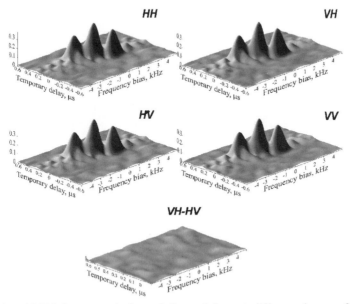

Fig. 14. Modules of BSM elements and of out-of-diagonal elements difference in case of noise-like signal
(variant 4)

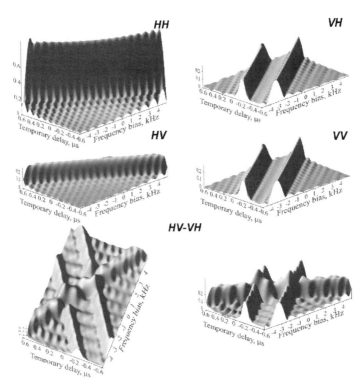

Fig.15 Modules of BSM elements and of out-of-diagonal elements difference in case of LFM signal
(variant 5)

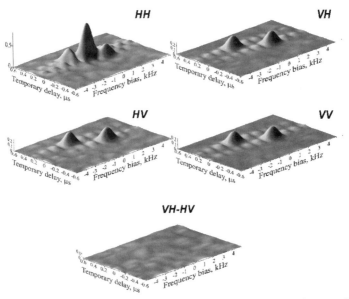

Fig. 16. Modules of BSM elements and of out-of-diagonal elements difference in case of noise-like signal
(variant 5)

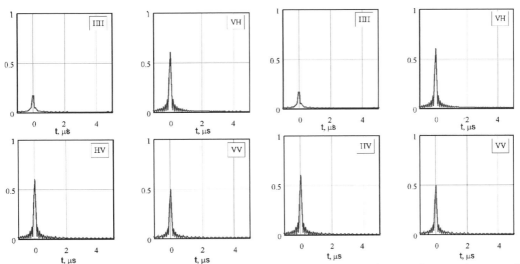

Fig. 17. Modules of BSM elements in case of   Fig. 18. Modules of BSM elements in case of
LFM signal. Cross-section at $\Omega = 0$ (variant 6)   noise-like signal. Cross-section at $\Omega = 0$
(variant 6)

Analysis of the results of modeling shows that in all cases of non-stable targets BSM estimation noise-like signals are better than LFM signals.

In the 1st and 2nd variants the module of difference $|J_{12} - J_{21}|$ of out-of-diagonal elements in case of LFM signals is comparable (by absolute value) with the elements J_{12} and J_{21}. In other words, the error of elements S_{12} and S_{21} estimation reaches 100%. That is unacceptable. The error of elements S_{12} and S_{21} estimation in case of PCM signals is much smaller and is defined by the level of ambiguity function side lobes and by mutual correlation of orthogonal signals. Comparing the general level of signals in the channels of matrix convolution for 1st and 2nd variants, it is necessary to point out, that when the velocity of the elementary reflectors increases the level of the PCM responses becomes smaller. It can be explained by the fact that the ambiguity body of the orthogonal PCM signals has "needle-like" appearance. So, the bigger is the interval of delay τ and Doppler frequencies Ω of the signals scattered by a spatially-distributed object, the larger is the area of the illuminated signal energy distribution in the process of scattering.

The general level of the response generalized function of LFM signal does not decrease. It can be explained by the fact, that the ambiguity functions of the orthogonal LFM signals are not symmetrical with the respect to the vertical axis (in co-ordinates time-frequency) and their prolonged combs coincide only in the vicinity of the point with co-ordinates $\tau = 0$, $\Omega = 0$. When Doppler frequency shifts have place, the responses in the plate $\Omega = 0$ run apart for every of orthogonal signals. So, the generalized responses of the elementary reflectors can appear inside the same sell of resolution of a radar for different combinations of these reflectors range and velocity. That is why there is not observed any decrease of the generalized function of the LFM signals.

For 3rd variant in case of correct BSM estimation the generalized function of response should consists of three separated peaks with co-ordinates ($\Omega_1 = 2V_1/\lambda_0 = -1,3$ kHz; 0), ($\Omega_2 = 0$; 0), ($\Omega_3 = +1,3$ kHz ; 0), correspondingly. It is seen that the estimates of BSM in case of

LFM signals (Fig. 11) are not correct. In the channels there are observed complicated responses not coinciding in cases of J_{12} and J_{21}.

The estimate in case of PCM signals (fig. 12) is correct and coincides with the theoretically expected one with accuracy up to the level of isolation of orthogonal signals. The out-of-diagonal elements difference of the received estimates $|J_{12} - J_{21}|$ is much smaller for PCM signals. The "surface of errors" for LFM signals allows to say, that LFM signals can't be used for estimation of polarization parameters of a set of moving objects.

Enlarging of the signal basis does not decrease the error of the moving radar object BSM estimation in case of LFM signal (variant 4, Fig. 13), while in case of PCM signal it does (Fig.14).

Comparison of Figs. 13, 15 with Figs. 14, 16 also shows great advantages of PCM signals (compared with LFM) in case of correct BSM estimation of the independently moving objects. Utilization of LFM signals with the basis about several thousand and time duration about 1 ms (variant 6, fig. 17 and 18) is possible only when the velocities of the objects are not high. BSM elements estimates at the moments of maximum responses by PCM and noise-like signals are in this case practically identical.

5. Simulation results for temporary weighted signals

In the process of simulation (modeling) there were formed in a digital form complex LFM and PCM signals modulated by an M-sequence. Both types have 8191 base, 1 ms duration and frequency band about 8 GHz. The signals were temporary-weighted by cos-like weighting function and after that the convolution was calculated, using (1). In the process of convolution there was estimated the maximal level of side lobes (along the axis of time).
Results of simulation are given in the Table 1. In case of PCM signals the form modulation of complex signals does not practically influence the level of side lobes, which stays low, much lower than in case of LFM signals. Only in case of symmetrical and fast falling down weighting function the level of LFM signal side lobes is as low as in case of PCM signals. But in order to provide the symmetry of the weighting function in time the system becomes more sophisticated.

6. Simulation results while frequency filtering of receiving signals is used

In the process of simulation there were used the same signals as in the previous case and the same kind of processing.

Results of simulation are given in the Table 2. The level of convolution function side lobes in case of LFM signals does not depend on the form of the receiver frequency response and is rather high. PCM signals keep their advantages over the LFM ones (the level of side lobes 20-25 dB lower) only when the frequency response is even. At the 0.5 level of the filter frequency response this advantage comes to 8-9 dB.

6. Conclusion

The correct estimation of non-stable spatially-distributed object BSM by the method utilizing LFM signals is possible only when velocities of the elementary reflectors, making up an object, have Doppler frequencies less than $1/T_0$, where T_0 is LFM signal duration.

The best of the investigated signals are PCM signals built by *M*-sequences. It is also possible to use noise-like signals, formed by phase modulation of the carrier frequency with the random noise.

When PCM are used the level of the radar response side lobes in case of simultaneous estimation of BSM elements is much lower than in case of LFM signals. Still, it is necessary to provide evenness of the receiver frequency responses.

Table 1. Maximal level of convolution function
(1) side lobes in the process of signal temporary weighting, dB

Form of the weighted function	LFM		PCM	
	Weighting only the received signal	Weighting the transmitted and received signals	Weighting only the received signal	Weighting the transmitted and received signals
	-14.5	-12.8	-34.5	-35.5
	-14.1	-14.2	-36.1	-37.6
	-26.0	-48.8	-36.4	-39.0
	-17.7	-23.9	-37.3	-38.7
Without weighing	-13.5	-13.5	-37.5	-37.5

Table 2. . Maximal level of convolution function (1) side lobes in case of signal filtering, dB

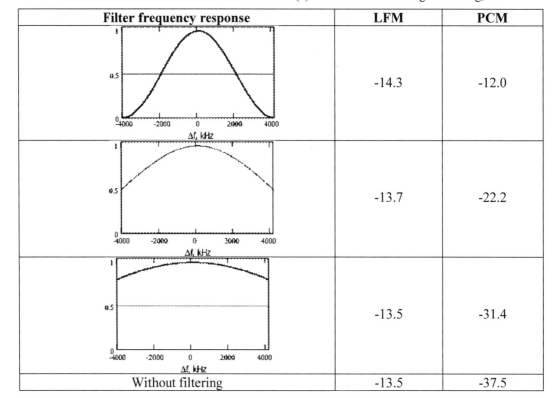

Filter frequency response	LFM	PCM
	-14.3	-12.0
	-13.7	-22.2
	-13.5	-31.4
Without filtering	-13.5	-37.5

References

[1] V.A.Klushov, L.P.Ligthart, G.S.Sharygin. "Principles of simultaneous measuring the radar objects' full scattering matrix", Proc. of MIKON'02, May 20-22, 2002, Gdansk, Poland, Vol.1, pp.155-158, 2002

[2] V.Khlusov, L.Ligthart, G.Sharygin. "Simultaneous measurement of radar objects scattering matrix all elements using complicated signals", Proc. of 8th Int. conf. "Radiolocation, Navigation, Communications", April 23-25, 2002, Voronedz, Russia, Vol. 3, pp. 1655-1667. (In Russian)

[3] V.A.Klushov, L.P.Ligthart, G.S.Sharygin. "Conditions to the radar vector signal providing simultaneous measurements of all back-scattering matrix elements", Proc. of MIKON'02, May 20-22, 2002, Gdansk, Poland, Vol. 2, pp.567-571, 2002

[4] A.Surkov, V.Khlusov, L.Ligthart, G.Sharygin. "Vector models of non-stable and spatially-distributed radar objects", see this issue.

[5] V.Khlusov, L.Ligthart, G.Sharygin, P.Vorobjev. "Polarization aspects of electromagnetic wave scattering by partly non-reciprocal media (Theory. Experimental modeling)", Proc. of SIBPOL-2002, October 4, 2002, Tomsk, Russia, pp. 22-31

Wave Propagation,Scattering and Emission in Complex Media
Edited by Ya-Qiu Jin
Science Press and World Scientific,2004

New Features of Scattering from
a Dielectric Film on a Reflecting Metal Substrate[*]
(Part I)

Zu-Han Gu[1], I.M.Fuks[2], Mikael Ciftan[3]

[1]Surface Optics Corporation
11555 Rancho Bernardo Road, San Diego, CA, 92127-1441, USA

[2]Zel Techologies, LLC and NOAA/ETL, 325 Broadway, Boulder, CO, 80305-3328, USA

[3]U.S. Army Research Office, P.O. Box 12211, Research Triangle Park, NC, 27709, USA

Abstract   We have recently observed several features from a randomly rough dielectric film on a reflecting metal substrate including the change in the spectrum of the light at the satellite peaks, the high order correlation and enhanced backscattering from the grazing angle.   In this paper we will focus on the enhanced backscattering phenomena.

The backscattering signal at small grazing angles is very important for vehicle re-entrance and subsurface radar sensing applications.   Recently, we performed an experimental study of the far-field scattering at small grazing angles, especially the enhanced backscattering at grazing angles. For a randomly weak rough dielectric film on a reflecting metal substrate, a much larger enhanced backscattering peak is measured.   Experimental results are compared with the theoretical predictions based on the two-scale surface roughness scattering model.

Key Words   *Enhanced backscattering effect, coherence effect, rough surface scattering.*

1. Introduction

Interference effects with diffuse light have been studied for a long time.   A recorded observation of the phenomenon was made by Newton about three centuries ago,[1] with a description of the appearance of a series of colored rings when a beam of sunlight falls on a concave, dusty back-silvered spherical mirror.   The phenomenon was explained later by Young[2] and Herschel by considering the interference between two streams of light: one scattered on entering the glass and the other scattered on emerging from the glass.

For scattering of light from a rough dielectric film on a reflecting substrate, there are three main kinds of trajectories that give rise to (a) Quetelet fringes, (b) Selenyi fringes, and (c) enhanced backscattering.   A typical Quetelet ring pattern consists of a series of elongated colored diffuse rings.   The white ring, corresponding to the zero order of interference, passes through both the specular and the backscattering directions, and as the angle of incidence is changed, new colors emerge from the center.

---

*   Correspondence: E-mail: zgu@surfaceoptics.com; Telephone: (858) 675-7404; FAX: (858) 675-2028.

specular and the backscattering directions, and as the angle of incidence is changed, new colors emerge from the center.

Besides the Quetelet-type ring, there is another kind of interference effect that has been observed in the scattering of light from dusty, backsilvered mirrors. The Selenyi rings[3] present a different kind of behavior under oblique illumination; there is no zero-order ring, and the rings are always centered about the normal to the sample. The occurrence of such rings has been explained in terms of the interference between waves scattered back directly from the top of the scattering layer without entering it, and waves reflected by the mirror after first having been scattered when entering the film.

One of the most interesting phenomena associated with the scattering of light from a randomly rough surface is that of enhanced backscattering. This is the presence of a well-defined peak in the retroreflection direction in the angular distribution of the intensity of the incoherent component of the light scattered from such a surface. It results primarily from the coherent interference of each multiply reflected optical ray with its time-reversed partner.

The scattering of light from a one-dimensional randomly rough dielectric film deposited on a flat reflecting substrate is studied.[4-5] In particular, the appearance of well-defined fringes in the angular distribution of the diffusely scattered intensity and their dependence on the angle of incidence, the roughness of the film, and the film's mean thickness is investigated. It is found that, for slightly rough films, the angle of incidence modulates the intensity of the fringes but has no effect on their angular position. For rougher films the contrast of the pattern decreases, and the fringes move with the angle of incidence in such a way that there are always bright fringes in the specular and backscattering directions. Eventually, for very rough films, the fringe pattern disappears, and a well-defined backscattering peak emerges in the retroreflection direction.

The measurement of the scattering of electromagnetic waves from a randomly rough surface at grazing angles of incidence presents a challenging problem.[6] This is due at least in part by the fact that if, say, a one-dimensional random surface is illuminated by a beam of finite width W, see Figure 1, its intercept with the mean scattering surface $\Delta = W/\cos(\theta_i)$, where θ_i is the angle of incidence measured counter clockwise from the normal to the mean scattering plane, increase to a very large value as θ_i approaches 90°. For example, if $\theta_i = 89°$, $\Delta = 57.3\ W$. We have to select a small beam size W about 1.5 mm and sample length L of the random surface should be large enough to compromise the grazing angle edge effect. L is chosen to be 200 mm.

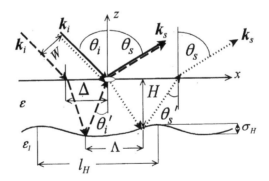

***Figure 1.* Physical schematic with indicatrix scattering.**

In this paper, we report the observation of enhanced backscattering at grazing angle. Part 2 will introduce the theoretical analysis followed by Part 3, the experimental results, and Part 4, the summary.

2.   Theoretical Analysis

Theoretical analysis of electromagnetic wave scattering from a randomly rough boundary of an arbitrary plane layered medium was performed in [7] and [8] by employing the small perturbation method. It was assumed that the surface roughness height $\varsigma(\mathbf{r})$ (where $\mathbf{r} = \{x, y\}$ is a radius-vector in the plane $z = 0$ of the medium boundary) is much smaller than the incident wave length λ.

The specific (referred to the unit area of surface $z = 0$) scattering cross-section $\sigma_{\alpha\beta}^0(\mathbf{k}_s, \mathbf{k}_i)$ of plane incident wave with wave vector $\mathbf{k}_i$ (see Fig.1) and polarization state $\beta = p, s$ [s polarization corresponds to the direction of electric vector is perpendicular to the plane of incidence xOz, and p polarization corresponds to electrical vector lies in this plane] into the scattered plane wave with the wave vector $\mathbf{k}_s = (k_0 \sin\theta_s \cos\varphi_s, k_0 \sin\theta_s \sin\varphi_s, k_0 \cos\theta_s)$ and the polarization state $\alpha = p, s$ [s polarization corresponds to electrical vector is perpendicular to the plane of scattering, that form the azimuthal angle φ with the plane of incidence xOz, and p polarization corresponds to electrical vector lies in this plane] can be written in the following form (see Eq.(5.3) in [8]):

$$\sigma_{\alpha\beta}^0(\mathbf{k}_s, \mathbf{k}_i) = \pi k_0^4 |\varepsilon - 1|^2 |f_{\alpha\beta}|^2 S_\varsigma(\mathbf{q}_\perp) \tag{1}$$

where $k_0 = 2\pi / \lambda$ is the wave number, $\varepsilon \equiv \varepsilon(0)$ is the upper limit value of the dielectric permittivity $\varepsilon(z)$ in the medium ($z<0$), $\mathbf{q} = \mathbf{k}_s - \mathbf{k}_i$ is a "vector of scattering", $\mathbf{q}_\perp$ is its projection on the plane $z=0$, and $S_\varsigma(\mathbf{q}_\perp)$ is a spatial power spectrum of surface roughness which can be introduced as a Fourier transformation of roughness auto-correlation function $W(\rho) = \overline{\varsigma(r + \rho)\varsigma(\rho)}$:

$$S_\varsigma(\mathbf{p}) = \frac{1}{(2\pi)^2} \iint W(\rho) e^{ip\rho} d\rho, \tag{2}$$

where the bar $\overline{...}$ denote the statistical averaging over ensemble of random function $\varsigma(\mathbf{r})$, and factors $f_{\alpha\beta}$ (that are proportional to the ``amplitude'' or ``length'' of scattering) are given by the expressions (5.6) - (5.9) from [8]:

$$f_{ss} = [1 + R_s(\theta_i)][1 + R_s(\theta_s)]\cos\varphi \tag{3}$$

$$f_{ps} = -[1 - R_p(\theta_s)][1 + R_s(\theta_i)]\cos\theta_s \sin\varphi \tag{4}$$

$$f_{pp} = \frac{1}{\varepsilon}\left[1 + R_p(\theta_i)\right]\left[1 + R_p(\theta_s)\right]\sin\theta_i \sin\theta_s -$$
$$- \left[1 - R_p(\theta_i)\right]\left[1 - R_p(\theta_s)\right]\cos\theta_i \cos\theta_s \cos\varphi \tag{5}$$

$$f_{sp} = \left[1 - R_p(\theta_i)\right]\left[1 + R_s(\theta_s)\right]\cos\theta_i \sin\varphi \tag{6}$$

Here R_s and R_p are the specular reflection coefficients from the lower ($z<0$) stratified media into the upper half-space ($z>0$) for s and p polarizations, correspondingly.

In a particular case of uniform (homogeneous) media with $\varepsilon(z) = \varepsilon = Const.$, when there is no volume scattering at all in the half-space $z<0$, coefficients R_p and R_s coincide with the usual Fresnel reflection coefficients R_{0p} and R_{0s} from the plane interface of two homogeneous media ($\varepsilon_0 = 1$, $z>0$) and (ε, $z>0$):

$$R_{op} = \frac{\varepsilon\cos\theta - \sqrt{\varepsilon - \sin^2\theta}}{\varepsilon\cos\theta + \sqrt{\varepsilon - \sin^2\theta}}; \qquad R_{os} = \frac{\cos\theta - \sqrt{\varepsilon - \sin^2\theta}}{\cos\theta + \sqrt{\varepsilon - \sin^2\theta}}. \tag{7}$$

In a general case of an arbitrary stratified media the reflection coefficients R_p and R_s can be represented in a form:

$$R_{p,s} = \frac{R_{0p,s} + R'_{p,s}}{1 + R_{0p,s}R'_{p,s}} \tag{8}$$

where $R'_{p,s}$ are the reflection coefficients from the buried layers ($R'_{p,s} = 0$ for the uniform $\varepsilon(z) = Const.$ homogeneous half-space $z<0$).

Here we apply this general theory to the simplest case of layered structure: the homogeneous dielectric layer of thickness H and permittivity ε lying on the homogeneous half-space ($z \leqslant H$) with a complex dielectric permittivity constant ε_1. The reflection coefficient $R_{p,s}(\theta)$ from this structure for every linear polarization (s,p) is given by the Equation (8), where $R'_{p,s}(\theta)$ can be written in the form $R'_{p,s}(\theta) = R_{1p,s}(\theta)\exp[i\varphi(\theta)]$, where $R_{1p,s}(\theta)$ is the Fresnel reflection coefficient from the interface of two media with dielectric permittivity constants ε and ε_1:

$$R_{1s} = \frac{\sqrt{\varepsilon - \sin^2\theta} - \sqrt{\varepsilon_1 - \sin^2\theta}}{\sqrt{\varepsilon - \sin^2\theta} + \sqrt{\varepsilon_1 - \sin^2\theta}}; \; R_{1p} = \frac{\varepsilon_1\sqrt{\varepsilon - \sin^2\theta} - \varepsilon\sqrt{\varepsilon_1 - \sin^2\theta}}{\varepsilon_1\sqrt{\varepsilon - \sin^2\theta} + \varepsilon\sqrt{\varepsilon_1 - \sin^2\theta}}. \tag{9}$$

and $\varphi(\theta) = 2k_0 H\sqrt{\varepsilon - \sin^2\theta}$.

The detailed analysis of the intensity spatial distribution pattern, originated from scattered by roughness $z = \varsigma(\mathbf{r})$ and multiple reflected (from planes $z=0$ and $z=-H$) wave interference, was done in [8]: the interference rings angular positions, their polarization dependence, periods of intensity oscillations as functions of parameters H and λ etc. These theoretical results are in a good agreement with experiments carried out with the perfect Fabry-Perot parallel-slided plates ($H = Const.$) and the very small surface settled scatterers (see, e.g., [9]). But in some experiments (see [10]) an essential disagreement with this theory was discovered. In [10] it was shown that the large-scaled roughness (LSR) can be the main reason of destroying the interference between some type of waves, that leads to the "surviving" only the one specific set of interference maxima and suppressing the others. The theoretical analysis conducted in [10] was restricted to consideration of interference of once reflected waves only. For very low grazing angle of incidence $\pi/2 - \theta_i$ the multiple wave reflections into the resonator formed by two planes $z=0$ and $z=-H$ can play the leading role in forming the scattered field intensity spatial distribution, and in particular, in forming the backscattering intensity peak

To investigate the effect of the LSR on the scattered intensity distribution, we assume that the successive wave reflections inside the layer ($0>z>-H$) every time take place from a horizontal plane (as it is depicted in Fig. 1) without changing the reflective angles θ_i' and θ_s' correspondingly before and after scattering by the rough patch of upper boundary), but the layer thickness H is different at the different points of reflection, as it shown in Fig. 1. The solution of this modeling problem can be represented in form (1) with amplitudes $f_{\alpha\beta}$, which can be obtained from those given by (3) through (6), with the following formal procedure. In $f_{\alpha\beta}$ representation by mentioned above equations, the factors $\left|1 \pm R_{s,p}\right|$ can be rewritten in the form, using (8):

$$1 \pm R = \frac{R_0 \pm 1}{R_0}\left[1 \pm \frac{R_0 \mp 1}{1 + R_0 R'}\right]. \tag{10}$$

Here, for short, we omit the arguments (θ_i and θ_s) or only their subscripts (i and s), and the polarization subscripts (p and s) in reflection coefficients, insomuch as it does not lead to confusion. Expand (10) in a series of $R' = R_1 \exp[i\varphi]$ powers, which is equivalent to field representation as a series of multiplicity reflection, we substitute instead $n\varphi$ the phase of the wave n-times reflected from the undulated interface between layer and substrate $n\varphi \Rightarrow \sum_{k=1}^{n} \phi_k$, where $\varphi_k(\theta) = 2k_0 H_k \sqrt{\varepsilon - \sin^2\theta}$, and $\{H_k\}$ is the set of layer thicknesses in the specular reflecting points:

$$1 \pm R \Rightarrow \frac{R_0 \pm 1}{R_0}\left[1 \pm (R_0 \mp 1)\sum_{n=0}^{\infty}(-R_0 R_1)^n \exp\left(i\sum_{k=1}^{n}\phi_k\right)\right]. \tag{11}$$

After this we can consider H as a random function of two variables (x,y), with a given average value $\langle H \rangle$, variance σ_H^2 and correlation length l_H.

Here we assume that there are no losses inside the dielectric layer, i.e., $\text{Im}\,\varepsilon = 0$ and present results only for the limiting case of extremely strong variations of layer thickness σ_H^2, when the corresponding Rayleigh parameter essentially exceeds the unity, i.e.,

$$\left\langle (\delta\phi_k)^2 \right\rangle = 4k_0^2 \sigma_H^2 \left(\varepsilon - \sin^2\theta_s\right) = \left(2k_1 \sigma_H \cos\theta_s'\right)^2 \gg 1, \tag{12}$$

where θ_s' and θ_s are related by Snell's refraction law: $\sin\theta_s' = \sin\theta_s / n$, $n = \sqrt{\varepsilon}$ is a refraction index, and $k_1 = nk_0$. If this inequality holds, it is possible to neglect the averaged value of exponents in (11), i.e., take $\langle \exp(i\phi_k) \rangle = 0$ in all equations that appears from (11).

The result of statistical averaging of scattered light intensity angular distribution over the set of random variables $\{H_k\}$, which we denote by the corner brackets $\langle ... \rangle$, strongly depends on the ratio of correlation length l_H and the distances Λ between the sequential specular reflecting points (see Fig. 1). If all distances Λ between every two arbitrary reflecting points from the substrate exceed essentially the correlation length l_H, then all subsequent specular reflections from the lower layer boundary can be considered as independent random events, and consequently the set of ϕ_k is the set of independent variables. Thus if inequality $\Lambda \gg l_H$ holds, then in all directions of scattering given by angles θ_s, φ, excluding the vicinity of backscattering direction ($\theta_s = \theta_i, \varphi = \pi$), the factors $1 \pm R(\theta_i)$ and $1 \pm R(\theta_s)$ are statistically independent and can be averaged separately. Statistical average value of the scattering cross section $\langle \sigma_{\alpha\beta}^0 \rangle$ over layer thickness fluctuations $\delta H = H - \langle H \rangle$ is proportional to the $\langle |f_{\alpha\beta}|^2 \rangle$, which for s polarization takes the form (see (3)):

$$\langle |f_{ss}|^2 \rangle = \langle |1 + R_s(\theta_i)|^2 \rangle \langle |1 + R_s(\theta_s)|^2 \rangle \cos^2\varphi \tag{13}$$

Compare the scattering cross section $\langle \sigma_{ss}^0 \rangle$, averaged over the layer thickness variations, with the corresponding value $\overline{\sigma}_{ss}$ for the homogeneous half-space bounded by the same rough surface. The explicit equation for $\overline{\sigma}_{ss}$ is given by (1) with f_{ss} from (3), where the reflection coefficients R_s have to be substituted by R_{0s}, i.e., if we put $R_s' = 0$ in all above equations. For ratio of these two cross-sections (which can be named, according to [11], as a contrast coefficient K_{ss}) we obtain:

$$K_{ss} = \frac{\langle \sigma_{ss}^0 \rangle}{\overline{\sigma}_{ss}} = C(\theta_i)C(\theta_s), \tag{14}$$

where function $C(\theta)$ is given by equation:

$$C(\theta) = \frac{1 + r_1^2(1 + 2r_0)}{1 - (r_0 r_1)^2}, \tag{15}$$

here, $r_0 = |R_{0s}|$ and $r_1 = |R_{1s}|$. In a special case of perfectly conducting substrate when $r_1 = 1$, (15) takes the form:

$$C(\theta) = \frac{2}{1 - r_0(\theta)}, \tag{16}$$

and we obtain the simple equation for the contrast coefficient (14):

$$K_{ss}(\theta_s, \theta_i) = \frac{4}{(1 - r_0(\theta_i))(1 - r_0(\theta_s))}. \tag{17}$$

It follows from this equation that $K_{ss}(\theta_s, \theta_i) \geqslant 4$, because $0 \leqslant r_0(\theta_{i,s}) \leqslant 1$, and in a particular case of $\varepsilon \gg 1$, when

$$1 - r_0(\theta) = \frac{2\cos\theta}{\sqrt{\varepsilon - \sin^2\theta} + \cos\theta} \cong \frac{2\cos\theta}{\varepsilon}, \tag{18}$$

we obtain for $K_{ss}(\theta_s, \theta_i)$:

$$K_{ss}(\theta_s, \theta_i) = \frac{\varepsilon}{\cos\theta_s \cos\theta_i} \gg 1. \tag{19}$$

It is seen that the average brightness of interference pattern due to the substrate can essentially exceed the one for homogeneous half-space (without substrate) even for very strong variations of layer thickness δH when all the interference maxima are utterly smoothen.

The backscattering case has to be considered separately because the dashed and dotted ray trajectories in Figure 2 are fully congruent in this case, and it is impossible to carry out the averaging over δH separately for each of them. When $\theta_s = \theta_i$ and $\varphi = \pi$, and all specular reflecting points for dashed and dotted rays coincide, we have to carry out the following averaging:

$$\langle \sigma_{ss}^0 \rangle \approx \langle |1 + R_s(\theta_i)|^4 \rangle, \tag{20}$$

where $1 + R_s(\theta_i)$ is represented as a sum (11) of independent specular reflections from the undulated substrate.

Skipping over the bulky derivations we present here only the final expression for the contrast coefficient K_{0ss} in backscattering direction as a result of averaging in the limiting case of very strong layer thickness fluctuations δH, when the inequality (12) holds:

$$K_{0ss} = \frac{\langle |1 + R_s|^4 \rangle}{|1 + R_{0s}|^4} = \left\{(1 + r_1^2)\left[(1 + r_1^2)(1 + A) + 8r_1^2 r_0\right] + 2r_1^2\left[(1 - A)^2 + 2A(3 - A)\right]\right\}(1 - A)^{-3}, \tag{21}$$

where $A = (r_0 r_1)^2$ and θ_i is supposed to be an argument for all reflection coefficients. Compare K_{0ss} given by (21) with the indicatrix contrast coefficient K_{ss} given by (14) for directions of scattering θ_s close to backscattering (i.e., putting there $\theta_s = \theta_i$ and $\mathbf{k}_s = -\mathbf{k}_i$, we can estimate the excess γ of backscattering peak of $\langle \sigma_{ss}^0(-\mathbf{k}_i, -\mathbf{k}_i) \rangle$ over the surrounding background:

$$\gamma = \frac{K_{0ss}}{K_{ss}}. \tag{22}$$

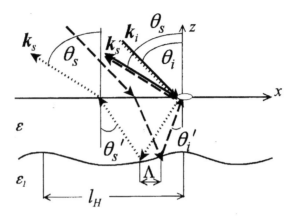

***Figure 2.* Physical schematic for near backscattering direction.**

Equation (21) takes the simpler form for the specific case of perfectly conducting substrate when $r_1 = 1$:

$$K_{0ss} = \frac{2(3-r_0)}{(1-r_0)^3}. \tag{23}$$

Compare this expression with the indicatrix contrast coefficient K_{ss} given by (17) for directions of scattering θ_s close to backscattering (i.e., putting there $\theta_s = \theta_i$:

$$K_{ss} = \frac{4}{[1-r_0(\theta_i)]^4}, \tag{24}$$

we can estimate the excess γ of backscattering peak of $\left\langle \sigma_{ss}^0(-\mathbf{k}_i, -\mathbf{k}_i \right\rangle$ over the surrounding background in this specific case :

$$\gamma = \frac{(3-r_0)}{2(1-r_0)}. \tag{25}$$

It is easy to see that the general phenomena of backscattering enhancement, in the specific problem under consideration, appears as a backscattering peak enhancement that for $(1-r_0) << 1$ can essentially (many times) exceed the background, in contrast to the well-known volume scattering problem, where this enhancement can achieve the value of several units only. In a specific case of low grazing angle or big layer dielectric permittivity ε, when inequality $\sqrt{\varepsilon - \sin^2 \theta_i} \gg \cos \theta_i$ holds, from (25) follows:

$$\gamma = \frac{\sqrt{\varepsilon - \sin^2 \theta_i}}{2 \cos \theta_i} \gg 1. \tag{26}$$

It is worth to emphasize that all above equations for contrast coefficients K_{ss} and K_{0ss}, as well as for backscattering peak enhancement γ, do not depend on the statistical parameters of upper surface roughness, and in particular, on its spatial power spectrum $S_\varsigma(\mathbf{q}_\perp)$. For non-absorptive dielectric layer with $\text{Im}\,\varepsilon = 0$ they do not depend also on the mean layer thickness $\langle H \rangle$ and its variance σ_H^2, if inequality (12) holds.

3. Experimental Results

A fully automated bidirectional reflectometer in Figure 3 is used to measure the fraction of incident light reflected by the sample into incremental angles over its field of view. It uses illumination from laser sources at 0.633 μm and enables measurements for any combination of incident and reflected angles over the entire plane, except for a small angle (about 0.5° away from the retroreflection direction) in which the source and detector mirrors interfere. A laser beam passes through a polarizer and is interrupted by a chopper and a half-wavelength plate, which enables rotation of the polarization of the beam. Then it is directed toward the sample by a folded beam system that collimates it into a parallel beam up to 25mm diameter. For the measurement, the beam size is set to 1.5 mm. The sample is viewed by a movable off-axis paraboloid that projects the light reflected by the sample onto the detector via a polarizer and a folding mirror. Four different polarization combinations of input and receiving beams are recorded. The reference standard used for these experiments is lab Sphere Gold, and the relative bidirectional reflectance is measured. The signal is recorded and digitized at each angular setting of interest throughout the angular range by an ITHACO lock-in amplifier and the data are stored in the memory of a personal computer (PC). The sample and the receiving telescope arm are separately mounted on two rotational stages run by two independent stepper motors that are controlled by the PC via a two-axis driver. Since the beam size is small, thus the average far-field speckle size is large. We have to average about 100 measurements to obtain the far-field scattering at each angle by scanning very small yaw and pitch directions of the sample.

The sample we used is a smooth aluminum that was coated with a dielectric film for high performance and protection. The thickness of the layer is approximately $H = 5.2$ μm. The complex permittivity ε_1 of Al at $\lambda = 0.6328$ μm is $\varepsilon_1 = -56.52 + 21.25i$. The refractive index of film is $n = 1.64$ (the dielectric constant is $\varepsilon = 2.69$). The rms height of the roughness of the film is about 60Å and $1/e$ correlation length is about 3000Å. The illuminating source in the experiment is a 15 mW He-Ne laser with $\lambda = 0.6328$ μm. Since the dielectric film is smooth, most energy goes to the specular direction thus a sensitive photomultiplier is used.

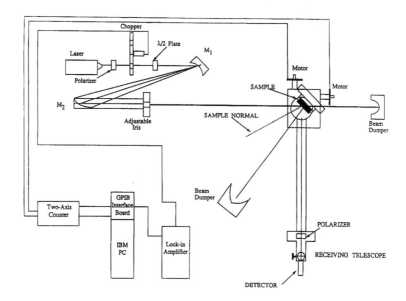

Figure 3. **Schematic of the bidirectional reflectometer.**

In the experiment, the beam size is set to W = 1.5 mm. Since the beam size is small, thus the average far-field speckle size is large. We have to average about 100 measurements to obtain the far-field scattering at each angle by scanning very small yaw and pitch directions of the sample.

Figure 4 (a) shows the experimental results for *p*-polarization, and Figure 4 (b) for *s*-polarization with $\theta_i = -89°$. There is a large enhanced backscattering peak at near grazing angle θ_s = 89°. The ratio of backscattering enhancement peak over the surrounding background at θ_s = 89° for *p*-polarization is about 19.6 and the ratio of backscattering enhancement peak over the surrounding background at θ_s = 89° for *s*-polarization is about 20.4. However in the experiment, the backscattering peak has a finite width and considerable uncertainty appears in determining the peak excess over the surrounding background.

The dependence of K_{0ss} on the angle of incidence θ_i is calculated for layer parameters, corresponding the experiment described above, and shown in figure 5. It is seen that for very low grazing angle $\pi/2 - \theta_i$, the backscattering contrast K_{0ss} can achieve values of several thousands. The plot of $\gamma(\theta_i)$ presented in Figure 6 shows the layer parameters corresponding to the experimental curve in Figure 4b. For $\theta_i = 89°$, the backscattering peak excess $\gamma \cong 20$, approximately coincides with the value observed in the experiment.

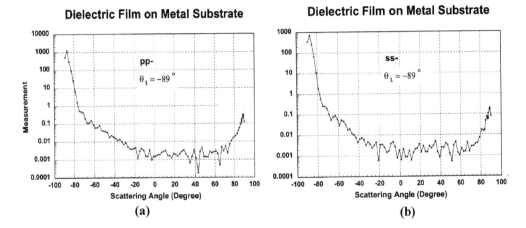

Figure 4.   **Experimental results for (a) p-polarization, and (b) for s-polarization.**

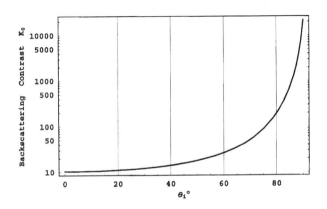

Figure 5. Backscattering contrast coefficient K_{oss}.

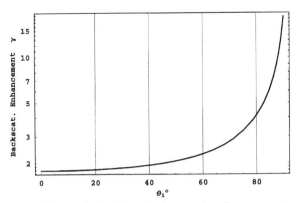

Figure 6.   **Backscattering peak enhancement γ.**

4. Summary

In conclusion, backscattering signals at small grazing angle are very important for vehicle re-entrance and lidar signature applications. For a randomly weak rough dielectric film on a reflecting metal substrate, a much larger enhanced backscattering at $\theta_s = 89°$ is measured which is compared with a theoretical calculation. Due to Quetelet's rings, the energy of diffusion is redistributed and a large portion of energy is attracted to the retro-reflection direction at grazing angle. That is why a large backscattering peak appears on the grazing angle.

Acknowledgements

The authors wish to express their gratitude to the U.S. Army Research Office for the support under Grants DAAD19-02-C-0056 and DAAD 19-02-1-0256.

References

1. Sir Isaac Newton, *Optics* (originally published in London, 1704; new edition by Dover, New York, 1952), pp. 289 ff

2. T. Young, "The Bakerian Lecture: On the theory of light and colors", Philos, Trans. R. Soc. London Part I, 12-48 (1802)

3. P. Selenyi, "Uber Lichtzerstreuung im Raume Wienerscher Interferenzen und neue, diesen reziproke interferenzerscheinungen", Ann. Phys. Chem. 35, 440-460 (1911)

4. Zu-Han Gu, M. Josse, and M. Ciftan, "Observation of Giant Enhanced Backscattering of Light from Weakly Rough Dielectric Films on Reflecting Metal Substrates", Opt. Eng. **35** (2), 370-375 (1996)

5. J.Q. Lu, J.A. Sanchez-Gil, E. Mendez, Zu-Han Gu, and A.A. Maradudin, "Scattering of Light from a Rough Dielectric Film on a Reflecting Substrate: Diffuse Fringes", J. Opt. Soc. Am. A, Vol. **15**, No. 1, 185-195 (1998)

6. Zu-Han Gu, I.M. Fuks, and Mikael Ciftan, "Enhanced Backscattering at Grazing Angles", Optics Letters Vol. **127**, No. 23, 2067-2069 (2002)

7. I.M. Fuks and V.G. Voronovich, "Wave diffraction by rough interfaces in an arbitrary plane-layered medium," Waves Random Media, **10**, 253-272 (2000)

8. I.M. Fuks, "Wave Diffraction by a Rough Boundary of an Arbitrary Plane-Layered Medium", *IEEE Trans. Antennas Propagat.*, **AP-49**, No. 4, pp. 630-639 (2001)

9. A.J. de Witte, "Interference in Scattered Light," Am. J. Phys. 35, 301-313 (1967)

10. Yu. S. Kaganovskii, V.D. Freilikher, E. Kanzieper, Y. Nafcha, and I.M. Fuks, "Light scattering from slightly rough dielectric films," Opt. Soc. Am. A. **16** (1999)

11.   A.I. Kalmykov, V.N. Tcymbal, I.M. Fuks, et al., "Radar observation of strong subsurface scatterers. Model of subsurface reflections," (in Russian), Preprint No. 93-6, Inst. Radiophys. And Electron. Nat. Acad. Sci. Ukraine, Kharkov (1993), English transl. in Telecommunications and Radio Engineering, **52**, 1-17 (1998)

Wave Propagation,Scattering and Emission in Complex Media
Edited by Ya-Qiu Jin
Science Press and World Scientific,2004

A Higher Order FDTD Method for EM Wave Propagation in Collision Plasmas

Shaobin Liu [1, 2], Jinjun Mo [1], Naichang Yuan [1]

1) Institute of Electronic Science and Engineering,
National University of Defense Technology, Changshai 410073, China
2) Basic Courses Department, Nanchang University, 330029, China
Email: liushaobin273@sohu.com

Abstract   A fourth-order in time and space, finite-difference time-domain (FDTD) algorithm is applied to study the electromagnetic propagation in homogeneous, collision and warm plasma. The approach can significantly minimize the dispersion errors while still maintaining minimal memory requirement. For the problem of three-dimensional propagation, the scheme requires only three additional memory cells per Yee cell over and above that of the generic FDTD scheme. To investigate the validation of the fourth-order FDTD algorithm, the reflection coefficient of a slab of non-magnetized collision plasma is calculated. Comparisons between the accurate data and the results of second-order or fourth-order FDTD methods are discussed.

Key Words   Fourth-order, FDTD, electromagnetic wave, plasma

1. Introduction

In recent years, the finite-difference time-domain (FDTD) algorithm has received considerable attention. The popularity of this algorithm stems from the fact that it is not limited to a specific geometry. And researchers can use the method for various materials (for example: isotropic material, anisotropic material [1,2], linearly dispersive media [3,4], and time-varying plasma media [5,6]). Furthermore, it provides a direct solution to problems with transient illumination. One of linear dispersive media is the isotropic collision plasma. For EM wave propagation in this medium, Young [7,8] recently proposed a direct integration finite-difference time-domain (DI-FDTD) algorithm that requires a minimal amount of memory and is second-order accurate. However, to obtain accurate solutions one must set the time step to be some arbitrarily small value (how small depends on the accuracy needed). Hence, Young [9] presented a fourth-order in time and space, finite-difference time-domain (FDTD) algorithm for radio-wave propagation in collisionless plasmas. The approach can significantly minimize the dispersion errors while still maintaining minimal memory requirements.

In this paper, we extended the fourth-order in time and space, FDTD algorithm to radio-wave propagation in collision, warm plasmas. One significant advance over the previous paper is that the fourth-order FDTD method is applied to loss plasmas. Thus, the attenuation of radio-wave by plasmas can be simulated using the fourth-order FDTD technique.

2. The FDTD Algorithm

A one-component, fluid model for electromagnetic wave propagation in a warm, isotropic, collision, plasma medium is assumed in the following discussion. And the motion of ions is neglected due to the ion' larger mass. Then the Maxwell's equations and the constitutive relation are given by (1)-(4)

$$\nabla \times \mathbf{H} = \varepsilon_0 \frac{\partial \mathbf{E}}{\partial t} + \mathbf{J} \tag{1}$$

$$\nabla \times \mathbf{E} = -\mu_0 \frac{\partial \mathbf{H}}{\partial t} \tag{2}$$

$$\frac{\partial \mathbf{J}}{\partial t} + \nu \mathbf{J} = \varepsilon_0 \omega_p^2 \mathbf{E} \tag{3}$$

Here, $\mathbf{E}$ is the electric field, $\mathbf{H}$ is the magnetic intensity, $\mathbf{J}$ the polarization current density, ε_0 the permittivity of free space, μ_0 the permeability of free space, $\omega_p^2 = n_e e^2 / m\varepsilon_0$ the square of plasma frequency, n_e is the electron concentration, ν the electron collision frequency, and e and m are the electric charge and mass of an electron, respectively.

For the ionosphere, the magnitude of the electron collision frequency for momentum loss was determined by the formula [10]

$$\nu = 8.3 \times 10^3 \pi a^2 \sqrt{T} N_m \tag{4}$$

where a is the radius of the molecule (the molecule is assumed to be rigid sphere. For air, $a = 1.2 * 10^{-10} m$ [10]), N_m is the number density of the molecules, T the plasma temperature.

The temporal derivatives and spatial derivatives (spatially-centered approximation) are discretized using the following fourth-order approximation [9]

$$\frac{\partial f(x,t)}{\partial t} \approx \frac{f(x, t + \Delta t/2) - f(x, t - \Delta t/2)}{\Delta t} - \frac{(\Delta t)^2}{24} \frac{\partial^3 f(x,t)}{\partial t^3} \tag{5}$$

$$\frac{\partial f(x,t)}{\partial x} \approx \frac{9}{8} \frac{f(x + \Delta x/2, t) - f(x - \Delta x/2, t)}{\Delta t} - \frac{1}{24} \frac{f(x + 3\Delta x/2, t) - f(x - 3\Delta x/2, t)}{\Delta x} \tag{6}$$

For the one-dimensional case, the electromagnetic wave is assumed to be propagation in x-direction, and consider only E_y and H_z components. Then the substitutions of (5)-(6) in (1)-(3), and using leap-frog integration give the following equations

$$\varepsilon_0 E_y\big|_i^{n+1} = \varepsilon_0 E_y\big|_i^{n} - \Delta t \left[\frac{9}{8} \left(\frac{H_z\big|_{i+1/2}^{n+1/2} - H_z\big|_{i-1/2}^{n+1/2}}{\Delta x} \right) - \left(\frac{H_z\big|_{i+3/2}^{n+1/2} - H_z\big|_{i-3/2}^{n+1/2}}{24\Delta x} \right) \right]$$

$$- \Delta t J_y\big|_i^{n+1/2} + \varepsilon_0 \frac{(\Delta t)^3}{24} \frac{\partial^3 E_y\big|_i^{n+1/2}}{\partial t^3} \tag{7}$$

$$\mu_0 H_z\big|_{i+1/2}^{n+1/2} = \mu_0 H_z\big|_{i+1/2}^{n-1/2} - \Delta t \left[\frac{9}{8} \left(\frac{E_y\big|_{i+1}^{n} - E_y\big|_i^{n}}{\Delta x} \right) - \frac{1}{24} \left(\frac{E_y\big|_{i+2}^{n} - E_y\big|_{i-1}^{n}}{\Delta x} \right) \right]$$

$$+ \mu_0 \frac{(\Delta t)^3}{24} \frac{\partial^3 H_z\big|_{i+1/2}^{n}}{\partial t^3} \tag{8}$$

$$J_y\big|_i^{n+1/2} = J_y\big|_i^{n-1/2} + \Delta t \varepsilon_0 \omega_p^2 E_y\big|_i^{n} - \nu \Delta t J_y\big|_i^{n} + \frac{(\Delta t)^3}{24} \frac{\partial^3 J_y\big|_i^{n}}{\partial t^3} \tag{9}$$

Here, n signifies the time $n\Delta t$, Δt is the time step, i signifies the space $i\Delta x$, Δx is the space step. After repeated use of (1)-(3), the terms of threes differentiations in (7)-(9) can be written as

$$\mu_0 \frac{\partial^3 \mathbf{H}}{\partial t^3} = -c^2 \nabla^2 (\nabla \times \mathbf{E}) + \omega_p^2 \nabla \times \mathbf{E} - \frac{\nu}{\varepsilon_0} \nabla \times \mathbf{J} \tag{10}$$

$$\varepsilon_0 \frac{\partial^3 \mathbf{E}}{\partial t^3} = c^2 \nabla \times \nabla^2 \mathbf{H} + v\varepsilon_0 \omega_p^2 \mathbf{E} - \omega_p^2 \nabla \times \mathbf{H} + \omega_p^2 \mathbf{J} + c^2 \nabla \times \nabla \times \mathbf{J} - v^2 \mathbf{J} \tag{11}$$

$$\frac{\partial^3 \mathbf{J}}{\partial t^3} = -c^2 \varepsilon_0 \omega_p^2 \nabla \times \nabla \times \mathbf{E} - \varepsilon_0 \omega_p^4 \mathbf{E} + v\omega_p^2 \mathbf{J} \tag{12}$$

Using (10)-(12), we generate the following FDTD equations for (7)-(9) as

$$E_y\Big|_i^{n+1} = \frac{48\varepsilon_0 + \varepsilon_0 (\Delta x)^3 v\omega_p^2}{48\varepsilon_0 - \varepsilon_0 (\Delta x)^3 v\omega_p^2} E_y\Big|_i^n$$

$$- \frac{48\Delta t}{48\varepsilon_0 - \varepsilon_0 (\Delta x)^3 v\omega_p^2} \left[\frac{9}{8} \left(\frac{H_z\big|_{i+1/2}^{n+1/2} - H_z\big|_{i-1/2}^{n+1/2}}{\Delta x} \right) - \left(\frac{H_z\big|_{i+3/2}^{n+1/2} - H_z\big|_{i-3/2}^{n+1/2}}{24\Delta x} \right) + J_y\big|_i^{n+1/2} \right]$$

$$+ \frac{2(\Delta t)^3}{48\varepsilon_0 - \varepsilon_0 (\Delta x)^3 v\omega_p^2} \left[\begin{array}{l} \dfrac{\omega_p^2}{\Delta x} \left(H_z\big|_{i+1/2}^{n+1/2} - H_z\big|_{i-1/2}^{n+1/2} \right) - \dfrac{c^2}{(\Delta x)^2} \left(J_y\big|_{i+1}^{n+1/2} - 2J_y\big|_i^{n+1/2} + J_y\big|_{i-1}^{n+1/2} \right) \\[2mm] - \dfrac{c^2}{(\Delta x)^3} \left(H_z\big|_{i+3/2}^{n+1/2} - 3H_z\big|_{i+1/2}^{n+1/2} + 3H_z\big|_{i-1/2}^{n+1/2} - H_z\big|_{i-3/2}^{n+1/2} \right) - v^2 J_y\big|_i^{n+1/2} \end{array} \right]$$

$$\tag{13}$$

$$\mu_0 H_z\Big|_{i+1/2}^{n+1/2} = \mu_0 H_z\Big|_{i+1/2}^{n-1/2} - \frac{9\Delta t}{8} \left(\frac{E_y\big|_{i+1}^n - E_y\big|_i^n}{\Delta x} \right) + \frac{\Delta t}{24} \left(\frac{E_y\big|_{i+2}^n - E_y\big|_{i-1}^n}{\Delta x} \right)$$

$$+ \frac{(\Delta t)^3}{24} \left(\begin{array}{l} \omega_p^2 \dfrac{E_y\big|_{i+1}^n - E_y\big|_i^n}{\Delta x} - \dfrac{v}{\varepsilon_0} \dfrac{J_y\big|_{i+1}^{n+1/2} - J_y\big|_i^{n+1/2}}{\Delta x} \\[3mm] - c^2 \dfrac{E_y\big|_{i+2}^n - 3E_y\big|_{i+1}^n + 3E_y\big|_i^n - E_y\big|_{i-1}^n}{(\Delta x)^3} \end{array} \right) \tag{14}$$

$$J_y\Big|_i^{n+1/2} = \frac{48 - 24v\Delta t + v\omega_p^2 (\Delta t)^3}{48 + 24v\Delta t - v\omega_p^2 (\Delta t)^3} J_y\Big|_i^{n-1/2} + \frac{48\Delta t \varepsilon_0 \omega_p^2}{48 + 24v\Delta t - v\omega_p^2 (\Delta t)^3} E_y\Big|_i^n$$

$$+ \frac{2(\Delta t)^3}{48 + 24v\Delta t - v\omega_p^2 (\Delta t)^3} \left(\begin{array}{l} c^2 \varepsilon_0 \omega_p^2 \dfrac{E_y\big|_{i+1}^n - 2E_y\big|_i^n + E_y\big|_{i-1}^n}{\Delta x^2} \\[3mm] - \varepsilon_0 \omega_p^4 E_y\big|_i^n \end{array} \right) \tag{15}$$

Here, the second-order accuracy spatial derivatives in (10)-(12) are computed due to the presence of the multiplicative term $(\Delta t)^3$, and shown as follows.

$$\frac{\partial f(x,t)}{\partial x} \approx \frac{f(x + \Delta x/2, t) - f(x - \Delta x/2, t)}{\Delta x} \tag{16}$$

$$\frac{\partial^2 f(x,t)}{\partial x^2} \approx \frac{f(x + \Delta x, t) - 2f(x,t) + f(x - \Delta x, t)}{(\Delta x)^2} \tag{17}$$

$$\frac{\partial^3 f(x,t)}{\partial x^3} \approx \frac{f(x + 3\Delta x/2, t) - 3f(x + \Delta x/2, t) + 3f(x - \Delta x/2, t) - f(x - 3\Delta x/2, t)}{(\Delta x)^3}$$

$$\tag{18}$$

3. Numerical Verification

To investigate the accuracy of the fourth-order FDTD method we compute the reflection coefficient of an unmagnetized collision plasma slab

$(\omega_p = 2\pi * 10 * 10^6 \, rad/s, \ \ v = 100 * 10^6 \, rad/s = 1.59 * 10^7 \, Hz)$

with a thickness of 12.5 m. (In the D layer of the isothere, the electron collision frequency for momentum loss is $v \sim 6 * 10^6 - 6 * 10^7$ while the number density of the air molecules is $N_m \sim 10^{15} - 10^{16}$ and the plasma temperature is $T = 300K$). In addition, we also compute the same coefficients using the second-order FDTD method. The computational domain is 125 m long and the plasma slab is defined by the region [56.25, 68.75] m. The reflection coefficient were computed by simulating the transient response of a normally incident plane wave on the plasma slab. The incident wave used in the simulation is a Gaussian pulse whose frequency spectrum peaks at 50 MHz and is 10dB down from the peak at 100MHz. For FDTD parameters, the spatial discretization, Δx, used in the simulations is 0.125 m (i.e., 48 points per free-space wavelength at $f = 50MHz$) and the time step $\Delta t (= 0.5\Delta x/c)$ is 0.208 ns. Then, the computational domain is subdivided into 1000 cells, plasma slab occupies cells 450 to 550, free space from 0 to 500 and 550 to 1000.

Figures 1 compare the reflection coefficients computed using the second-order FDTD and the fourth-order FDTD methods with those of the analytical solution for a plasma with collision frequency $v = 100 * 10^6 \, rad/s$. The Fourier transforms (FFT) were used when the reflection coefficients of the plasma slab from the air-to-plasma and plasma-to-air interfaces were computed using the second-order or fourth-order FDTD. And the second-order or fourth-order perfectly match layers [11,12] were used to absorb the electromagnetic waves traveling towards boundaries, respectively. The exact solution of reflection coefficients is the following [13]:

$$R = \frac{\left(\dfrac{n-1}{n+1}\right)\left[1 - \exp\left(-2j\left(\dfrac{\omega}{c}\right)nd\right)\right]}{1 - \left(\dfrac{n-1}{n+1}\right)^2\left[1 - \exp\left(-2j\left(\dfrac{\omega}{c}\right)nd\right)\right]} \qquad (19)$$

where R is the reflection coefficient of EM wave, ω is the incident EM wave frequency, d is the thickness of the plasma slab, n is the refractive index of the plasma and has the form

$$n = \sqrt{1 - \frac{\omega_p^2}{\omega^2 + v^2} - j\frac{v}{\omega}\frac{\omega_p^2}{\omega^2 + v^2}} \qquad (20)$$

Upon the comparison of those two plots we observe that the bandwidth associated with the fourth-order scheme is wider than its second-order counterpart. Furthermore, at the higher frequencies the frequency shift [9] in the nulls of the reflection coefficient doesn't occur.

In the F layer of the isothere, the electron collision frequency for momentum loss is $v \sim 10^4 \, Hz$ while the number density of the air molecules is $N_m \sim 10^6$ and the plasma temperature is $T = 300K$ [10]. Hence, we computed the reflection coefficients of a plasma slab with collision frequency $v = 2\pi * 10^4 \, rad/s$ and $v = 10^9 \, rad/s$ using the fourth-order FDTD technique and exact solution (Figure 2). Figures 1-2 show that the magnitude of the reflection coefficient can be decreased by increasing the plasma collision frequency. Upon the comparison of Fig. 2, we can observe that the bandwidth associated with the higher collision frequency is wider than its low collision frequency counterpart.

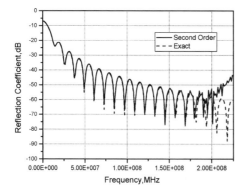 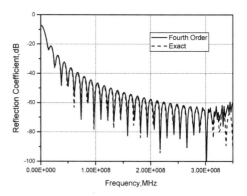

Fig. 1. A frequency domain plot of the reflection coefficient for $v = 100*10^6\,rad\,/\,s$. Comparison is between the exact data and the data obtained from the second-order FDTD (fourth-order FDTD).

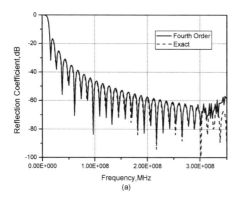 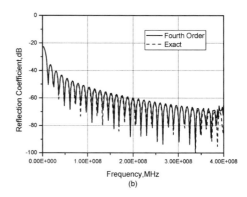

Fig. 2. A frequency domain plot of the reflection coefficient for (a) $v = 2\pi*10^4\,rad\,/\,s$, (b) $v = 10^9\,rad\,/\,s$. Comparison is between the exact data and the data obtained from the fourth-order FDTD.

4. Summary

In this paper, we have introduced a fourth-order FDTD formulation for predicting the propagation characteristics of an electromagnetic wave in collision plasma. For three-dimensional propagation, the scheme requires only three additional memory cells per Yee cell over and above that of the generic FDTD scheme. With respect to accuracy, the scheme is accurate to fourth-order in both time and space. And the approach can significantly minimize the dispersion errors while still maintaining minimal memory requirements. In particular, the fourth-order FDTD algorithm is well fit for simulating the electromagnetic problems of those devices with bigger structures

The study of the reflection coefficient of collision plasmas is of practical significance. It is well known that collision plasmas can attenuate the energy of incident EM wave. Hence, in some specific cases collision plasmas can be used as EM-wave absorbers [13-14, 16]. The reduction of the radar cross section (RCS) of a target surrounded by collision plasma is an example.

References
[1] J. Schneider and S. Hudson. IEEE Trans. A. P. 1993, 41(7): 994-999
[2] S. D. Gedney. IEEE Trans. A. P. 1996, 44(12): 1630-1639
[3] R. J. Luebbers, F. Hunsberger and K. S. Kunz. IEEE. Trans. A. P. 1991, 39(1): 29-34
[4] J. L. Young and R. O. Nelson. IEEE Antennas and Propagation Magazine, 2001,

43(1): 61-77

[5] J. H. Lee, D. K. Kalluri IEEE Trans. A. P. 1999, 47(7): 1146-1151

[6] J. H. Lee, D. K. Kalluri and G. C. Nigg J. of Infrared and Millimeter Waves, 2000, 21(8): 1223-1253

[7] J. L. Young. IEEE Trans. A. P. 1995, 43(3): 422-426

[8] J. L. Young. Radio Science, 1994, 29(6): 1513-1522

[9] J. L. Young. IEEE Trans. A. P. 1996, 44(9): 1283-1289

[10] V. L. Ginzburg. The Propagation of Electromagnetic Waves in plasmas, 2nd ed., Pergamon, 1970

[11] J. P. Berenger. J. of Computational Physics, 1994, 114: 185-200

[12] A. Taflove. Advances in Computational Electrodynamics: The Finite-Difference Time-Domain Method, Artech House, Boston, London. 1995: 63-105

[13] J. H. Yuan and D. H. Mo, Waves in Plasmas (in Chinese), UESTC press, 1990: 153-156

[14] J. J. Mo, S. B. Liu and N. C. Yuan Chinese Journal of Radio Science (in Chinese), 2002, 17(1): 69-73

[15] S. B. Liu, J. J. Mo and N. C. Yuan Chinese Journal of Radio Science (in Chinese), 2002, 17(2): 134-137

[16] S. Liu, J. Mo and N. Yuan. J. of Infrared and Millimeter Waves, 2002, 23(12)

V. Radiative Transfer and Remote Sensing

Wave Propagation,Scattering and Emission in Complex Media
Edited by Ya-Qiu Jin
Science Press and World Scientific,2004

303

Simulating Microwave Emission from Antarctica Ice Sheet with a Coherent Model

M. Tedesco , P. Pampaloni

Institute of Applied Physics 'Carrara' – CNR - Italy

mail to: M.Tedesco@ifac.cnr.it

Abstract A multi-layer electromagnetic model (Microwave Model Emission Snow Ifac - Multilayer Coherent, MIMESI-MC) based on Wave Approach, Strong Fluctuation Theory and Fluctuation-Dissipation Theorem was employed to investigate the microwave emission features of Antarctica ice sheet at DOME C. The inputs to the electromagnetic model were derived from a glaciological model describing the firn properties, such as density, mean grain size and temperature, as a function of depth. Obtained results were compared with space-borne observations, and a sensitivity analysis was performed. Lastly, L-band brightness temperature was computed, as a function of the observation angle and of the day of the year. The latter simulation pointed out the high stability of the low frequency emission.

Introduction

Antarctica is one of the most interesting and challenging natural laboratories of the Earth, playing a fundamental role in the hydrological and meteorological cycles. It has an area of approximately 14 million km^2 almost completely covered by ice and snow. Air temperature varies from –10 °C to –80 °C throughout the year. Antarctica contains about 90 % of the world's ice (the average thickness is approximately 2100 m with a maximum up to 4500 m).

The DOME C, located at 75°06'06''S and 123°23'42''E on the polar plateau of East Antarctica (Figure 1) with an altitude of 3280 m a.s.l, is one of the coldest and highest place on Earth with its –50.4 °C of average temperature (the minimum recorded temperature was – 81 °C !). This location is spatially homogeneous, with small surface slopes. The mean accumulation rate is about 3.7 cm/year and the mean wind speed is 2.7 m/s. Recent studies have shown that DOME C ice is one of the oldest of the planet: indeed the ice up to 800 meters below the surface is about 45000 years old.

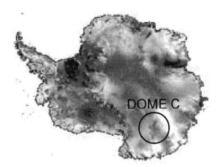

Figure 1 DOME C location

In addition to the interests related to climatic changes, to glaciological and hydrological applications, there is a growing interest for this zone from the microwave remote sensing community because the size, structure, spatial homogeneity and thermal stability of this area makes it a good candidate for external calibration and data   validation of satellite-borne microwave radiometers such as SMOS (Soil Moisture and Ocean Salinity).

The purpose of this research was to investigate the microwave emission features of the Dome C ice sheet by using simulations obtained with a coherent electromagnetic model.

In the first section of this paper we describe the electromagnetic model and the formulas employed for the brightness temperatures. The second section regards the glaciological properties of the DOME C site and the inputs to the electromagnetic simulations. In the third section simulated brightness temperatures are compared with those measured by SMMR and SSMI and a sensitivity analysis is performed. Lastly, L-band brightness temperatures, computed as a function of the observation angle and of the day of the year, are shown.

The Electromagnetic Model

Behind being an important tool for interpreting experimental data, modelling represents an attractive approach to infer radiometric properties of ice in those regions where extreme environmental conditions make it difficult to perform local measurements for long time. A coherent electromagnetic multi-layer model (MIMESI-MC) based on the Strong Fluctuation Theory (SFT) and Wave Approach (WA) was implemented and employed to simulate the brightness temperature of snow pack in DOME C.

In this model the reflection and transmission wave amplitudes were computed for a layered medium using the wave approach [1]. Any incident polarization was expressed as a linear combination of the horizontal (TE) and vertical (TM) polarizations. The medium under observation was divided into n layers with boundaries at z = -d_0, -d_1, -d_2, ..., -d_n, with d_0=0. The $(n+1)^{th}$ region was considered semi-infinite and it was labelled t, t = n+1. The reflection coefficient was expressed in a recurrent relation and the amplitudes of waves in each layer (A_l, B_l, C_l and D_l with l=0,..,n+1) were obtained by using the propagating matrices [1].

For each of the n+1 boundaries there were two equations. Therefore, we had 2n+2 equations to solve the 2n+2 unknowns A_l and B_l, l=1,2,...,n and A_0 and B_t. Expressing A_l and B_l in terms of A_{l+1} and B_{l+1}, and forming the ratio of the formulas we obtained the reflection coefficient for horizontal polarization in region l at the interface separating regions l and l+1. Thus it was possible to calculate recurrently the reflection coefficient for the stratified medium R_h= A_0/B_0. The solutions for the vertically polarized incident wave were obtained by duality with the replacements $\overline{E} \rightarrow \overline{H}, \overline{H} \rightarrow -\overline{E}, \mu \leftrightarrow \varepsilon$.

Using the result of Fluctuation- Dissipation Theorem [2], the *p*-polarized (*p*=*v,h*) brightness temperatures of multi-layers media were expressed as shown in Eq.1 [1-2].

$$T_{Bp}(\theta) = \frac{k}{\cos\theta} \sum_{l=1}^{n} \frac{\varepsilon_l'' T_l}{2\varepsilon_0} \{ \frac{|A_l|^2}{k_{lz}''} (e^{2k_{lz}''d_l} - e^{2k_{lz}''d_{l-1}}) - \frac{|B_l|^2}{k_{lz}''} (e^{-2k_{lz}''d_l} - e^{-2k_{lz}''d_{l-1}})$$

$$+ i\frac{A_l B_l^*}{k_{lz}'} (e^{-i2k_{lz}'d_l} - e^{-i2k_{lz}'d_{l-1}}) - i\frac{A_l^* B_l}{k_{lz}'} (e^{i2k_{lz}'d_l} - e^{i2k_{lz}'d_{l-1}}) \} + \frac{k}{\cos\theta} \frac{\varepsilon_t'' T_t}{2\varepsilon_0 k_{tz}''} |T^p|^2 e^{-2k_{tz}''d_n} \qquad (1)$$

For details about the quantities appearing in the formula, please see [1-3]. The brightness temperature of the layered ice sheet was obtained through the following steps: the number of layers and the thickness of the single layer in which the medium had been divided were given as inputs along with the frequency, surface temperature, observation angle and correlation length. The fractional volume, temperature and correlation length were computed at different depths employing the formulas obtained with the glaciological model. The permittivity of ice was computed at each single layer as a function of frequency and temperature by employing the formula proposed by Hufford [4]. The effective permittivity for each single layer was, then, computed by using the Strong Fluctuation Theory. Finally, the brightness temperatures were computed by using the Fluctuation-Dissipation Theorem.

The Glaciological Model and the Inputs for Electromagnetic Model at Dome C

Many experimental campaigns have been carried out and are planned to study several aspects of Dome C region. For example, in the European Project for Ice Coring in Antarctica (EPICA) an analysis of ice cores properties such mean particle size, density and impurities was performed from drilled ice cores. The data provided by this and other experiments were fundamental to express firn properties, such as grain size, density and temperature of each layer in which the medium was divided. In turn these data were employed to compute the electromagnetic properties of the layered medium.

Grain size

The size and distribution of grains, as well as their shapes and topological characteristics are factors that strongly influence microwave emission from snow and firn [5]. The definition of grain "size" is not trivial and different size estimation methods can lead to very different results [6]. As for many crystalline materials at relatively high temperature, polar ice experiences grain growth through time [7]. This is sometimes expressed as a parabolic growth law for the mean grain "size" D^2 that can be expressed as a function of the depth z:

$$D^2 = D_0{}^2 + Kz \qquad (2)$$

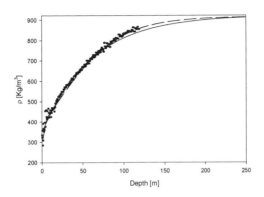

Figure 2 Measured values of density (dots) as a function of depth and fitting curves obtained with Eq. 4 (cont. line) and Eq. 5 (dashed line)

where D_0 is the initial mean cross-sectional dimension and K a constant. Experimental data have shown that the mean cross-sectional area measured at DOME C increased as a function of depth down to around 350 m and decreased between 430 m and 500 m and then increased again down to 800 m. This trend was associated to the Holocene/Last Glacial transition. Obtained values in [6] show that K varies with the method used to evaluate the mean squared grain size up to a factor of about 2, ranging from $0.6*10^{-2}$ to $1.68*10^{-2}$ mm^2/m.

Fractional volume

As for grain size, density is another factor that strongly influences microwave emission of snow. In general, firn density $\rho(z)$ can be assumed to increase exponentially with depth as in the following Eq. 3 [7,8], which is valid for average grain dimension:

$$\rho(z) = 0.922 - 0.564\exp(-0.0165z) \qquad (3)$$

where $\rho(z)$ is expressed in g/cm^3 and the coefficients are specifically obtained from DOME C ice cores. On the other hand, in [8] Barnes suggested for the same relationship the following formula:

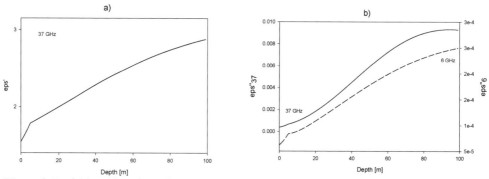

Figure 3 Real (a) and imaginary (b) part of the permittivity of the layered medium as a function of depth

$$\rho(z) = \frac{\rho_{ice}}{1 + e^{-g \cdot z + b}} \tag{4}$$

where g and b have been fitted with data obtained during the EPICA project. In [9] three stages of densification are taken into account; initially a settling and packing grains in the upper layers of snow leading to rapid densification down to a critical depth. Below this level the density increases more slowly leading to the closing off of interconnecting air ways and further densification continues with the compression of the air bubbles formed after close off. The critical depth occurs at 5 m in this core, which corresponds to a density of 421 kg m-3 and a volume fraction of 0.46. The computed values for g and b are, respectively: $g = 0.0967$ m^{-1} and $b = -0.6503$ for a depth up to 5 m and $g = 0.0253$ m^{-1} and $b = -0.2910$ from 5 to 120 m.

Figure 2 shows measured and computed values of density as a function of snow depth. It should be noted that both Eq. 3 and Eq. 4 well reproduce the density profile even if the latter equation works better for values down to 5 meters below the surface. This can be important when dealing with high frequencies and only the upper part of the medium is involved into the microwave radiometric response. In the following the fractional volume will be computed at different depths by using Eq. 4.

Note that the permittivity of the medium is a growing function of depth, and therefore of fractional volume. Figure 3a shows the real part of the permittivity of the layered medium at 37 GHz (the behaviour at 6 GHz is almost the same) where Figure 3b shows the imaginary part at 6 and 37 GHz as a function of depth. Note that the scales in Figure 3b are different for the two frequencies and the values of the permittivity differ by several orders of magnitude.

Thermal properties

The conventional heat conduction theory was employed to calculate the vertical temperature profile given by [8]:

$$T(z,t) = T_m + T_a \cdot e^{-z \left(\frac{\omega}{2k}\right)^{0.5}} \cdot \sin(\omega(t - \Phi) - z \cdot \left(\frac{\omega}{2k}\right)^{0.5}) \tag{5}$$

where T_m and T_a are, respectively, the mean annual temperature and seasonal amplitude, k is the thermal diffusivity of snow which depends on the thermal conductivity Ks, density ρ, and specific heat capacity γ. The values of T_m, T_a, ω and Φ were computed by fitting the temperatures recorded from the beginning of 1984 to the end of 1995 at DOME C by an Automatic Weather Station (AWS) installed by the National Science Foundation Office of Polar Programs. The obtained parameters are T_m = -50.72 °C, T_a = 16.73 °C, ω = $2\pi/364.8$ and Φ = -91.26 days. Figure 4 shows the daily average temperature measured by AWS (dots) and fitting curve obtained with the Eq. 5.

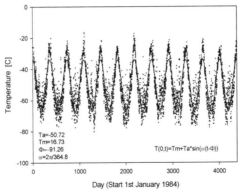

Figure 4 Daily average temperature measured by AWS (dots) and fitting curve obtained with Eq.6.

Simulation Results

For the Dome C region, temperature, density and mean grain size stratigraphic profiles were modelled by means of semi-empirical formulas obtained by examining ice cores from DOME C and considering general relationships regarding the polar ice. In order, to test the validity of MIMESI-MC, the simulated brightness temperatures at 37, 19, 10 and 6 GHz were compared with data acquired over DOME C by SMMR from 1979 to 1987 and by SSMI from 1988 to 2000. Experimental data were derived by fitting SMMR and SSMI measurements collected over more than ten years with a sinusoidal function, as already done for the surface temperature. The input parameters employed for the simulations were the surface temperature and the mean particle size (taken into account throughout the correlation length), along with the number of layers and the thickness of each single layer. The evolution of the fractional volume, (assumed to vary linearly with the depth, as the mean particle size) temperature and correlation length were described by using the equations proposed in the section regarding glaciological model.

The comparison of the annual average trend of measured and simulated brightness temperatures were in good agreement. Figure 5 shows the behaviour of 37 and 10 GHz simulated (line) and measured (dots) brightness temperatures (V polarization) as a function of the day of the year. The agreement was better for V than for H polarization. A reason for this discrepancy could be the assumption of spherical snow particles in contrast with the real shape of the measured particle, which were flattened along the horizontal plane [6], with the flattening increasing with depth.

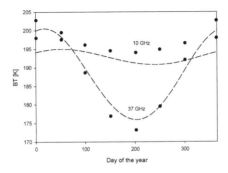

Figure 5 Simulated BT at 37 and 10 GHz V pol (lines) as a function of the day of the year and values from fitted SMMR and SSMI data (dots)

In order to evaluate the effects of ice layering, a sensitivity analysis of the model was performed: Figure 6 and Figure 7 show, respectively, the simulated brightness temperature (H pol.) and the horizontal reflectivity $|R_{TE}|^2$ (θ =53 ° , freq = 6, 19 and 37 GHz) as a function of the number of layers. Figure 6a refers to a total depth equal to 5 m with the number of layers ranging from 2 to 150, corresponding to a minimum thickness L_{min} of 3.3 cm (150 layers) and L_{max} of 2.5 m (2 layers); the Figure 6b refers to a total depth of 40 m, a minimum thickness of each single layer of 4 cm (100 layers) and a maximum of 20 m (2 layers). In the same way, the Figure 7a shows the horizontal reflectivity with a total depth of 5 m, L_{min} of 3.3 cm (150 layers) and L_{max} of 2.5 m (2 layers); the Figure 7b refers to a total depth of 100 m, L_{min} of 1 m   (100 layers) and L_{max} of 50 m (2 layers). The choice of the H polarization was related to the highest sensitivity shown by this polarization at an observation angle of 53°. From Figures 6 and 7, we see that the sensitivity of the electromagnetic quantities depends on the total depth considered for the simulations. For example, in Figure 7b if we consider only two layers, the 37 GHz emission is mostly influenced by the first layer whereas the 6 GHz emission is generated by the two layers having very different permittivity values. As the number of layer increases, the discontinuity of permittivity between contiguous layers decreases and the snow pack tends to a continuous medium.

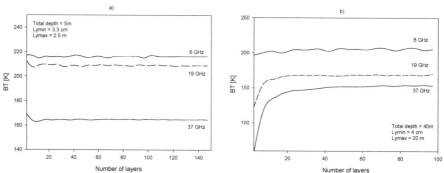

Figure 6 Simulated brightness temperature (H pol.) as a function of the number of layers
(see text for details)

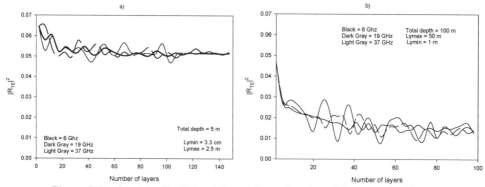

Figure 7 Horizontal reflectivity at θ = 53 ° as a function of the number of layers
at 6 (black), 19 (gray dashed) and 37 GHz (cont. light gray) (see text for details)

As already said it is particular interesting to investigate the behaviour of low frequency emission, either because it presumably gives information on deep ice layers or because it can be used as reference for L-band satellite radiometers such as SMOS. Figure 8 shows the DOME C L-band simulated brightness temperatures as a function of observation angle where Figure 9 shows the L-band temperature annual trend. As expected the variations along the year of the L-band brightness temperature are very small, confirming the hypothesis of high stability of the DOME C snow-pack.

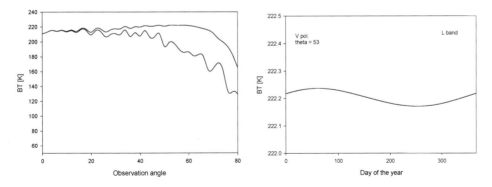

Figure. 8 L-band simulated T$_B$ vs observation angle   **Figure 9** Annual simulated trend of L band T$_B$

Conclusions

In this study a multi-layer electromagnetic model was employed together with a glaciological model to simulate microwave radiometric responses of the DOME C, in Antarctica.

A comparison between simulated and space-borne measured brightness temperatures showed good agreement in the case of V polarization whereas it pointed out some discrepancies at H polarizations. This could be due to the approximation assumed for the particle shape.

The very small variation (< 0.1K) of L-band simulated emission during the yearly cycle confirmed that the contribution to the low frequency microwave emission mostly comes from the deeper layers of ice, where temporal changes of   temperature and permittivity are very small.

Acknowledgments

The authors wish to thank Hugh Corr and Eric W. Wolff (British Antarctic Survey) for providing useful information about the EPICA project and for the constructive discussions; Mark Drinkwater and Nicolas   Floury (ESA –ESTEC) for providing the temporal trends of SMMR and SSMI data over DOME C and for the exchange of experience in modelling the DOME C microwave response; Giovanni Macelloni (IFAC – CNR) for the many discussions and suggestions. This study was supported by the Italian Space Agency (ASI).

References

[1] Kong, J.A. *Electromagnetic Wave Theory.* New York: Wiley-Interscience, 1990

[2] Jin Y.-Q., *Wave approach to brightness temperature from a bounded layer of random discrete scatterers*, Electromagnetics, 1984, Vol. 4, pp 323-341

[3] Tsang L., J. A Kong, K.-H. Ding, *Scattering of Electromagnetic Waves: Theories and Applications,* Wiley and Sons, 2000, USA

[4] Hufford G., *A model for the complex permittivity of ice at frequencies below 1 THz,*Int. Journ. Of Infrared and MM Waves, Vol. 12, No. 7 , 1991

[5] Macelloni, G., S. Paloscia, P. Pampaloni and M. Tedesco, *Microwave Emission from Dry Snow: A Comparison of Experimental and Model Result.* Transactions on Geoscience and Remote Sensing , 2001 vol. 39, n. 12, 2649-2656

[6] Gay M. and J. Weiss, *Automatic reconstruction of polycrystalline ice microstructure from image analysis: application to the EPICA ice core at Dome Concordia, Antarctica*, Journal of Glaciology, Vol. 45, 1999, pp 547-554

[7] Alley R. B. , Bolzan J. and Whillans I., *Polar firn densification and grain growth*, Ann. Glaciol., vol. 3, pp. 7-11, 1982

[8] Bingham, A.W.; Drinkwater M.R.;, *Recent changes in the microwave scattering properties of the Antarctic ice sheet*, IEEE Transactions on Geoscience and Remote Sensing, Volume: 38 Issue: 4 , Jul 2000 ,1810 -1820

[9] Barnes P. R. F., *The location of impurities in polar ice*, PhD Thesis, Open University, Sept. 2002

Wave Propagation,Scattering and Emission in Complex Media
Edited by Ya-Qiu Jin
Science Press and World Scientific,2004

Scattering and Emission from Inhomogeneous Vegetation Canopy and Alien Target by Using Three-Dimensional Vector Radiative Transfer (3D-VRT) Equation

Ya-Qiu Jin, Zichang Liang
Center for Wave Scattering and Remote Sensing
Fudan University, Shanghai 200433, China

Abstract   To solve 3D-VRT equation for the model of spatially inhomogeneous scatter media, the finite enclosure of the scatter media is geometrically divided, in both the vertical z and horizontal (x,y) directions, to form very thin multi-boxes. The zero-th order emission, first-order Mueller matrix of each thin box and an iterative approach of high-order radiative transfer are applied to deriving high-order scattering and emission of whole inhomogeneous scatter media. Numerical results of polarized brightness temperature at microwave frequency and under different radiometer's resolutions from inhomogeneous scatter model such as vegetation canopy and embedded alien target are simulated and discussed.

Key Words   inhomogeneous media, 3D-VRT, scattering and emission.

1. Introduction

To describe multiple scattering, absorption and emission of radiance intensity, the vector radiative transfer (VRT) equation of the Stokes vector $\bar{I}$ has been studied and applied in broad areas. Conventional VRT is usually for the models of parallel-layered media, i.e. one-dimensional (1D) VRT equation ($d\bar{I}/dz$)[1, 2]. There have been some studies of 2D or 3D scalar RT equations, such as the Monte Carlo method and multi-modes approach for radiance of atmospheric discrete clouds [3~6], the discrete-ordinate and finite difference method for the problems of heat transfer[7,8] and neutron transport etc. However, in all of these approaches it has been necessary to assume that the host medium itself is homogeneous, i.e. extinction, scattering and phase functions are considered to be independent of location within the medium. This fact simplifies the problem somewhat as the inhomogeneity in RT and then requires only a suitable treatment of the lateral sides of the medium and their associated boundary conditions. In advances of polarimetric and radiometric observation in remote sensing and the imaginary technology with an improved high resolution, development of 3D-VRT ($d\bar{I}/dx$, $d\bar{I}/dy$, $d\bar{I}/dz$) model for scattering simulation and detection technology from spatially inhomogeneous scatter media such as vegetation canopy, embedded target etc. becomes of great interest [9~11]. However, modeling and numerical solution of 3D-VRT for inhomogeneous scatter media remains to be studied.

To solve 3D-VRT equation for spatially inhomogeneous scatter media and obtain high-order scattering and emission, the finite enclosure of the scatter media is first geometrically divided into many thin boxes, slicing the media in both the vertical z and horizontal (x,y) directions. The zero-th order emission of each thin box and an iterative approach of high-order radiative transfer via the Mueller matrix solution of VRT in all media boxes are applied to deriving high-order scattering and emission of whole inhomogeneous scatter media. High order scattering and brightness temperature of the inhomogeneous scatter media in 3D geometry can be numerically calculated. According to the spatial resolution of the observation to divide the finite enclosure of scatter media, this approach presents numerical method to obtain high-order scattering and emission of the 3-D models for inhomogeneous scatter media such as heterogeneous vegetation canopy and embedded alien target.

2. The 3D – VRT equation

A geometric model is shown in Fig.1, where non-spherical particles are randomly and non-uniformly distributed within random limited space, and an alien target lays on the ground surface. It can be the model for tree canopy and alien target embedded. Suppose that the finite enclosure of the media has the length and width as W_1 and W_2, in the x and y directions respectively, and the height d in the z direction. Note that in this model the top and lateral boundaries of the enclosure are not solid, and the underneath of the bottom surface is a homogeneous half-space (land-) medium. The alien target can have different dielectric constant or physical temperature from the underlying surface.

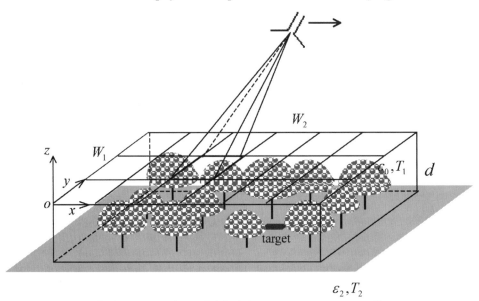

Fig.1. A geometric model for inhomogeneous scatter media.

The 3D-VRT [8,10,12] of the random scatter media is written as

$$(\cos\theta\frac{\partial}{\partial z}+\sin\theta\cos\varphi\frac{\partial}{\partial x}+\sin\theta\sin\varphi\frac{\partial}{\partial y})\bar{I}(\theta,\varphi,x,y,z)$$

$$=-\overline{\overline{\kappa}}_e(\theta,\varphi,x,y,z)\cdot\bar{I}(\theta,\varphi,x,y,z)+\overline{\overline{\kappa}}_a(\theta,\varphi,x,y,z)\cdot C\overline{T}_1(x,y,z) \qquad (1)$$

$$+\int_0^{2\pi}d\varphi'\int_0^\pi\overline{\overline{P}}(\theta,\varphi,\theta',\varphi',x,y,z)\cdot\bar{I}(\theta',\varphi',x,y,z)\sin\theta'd\theta' \quad ,$$

where $\overline{\overline{\kappa}}_e$, $\overline{\overline{\kappa}}_a$ are, respectively, the extinction and absorption matrices, $\overline{\overline{P}}$ is the phase matrix, and $\overline{T}_1$ is the physical temperature of the scatter media.

The boundary conditions at the top ($z=0$) and bottom ($z=-d$) surfaces are written as

$$\bar{I}(\pi-\theta,\varphi,x,y,z=0)=0, \qquad (2a)$$

$$\bar{I}(\theta,\varphi,x,y,z=-d)=\overline{\overline{R}}_{12}(\theta)\cdot\bar{I}(\pi-\theta,\varphi,x,y,z=-d)+\overline{\overline{T}}_{12}(\theta)\cdot C\overline{T}_2, \qquad (2b)$$

where $\overline{\overline{R}}_{12}(\theta)$, $\overline{\overline{T}}_{12}(\theta)$ are the reflectivity and transmittivity matrices of the bottom surface, respectively. Here, the subscript 12 denotes "between" the random media (1) and underlying medium (2). T_2 is the physical temperature of the bottom medium. Eq.(2a) means that there is no downwards emission incidence at $(\pi - \theta, \varphi)$ from up-space ($z > 0$) to the top surface ($z = 0$). Eq. (2b) indicates that the upwards radiance at the bottom ($z = -d$) is contributed by reflection of downwards radiance of the random media and transmitted emission from the underlying medium.

The boundary conditions at the lateral surfaces at $x = 0$, $x = W_1$, $y = 0$ and $y = W_2$ are respectively written as

$$\begin{cases} \bar{I}(\pi - \theta, \varphi, x = 0, y, z) = 0 \\ \bar{I}(\theta, \varphi, x = 0, y, z) = \bar{\bar{T}}_{12}(\theta) \cdot C\bar{T}_2 \end{cases} \quad 0° < \varphi < 90°, 270° < \varphi < 360°, \quad (2c)$$

$$\begin{cases} \bar{I}(\pi - \theta, \varphi, x = W_1, y, z) = 0 \\ \bar{I}(\theta, \varphi, x = W_1, y, z) = \bar{\bar{T}}_{12}(\theta) \cdot C\bar{T}_2 \end{cases} \quad 90° < \varphi < 270°, \quad (2d)$$

$$\begin{cases} \bar{I}(\pi - \theta, \varphi, x, y = 0, z) = 0 \\ \bar{I}(\theta, \varphi, x, y = 0, z) = \bar{\bar{T}}_{12}(\theta) \cdot C\bar{T}_2 \end{cases} \quad 0° < \varphi < 180°, \quad (2e)$$

$$\begin{cases} \bar{I}(\pi - \theta, \varphi, x, y = W_2, z) = 0 \\ \bar{I}(\theta, \varphi, x, y = W_2, z) = \bar{\bar{T}}_{12}(\theta) \cdot C\bar{T}_2 \end{cases} \quad 180° < \varphi < 360°. \quad (2f)$$

Eq.(2c) indicates that there is no downwards incidence from outside (defined by the azimuth φ region: $0° < \varphi < 90°, 270° < \varphi < 360°$) to enter the lateral side ($x = 0$) of the random media, and the upwards radiance from outside comes from the transmitting emission from the underlying half-space medium. Same physical meanings are described by Eqs.(2d,e,f) for other lateral sides.

Generally, $\bar{\bar{\kappa}}_e$ is non-diagonal for non-uniformly (the Euler angles β, γ)[1] oriented scatterers. To find the matrix $\bar{\bar{E}}$ and its inverse $\bar{\bar{E}}^{-1}$, $\bar{\bar{\kappa}}_e$ can be diagnosed for random scatterers as [1, 2, 13]:

$$\bar{\bar{\beta}}(\theta, \varphi, x, y, z) = \bar{\bar{E}}^{-1}(\theta, \varphi, x, y, z) \cdot \bar{\bar{\kappa}}_e(\theta, \varphi, x, y, z) \cdot \bar{\bar{E}}(\theta, \varphi, x, y, z), \quad (3)$$

where the ii-th elements of the diagonal $\bar{\bar{\beta}}$ is denoted as β_i, $i = 1,2,3,4$. It can be known that β_i is the eigen-values of $\bar{\bar{\kappa}}_e$, and $\bar{\bar{E}}$ is composed by the eigen-vectors of $\bar{\bar{\kappa}}_e$. All formulations of $\bar{\bar{\kappa}}_e$, $\bar{\bar{\beta}}$ and $\bar{\bar{E}}$ can be found in Refs. [1,2].

Left-multiplying $\bar{\bar{E}}^{-1}$ on the both sides of Eq.(1), it yields

$$(\cos\theta \frac{\partial}{\partial z} + \sin\theta\cos\varphi \frac{\partial}{\partial x} + \sin\theta\sin\varphi \frac{\partial}{\partial y})\bar{I}^E(\theta, \varphi, x, y, z)$$

$$= -\bar{\bar{\beta}}(\theta, \varphi, x, y, z) \cdot \bar{I}^E(\theta, \varphi, x, y, z) + \bar{\bar{\kappa}}_a^E(\theta, \varphi, x, y, z) \cdot C\bar{T}_0^E(x, y, z) \quad (4)$$

$$+ \int_0^{2\pi} d\varphi' \int_0^{\pi} \bar{\bar{P}}^E(\theta, \varphi, \theta', \varphi', x, y, z) \cdot \bar{I}^E(\theta', \varphi', x, y, z)\sin\theta' d\theta' ,$$

where

$$\bar{I}^E(\theta, \varphi, x, y, z) = \bar{\bar{E}}^{-1}(\theta, \varphi, x, y, z) \cdot \bar{I}(\theta, \varphi, x, y, z), \quad (5a)$$

$$\bar{T}_0^E(x, y, z) = \bar{\bar{E}}^{-1}(\theta, \varphi, x, y, z) \cdot \bar{T}_0^E(x, y, z), \quad (5b)$$

$$\bar{\bar{P}}^E(\theta, \varphi; \theta', \varphi', x, y, z) = \bar{\bar{E}}^{-1}(\theta, \varphi, x, y, z) \cdot \bar{\bar{P}}(\theta, \varphi; \theta', \varphi', x, y, z) \cdot \bar{\bar{E}}(\theta, \varphi, x, y, z), \quad (5c)$$

$$\bar{\bar{\kappa}}_a^E(\theta, \varphi, x, y, z) = \bar{\bar{E}}^{-1}(\theta, \varphi, x, y, z) \cdot \bar{\bar{\kappa}}_a(\theta, \varphi, x, y, z) \cdot \bar{\bar{E}}(\theta, \varphi, x, y, z). \quad (5d)$$

For convenience, all notations of E would not be especially indicated in next derivations.

Let's slice the media enclosure into many thin-slabs with the thickness Δd along the z direction, and denote the slabs by the subscripts $\ell = 1,2,\cdots,L$ from the top to the bottom. According to the spatial resolution Δh, dividing the (x,y) plane of the media enclosure to form many thin rectangular boxes with the length-width Δh and thickness Δd as illustrated in Fig.2. Because all boxes are very thin, we can assume that the medium within each box is homogeneous. But different boxes can be different, e.g. with different particles, different particle's fractional volumes, or different physical temperatures, etc. It is noted that as the $\overline{\overline{\kappa}}_e$ and $\overline{\overline{P}}$ of random particles have been calculated, the approach of VRT of random discrete particles is the same as one of continuous random media no matter what size the particles are comparing with Δd.

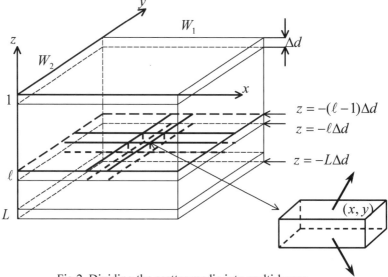

Fig.2. Dividing the scatter media into multi-boxes.

As illustrated in Fig.2, the radiance intensity of the ℓ-th slab is defined as

$$\bar{I}_\ell(\theta,\varphi,x,y) = \begin{cases} \bar{I}(\theta,\varphi,x,y,z=-(\ell-1)\Delta d) & 0° < \theta < 90° \\ \bar{I}(\theta,\varphi,x,y,z=-\ell\Delta d) & 90° < \theta < 180° \end{cases} \quad (6)$$

where $\bar{I}_\ell(\theta,\varphi,x,y)$ is the radiance intensity in the direction (θ,φ) from the box of the ℓ-th slab whose center is defined at (x,y).

3. Scattering and radiative transfer of 3D-thin boxes

Because Δd is very small ($\Delta d << \Delta h$), change of the radiance intensity through the lateral sides is always much smaller than one through the top and bottom surfaces. Thus, based on VRT Eq.(1), the changes of up-going ($0° < \theta < 90°$) and down-going ($90° < \theta < 180°$) radiance intensities through the box of the ℓ-th slab whose center is defined at (x,y) are derived as follows

$$\bar{I}_\ell^{(n)}(\theta,\varphi,x+\Delta x,y+\Delta y) = \bar{I}_{\ell+1}^{(n)}(\theta,\varphi,x,y)\exp[-\bar{\beta}_\ell(\theta,\varphi,x,y)\Delta d \sec\theta]$$
$$+ \bar{I}_{\ell s}^{(n)}(\theta,\varphi,x+\Delta x,y+\Delta y) , \quad (7a)$$

$$\bar{I}_\ell^{(n)}(\theta,\varphi,x+\Delta x,y+\Delta y) = \bar{I}_{\ell-1}^{(n)}(\theta,\varphi,x,y)\exp[\bar{\beta}_\ell(\theta,\varphi,x,y)\Delta d \sec\theta]$$
$$+ \bar{I}_{\ell s}^{(n)}(\theta,\varphi,x+\Delta x,y+\Delta y) , \quad (7b)$$

where $\bar{I}_{\ell s}^{(n)}$ is the radiance intensity (zero-th order emission and multiple scattering) happening in the box self, where the superscript n indicates the iteration number.

$\bar{\beta}_\ell = [\beta_{1\ell}, \beta_{2\ell}, \beta_{3\ell}, \beta_{4\ell}]$ is the vector of diagonal $\bar{\bar{\beta}}$ of the ℓ-th slab.

Radiance transfering from the $(\ell-1)$-th and $(\ell+1)$-th slabs can present the following relations:

$$\Delta x = \tan\theta \cos\varphi \cdot \Delta d, \quad \Delta y = \tan\theta \sin\varphi \cdot \Delta d. \tag{8}$$

The zero-th order emission of the box self is

$$\bar{I}_{\ell s}^{(n=0)}(\theta, \varphi, x, y) = \frac{1 - \exp(-\bar{\beta}_\ell \Delta d | \sec\theta |)}{\bar{\beta}_\ell} \bar{\bar{\kappa}}_{a\ell}(\theta, \varphi, x, y) C \bar{T}_{0\ell}(x, y). \tag{9a}$$

Multiple scattering from all directions under up-going and down-going incidences upon the box is

$$\bar{I}_{\ell s}^{(n>0)}(\theta, \varphi, x, y)$$
$$= \int_0^{2\pi} d\varphi' \int_{\pi/2}^{\pi} \bar{\bar{M}}_\ell(\theta, \varphi, \theta', \varphi', x', y') \cdot \bar{I}_{\ell-1}^{(n-1)}(\theta', \varphi', x', y') \sin\theta' \, d\theta' \tag{9b}$$
$$+ \int_0^{2\pi} d\varphi'' \int_0^{\pi/2} \bar{\bar{M}}_\ell(\theta, \varphi, \theta'', \varphi'', x'', y'') \cdot \bar{I}_{\ell+1}^{(n-1)}(\theta'', \varphi'', x'', y'') \sin\theta'' \, d\theta''$$

and the relations are also given as

$$x' = x + \frac{\Delta x}{2} + \frac{\Delta d \tan\theta' \cos\varphi'}{2}, \quad y' = y + \frac{\Delta y}{2} + \frac{\Delta d \tan\theta' \sin\varphi'}{2}, \tag{10a}$$

$$x'' = x + \frac{\Delta x}{2} - \frac{\Delta d \tan\theta'' \cos\varphi''}{2}, \quad y'' = y + \frac{\Delta y}{2} - \frac{\Delta d \tan\theta'' \sin\varphi''}{2}. \tag{10b}$$

The Mueller matrix of the ℓ-th slab, $\bar{\bar{M}}_\ell$, is approximated by the first order Mueller matrix as:

$$M_{ij\ell}(\theta, \varphi, \theta', \varphi', x, y) \approx M_{ij\ell}^{(1)}(\theta, \varphi, \theta', \varphi' x, y)$$

$$= \frac{P_{ij\ell}(\theta, \varphi, \pi-\theta', \varphi' x, y) \sec\theta}{\beta_{j\ell} \sec\theta' + \beta_{i\ell} \sec\theta} \begin{cases} 1 - \exp[-\Delta d(\beta_{j\ell} \sec\theta' + \beta_{i\ell} \sec\theta)] \\ \quad 0° < \theta < 90°, 0° < \theta' < 90° \\ \exp(-\beta_{j\ell}\Delta d \sec\theta') - \exp(\beta_{i\ell}\Delta d \sec\theta) \\ \quad 90° < \theta < 180°, 0° < \theta' < 90° \\ \exp(\beta_{j\ell}\Delta d \sec\theta') - \exp(-\beta_{i\ell}\Delta d \sec\theta) \\ \quad 0° < \theta < 90°, 90° < \theta' < 180° \\ 1 - \exp[\Delta d(\beta_{j\ell} \sec\theta' + \beta_{i\ell} \sec\theta)] \\ \quad 90° < \theta < 180°, 90° < \theta' < 180° \end{cases}, \tag{11}$$

where the subscripts $i, j = 1, 2, 3, 4$.

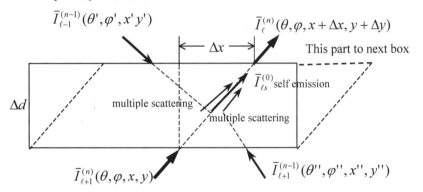

Fig.3. Illustration of radiative transfer through a box in the zox section.

From Fig.3 and Eq.(7a), it can be seen that the radiance $\bar{I}_{\ell+1}^{(n)}(\theta,\varphi,x,y)$ of the box at the center (x,y) becomes $\bar{I}_{\ell}^{(n)}(\theta,\varphi,x+\Delta x,y+\Delta y)$ after propagating and attenuating through this box medium. $\bar{I}_{\ell}^{(n)}(\theta,\varphi,x+\Delta x,y+\Delta y)$ will be distributed and enter into the boxes centers of the (ℓ-1)-th slab based on the ray-projected areas.

The thin-lines with arrow in Fig.3 indicate the zero-th order emission $\bar{I}_{\ell s}^{(0)}$ and multiple scattering from the up and bottom interfaces of the box.

Suppose $\Delta h=100cm$, $\Delta d=1cm$, the calculation of Eq.(8) yields $\Delta x=16.5cm$ and $\Delta y=9.5cm$ for the direction $\theta=87°$ and $\varphi=30°$ (noted that the maximum $\theta_M=\cot^{-1}(2\Delta d/\Delta h)\approx 89°$). The result shows that 0.756 of $\bar{I}_{\ell}^{(n)}(\theta,\varphi,x+\Delta x,y+\Delta y)$ remained within the box of the center (x,y), and only 0.079, 0.016, 0.149 of it enter into the neighbor boxes of the centers at $(x,y+\Delta h)$, $(x+\Delta h,y+\Delta h)$, $(x+\Delta h,y)$, respectively. Thus, the radiance intensity through the lateral sides of each box is much smaller than one through the top and bottom interfaces due to very small Δd, and $\bar{I}_{\ell s}^{(n)}$ of Eq.(7) can be calculated by Eqs.(9-a,b).

Now the steps to calculate brightness temperature emitted from the top interface $\bar{T}_B(\theta,\varphi,x,y,z=0)$ is summarized as follows[14]:

(i) By using Eqs.(9a,11), the zero-th order emission $\bar{I}_{\ell s}^{(n=0)}(\theta,\varphi,x,y)$ and the Mueller matrix $\overline{\overline{M}}_{\ell}(\theta,\varphi,\theta',\varphi',x,y)$ of each box are calculated;

(ii) By using the boundary condition (2), calculate the radiative transfer Eq.(7) from the top surface, the slabs $\ell=1,2,\cdots,L$ sequentially, and finally to the bottom surface. Adding the emission and reflection of the bottom surface, calculate the radiative transfer Eq.(7) from the bottom surface, the slabs $\ell=L,L-1,\cdots,1$ sequentially, and finally to the top surface. Thus, $\bar{I}_{\ell=1}^{(n=0)}(\theta,\varphi,x,y)$ is obtained;

(iii) By using Eq.(9b), calculate the (n+1)-th (n=0,...) order iteration to obtain $\bar{I}_{\ell s}^{(n+1)}(\theta,\varphi,x,y)$ of each box;

(iv) Repeating the steps of (2) and (3) to the N-th iteration, the calculation is finished when $\bar{I}_{\ell=1}^{(N)}(\theta,\varphi,x,y)$ is small enough.

Let $\bar{I}_{\ell=1}(\theta,\varphi,x,y)$ is the sum of all iterations $\bar{I}_{\ell=1}^{(n)}(\theta,\varphi,x,y)$ ($n=0,1,\cdots,N$), the brightness temperature observed in the up-space is

$$\bar{T}_B(\theta,\varphi,x,y,z=0)=\frac{1}{C}\bar{I}_{\ell=1}(\theta,\varphi,x,y), \qquad 0°<\theta<90°. \qquad (12)$$

4. Numerical results

4.1 Homogeneous media

First, compare the results of 3D-VRT (as scatterers are uniformly distributed within the rectangular enclosure) with conventional 1D-VRT for a homogeneous scatter medium. Suppose that the radiometer's frequency is 3GHz, the scatter particles are prolate spheroids with the semi-radii $a=b=0.1$ cm and $c=2.5$ cm, the dielectric constant of the particle is $\varepsilon_s=22+5i$, the fractional volume is 0.0075, the dielectric constant of the bottom medium is $\varepsilon_2=8+1i$. The physical temperatures $T_1=T_2=284$ K.

Both the length and width of the rectangular enclosure are 15m, and the depth d=1m, respectively. Divide the rectangular enclosure to form thin multi boxes with $\Delta h=1m$,

$\Delta d = 0.01$ m (i.e. 100 thin slabs), and take calculations at discrete angles $\Delta \theta = 9°$, $\Delta \varphi = 18°$.

From Eqs.(7,9), it can be seen that rigorous calculation requires to storage all radiance at different angles (θ, φ) and different centers (x, y) of the boxes. It needs the storage memory about 144M. When the media enclosure becomes large, the storage memory would be tremendously increased. To reduce such requirement, we propose to use a parabola line to approximately match the radiance of each slab along all direction (θ, φ), and only storage some coefficients of the matching line.

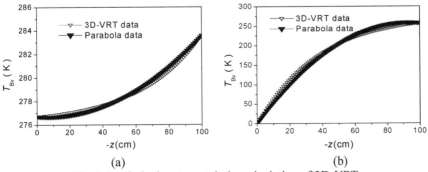

(a)   (b)

Fig.4. Parabola data to match the calculation of 3D-VRT.

As shown in Figs. 4(a,b), three matching points of the parabola line are chosen at $z = 0$, $-d/2$, $-d$. It only needs to storage 3 coefficients from 100 data, and significantly reduces the storage memory to 4M. For example, Figs. 4(a,b) show the data of 3D-VRT from the slabs, respectively, along the line $(\theta = 58°, \varphi = 9°)$ and $(\pi - \theta = 58°, \varphi = 9°)$, and good matching by a parabola line.

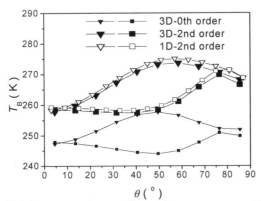

Fig.5. Brightness temperature from a homogeneous scatter medium.

Fig.5 presents the zero-th order (small black points) and $N=2$ (big black points) iterative brightness temperatures observed towards the center ($x = 7.5$ m, $y = 7.5$ m, $z = 0$). Calculation shows that $N=2$ iteration is good enough in this 3 GHz case. Difference of the radiance of the zero-th order and $N=2$ results is significant as about 10K. Because the brightness temperature observed at this center won't be significantly affected by the radiance afar from the lateral sides, as we believed, the result of 3D-VRT is well matched to the result of conventional 1D-VRT [13] (1D means that the enclosure becomes infinite).

It numerically validates our 3D-VRT codes.

4.2 High-orders scattering and emission from inhomogeneous scatter media

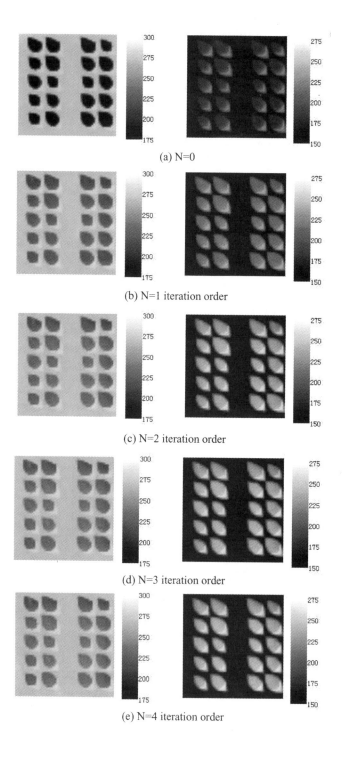

(a) N=0

(b) N=1 iteration order

(c) N=2 iteration order

(d) N=3 iteration order

(e) N=4 iteration order

Fig.6 High-orders from N=0 to N=4 scattering and emission at 5GHz from inhomogeneous canopy.

Now suppose the rectangular enclosure of Fig.1 has both the length and width 55 m, and the depth d=4m. Spheroid particles are randomly distributed within geometrical cones. The radius and height of the cones are randomly takes the values of $3 \sim 4$ m from Monte Carlo realization, and the cone radius is equal to height only for simplicity. Distance between the cones is assumed be 10m. The spatial orientation distribution of random particles is uniformly over $\beta \in (30^\circ, 60^\circ)$ and $\gamma \in (0^\circ, 360^\circ)$. Both the physical temperatures of random scatterers and ground surface are 284 K. Other parameters are the same as Fig.5.

It is noted that by using the parabola line to reduce the storage (about 58M), the matching points should be chosen at those locations where there are scatter particles.

Suppose that the radiometer at 5GHz with the spatial resolution Δh=1m at the angle $\theta = 58^\circ, \varphi = 315^\circ$ observes each box one by one. Figs. 6(a-e) show the T_{Bv}, T_{Bh} solution from the zero-th order to N=4 iteration, sequentially.

It can be seen that higher-order iterative brightness temperature from those locations of random scatter particles significantly enhances the zero-th order emission. Brightness temperature from some parts within the cone locations are higher than the radiance from bottom surface medium because the emission is contributed by random particles, scattering reflected from the bottom surface and the emission from the bottom medium. However, there are also some parts within the cone locations where brightness temperatures become lower because stronger back-scattering of random particles darken the emission from the bottom medium. On the other hand, because the horizontally polarized brightness temperature from flat bottom surface is low, random particles always enhance brightness temperature within those cone locations.

It is interesting to note that the maximum increase of the 1-st order iteration can reach 50K, and then each order iterative solutions sequentially increase as 15K, 5K, and 1K. Higher iteration is not necessary.

Calculation of the zero-th order solution by using the Pentium 1.8 GHz PC takes about 10 min, but higher order solutions would take much more as 10 hr.

4.3 Scattering and emission from an embedded alien target

Let's consider that there is an alien target laying over the ground and embedded under inhomogeneous tree canopy.

Suppose the radiometer's frequency 3GHz, the sizes of oblate spheroids for tree leaves model are $a = b = 1.5$ cm, $c = 0.05$ cm, fractional volume 0.001, dielectric constant $\varepsilon_s = 15 + 5i$, spatial orientations are uniform over $\beta \in (30^\circ, 60^\circ)$, $\gamma \in (0^\circ, 360^\circ)$. The crowns of tree canopy are modeled as semi-spheroids, whose radii of the bottoms randomly take the value within 4~6m in Monte Carlo realization, and tree heights are assumed as $3/4$ of respective bottom radius. All tree bottoms are simply assumed on the same level over the ground at height 2m. Physical temperature of tree canopy and ground surface are 293K. The observation angle is $\theta = 49.5^\circ, \varphi = 315^\circ$. In computation, both the length and width of each small box are 50 cm, and the height 1.5cm. The resolution of the radiometer is 1m, i.e. is coarser than the small box size. Scatterings from small boxes within the resolution are averaged. The canopy enclosure covers the area 55m×55m, dielectric constant of ground surface is $\varepsilon_2 = 15 + 2i$.

An alien target with the rectangular sizes 3m×3m lays over the ground, and its dielectric constant $\varepsilon_{t\,\mathrm{arg}} = 80 + 10i$ is much different from the ground surface, e.g. water or conductor.

Figs. 7(a,b,c) show the T_{Bv}, T_{Bh} solution from the zero-th to $N=2$ iteration, sequentially.

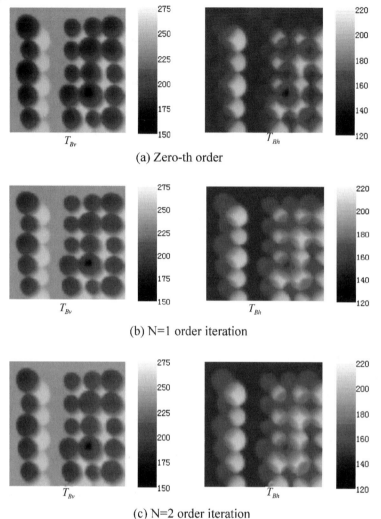

(a) Zero-th order

(b) N=1 order iteration

(c) N=2 order iteration

Fig. 7. Brightness temperature from tree canopy and an alien target at 3 GHz, N=0~2 orders.

It can be seen that bright spots come from stronger reflected emission from interaction of tree canopy and ground surface. Much lower emission from the target area, where the dielectric constant is much larger than surrounding areas, can be well identified at this spatial resolution 1m, and be darken by high-order scattering.

4.4 Different Resolutions for target detection

To see the resolution in observations, we consider another example when the alien target has much higher physical temperature 30K than the surrounding ground surface. The radiometer's frequency is 2GHz, the T_{Bv}, T_{Bh} solution only need $N=1$ iteration order. The ground surface has $\varepsilon_2 = 6 + 1i$. Other parameters are the same as Fig.7.

Figs. 8(a,b,c) give the T_{Bv}, T_{Bh} solutions at N=1 order for the radiometer's resolution Δh=1m, 2m, and 4m, respectively. It can be seen that variation of brightness temperature in whole area becomes a little smooth because coarser resolution takes the average of brightness temperature over a larger resolution area and whole image becomes blurred. Actually, at much coarser resolution Δh=4m, the target cannot be identified.

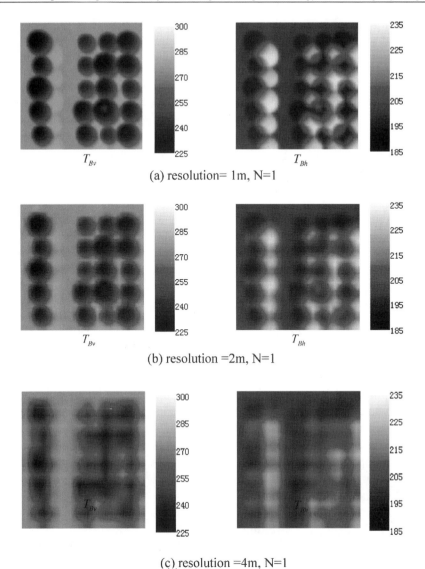

(a) resolution= 1m, N=1

(b) resolution =2m, N=1

(c) resolution =4m, N=1

Fig.8 Brightness temperature for different resolution at 2 GHz, N=1 order iteration.

To study the visibility, two images at Δh=1m with and without the target are subtracted as shown in Figs. 9(a,b).

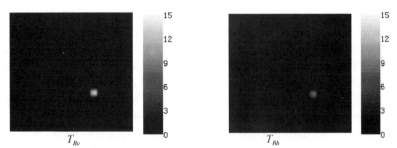

Fig.9 subtracting two images at Δh=1m with and without the target.

Define the visibility as the maximum of the brightness temperatures difference ΔT_B. Fig. 9(a) shows that identification of ΔT_B becomes worse when the resolution is coarser.

It is noted that at coarser resolution, the pixels number becomes less. For example, there are 16 pixels of a picture within 1m resolution, but only 1 pixel within 4m resolution. Another way to define the visibility is to sum all ΔT_B within the resolution. Fig. 9(b) shows $\sum \Delta T_B$ vs. resolution Δh.

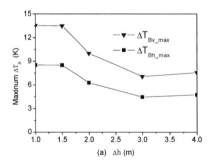 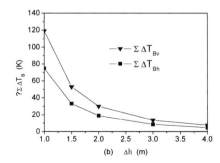

Fig.9 Visibility vs resolution

5 Conclusions

In this paper, a novel approach to solving 3-D VRT equation for spatially inhomogeneous scatter media is developed. The finite enclosure of the scatter media is divided into many thin boxes. The Mueller matrix solution of each thin box is employed for iterative algorithm of radiative transfer of whole scatter media. Our approach of 3D-VRT is first validated by conventional 1D-VRT for the model of a homogeneous scatter medium. Numerical results of high order scattering and brightness temperature of the inhomogeneous scatter media are discussed.

According to the spatial resolution of the observation to divide the finite enclosure of scatter media, this approach presents numerical solution of high-order scattering and emission of the 3-D inhomogeneous scatter media model, e.g. for vegetation canopy, embedded target on the ground etc.

Acknowledgments

This work was supported by the China State Key basic Research Project 2001CB309401~5 and CNSF 60171009.

References

[1] Y. Q. Jin, *Electromagnetic Scattering Modeling for Quantitative Remote Sensing*, Singapore: World Scientific, 1994

[2] L. Tsang, J.A. Kong, *Polarimetric Remote Sensing* , New York: Elsevier, 1990

[3] H.W. Barker, "Solar radiative transfer through clouds possessing isotropic variable extinction coefficient", *Q. J. Roy. Meteor. Soc.*, 118: 1145-1162, 1992

[4] K.F. Evans, "Two-dimensional radiative transfer in cloudy atmospheres-the soherical harmonic spatial grid method", *J. Atmos. Sci.*, 50: 3111-3124, 1993

[5] O.J. Dittmann, "The influence of cloud geometry on continuum polarization", *J. of Quantitat. Spectros. Radiat. Transfer*, 57: 249-263, 1997

[6] G.L. Stephens, "Radiative transfer in spatially heterogeneous two dimensional anisotropically scattering media", *J. of Quantitat. Spectros. Radiat. Transfer*, 36: 51-67, 1986

[7] J.S. Truelove, "3-D Radiation in absorbing-emitting-scattering media using the discrete-ordinates approximation", *J. of Quantitat. Spectros. Radiat. Transfer*, 39: 27-31, 1988

[8] S. Maruyama, Y. Takeuchi, S. Sakai, Z. Guo, "Improvement of computational time in

radiative heat transfer of three-dimensional participating media using the radiation element method", *J. of Quantitat. Spectros. Radiat. Transfer*, 73: 239-248, 2002

[9] G. Sun, K.J. Ranson, "A three-dimensional radar backscatter model of forest canopies", *IEEE Trans. On Geosci. Rem. Sens.*, 33: 372-382, 1995

[10] J.P. Gastellu-Etchegorry, V. Demarez, V. Pinel, and F. Zagolski, "Modeling Radiative Transfer in Heterogeneous 3-D Vegetation Canopies", *Remote Sensing of Environment*, 58: 131-156, 1996

[11] Z. She, "Three-dimensional space-borne synthetic aperture radar (SAR) imaging with multiple pass processing", *Int. J. Remote Sensing*, 2002, 23: 4357-4382

[12] S. Chandrsekhar, *Radiative Transfer*, New York: Dover Pub. 1960

[13] Z.C. Liang, Y.Q. Jin, "Iterative approach of high-order scattering solution for vector radiative transfer of inhomogeneous random media", *J. of Quantitat. Spectros. Radiat. Transfer*, 77: 1-12, 2003

[14] Y.Q. Jin and Z. Liang, "An approach of the three-dimensional vector radiative transfer equation for inhomogeneous scatter media", *IEEE Transaction on Geoscience and Remote Sensing*, in press, 2003

Wave Propagation,Scattering and Emission in Complex Media
Edited by Ya-Qiu Jin
Science Press and World Scientific,2004

324

Analysis of land Types Using High-resolution Satellite Images and Fractal Approach

H.G. Zhang, W.G. Huang, C.B. Zhou, D.L. Li, Q.M. Xiao

Lab. of Ocean Dynamic Processes & Satellite Oceanography,Second Institute of Oceanography, State Oceanic Administration, Hangzhou, 310012,China,

E-mail: zhanghg@163.com

Abstract   The purpose of this work is to study the land cover and land types of Nanji Island. A scene IKONOS image was taken to classify the land types including village, farmland, shrubbery, meadow, reservoir, sands and so on. Then several models were built base on fractal theory to analyze the land types. Condition of the land cover and land use was analyzed at three aspects as following: 1) effects of patch area; 2) fractal characters of land types; 3) test of difference of fractal character between every two land types. The results show that the values of D of meadow and shrubbery are higher, and those of farmland and village are smaller, and that the fractal characters are determined by the degree of interferes of human activities.

Key Words   IKONOS, Fractal, Land cover, Land type, Remote sensing, GIS, Nanji Island

1. Introduction

Land types refer to a mosaic of heterogeneous vegetation types, shapes and land use. The relevance of the land types structure to bio-diversity is well accepted today due to the extensive literature on habitat fragmentation. According to Noss and Harris (1986), ecological studies in natural areas revealed that every land type has indefinable patterns of disturbance and recovery processes. The natural disturbance regime intrinsic to a landscape interacts with vegetation and habitat variables to produce a mosaic of land type patches of different sizes and in different phases of post disturbance regeneration. The expansion of human activities during the past decades has led to a reduction of the natural vegetations in most parts of the world. Especially, in the study area of this work as a marine natural reserve, this reduction will cause a fragmentation process, and then the natural environment were disturbed and separated by humans. This process gave rise to natural patches of different areas, shapes, isolation levels, types of neighborhoods and disturbance history (Viana, 1990).

Remote sensing (Giles et al., 2000; Gupta et al., 1998; and Luiz et al., 1998) and fractal theory (Gupta et al., 1998; Luiz et al., 1998; and Ma et al., 2000) were widely used in land covers and land uses analysis. The goal of the present project was to study the fragmentation of land types in an island of the marine natural reserve. The specific aims were: 1) to survey the land uses and land covers of the study area by using IKONOS images; 2) to evaluate edge effects caused by habitat fragmentation by using remote sensing and fractal approach: 3) to develop and adjust fractal models that may help in the land types structure analysis. The results of this study are intended to aid in the planning of future actions concerning the island development.

2. Data and methodology

2.1 Study area and data

The study area is Nanji Island in East China Sea, which is the main island of Nanji

Islands National Marine Natural Reserve, and which also is the member of Biosphere Reserve Designated by UNESCO (United Nations Educational, Scientific, and Cultural Organization). It covers an area of 7.64 km^2.

The data used in this paper were from a digital map of Nanji Island at scale 1:10000 and IKONOS images. IKONOS data were acquired on July 22, 2001, and which consist of a panchromatic band (0.45 to 0.90 μm) with 1 m spatial resolution and four multispectral bands: blue (0.45 to 0.53 μm), green (0.52 to 0.61 μm), red (0.64 to 0.72 μm), near infrared (0.77 to 0.88 μm) with 4 m spatial resolution.

2.2 IKONOS image processing

A simple computer program was designed to process the image, so that different land uses and land covers types could be distinguished. After noise was eliminated and linear contrast was increased, several false color compositions were tested. The best RGB colored composition was: Near Infrared (NIR), presented in red, red band, in green; and green band, in blue. Then another color image with 1 m resolution was obtained by fusion using NIR band, red band, green band and panchromatic bands using HIS (Hue, Intensity, Saturation) transform method.

2.3 Land types derived from IKONOS images

To classify the land types from IKONOS images, the threshold method, vegetation Index and others were employed (Zhou, et al., 1999). Shoreline was derived from NIR band with threshold method, and confirmed using digital map and false color image in the place where was covered by cloud. In accordance with Frank (1988), the following normalized difference vegetation indices $NDVI=(R_{NIR}-R_{RED})/(R_{NIR}+R_{RED})$ were obtained. The vegetation patterns could be derived from the NDVI gay image. Texture, spatial position and shape were taken account to classify the village, farmland, sands from the false color images.

However, the results of this digital classification were difficult to analyzing. Thus, classified image was transferred to the GIS. After topology process, polygons were then generated.

2.4 Indicators of the land types structure

The condition of land cover and land use that characterizes a given area may be described as a function of the varying size, shape and distribution of patches. Some groups of statistics studied by Ripple et al. (1991) were used to quantify fragmentation of the land types in the study area. Patch size was expressed in terms of the average patch area Distribution of patch size was given as the frequencies ratio of size classes. Patch shape was measured by fractal dimension (D).

2.5 Fractal method

Fractals are now widely used for measuring forms and processes, and are attractive as a spatial analytical tool in the mapping sciences. Fractals were derived mainly because of the difficulty in analyzing forms and processes by classical geometry. In fractal geometry (Mandelbrot, 1977; and Mandelbrot, 1983), the dimension (D) of a curve can be any value between I and 2, and a surface between 2 to 3, according to the complexity of the curve and surface. The key concept underlying fractals is self-similarity. Fractal can well describe the anomalous, unstable and complex phenomena in order to help understand the characteristics of those phenomena and their mechanism. So far, very perfect results have been achieved (Li, 2000; Liu et al., 2000; Mladenoff et al., 1993; and Sugihara et al., 1990).

There are several definitions about fractal dimension of a polygon, and the approaches for estimating fractal dimension are more. Structured walk method, Douglas-Pueker method, Box counting method, Dilation method, Euclidean distance mapping (EDM) method (Dominique et al., 1999) and others were employed for estimating D of land cover patches. In this article, perimeter-area method and measure-perimeter-area method were used.

2.6 Perimeter-area method

A perimeter-area method was used to calculate the fractal dimension (D). The perimeter and area of patches were obtained from GIS coverage data. The method was very useful for estimating entire fractal dimension of a group of patches. According to Mandelbrot (1983), the relationship of patch perimeter (P) and patch area (A) can be expressed as

$$P = C_0 A^{D/2} \tag{1}$$

where D is the fractal dimension and C_0 the constant. After logarithmic operation both sides we have

$$\ln P = C + D \ln A^{1/2} \tag{2}$$

where $C = \ln C_0$. D of a group of patches can be estimated from a set of $A^{1/2}$ and P data of using equations (1)-(2) as long as there are enough number patches.

2.7 Measure-perimeter-area method

To estimate the fractal dimension for a single patch, Wang et al (1998) developed measure-perimeter-area method, which also was based on perimeter-area method. The relation of measure, perimeter and area is as follows:

$$P^{1/D} = a_0 S^{(1-D)/D} A^{1/2} \tag{3}$$

where D is the fractal dimension, a_0 the shape parameter, S the size of measurement, P and A are the perimeter and area of the polygon respectively. After logarithmic operation both sides equation (3) becomes

$$\ln(P/S) = D \ln a_0 + D \ln(A^{1/2}/S) \tag{4}$$

For a patch, D and a_0 are constants. Thus equation (4) can be written as

$$\ln(P/S) = C + D \ln(A^{1/2}/S) \tag{5}$$

where

$$C = D \ln a_0 \tag{6}$$

The fractal dimension D of a patch can be estimated from a set of S, A and P data using equations (4)-(6).

However, some investigators have accumulated evidence that many natural phenomena do not display self-similarity over all scales (Goodchild, 1980; and Lin, 1998). Thus, changes in D cannot be ignored. Therefore, in the present work, the hypothesis that D varies over scales has been assumed and an attempt was made to determine the scale magnitudes in which D remains unchanged. For this purpose, a program was developed to determine the range of measures, in which the measures was used to calculate D.

3.   Results and analysis

3.1 Statistics of land types from IKONOS images

Six main land types were observed on Nanji Island, including village, farmland, shrubbery, meadow, reservoir, sands and so on. Aiming to analyze the condition of distribution of land types, the information of all land types was listed in table 1.

Table 1 Land types and their areas and percents in relation to the total study area obtained from IKONOS images

Land type	Area (m^2)	Percent $(\%)$	Patch number	Average patch area (m^2)
Meadow	4193970.986	55.81	26	161306.576
Shrubbery	2538597.763	33.78	25	101543.910
Farmland	295802.259	3.94	16	18487.641
Village	439222.274	5.84	32	13725.696
Reservoir	4403.178	0.06	2	2201.589
Sands	42701.336	0.57	2	21350.668
Total	7514697.796	100	103	72958.231

From table 1, it was reflected that shrubbery and meadow types occupied about 89.6% of the total study area and their patch sizes are larger than others. Thus, the natural vegetation types are dominant on Nanji Island. The peach number of village type is biggest, but their patch sizes are small, which means that the villages are sporadic. Reservoir and sands both have only two patches and their sizes are small.

3.2 Effects of patch area

In this section, discussions of patch area effects on frequencies, fractal dimension (estimated by measure-perimeter-area method) and patch perimeter were carried out. Fig 1(a) shows the distribution of land types patches as a whole in size classes. The plot reveals that the frequencies of small patches are higher than that of larger patches. Fig.1 (b) is the distribution of fractal dimension in patch size (expressed by patch area in logarithms reference, lnA). According to Knimmel et al. (1987), it was assumed that in the perimeter-area relationship, the patches of all land types as a whole have well-defined length scales, then one can compare large and small characteristic lengths. This would lead to different perimeter-area relationships for large and small structures and thus to different values of D, indicating scale distinctions in the underlying processes that generate the land type patch shapes. From the Fig.1 (b), large patches have higher values of D than small patches. The relationship of patch perimeter (P) and patch area (A) both is strongly linear, but relationship of lnP and $\ln A^{1/2}$ is more stronger (see Fig1(c) and Fig.1 (d)). Therefore, Fractal theory is more propitious to describe the land cover patches structures.

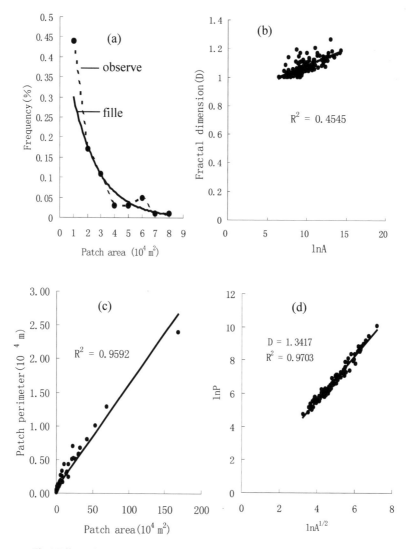

Fig,1 Effect of patch area, (a) distribution of patch number in classes of size;(b) D in relation to lnA;(c)behavior of patch perimeter as a relation to patch area;(d) natural logarithms of patch perimeter in relation to patch area and the slope line of corresponding perimeter-area relationship of the all study patches.

3.3 Fractal characters of land types

According perimeter-area method, values of D of main land types as a whole were estimated (excluding reservoir and sands because patch number is too small too calculate it). Fig.2 (a), Fig.2 (b), Fig.2(c) and Fig.2 (d) show lnP-lnA$^{1/2}$ relationships of meadow, shrubbery, farmland and village respectively. The value of D of meadow, shrubbery, farmland and village are 1.3433, 1.3728, 1.2563 and 1.2156 respectively. On the other hand, D of each patch was calculated using measure-perimeter-area method. Average values of D of meadow, shrubbery, farmland, village, reservoir and sands are 1.096, 1.102, 1.05, 1.062, 1.002 and 1.047 (see Table 2). The values of D from two methods are different, but they both reveal that the patches structures of meadow and shrubbery are more complex than that of farmland and village. The main reason is that the meadow and shrubbery are formed under natural processes, whereas farmland and village are planned by human usually. In addition, another reason probably is that patches of meadow and shrubbery are large, so that they may contain some patches of other land types, which makes the patch boundary more complex.

Table 2 Statistics of fractal dimension in classes of land type obtained by measure-perimeter-area method

Land type	Patch number	Minimum D	Maximum D	Average D
Meadow	26	1.032	1.19	1.096
Shrubbery	25	1.008	1.268	1.102
Farmland	16	1.001	1.123	1.05
Village	32	1.001	1.198	1.062
Reservoir	2	1.002	1.002	1.002
Sands	2	1.043	1.052	1.047

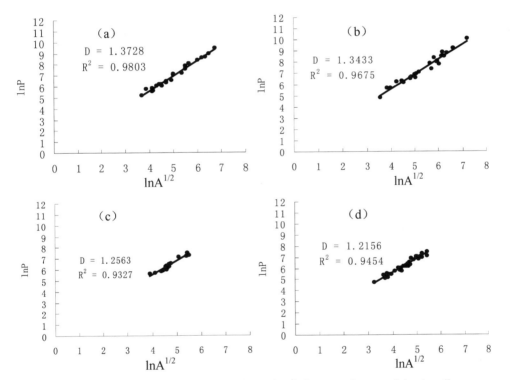

Fig. 2 Natural logarithms of patch perimeter in relation to patch area and the slope line of corresponding perimeter-area relationship of the main four land types: (a) meadow;(b) shrubbery;(c) farmland;(d) village.

3.4 Test of difference of fractal character between every two land types

The results of the above section show that the fractal characters of land types are different from each other. To analyze the difference between land types, T-test is carried out by using the values of D of all patches (excluding reservoir and sands) classified in land types. Results listed in Table 3 indicated that meadow and shrubbery are both different to village and farm- land distinctly, but that the difference between meadow and shrubbery and that between village and farmland are indistinct. So the results confirmed the conclusions of above section again.

Table 3 T-Test of difference between every two land types

Land type	Meadow	Shrubbery	Farmland	Village
Meadow	—	n	0.005	0.01
Shrubbery	—	—	0.005	0.005
Farmland	—	—	—	n
Village	—	—	—	—

Note: n means indistinct.

4.   Discussions and conclusions

Having analyzed the results obtained in the present investigation, it may be concluded that:

1.   The land cover and land use are simple on Nanji Island. The main land types are meadow, shrubbery, farmland, village, reservoir and sands. Meadow and shrubbery are the dominant types, and occupy about 89% of all area of the study area.

2.   Patch number, D of patch and patch perimeter are related to patch area. When patch area increases, the frequency rate will be down along a exponential curve, but values of D will increases slowly. The relationship of lnP and lnA$^{1/2}$ are more strongly linear than that of P and A. Thus, according fractal theory, patches of land types are fractal objects, but classical geometry objects.

3.   The values of D of meadow and shrubbery are higher, and those of farmland and village are smaller. Thus it is indicated that patch structure of meadow and shrubbery are more complex than farmland and village. Further analysis show that the fractal characters are determined by the degree of interferes of human activities.

Acknowledgments

We would like to thank the Development Project of National High Technical Industry of P.R. China (No.1999-2062) and Specific Fund of Marine Development and Management of Zhejiang Province, P.R. China (No.00-2-6-1-02) for support this project. We are grateful to Beijing Yimi Space Image Technology Corp. for IKONOS images.

References

[1] Dominique B., M. JeÂbrak., 1999, High precision boundary fractal analysis for shape characterization. Computers & Geosciences. Vol. 25, pp: 1059~1071

[2] Frank, T.D., 1988, Mapping dominant vegetation communities in the Colorado Rocky Mountain front range with Landsat thematic mapper and digital terrain data [J]. Photogr. Engin. Remote Sens. Vol.54, pp: 1727~1734

[3] Giles M. Foody, 2000, Mapping Land Cover from Remotely Sensed Data with a Softened forward Neural Network Classification [J]. Journal of Intelligent and Robotic Systems, No.29, PP: 433~449

[4] Goodchild, M. F., 1980, Fractals and the accuracy of geographical measures [J]. Mathematical Geology. Vol.12, pp: 85~98

[5] Gupta R.K., Prasad S. and P.V. K. Rao., 1998, Evaluation of Spatial Upscaling Algorithms for Different Land Cover Types. Adv [J]. Space Res. Vol.22, No.5, pp: 625~628

[6] Knimmel, J.R., Gardner, R.H., Sugihara. G., O'Neill. R.V., Cole-man, P.R., 1987, Landscape patterns in a disturbed environment [J]. Oikos. Vol.48，pp: 321~324

[7] Li B., 2000. Fractal geometry applications in description and analysis of patch patterns and patch dynamics [J]. Ecological Modelling. Vol.132 (1), pp:33~50

[8] Lin H., Li Y., 1992, Fractal theory—exploring of oddity [M]. Beijing: Beijing Sci. Tech. Univ. Press (In Chinese)

[9] Liu C., Chen L., 2000, Landscape Scale Fractal Analysis of Patch Shape in the Vegetation of The Beijing Region [J]. Acta Phytoecologica Sinica, Vol.24, No.2, pp:129~134 (In Chinese)

[10] Luiz A., Blanco J., Gilberto J.G., 1997, A study of habitat fragmentation in Southeastern Brazil using remote sensing and geographic information systems (GIS) [J]. Forest Ecology and Management, Vol.98, pp:35~47

[11] Ma K., Y. Zu, 2000, Fractal Properties of Vegetation Pattern [J]. Acta Phytoecologica Sinica, Vol.24, No.1, pp.111-117 (In Chinese)

[12] Mandelbrot, B.B., 1977, Fractal Form. Chance and Dimension [M]. Freeman. New York, pp: 365

[13] Mandelbrot, B.B., 1983, The Fractal Geometry or Nature [M]. Freeman, New York, pp: 468

[14] Mladenoff, D. J., M. A. White, J. Pastor and T. R. Crow, 1993, Comparing spatial pattern in unaltered old-growth and disturbed forest landscapes [J]. Ecological Applications. Vol.3, pp: 294~306

[15] Noss, R.F., Harris. L.D., 1986, Nodes, networks, and mums: preserving diversity at all scales [J]. Environ. Manage. No. 10, pp: 299-309

[16] Ripple, W.J.. Bradshaw. G.A., Spies, T.A.. 1991. Measuring forest landscape patterns in the cascade range of Oregon, USA. Biol. Conserv. Vol.57, pp:73~88

[17] Sugihara, G. and R. M. May, 1990, Applications of fractals in ecology [J]. Trends in Ecology and Evolution. No.5, pp: 79~86

[18] Wang Q., Wu H., 1998, Study on fractal description of geographic information and automatic generalization [M]. Wuhan: Wuhan Tech. Univ. of Survey. Mapp. Press (In Chinese)

[19] Zhou C., Luo J. and Yang X. et al., 1999, Geosciencs analysis of remote sensing images [M]. Bejing: Science Press. pp: 59~60 (In Chinese)

Data Fusion of RADARSAT SAR and DMSP SSM/I for Monitoring Sea Ice of China's Bohai Sea

Ya-Qiu Jin

Center for Wave Scattering and Remote Sensing

Fudan University, Shanghai, China 200433

Abstract   Sea ice at the middle latitudes demonstrates high variability and inhomogeneity. By using image data from both active RADARSAT SAR and passive DMSP SSM/I measurements over the Bohai Sea ice in January 23, 1999, we present a data fusion analysis for sea ice detection and classification. Numerical modeling of scattering and thermal radiative transfer for the Bohai Sea shore ice is studied. The scattering index (SI), polarization index (PI), and the polarization ratio (PR) from passive observation of the brightness temperature (T_B) are defined to detect the existence of sea ice. Further, by using the co-horizontally polarized back-scattering coefficient (σ_{hh}) measured by the RADARSAT high resolution SAR, we can further classify the sea ice of the Bohai Sea. Temporal and spatial variations of these indexes are discussed and applied to monitor the Sea ice of the Bohai Sea.

Key Words   SAR and SSM/I, sea ice, Bohai Sea, characteristic indexes

1. Introduction

During recent decades, a significant advance of satellite-borne microwave remote sensing has been the continuous monitoring of global physical and hydrological parameters. Active and passive microwave remote sensing, such as the synthetic aperture radar (SAR) and radiometry technologies, has greatly promoted the sea ice study [Lythe M. et al. 1997; Sandven S. et al. 2001; Bertoria S.V. 2001; Shokr M E. et al. 1999; Carlstrom A. et al 1997]. The detection and specification of sea ice is of great importance to studies of global change, the radiation budget of the earth's hydrosphere, polar climate, and other characteristics. The main issues currently involve understanding and quantitatively retrieving the physical and hydrological information from those multi-channels, multi-polarized, and multi-sensor measurements.

For example, the Canadian RADARSAT SAR launched in November 1995 presented measurements of the co-horizontally polarized backscattering coefficient (σ_{hh}) at the C band. One of most successful programs of passive microwave remote sensing since 1989 is the DMSP (Defense Meteorological Satellite Program) SSM/I (Specific Sensor Microwave /Imager) operated at seven channels (19, 37, 85 GHz with dual polarization, and 22 GHz with vertical polarization only). The brightness temperatures (T_B) in these seven channels are measured. Analysis of both active and passive observations of sea ice have been also discussed [e.g. Kwok R et al. 1998; Lythe M. et al. 1999; Kwok R. et al. 1998; Johannessen O. M et al. 1996]. Monitoring regional sea ice is of special interest to operations involved in the exploration of mineral and petroleum resources as well as in navigation. However, most sea ice research has focused on the polar region at high latitudes, where most sea ice is multi-year (MY) ice. Shore ice at the middle latitudes, such as at China's Bohai Sea or Huang Hai Sea, lasts only a few months during the winter season. Its physical constitution and high temporal and spatial variability are significantly different from ice at the polar regions. Therefore, it is especially necessary to collect data of both active and passive remote sensing and develop analysis and algorithms for shore ice at these regions.

By using both active and passive data from the RADARSAT SAR and DMSP SSM/I observations over the Bohai Sea ice on the same day (January 23, 1999 as a representation), we present a data fusion analysis for sea ice detection and classification. Sea ice is usually modeled as a component medium of ice/brine background embedded randomly by air bubbles. By using numerical simulations of scattering and thermal radiative transfer of a sea ice layer, and comparing these with valid data from the RADARSAT and SSM/I observations, we can study the characteristic variation of the Bohai Sea shore ice. The scattering index (SI), polarization index (PI), and polarization ratio (PR) from passive T_B observations are defined to categorize the existence of sea ice. Further, by using active σ_{hh} measurements from the RADARSAT high-resolution SAR, we can classify the sea ice of the Bohai Sea. Temporal and spatial variations of these indexes can be effectively used to monitor the sea ice of the Bohai Sea.

2. Scattering and Emission of Sea Ice

Sea ice is usually classified and young ice, new ice, first-year ice (FY), multi-year (MY) ice, and so forth [Ulaby et al., 1985]. The ice thickness might be listed, accordingly, in a similar order, from a few centimeters to one meter or more. The sea ice of China's Bohai Sea usually lasts from the end of November to the beginning of March of the next year. The ice thickness is usually from 30 or 40 cm (shore ice) to a few centimeters towards the ocean. The temporal and spatial variability of the shore ice is quite high compared to that of MY ice in the polar region. The dielectric constant (ε_b) of the ice/brine medium, as a background body of sea ice, has been discussed in the literature [Ulaby et al.1985; Jin, 1994], and is taken as $\varepsilon_b = (3.2 + i0.2)\varepsilon_0$ in our calculations. The subscript b denotes the sea ice background. The fractional volume of air bubbles (f_a) is generally less than 0.05 for FY ice. It yields a specific gravity of about $0.87 \, g/cm^3$. In contract, MY ice is usually multi-layered with a f_a of about 0.2, and a specific gravity of $0.7 \, g/cm^3$.

Sea ice can be modeled as a strongly fluctuating random medium [Jin, 1994]. The dielectric fluctuation is described by the variables of

$$\xi_a = 3\frac{\varepsilon_g}{\varepsilon_0}(\frac{\varepsilon_b - \varepsilon_g}{\varepsilon_b + 2\varepsilon_g}) \text{ and } \xi_a = 3\frac{\varepsilon_g}{\varepsilon_0}(\frac{\varepsilon_a - \varepsilon_g}{\varepsilon_a + 2\varepsilon_g}) \tag{1}$$

where the effective permittivity ε_g can be determined by solving

$$f_a\xi_a + (1 - f_a)\xi_b = 0 \tag{2}$$

Note that the dielectric constants for air bubbles, $\varepsilon_a = \varepsilon_0$, and for ice/brine, ε_b, are almost constant with no frequency dependence.

As a model of one-layer sea ice, the co-polarized pp (=vv co-vertically, and hh co-horizontally)-backscattering coefficient has three contributions:

$$\sigma_{pp}(\theta) = \sigma_{ICE}(\theta) + \sigma_{SUR1}(\theta) + \sigma_{SUR2}(\theta)\exp(-2k_g''d\sec\theta) \tag{3}$$

where the first term on the right-hand side (RHS) is the volumetric scattering due to air bubbles in the ice medium, and the second and third terms are the surface scattering from the top and underlying interfaces of the ice layer, respectively. Attenuation through the ice layer with thickness d is taken into account in the exponential term, where $k_g'' = \text{Imag}(k_g)$.

According to the strong fluctuation theory [Jin, 1994], the first term of the RHS in hh-polarization is written as

$$\sigma_{ICE,hh}(\theta) = k_0^4 W \left| \frac{X_{01i}}{D_{2i}} \right|^2 \Phi_{hh} \tag{4}$$

where k_0 is the wavenumber of free space, $|X_{01i}/D_{2i}|^2$ takes into account multi-reflections between the top and underlying interfaces,

$$W = \frac{1}{3}a^3(f_a |\xi_a|^2 + f_b |\xi_b|^2) \tag{5}$$

and

$$\Phi_{hh} = \frac{1}{4k_{gzi}''}(1 - e^{-4k_{gzi}''d})(1 + |R_{12i}|^4 e^{-4k_{gzi}''d}) + 4d|R_{12i}|^2 e^{-4k_{gzi}''d} \tag{6}$$

and where a in Eq.(5) is the radius of the spherical air bubble scatter, R_{12i} is the h-polarized reflection coefficient of the underlying smooth interface, and

$$k_{gzi}'' = \text{Imag}(k_{gzi}), \text{ and } k_{gzi} = \sqrt{k_g^2 - k_0^2 \sin^2\theta}, \quad k_g = k_0\sqrt{\varepsilon_g/\varepsilon_0}.$$

The rough surface scattering is calculated by using the two-scale model [Ulaby et al., 1985] as

$$\sigma_{SURn} = \sigma_n^{KA} + <\sigma_n^{SPA}> \tag{7}$$

where n = 1 (top interface) or 2 (bottom interface), and KA and SPA indicate the Kirchhoff (large-scale) and small perturbation (small-scale) approximations, respectively. Specifically,

$$\sigma_{pp}^{KA} = \frac{|R_p(0)|^2}{2s^2\cos^4\theta}\exp(-\frac{\tan^2\theta}{2s^2}), \quad p=h \text{ or } v \tag{8}$$

where $R_p(0)$ is the p-polarized reflection coefficient of smooth interface at $\theta = 0$, and the Gaussian surface slope $s = \sqrt{2}\delta/\ell$, where δ and ℓ are, respectively, the square root of the surface variance and correlation length in large scale. Further,

$$\sigma_{pp}^{SPA}(\theta) = 8k_0^4\delta_1^2\cos^4\theta |A_p|^2 W(2k_0\sin\theta,0), \quad p=v \text{ or } h \tag{9}$$

where $W(k_\perp,k_z)$ is the Fourier transform of the correlation function of surface height with the square root of the variance δ_1 and correlation length ℓ_1, and the function A_p is related to the p-polarized reflection coefficient of the smooth interface at θ. Equation (9) is averaged over the large-scale undulation as [Ulaby et al., 1985]

$$<\sigma_{pp}^{SPA}(\theta)> = \frac{1}{2\pi}\int_{\pi/2}^{-\pi/2} d\phi_n [\sigma_{pp}^{SPA}(\theta')|_{\theta_n=\tan^{-1}s} + \sigma_{pp}^{SPA}(\theta')|_{\theta_n=-\tan^{-1}s}] \tag{10}$$

where $\theta' = \cos^{-1}(\cos\theta_n\cos\theta + \sin\theta_n\sin\theta\cos\varphi_n)$.

Figure 1 shows simulated σ_{hh} at the C band and incident angle of 35° for increasing ice thickness d. Because the air bubble's f_a in FY ice is always small, scattering enhancement is mainly contributed by surface scattering of σ_{SUF1} and σ_{SUR2} (because d very small). As d keeps increasing, σ_{SUR2} gradually darkens due to attenuation through the ice layer, and scattering is dominated by σ_{SUF1}. Sea surface wind can drive a free sea surface to rougher and also enhance scattering. Thus, without any prior knowledge, it is not appropriate to use a single instance of increasing σ_{hh} to classify sea ice existence.

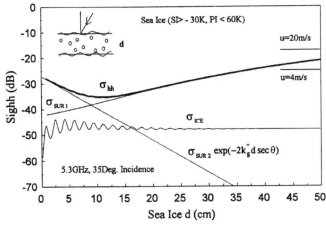

Fig.1. Back   air bubbles: $f_a = 0.05$, radius a = 0.025cm; d<50 cm. Roughness of the top surface: $\delta = 2 + (d-1)/49\,cm$, $\ell = 15\,cm$, $\delta_1 = 0.1\,cm$, $\ell_1 = 5\,cm$. Roughness of the bottom surface: $\delta = 2\,cm$, $\ell = 20\,cm$, $\delta_1 = 0.1\,cm$, $\ell_1 = 5\,cm$.

The vector radiative transfer (VRT) equation for calculation of $\overline{T}_B(= T_{Bv}, T_{Bh})$ is written [Jin, 1994; Jin and Zhang, 1999]:

$$\cos\theta \frac{d}{dz}\overline{T}_B(\theta, z) = -\kappa_a(\theta)\overline{T}_B(\theta, z) + \int_0^\pi d\theta' \sin\theta' \overline{\overline{P}}(\theta, \theta') \cdot \overline{T}_B(\theta', z) + \kappa_a \overline{T} \qquad (11)$$

By solving the VRT equation (11) with the boundary conditions, we can calculate the brightness temperature $T_{Bp}(p = v, h)$ at SSM/I channels. Definitions of the extinction coefficient κ_e, absorption coefficient κ_a, and phase matrix $\overline{\overline{P}}$ of small air bubble scatters, and the numerical approach of the VRT equation are given by Jin (1994).

Due to the existence of sea ice, the T_B of all SSM/I channels is increased, and the difference between different channels becomes smaller. Because atmospheric absorption at 22 GHz and 85 GHz channels is fairly close, compared to other channels, the scattering index is defined as $SI = T_{B22v} - T_{B85v}$ to indicate a reduced T_B difference in high- and low-frequency channels. SI is sensitive to the existence of sea ice. When no sea ice exists, it has SI << 0, e.g. SI < - 40K. When sea ice appears, the index SI gradually increases (> -30K). Meanwhile, the T_B difference between vertical and horizontal polarization also becomes smaller. The polarization index is defined as $PI = T_{B19v} - T_{B19h}$.

Figure 2 shows numerical simulations of SI-PI for different situations of increasing ice thickness from 0 to 15 cm obtained by solving the VRT equation (11). The solid points and line indicate the numerical results, and the circle points are some typical SSM/I data selected from observations over the Bohai Sea during October 1995-April 1996. Figure 2 roughly indicates a rule of ice detection as SI > -30 K, PI < 60 K [Jin, 1998].

Figure 3 is an image of the brightness index (DN) from observation of RADARSAT SAR data on January 23, 1999, over the Bohai Sea. Following the instructions of the RADARSAT Data Products Specification (1997), the backscattering coefficient σ_{hh} (dB) at the j^{-th} pixel can be obtained from the DN dB value as

$$\sigma_{hh}(j) = \beta(j) + 10 \times lg_{10}(\sin\theta_j) \tag{12a}$$

$$\beta(j) = 10 \times lg_{10}[DN(j)^2 + A_3 / A_2(j)] \tag{12b}$$

where $\beta(j)$ is the radar brightness of the j^{-th} pixel, $A_2(j)$ is the scaling gain value and A_3 is a fixed offset, and $\theta(j)$ is the local incidence angle to the j^{-th} pixel, determined by the satellite altitude, the earth radius, and the slant range from the satellite to the j^{-th} pixel.

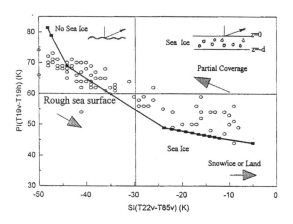

Fig. 2 The SI-PI graph and VRT results for detection of sea ice.

Ocean: $T_2 = 273\,K$; Sea ice: $T_1 = 268\,K$; Air bubble: a=0.02 cm, $f_a = 0.08$.

Atmosphere: $T_a = 278\,K$, Atmospheric opacity: $\tau(19) = 0.06$, $\tau(22) = 0.12$, $\tau(37) = 0.08$, $\tau(85) = 0.22$.

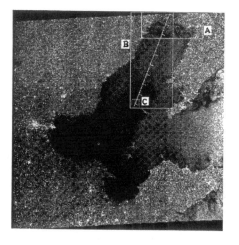

RADARSAT SAR DN Image

Fig. 3 Image of DN values from RADARSAT SAR.

Figures 4a and 4b show the σ_{hh} values of the areas circled by frames A and B in Fig. 3, respectively. Figure 4c shows the variation of σ_{hh} along line C in Fig. 3. The numbers on the horizontal axis indicate the pixel location from the north (value 0) to the south.

It can be seen from Fig. 4c that σ_{hh} is, in comparison with the sea water area, significantly increased (from a low of -30 to -18dB), due to the north shore ice of the

North Gulf of (locations 0 to 500. The large fluctuation between locations 300 and 1000 correspond to the heterogeneous distribution of ice type and thickness. Location 1000 seems to be the border between sea ice and seawater. After location 1000, the value of σ_{hh} remains within -35dB to -25dB; these locations have no ice, and scattering is coming from the rough sea surface. Without a priori knowledge or the variations of the spatial correlation of σ_{hh}, it would not be easy to distinguish the sea ice from the water areas. After location 3500, σ_{hh} decreases as a typical value of flat- surface sea water e.g. less than –40 dB.

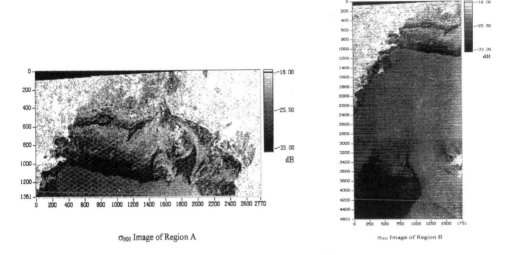

σ_{HH} Image of Region A

σ_{HH} Image of Region B

Fig. 4a σ_{hh} for frame A of Fig. 3.   Fig. 4b σ_{hh} for frame B of Fig. 3.

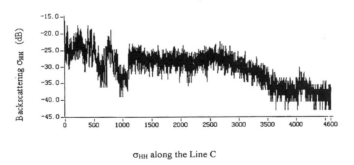

σ_{HH} along the Line C

Fig. 4c Variation of σ_{hh} along line C of Fig. 3.

Figures 5a and 5b present, respectively, the images of the SI and PI indexes from the SSM/I T_B observations on the same day and in the same area as Figs. 3 and 4. The rule of ice detection using SI and PI as discussed above seems direct for finding the shore ice, especially near Ying Kou City. Figure 5c gives the SI and PI values along line C of Fig. 3. It can be seen that sea ice exists from location 0 to location 1000. Further, from the high resolution σ_{hh} images of SAR in Figs. 4a and 4b, and its spatial variation, the physical status of sea ice, such as thickness, surface roughness, and snow cover, can then be detailed.

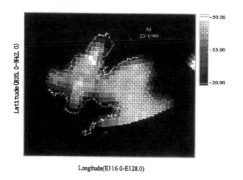

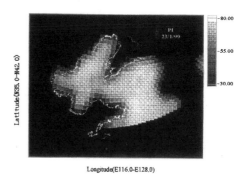

Fig. 5a. SI from SSM/I T_B data.   Fig. 5b. PI from SSM/I T_B data.

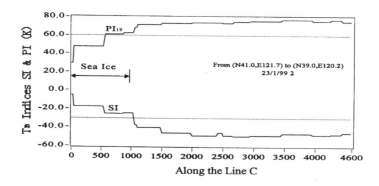

Fig. 5c SI and PI indexes along line C.

Due to the spatially heterogeneous distribution of sea ice and seawater, an index of the polarization ratio (PR) at the SSM/I 37GHz channel was proposed to describe ice coverage percentage (concentration) [Haggar et al., 1998]:

$$PR = \frac{T_{B37v} - T_{B37h}}{T_{B37v} + T_{B37h}} \tag{13}$$

Empirically, at PR=0.1851, the surface is 100% water, and at PR=0.0157, it is 100% ice. The usefulness of these empirical values has not yet been validated. Actually, the radiative transfer equation of the atmosphere and underlying sea ice/water can be simply written as

$$T_{Bp} = e_p T_s e^{-\tau} + (1 - e^{-\tau})(1 + r_p e^{-\tau})T_a + s_p T_s e^{-\tau}, \quad p = v, h \tag{14}$$

where e_p and τ_p are, respectively, the p (v,h)-polarized surface emissivity and reflectivity, with $e_p = 1 - r_p$; τ is the atmospheric opacity; T_s and T_a are, respectively, the surface physical temperature and atmospheric effective temperature.. On the RHS of equation (14), the first term represents the emission from sea ice to water. The second term is the atmospheric emission, and the third term is the additional contribution to T_{Bp} due to surface roughness, which is taken into account by using the empirical parameter s_p (p = v, h). By letting $T_a \approx T_s$, we obtain

$$PR = \frac{e^{-\tau}[(r_h - r_v)e^{-\tau} + (s_v - s_h)]}{[2 - (r_h + r_v)]e^{-\tau} + (1 - e^{-\tau})[2 + (r_h + r_v)e^{-\tau}] + (s_v + s_h)e^{-\tau}} \tag{15}$$

Thus, PR has a functional dependence on atmospheric opacity τ and surface reflectivity r_p. The reflectivity of an ice layer over water is calculated as

$$r_p = \left| \frac{R_{01p} + R_{12p}\exp(ik_{gz}d)}{1 + R_{01p}R_{12p}\exp(ik_{gz}d)} \right|^2 \quad , p = v, h \tag{16}$$

where R_{01p} and R_{12p} are, respectively, the p-polarized reflection coefficients of the interfaces of air-ice and ice-water.

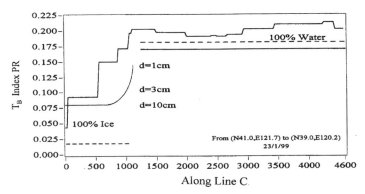

Fig. 6. Index PR along line C of Fig. 3 compared with Eq.(15) and empirical values.
$\tau = 0.05$, $s_v = 0.1$, $s_h = 0.01$.

Figure 6 shows the PR values of equation (13) from SSM/I data along line C of Fig. 3, where the solid lines near the 100% indications for water and ice give the PR values from equation (15), and the dotted lines are the empirical values given by Haggar et al.(1998). Figure 6 shows that the empirical value for a 100% ice surface is underestimated because it is even less than the value for land at location 0. Considering partial ice coverage within the spatial resolution, the detection rule of the SI and PI indexes should be empirically modulated.

We propose that the SI, PI, and PR indexes be employed to categorize the existence of sea ice, and the variation of σ_{hh} with high spatial resolution then be used to determine the sea ice status. Further collection of ground truth data should be of much help to improve validation and classification. A study of data from infrared and visible light observations and a comparison with microwaves data are also suggested.

4. Conclusions

The shore ice of China's Bohai Sea, which lasts for a few months during the winter season, is different from polar MY ice. The rule for ice detection and classification for the Bohai sea should be specifically studied. By using the image data from active and passive measurements of the RADARSAT SAR and DMSP SSM/I over the Bohai Sea ice, we have developed a correlated analysis for the Bohai Sea ice. By using numerical modeling of scattering and radiative transfer of a layer of sea ice, we demonstrate the characteristic variation of the Bohai Sea shore ice. The T_B indexes, SI, PI, and PR, are defined to detect the existence of sea ice. Further, using back-scattering σ_{hh} measured by high-resolution SAR, the sea ice of the Bohai Sea can be further classified. We propose to collect more ground truth data to determine uncertain parameters. Temporal and spatial variations of these indexes can then be applied to monitor the sea ice of the Bohai Sea. In future work, more collection of ground truth data, such as surface roughness, snow cover, atmospheric absorption, and other parameters should be quite helpful in improving sea ice detection and classification. The correlation of microwave active and passive measurements, along with data fusion of infrared and visible light observations, can effectively monitor the Bohai Sea ice.

Acknowledgments

This work was supported by the China National Key Basic Research Program 2001CB309401, and NSFC 60171009.

References

Bertoria S V (2001), Special issue-stuudy of multi-polarization C-band backscattering signatures for Arctic Sea ice mapping with future satellite SAR, *Canadian Journal of Remote Sensing*, 27(5): 387-402

Carlstrom A (1997), A microwave backscattering model for deformed first-year sea ice and comparisons with SAR data, *IEEE Trans. on Geosci. Rem. Sens.*, 35(2): 378-391

Haggar C E, Garrity C, and Ramseier R O (1998), The modeling of sea ice melt-water ponds for the high Arctic using an airborne line scan camera and applied to SSM/I, *Internat. J. Rem. Sens.* 18(12): 2373-2394

Haverkamp D, Soh L K, Tsatsoulis C (1995), Automated approach to determining sea ice thickness from SAR data, *IEEE Trans. Geosci. Rem. Sens.* 33(1): 46-50

Jin Y Q (1994), Electromagnetic scattering modelling for quantitative remote sensing. Singapore: World Scientific

Jin Y Q (1998), Monitoring regional sea ice of China's Bohai Sea by using SSM/I scattering indexes, *IEEE J. Ocean. Engin.* 23(2):141-144

Jin Y Q and Zhang N (1999), Correlation of the ERS and SSM/I observations over snowpack and numerical simulation, *International Journal of Remote Sensing*, 20(15/16): 3009-3018

Johannessen O M, Sandven S, Pettersson L H, and others (1996), Near-real-time sea ice monitoring in the Northern Sea Route using ERS-1 SAR and DMSP SSM/I microwave data, *Acta Astronautica*, 38(4/8): 457-461

Kwok R, Schweiger A, Rothrock D A and others (1998), Sea ice motion from satellite passive microwave imagery assessed with ERS SAR and buoy motions, *Independent Energy*, 28(4): 8191-8213

Lythe M, Hauser H, Wendler G (1999), Classification of sea ice types in the Ross Sea, Antarctica from SAR and AVHRR imagery, *International Journal of Remote Sensing*, 20(15/16): 3072-3078

Lythe M, Hauser A, Wendler G (1997), Model-based interpretation of ERS-1 SAR images of Arctic sea ice, *International Journal of Remote Sensing*, 18(12): 2483-2503

RADARSAT DATA Products Specifications (1997), RADARSAT International, Canada: Richmond

Sandven S, Dalen O, Lundhaug M and others (2001), Special Issue-sea ice investigations in the Laptev sea area in late summer using SAR data, *Canadian Journal of Remote Sensing*, 27(5): 501-515

Shokr M E, Jessup R, Ramsay B (1999), An interactive algorithm for derivation of sea ice classifications and concentrations from SAR images, *Canadian Journal of Remote Sensing*, 25(1): 70-79

Ulaby F T, Moore R K, and Fung A K (1985), Microwave Remote Sensing, 3 Vols. Mass: Artech Press

Wave Propagation, Scattering and Emission in Complex Media
Edited by Ya-Qiu Jin
Science Press and World Scientific, 2004

Retrieving Atmospheric Temperature Profiles from Simulated Microwave Radiometer Data with Artificial Neural Networks

Zhigang Yao[1], Hongbin Chen[2]

1 Institute of Meteorology, PLA University of Science and Technology,
Nanjing 211101，China

2 LAGEO, Institute of Atmospheric Physics, CAS, Beijing 100029, China

Abstract   Artificial neural networks are used to retrieve vertical profiles of atmospheric temperature from simulated microwave radiometer data. The global and local retrieval algorithms are compared. The global retrieval experiments show that the overall root mean square error in the retrieved profiles of a test dataset is about 7% better than the overall error using a linear statistical retrieval，and the retrieval errors are about 0.5K better at 200- and 250-hPa levels. The local experiments show that the differences between the results of neural network and linear statistical retrieval approaches are different from region to region but are not remarkable. In comparison with the global retrieval, the local retrieval is well.

Key Words   artificial neural networks，microwave remote sensing，temperature profiles，retrieval and simulation

1. Introduction

During recent years the radiosonde network has been the primary observing system for monitoring atmospheric temperature. However, radiosonde instruments provide sounding information over limited land areas, while vast oceanic areas are void of this data. In this sense, the retrievals of temperature profiles from satellites remote sensing for applications such as weather analysis and data assimilation in numerical weather prediction models has become of great importance. Meeks and Lilley (1963) suggested that atmospheric temperature profiles could be retrieved from the radiances in oxygen absorbing bands measured by a vertically pointing microwave radiometer. In their study, the microwave weighting functions were calculated for the first time. The primary advantage of the remote sensing of atmospheric temperature profiles using microwave radiometers is that the microwave radiance can penetrate non-precipitating clouds, especially cirrus. Although infrared sounding data has been applied to retrieve atmospheric temperature profiles, it is impossible to obtain the atmospheric temperature information below the clouds. Consequently, in addition to developing infrared remote sensing technology, it is imperative to develop microwave remote sensing technology to improve the temperature sounding accuracy, reduce the clouds effects on sounding temperature profiles and enhance the capability of remote sensing of atmospheric temperature profiles in the whole day and night.

In order to obtain atmospheric temperature profiles from the radiances measured by a vertically pointing instrument, the conventional physical retrieval methods have been studied (Chahine 1970; Smith 1970), which generally requires a good initial guess and accurately direct transfer model. Furthermore, the methods easily lead the results to a local minimum due to that the equation is nonlinear. One other technique for retrieving temperature profiles from microwave radiances is a linear statistical inversion (Strand and Westwater 1968; Hogg et al. 1983). Compared with the physical retrieval methods, the statistical retrieval method is easy and practical although it is lack of the capability of retrieving temperature profiles in extreme cases. A linear statistical inversion is used to obtain continuous soundings of temperature in real time (Hogg et al. 1983). The root mean square difference between these temperature profiles and the U.S. National Weather Service operational radiosonde profiles at the same site is generally 1-3K throughout the troposphere (Westwater et al. 1984).

Neural networks have been broadly applied to estimation of high quality geophysical parameters (information about physical, chemical, and biological properties of the oceans, atmosphere, and land surface) from remote (satellite, aircraft, etc.) measurements (Vladimir M. Krasnopolsky et al. 2002). There are several similarities between linear statistical inversion and neural network retrieval (Churnside et al. 1994). In both cases, some input vector is operated on to produce an output vector. For our purposes, the input vector is a set of microwave radiances, and the output vector is the vertical profile of temperature. In both cases, the operator is generated using an existing set of paired input and output vectors. In the linear statistical inversion, this set is used to obtain the inversion coefficients. In the neural network retrieval, this set is used to train the network. Some groups have used neural networks to retrieval temperature profiles from microwave radiometer data. Churnside et al. (1994) applied neural networks to obtain vertical profiles of temperature from simulated microwave radiometer data. In their study, the overall rms error in the retrieved profiles of a test dataset was only about 8% worse than the overall error using a statistical retrieval. But the data used were taken only from routine National Weather Service radiosonde data at a single radiosonde station, and the results were not representative. Charles et al. (1996) obtained good results with a neural network in comparison with an operational linear -regression algorithm. Although the data used were collected from a 30-day period observation in northern hemisphere winter, only a global retrieval was demonstrated. In order to fully understand the performance of retrieving temperature profiles using neural networks, the data are taken from different radiosonde stations in the world area, and the global and local retrieval neural networks are considered respectively in our study.

2. Neural Network Approaches

A schematic diagram of the type of neural network that is used for this work is presented in Fig. 1 (Hornik et al., 1989). The network has an input layer of L nodes to which an input vector $\mathbf{X}$ of length L is applied. The input layer has no function of computation and is only used to enter the input vector. Each input node is connected to all M nodes in a hidden layer. Each node in the hidden layer performs a weighted sum over all the input values to produce an output vector $\mathbf{Y}$. Each node in the hidden layer is connected to each node in an output layer, which performs a weighted sum over all of the results of the hidden-layer calculations. The N values from the output-layer nodes create the output vector $\mathbf{Z}$. For the i-th node in the hidden layer, a neuron performs a weighted sum of its inputs and applies a sigmoidal function to the result to produce an output, which can be expressed as

$$Y_j = S(\sum_{i=1}^{L} w_{ij} X_i + b_j),\tag{1}$$

where S is the sigmoidal function (Fig. 2) in the form

$$S(a) = \frac{1}{1+e^{-a}}.\tag{2}$$

In Eq. (1), w_{ij} is the weighting of the connection between the ith input neuron and the j-th hidden neuron; b_j is the bias in the j-th neuron of the hidden layer. The purelin linear function is applied in the output layer. As a result, the output values can be arbitrary.

The weights w_{ij} and the biases b_j are determined during the training process. They are obtained using a back-propagation algorithm that is described in detail by Rumelhart et al. (1986). This algorithm adjusts the weights and biases iteratively to reduce the difference between the actual training set output vectors and the estimated output vectors calculated by the network using the input vectors of the training set.

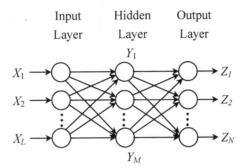

Fig. 1.   Diagram of the neural network employed

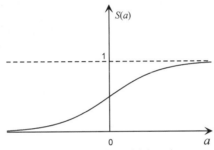

Fig. 2.   The sigmoid function

3. Simulating Brightness Temperatures

For a non-scattering plane-parallel atmosphere in local thermodynamic equilibrium, the brightness temperature $T_B(\theta)$ observed by a satellite-borne microwave radiometer can be expressed as (Ulaby, 1981; Chen, 1991)

$$T_B(\theta) = T_{UP} + T_{BS} + T_{DN} + T_{EXE},\tag{3}$$

where T_{UP} is the contribution from the total upward atmospheric radiance to $T_B(\theta)$; T_{BS} is the contribution from the radiance emitted by the surface and attenuated by the atmosphere; T_{DN} is the contribution from the downward atmospheric radiance reflected by the land or sea surface and then attenuated by the atmosphere; T_{EXT} is the contribution from the cosmic background radiance which penetrate the atmosphere and is reflected by the surface and penetrate the atmosphere again. According to the solution of the radiative transfer equation, in the low frequency band ($f < 100GHz$), the four terms are respectively given by

$$T_{UP}(\theta) = \int_0^\infty k_a(z)T_a(z)e^{-\int_z^\infty k_a(z')dz'}dz/\cos\theta,\tag{4}$$

$$T_{BS} = \varepsilon(\theta,p)T_{SN}e^{-\tau/\cos\theta},\tag{5}$$

$$T_{DN} = [1-\varepsilon(\theta,p)]e^{-\tau/\cos\theta}\int_0^\infty k_a(z)T_a(z)e^{-\int_0^z k_a(z')dz'}dz/\cos\theta,\tag{6}$$

$$T_{EXT} = T_{\cos}(1-\varepsilon)e^{-2\tau/\cos\theta},\tag{7}$$

where $k_a(z)$ is the absorption coefficient at the level z; $T_a(z)$ is the atmospheric physical temperature at the level z; $\varepsilon(\theta, p)$ is the land or sea surface emissivity, which depends on the nadir angle θ and the polarization p (vertical or horizontal); $[1 - \varepsilon(\theta, p)]$ is the surface reflectivity; T_{cos} is the cosmic background with an approximate value of 2.7K, since the brightness temperature contribution T_{EXT} from the cosmic background is the product of the reflectivity, the square of the transmittance from the surface, it is very small and can be ignored. According to what has been analyzed above, the equation of microwave remote sensing from satellite-borne radiometer can be expressed as

$$T_B(\theta) = \varepsilon(\theta, p) T_{SN} e^{-\tau/\cos\theta} + \int_0^\infty J(\theta, p) k_a(z) e^{-\int_z^\infty k_a(z')dz'} dz / \cos\theta, \qquad (8)$$

where $J(\theta, z, p)$ is given by

$$J(\theta, z, p) = \{(1 + [1 - \varepsilon(\theta, p)] e^{-2\int_0^z k_a(z')dz'}\} T_a(z). \qquad (9)$$

The weighting function is defined as

$$K(z) = \{1 + [1 - \varepsilon(\theta, p)] e^{-2\int_0^z k_a(z')dz'}\} k_a(z) e^{-\int_z^\infty k_a(z')dz'}, \qquad (10)$$

which provides weights of the contribution of $T_a(z)$ to $T_B(\theta)$.

For the clear plane-parallel atmosphere, the microwave brightness temperatures and the weighting functions are calculated by dividing the atmosphere into N thin layers; the atmospheric parameters and the absorption coefficient for a certain thin layer are assumed to be uniform, then the upward radiance and the downward radiance and the attenuated radiance for each layer can be calculated one by one; the contribution from the whole atmosphere including the surface to the radiance measured by a radiometer is accumulated from layer to layer. Due to the high variability and high emissivity values over land, the retrieval of atmospheric temperatures is more difficult. Compared to the land surfaces, the variability of the ocean-surface emissivity and its value are small over oceans. That is to say that the ocean surface is a cold background. So the case over the ocean surface is only considered in our study.

A non-scattering microwave radiative transfer model has been developed by Chen (1991). Liebe's microwave propagation model is applied (Liebe, 1989), and Rosenkranz's oxygen absorption model is referenced in this study (Rosenkranz, 1993). For ocean surface emissivity, Wentz's model is used (Wentz, 1983), which is simple and is continuous at the wind speed 7m/s. In our study, the nadir angle is assumed to be zero. The sea surface temperature, the sea surface wind speed and the sea surface water salinity are the three input parameters of the sea surface emissivity model (Wentz, 1983). In order to simulate all kinds of conditions over the sea surface, the sea surface temperature is assumed to be the sum of the surface air temperature and a random value from −5- to +5-K, and the wind speed over the sea surface is assumed to be a random value from 0- to 25-m/s. The salinity varies from about 32 to 37 ‰ in the open ocean. This salinity variation translates into a brightness temperature variation of the sea surface, which is small in the oxygen absorption band used. Hence we simply assume a nominal value of 34‰ for all computations.

4.   Retrieval Experiments and Results

In this paper, the atmospheric temperature retrievals reported are obtained using simulated brightness temperatures, which are at the seven frequencies in the 50- to 60-GHz oxygen absorption band (50.50-, 53.20-, 54.35-, 54.90-, 58.825-, 59.40- and 58.40-GHz). The peaks of the weighting functions of the former four channels are in the troposphere, and the ones of the latter three channels are in the stratosphere. The seven channels can be utilized to retrieve atmospheric temperature profiles from the 1000- to the 10-hPa levels.

Tab. 1.   Number of atmospheric profiles per month contained in the training and testing sets

Station Number	70200	47807	48698	94975
Latitude	64.50	33.58	1.36	-42.83
Longitude	165.43	130.38	103.98	147.50
Jan	71	113	103	45
Feb	77	96	89	42
Mar	10	93	92	48
Apr	108	108	90	48
May	124	105	88	22
Jun	118	93	108	51
Jul	87	80	81	44
Aug	126	103	103	62
Sep	112	103	99	51
Oct	87	105	95	49
Nov	69	114	93	67
Dec	81	135	75	64
Training	760	832	744	396
Testing	380	416	372	197

Here the radiosonde data from four representative radiosonde stations are used to simulate the brightness temperatures. The detailed information about the distribution of the datasets is listed in Tab. 1. The four stations stand in the different areas, the station numbers of which are 70200, 47807, 48698 and 94975 respectively. According to the latitudes and longitudes of the stations in Tab. 1, it is known that the four stations stand in the high-latitude area of the Northern hemisphere, the mid-latitude area of the Northern hemisphere, the low-latitude area of the Northern hemisphere, and the area of the Southern Pacific Ocean respectively. In this paper, the four stations represent the different areas. Because the weighting functions can reach the altitude of 10-hPa, only the radiosonde data that includes the parameters at 10-hPa are collected. The parameters of the atmosphere at the altitudes above 10-hPa are substituted by the parameters of the American standard atmosphere. Furthermore, for reflecting the actual vertical distribution of the atmospheric parameters, only the high vertical resolution radiosonde data are used, which at least have 40 sounding points. There are 4097 examples that are satisfied with the requirements mentioned above, which are from the radiosonde observations at the four stations in 2000-2002. The examples include all kinds of the vertical distributions of the atmospheric parameters in the four different areas and in the different seasons. In the experiments, the 2732 examples (about two thirds) are used to train neural networks, and the 1365 (about one third) examples are used to test neural networks.

In practice the accuracy of microwave atmospheric retrievals is limited in part by the presence of noise in the measured spectra. The major source of noise is the thermal noise produced by radiometer itself. To account for instrument noise, a zero-mean Gaussian random error with 1.0-K standard deviation is added to each calculated brightness temperature.

In our study, the three-layer neural networks are utilized to retrieve atmospheric temperature profiles. The input vector comprises seven elements, which are the simulated brightness temperatures at the seven frequencies of the radiometer. There are 20 nodes in the hidden layer. The output vector is an 18 elements vertical profile of the temperature including the sea surface water temperature, the sea surface air temperature, and air temperatures at the 16 levels from 1000- to 10-hPa (1000-, 925-, 850-, 700-, 500-, 400-, 300-, 250-, 200-, 250-, 100-, 70-, 50-, 30-, 20-, 10-hPa).

4.1 Global Retrieval

In order to evaluate the capability of retrieving atmospheric temperature profiles using neural networks, the multiple linear regression method is used to retrieve temperature profiles at the same time. The 2732 examples and the corresponding simulated brightness temperatures are used to train the neural networks. After training, 1365 simulated brightness temperature vectors are inputted to the neural network, the corresponding output vectors are the retrieved temperature profiles. Additionally, we use the same datasets to obtain the regression coefficients and test the statistical retrieval method. The overall rms errors of the neural network and linear statistical retrieval approaches are 2.63K and 2.84K respectively. The rms errors at the different pressure levels (from 1000- to 10-hPa) are shown in Fig. 3 and Fig. 4. Fig. 3 shows the results of the whole data from the four stations，while Fig 4 show the results of the data from the four different stations respectively.

The global retrieval experiments show that the overall rms error in the retrieved profiles of a test dataset using the neural network is about 7% better than the overall error using the linear statistical retrieval. In Fig. 3, one can clearly see that the retrieval errors using the neural network approach are about 0.5K better at 200- and 250-hPa levels than ones using the linear statistical approach. In Fig. 4b, it can be seen that the capabilities of the two global retrieval approaches have no much differences in the mid-latitude area of the Northern hemisphere; In Fig. 4c, it can be seen that the results of the global neural network retrieval approach below the 100-hPa pressure level is better than ones of the global statistical retrieval approach in the low-latitude area of the Northern hemisphere; In Figs. 4a and 4d, one can see that the results of retrieving atmospheric temperatures in the vicinity of the tropopause with the global neural network retrieval approach are apparently better than the results with the global statistical retrieval approach in the high-latitude area of the Northern Hemisphere and the area of the Southern Pacific Ocean. According to the above results, we can draw a conclusion that when a global retrieval approach is applied, the neural network retrieval will be better than the linear statistical approach in the whole world area, while the performances in the different areas with the global retrieval approach are slightly different.

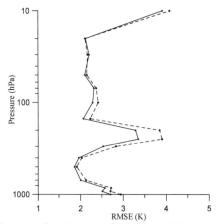

Fig. 3.   Comparison of total rms retrieval errors of neural (solid) and statistical approaches (dashed)

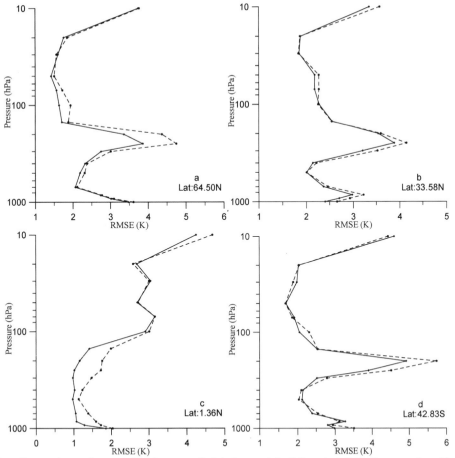

Fig. 4.   Comparison of rms retrieval errors of global neural (solid) and statistical approaches (dashed) at four individual sites

4.2 Local Retrieval

In order to further understand the capability of retrieving temperature profiles using neural networks, we construct the independent neural networks with the data taken from the four different areas respectively. As a result, there are four different neural networks in this section. In order to differ from the neural network in section 4.1, we define the neural networks in this section as the local neural networks, and correspondingly we define the retrieval approaches as the local retrieval approaches. As being done in section 4.1, the neural network approach and the linear statistical approach are taken at the same time in order to evaluate the performances of retrieving temperature profiles using neural networks. The whole rms errors of the testing datasets taken from four different areas using the corresponding four local neural networks are listed in Tab. 2, in which the whole rms errors of the testing datasets using the same global neural network in section 4.1 are also shown. The rms errors at the different levels in the different areas using the local retrieval approaches are shown in Fig. 5.

In Table 2, one can see that the capability of retrieving temperature profiles with two different approaches is slightly different from region to region. One can also see that the local neural network retrieval approach is the best in the high- and mid-latitude areas of the Northern hemisphere, and the local linear statistical retrieval approach is the best in the low-latitude area of the Northern hemisphere and the area of the Southern Pacific Ocean respectively. As a whole, compared with the global retrieval approaches, the local retrieval approaches are better. Consequently applying the local retrieval approach is worth in a given area.

Table 2.   Total retrieval errors of linear statistical and neural network approaches

Approaches	RMSE (K)			
	70200	47807	48698	94975
Global statistical retrieval	3.04	2.81	2.41	3.24
Global network retrieval	2.79	2.66	2.21	2.97
Local statistical retrieval	2.76	2.56	2.12	2.65
Local network retrieval	2.67	2.52	2.44	2.76

In Fig. 5, one can also see that the local neural network retrieval approach obtain the slightly better results than the local statistical retrieval approach in the high-latitude area of the Northern hemisphere, which is due to that the profile variation in the area is large in different quarters, which lead to that the non-linearity is more obvious in the microwave transfer equation. It is known that the most inportant characteristics of a network for solving inversion problems, the class of problems represented by profile retrieval, are that it is capable of modeling highly nonlinear data from examples and is able to generalize or interpolate. The ability to generalize guarantees that each profile retrieved by the network during actual operation is a unique, new function, and not simply the learned training example most closely approximating the required profile. In the low-latitude area of the Northern hemisphere, the variation of temperature profiles is relatively small, so the linear statistical approach can obtain slightly better results, and the rms errors at the altitudes below 200-hPa are less than 1K. Especially, it should be noticed that the local neural network can obtain better results at the levels in the vicinity of the tropopause in the high- and mid-latitude areas of the Northern hemisphere. According to what has been discussed, there are slightly differences between the results of the local neural network and linear statistical retrieval approaches.

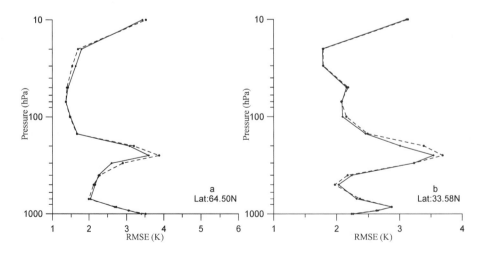

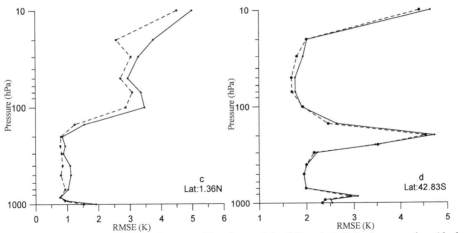

Fig. 5.   Comparison of rms retrieval errors of local neural (solid) and statistical approaches (dashed)

5. Conclusions

In this study, neural networks are used to retrieve vertical profiles of atmospheric temperature from simulated radiometer data. One hidden layer is adopted in the neural networks. Twenty neurons in the hidden layer are used. Increasing the number of hidden layers or neurons did not seem to increase the performance significantly. The atmospheric temperature profiles used in the simulation experiments are taken from different areas in the different seasons. Although the data used here are still limited, some interesting results have been obtained. In comparison with the linear statistical approach, we can draw the following conclusions:

(1) When a global retrieval approach is applied, the neural network retrieval can obtain better results in the whole world area, while the performance in the different areas with the global retrieval approach is slightly different.

(2) When a local retrieval approach is applied to a certain area, the capability of the neural network retrieval is different from region to region. In the areas where the variation of temperature profiles is large, the neural network retrieval can obtain slightly better results than the linear statistical retrieval. In the areas where the variation of temperature profiles is relatively small, such as over the low-latitude area of the Northern hemisphere, the linear statistical retrieval can obtain slightly better results.

(3) Generally, applying the local retrieval approach can obtain better results than the global retrieval approach.

Acknowledgements

The authors would like to thank Prof. Y. Q. Jin of Fudan University for his constructive discussions and valuable comments. The first author would also like to thank Prof. W. Yan of the PLA University of Science and Technology for his help on artificial neural networks.

This work was supported by the China State Key Basic Research Project (2001CB309402).

References

Chahine M T. (1970), Inverse problems in radiative transfer: Determination of atmospheric parameters. *J. Atmos. Sci.*, **27**: 960-967

Charles T. Bulter, R. V. Z. Meredith (1996), Retrieving atmospheric temperature parameters from DMSP SSM/-1 data with a neural network. *J.Geophys.Res.*, **101:** 7075-7083

Chen H B. (1991), Simulation of the microwave brightness temperature; Study of the possibility to retrieve some cloud parameters from the MMS 5 channel measurements [Ph. D. Thesis]. Lille: USTL

Churnside J H, Stermitz T A, and Schroeder J A. (1994), Temperature profiling with neural network inversion of microwave radiometer data. *J. Atmos. Oceanic Technol.*, **11(1):** 105-109

Hogg D C, Decker M T, Guiraud F O, Earnshaw K B, et al. (1983), An automatic profiler of the temperature, wind, and humidity in the troposphere. *J. Appl. Meteor.*, **22:** 807-831

Hornik K M, Stinchcombe M, and White H. (1989), Multilayer feedforward networks are universal approximators. *Neural Networks*, **2:** 359-366

Lee J, Weger R C, Sengupta S K, and Welch R M. (1990), A neural network approach to cloud classification. *IEEE Trans. Geosci. Remote. Sens.*, **26:** 846-855

Liebe H. (1989), MPM – an atmospheric millimeter-wave propagation model. *Inter. J. IR and Millimeter Wave.*, **10:** 630-650

Meeks M J, and Lilley A. E. (1963), The microwave spectrum of oxygen in the earth's atmosphere., *J. Geophys. Res.*, **68:** 1683-1703

Rosenkranz P W. (1993), Absorption of microwaves by atmospheric gases, in *Atmospheric Remote Sensing by Miceorave Radiometry*, M. A. Janssen, Ed. New York: Wiley, Ch.2

Rumelhart D E, Hinton G, and Williams. (1986), Learning internal represcntation by error propagation. *Parallel Distributed Processing: Explorations in the Microstructure of Cognition*, vol. 1: Foundations. Cambridge: MIT Press, 318-362

Smith W L. (1970), Interative solution of the radiative transfer equation for the temperature and absorbing gas profile of an atmosphere. *Appl. Opt.*, **9:** 1993-1998

Strand, O. N., and E. R. Westwater (1968), Minimum rms estimation of the numerical solution of a Freholm integral equation of the first kind. *J. Numer. Anal.*, **5:** 287-295

Ulaby F T, Moore R K, and Fung A K. (1981), Microwave remote sensing: active and passive. Vol.1: Microwave remote sensing fundamentals and radiometry. Massachusettes: Addison-Wesley Publ. Com, 145-148, (in Chinese)

Vladimir M. Krasnopolsky, Helmut Schiller (2003), Some neural network applications in environmental sciences forward and inverse problems in geophysical remote measurements. *Nueral Networks*, **16:** 321-348

Wentz F J. (1983), A model function for ocean microwave brightness temperature. *J.Geophys.Res.*, **88:** 1892-1907

Westwater, E. R., and N. C. Grody (1980), Combined surface- and satellite-based microwave temperature profile retrieval. *J. Appl. Meteor.*, **19:** 1438-1444

VI. Wave Propagation and Wireless Communication

Wave Propagation,Scattering and Emission in Complex Media
Edited by Ya-Qiu Jin
Science Press and World Scientific,2004

353

Wireless Propagation in Urban Environments: Modeling and Experimental Verification

Danilo Erricolo, Piergiorgio L.E. Uslenghi(*), Ying Xu, Qiwu Tan
Department of Electrical and Computer Engineering
University of Illinois at Chicago
851 S. Morgan St, Chicago, IL, 60607-7053, USA
email: derricol@ece.uic.edu, uslenghi@uic.edu

1   Introduction

Wireless communication systems play an important role in our society not only for voice transmission but also for data communications. New applications are constantly being developed and they usually result in a demand for higher transfer data rates. Therefore the design of reliable communication systems must be made with advanced engineering tools that account for the complex environment where communications take place so that higher data rates can be guaranteed. To achieve these data rates, one of the tasks that need to be accomplished is the proper characterization of the communication channel.

In order to obtain a characterization of the communication channel, one must be able to describe the propagation of electromagnetic waves in the environment under consideration. At the beginning of radio communications, one of the main concerns was propagation past large obstacles, such as hills or mountain ridges, and around the earth. Diffraction is the mechanism that explains the propagation around obstacles. In order to take advantage of the classical solution for the problem of diffraction past a half plane, many earlier works have assimilated large obstacles to knife edges. Later, propagation problems have been successfully studied with a simplification of the wave equation that leads to the parabolic-equation method. These different methods are described in, among other works,[1], [2], [3],[4], [5], [6],[7].

In this presentation, using ray-tracing methods, we examine the efforts that have been taken by our research group in order to provide effective models to characterize the propagation of electromagnetic fields inside urban environments. We first describe our polygonal line simulator for the analysis of electromagnetic wave propagation in a vertical plane. We then discuss its accuracy by showing the results of measurements that were taken in the frequency-domain and in the time-domain. Next the polygonal line simulator is compared with the methods of Hata , Zhang and COST-231. Then a general method to address propagation in a horizontal 2D environment is explained. Finally, we present a new 3D propagation method that is currently under development within our research group.

2 The polygonal line simulator

This two-dimensional simulator analyzes the path loss experienced by electromagnetic waves that propagate through the obstacles of an urban environment using ray-tracing methods. Because it is a two-dimensional simulator, only the trajectories that are contained in a vertical plane passing through the transmitter and the receiver are considered. In addition, it is assumed that the vertical plane is normally incident on the walls of the buildings between the antennas. The profile of the obstacles that are cut by the vertical plane is represented by a polygonal line, so that both variations of the height of the terrain and complex building shapes are taken into account. The new algorithm introduced into this simulator calculates all the rays that propagate along a vertical plane from the transmitter to the receiver, neglecting any backscattered ray.

Reflections can occur either on the terrain or along building surfaces. Diffractions are calculated using the Uniform Theory of Diffraction (UTD) [8] and its extensions designed to account for scattering by impedance wedges. Each segment of the polygonal profile is associated with its own value of surface impedance so that varying electrical characteristics for both terrain and building surfaces are accounted for. One advantage of using a polygonal

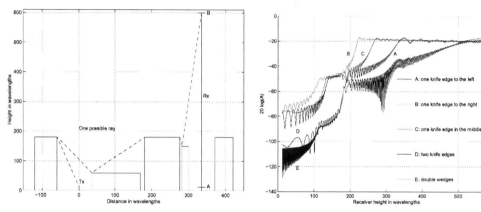

(a) Profile for buildings approximated using double wedges

(b) Overall attenuation for the profile of Fig. 1(a) using different building models in the case of hard boundary. Results are shown for: A) One knife edge to the left; B) One knife edge to the right; C) One knife edge in the middle; D) Two knife edges; E) Double wedges.

Figure 1: Rectangular shapes vs knife edges to approximate buildings

line is that buildings may be modelled using a rectangular shape [9]. The advantage of using a rectangular shape over a knife edge becomes clear as shown in Fig. 1. In fact, when one of the buildings of Fig. 1(a) is replaced with a knife edge, some possible choices are A) locate a knife edge at the left wall of each building; B) locate a knife edge at the right wall of each building; C) locate a knife edge at the center of each building; and D) locate two knife edges one at the left wall and the other at the right wall of each building. All knife edges have the same height of the building that they approximate. It is worth noting that choices A) and

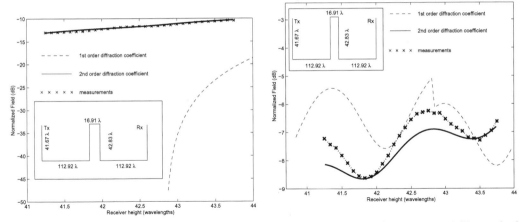

(a) soft polarization. Mean error: 0.14 dB, standard deviation 0.08 dB

(b) hard polarization. Mean error: 0.36 dB, standard deviation 0.27 dB

Figure 2: Advantages of the use of double diffraction coefficients

B) violate the reciprocity principle, since if T_x is switched with R_x the new situation is not reciprocal of the previous one. An automatic ray tracing for the profile shown in Fig. 1(a) was carried out and the receiver height was varied from A to B at small increments of $\lambda/10$ each. The results for perfect electrical conductor buildings and hard boundary are shown in Fig. 1(b).

The choice of rectangular shapes to approximate buildings must go together with the use of appropriate diffraction coefficients [10]. Specifically, one of the advantages of this simulator is the implementation of the diffraction coefficients for double wedges developed by Albani et al. [11] and those for impedance wedges of Herman and Volakis [12]. The importance of correctly characterizing double wedges is seen referring to Fig. 2, which shows the comparison among measurements and predictions computed using cascaded single order diffraction coefficients and double wedge diffraction coefficients. These results point out that the use of cascaded diffraction coefficients for soft polarization, Fig. 2(a), and hard polarization, Fig. 2(b), leads to errors, particularly for soft polarization.

2.1 Anechoic chamber experiments to measure the accuracy of the predication

The novelty of these studies is that the measurements are conducted inside the anechoic chamber facility at the University of Illinois at Chicago, using scaled models of urban environments and appropriate antennas. The advantage of this approach is that all parameters such as physical dimensions of the model and material properties are known with very low tolerances. Therefore, it is possible to make extremely accurate comparisons with the theoretical prediction obtained using a ray-tracing method. The experiments must approximate, as closely as possible, the two-dimensional assumption. In particular, the trajectories considered by the polygonal line simulator are contained in a vertical plane. As a consequence,

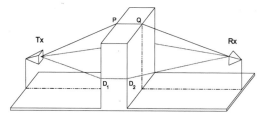

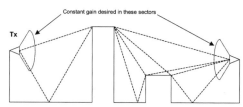

(a) Example of three-dimensional propagation. The trajectory $T_x \to P \to Q \to R_x$ that is contained in the vertical plane is of interest for the experiments; however, the trajectory $T_x \to D_1 \to D_2 \to R_x$ is an undesired one.

(b) Example of the angular sectors within which the directive gain must be as constant as possible to guarantee that all trajectories of interest are equally weighted.

Figure 3: 3D trajectories and vertical pattern requirements

in order to experimentally emphasize only those trajectories that are contained in the vertical plane, such as trajectory $T_X \to P \to Q \to R_X$ of Fig. 3(a), the patterns of both antennas in the horizontal plane must be narrow and directed along the line connecting the antennas. Furthermore, for both antennas the directivity patterns in the vertical plane must be isotropic, so that the contributions from all possible trajectories are equally weighted. At least within the angular sector of interest shown in Fig. 3(b), the directivity gain must be reasonably constant. The experiments measure the field $E(R_x)$ that reaches R_x after propagating past the scaled building models. During the experiments, T_x is fixed while R_x is manually moved vertically. Measurement data are plotted in terms of the normalized field $E_0(R_x)$, i.e the measured field $E(R_x)$ divided by the magnitude of the field measured for the same position of the antennas in free space $E_{\text{free space}}(R_x)$:

$$E_0(R_x) = \frac{E(R_x)}{|E_{\text{free space}}(R_x)|} \ . \tag{1}$$

Two polarizations are considered: hard (Neumann boundary condition and soft (Dirichlet boundary condition). The geometrical shapes of the scaled models are simple but effective to test the performance of the polygonal line simulator. Further details of this investigation were given in [13].

2.1.1 Frequency domain measurement

As an example of these measurements we consider the two-building profile that may be regarded as the simplest case of multiple rows of buildings having nearly uniform height, which is a situation of practical interest, especially in suburban areas. The case of two buildings of almost equivalent height is challenging for any ray-tracing method.

Referring to the inset of Fig. 4, the challenge comes from the trajectories that illuminate the building to the left and are diffracted towards the building to the right. These trajectories are in the transition zone of the edges and, therefore, the application of the ray-tracing method must be verified. In particular, the most difficult case occurs when the transmitter is below the building rooftop since the field at the receiver is only due to diffraction mechanisms, which have to be carefully computed. This explains why the configurations that were

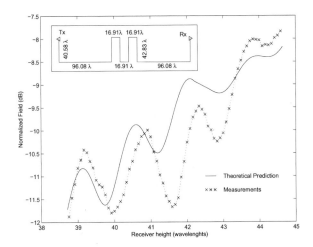

Figure 4: Propagation with Tx below rooftops. Hard polarization case for grazing observation aspect with an incidence angle is 1.34°. For these data, the mean error is 0.59 dB and the standard deviation is 0.50 dB.

measured consider the transmitter either below the rooftop or slightly below the rooftop to cause grazing incidence.

In particular, here we examine the configuration depicted in the inset of Fig. 4 that causes the transmitter to create grazing incidence. In fact, in Fig. 4, the trajectory that impinges the left edge of the left building makes an angle of 2.18° with the line through the building rooftops, whereas in Fig. 4 the same angle is reduced to 1.13°. Fig. 4 shows the details of this comparison and corresponds to a mean error of 0.59 dB and a standard deviation of 0.50 dB. Improvements could be achieved using higher order diffraction coefficients, but the benefits would be far too small when compared to the increased computational effort.

2.1.2 Time domain measurement

The purpose of the TD analysis is the measurement and identification of the multipath components that actually contribute towards the received field in a selected frequency band. This requires the ability to resolve two closely time spaced multipath components, which we will refer to as *response resolution*, and the ability to locate the peak, in time, of a single multipath component, which we will refer to as *range resolution*. The experiments consist of pulses that are launched from T_x and measured at R_x.

For a given scaled model of an urban environment, different configurations are considered. Each configuration is characterized by the positions of the transmitter and receiver with respect to the scaled building model. Given a certain configuration as input, the PL simulator computes the complex value of the EM field for a pre-determined set of frequencies. Both the experiments and the theoretical predictions return results in the frequency domain. The TD analysis was accomplished via post-processing using MATLAB.

Similarly to what we already examined in the frequency-domain case,the TD analysis of the multipath components is now applied to the case of the two-building profile. This profile

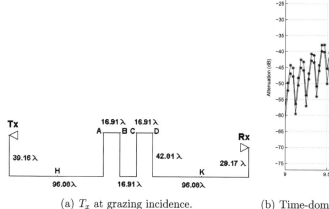

(a) T_x at grazing incidence.

(b) Time-domain analysis for hard polarization. The curves show the PL simulator prediction (black line) and the measurements (blue line).

Figure 5: Time-domain analysis

represents the simplest case of rows of parallel buildings. In addition, since the two buildings have exactly the same height, it provides a way to challenge the ray-tracing algorithm because of the intrinsic grazing incidence condition created by this particular shape. Further details are given in [14]. As an example, Figure 5(a) shows the geometry for the configuration of the T_x at grazing incidence. For this configuration, the trajectories that provide the most important contributions are:

1. $T_x \rightarrow A \rightarrow B \rightarrow C \rightarrow D \rightarrow R_x$

2. $T_x \rightarrow A \rightarrow B \rightarrow C \rightarrow D \rightarrow K \rightarrow R_x$

3. $T_x \rightarrow H \rightarrow A \rightarrow B \rightarrow C \rightarrow D \rightarrow R_x$

The TD analysis is shown in Figure 5(b), which corresponds to a hard-polarization case. Hard polarization was chosen because it results in stronger values for the received field and therefore makes the measurements easier.

Table 1: Configuration 3: Comparison between measured data and predictions given by the PL simulator.

Trj	Measured Data		PL Simul	
	Delay (ns)	Attn (dB)	Delay (ns)	Attn (dB)
1	9.750	-21.25	9.763	-21.36
2	10.655	-35.40	10.680	-36.84
3	10.934	-38.65	10.964	-41.07

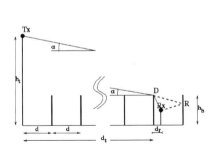

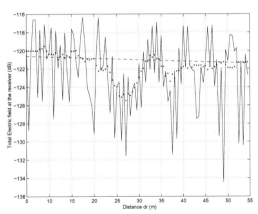

(a) Geometry for a simplified urban environment where buildings are replaced using knife edges. The continuous line represents a diffracted path, whereas the dashed line represents a diffracted-reflected path.

(b) Comparison with COST-231 Walfisch-Ikegami model. The solid line represents the polygonal line simulator; the dashed line is the COST-231 Walfisch-Ikegami model; and the dotted line represents the local average of the polygonal simulator results. Simulation data for this comparison are given in Table 2. d_r varies at increments of 0.5 m, which corresponds to 3.6 wavelengths.

Figure 6: Comparison with COST-231 for knife-edge buildings

The main peak of Figure 5(b) is easily identified with trajectory 1 of Table 1. The agreement with the theoretical prediction is very good both for the main and the secondary peaks. Also, trajectories 2 and 3 provide similar contributions.

2.2   Comparison with Cost-231, Zhang and Hata

Many models for propagation in urban environment represent building obstructions using the knife edge approximation, as reported, for example, in [15] [16] [17] [18] [19]. Therefore the geometry of Fig. 6(a) is considered herein to compare different propagation prediction methods because it represents an urban environment where parallel rows of buildings are modelled using knife edges. In the configuration of Fig. 6(a), the transmitter Tx is always above the rooftop height and the receiver is always in an obstructed area. Further details are provided in [20]. A comparison of the prediction obtained with the polygonal line simulator and the COST-231 Walfisch-Ikegami model is shown in Fig. 6(b). For this comparison, as well as for those shown in Fig. 7(a) and 7(b), the geometry of the environment under study is shown in Fig. 6(a) and the parameters for the simulation are shown in Table 2. The comparison is carried out by computing the total electric field at the receiver, while the receiver moves horizontally and d_r measures its distance from the knife edge to its left. The total field is calculated assuming an isotropic source with transmitted power $P_t = 1W$ and vertical polarization. Referring to Fig. 6(b), the prediction obtained using the COST-231 Walfisch-Ikegami model is in agreement with the one of the polygonal line simulator. In fact, the average value of the difference between the two curves is 1.46 dB and the standard

Table 2: Parameters for the geometry of Fig. 6(a)

Parameter	Value
frequency	f=2.154 GHz
transmitter height	$h_t = 12$ m
building separation	d=60 m
horizontal distance between T_x and D	d_t=1020 m
distance from mobile to left building	$5 \leq d_r \leq 55$ m
mobile height	h_r=1.6 m
building height	h_b=10 m
relative dielectric permittivity	ε=5
knife edges between T_x and R_x	n=17

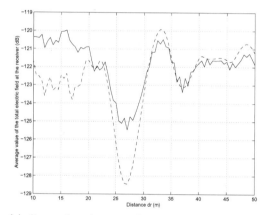 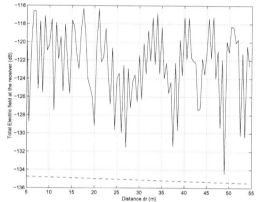

(a) Comparison between the average values of the two-dimensional simulator (continuous line) and Zhang's method (dashed line).

(b) Comparison with Hata's model. The solid line is the polygonal line simulator result; the dashed line is Hata's model result.

Figure 7: Comparison with Hata's method for knife-edge buildings

deviation of this difference is 4.0 dB. In this comparison, the correction to the COST-231 Walfisch-Ikegami model described in [21] was introduced. The next comparison is with the method of Zhang and the corresponding result is shown in Fig. 7(a). The parameters for this comparison are reported in Table 2 and for the simplified geometry under examination, there is good agreement between the two predictions: the difference between the average values of the two predictions never exceeds 3 dB.

Finally a comparison is given with Hata's model [22] in Fig. 7(b). The average difference between the two curves is 12.6 dB and the standard deviation of the difference is 4.0 dB, a result that shows Hata's prediction is too pessimistic. The difference between Hata's prediction and the polygonal line simulator may be explained on the basis that Hata's model was obtained by fitting the experimental data measured by Okumura [23] and using only a few parameters to describe the environment.

3   A general method to study 2D propagation

Electromagnetic wave propagation in urban environments among high-rise buildings may be described using trajectories that propagate around buildings instead of over them. This problem may be simplified by modelling the environment with polygonal cross-section buildings that are infinitely high and by assuming that propagation takes place in a plane perpendicular to the buildings. With these simplifications, the problem is reduced to the scattering of a cylindrical wave incident on a group of two-dimensional polygonal cylinders. The goal of this study is the evaluation of the field intensity at an arbitrary observation point, while accounting for the interaction of the cylindrical wave with the polygonal obstacles. The study is carried out using ray tracing methods, which implies the determination of all the trajectories connecting the source point to the observation point. The approach taken consists of:

- Preprocessing the information related to the position of the polygonal cylinders. This operation is carried out to establish the "local" visibility that each polygon has in relation to its neighbors.

- The construction of an overall visibility graph using the results found in the previous step.

- The extraction of the ray trajectories connecting the source point to the observation point by means of a traversal algorithm that is applied to the overall visibility graph.

For each ray trajectory, it must be verified that both geometrical optics (i.e., reflection law) and geometrical theory of diffraction (i.e., edge diffraction law) are fulfilled. Then the field strength at the observation point is computed by coherent superposition of the field strength associated with each ray trajectory. The field intensity of each ray trajectory is calculated by introducing the proper reflection and diffraction coefficients along the path. The electromagnetic properties of the polygonal cylinders are given in terms of surface impedance, which makes this study quite flexible in terms of the variety of situations that can be represented. Finally, the algorithm is implemented in C++ language to take full advantage of the methodologies offered by object-oriented programming. Further details are found in [24].

3.1   The Method

Free space propagation and reflection are easily treated with ordinary ray tracing methods. Diffraction cannot be studied using an ordinary ray tracing approach because of the law of edge diffraction according to which a ray incident on an edge creates a whole cone of diffracted rays. Therefore, an inverse method is used to avoid the problem of dealing with a cone of an infinite number of diffracted rays. The idea is to consider two points, the source and the observation, and to connect them with all the trajectories that it is possible to draw according with the constraints imposed by the environment. The polygons can be concave or convex and both their faces and vertices will be referred to as the scattering elements of the environment. Rays that propagate from the source to the observation point interact with the environment at the scattering elements. The interactions at the scattering elements are either reflections or diffractions. For a given ray trajectory, the number of interactions at the scattering elements determines the order of the ray. Therefore, a direct ray from source to observation is a zero order path, whereas a ray that undergoes either one reflection or one diffraction during its propagation will be referred to as a first order path. The algorithm that will be presented can compute the trajectories under the assumption that a given path must not pass twice through the same point. The algorithm determines the trajectories starting from order zero up to a specified order. The first order trajectory is computed using the following steps:

1. All scattering elements that are visible from the source (i.e. all elements that have a line of sight with the source) are determined and collected together in what will be referred to as the visibility list of the source.

2. Each element of the visibility list that has a line of sight with the observation point is selected to become part of a candidate trajectory.

3. The candidate trajectory is composed of the source, the scattering element, and the observation point. If the scattering element is a vertex, where a diffraction occurs, then the trajectory passing through the scattering element is a physical trajectory. However, if the scattering element is a segment, one has to establish whether the angles of incidence and reflection satisfy the reflection law.

3.2   Determination of the visibility

One of the most difficult tasks to solve in this problem is the determination of the visibility in an environment with a potentially large number of objects. A statistics of ray tracing applications shows that, in order to render images where thousands of objects are present, more than 95% of the time is spent performing intersection calculations between rays and objects [25], if an exhaustive approach is taken. An exhaustive implementation is prohibitive and before implementing any ray tracing algorithm it is necessary to become acquainted with acceleration techniques [26]. The first step to accelerate intersection computations consists of enclosing each polygon into a rectangular bounding box. A further improvement consists of creating a hierarchical structure of bounding boxes that contain groups of bounding boxes. In this way, only the bounding boxes that are pierced by the rays are considered, therefore greatly reducing the number of intersection calculations. A final improvement consists of introducing the so called painter's algorithm that simply consists of keeping track of the objects that have a line of sight with the point of view and neglecting those that are shadowed by the closest ones.

3.3   The graph theoretical model

For the environment shown in Figure 8(a) one could draw a graph that collects all the visibility information obtained for each scattering element. The information is presented in Figure 8(b) and it represents a collection of elements that have a line of sight among each other. The process of the determination of the trajectories can be thought of as the extraction of the paths of Figure 8(b) that connect S with O. Because of the reasons

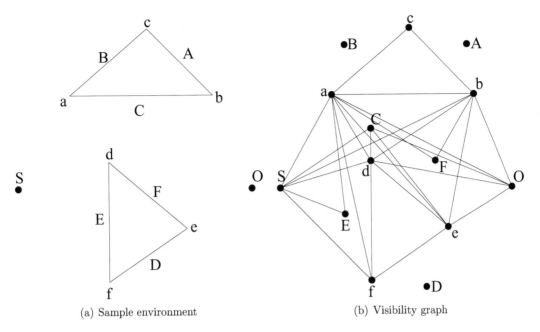

(a) Sample environment   (b) Visibility graph

Figure 8: Sample 2D environment and corresponding visibility graph

previously explained, all paths containing at least one reflection need to be checked to assess whether they satisfy the reflection law.

4 A new 3D propagation method

In 3D situations, ray tracing methods [27] [28] become very complicated because of the Keller's diffraction cone [8]. On the other hand, integral methods don't need tracing rays, but the computation increases tremendously when many buildings are present. However, using integral methods diffraction is usually considered only up to order 4, which makes it easier to implement them. Diffractions of higher orders usually become too weak to be important. Generally speaking, integral methods include solving parabolic equation (PE) with split-step Fourier transform [29], computing Fresnel-Kirchhoff integral by interpolation [30], [31], and applying asymptotic path-integral technique with repeated integral of the error function. The path-integral is equivalent to the Fresnel-Kirchhoff integral, but the repeated integral of error function is more time-consuming than the interpolation method. Here Fresnel-Kirchhoff integral is used to compute path loss from buildings in 3D urban and suburban environment (except downtown area), both diffraction and reflection are considered. An example of a perfectly conducting building as shown in Figure 9(a) is considered. The coordinates of the transmitter are (25.2cm, 0, 46cm). The building first intersects the x axis at x=68.9cm, and ends at x=68.9cm+8.7cm. The width and height of the building is 30cm and 50cm, respectively. The receiver moves along an arc centered at the origin with a radius of 114.8cm. The height of the receiver is 40cm. In the case of vertical polarization,

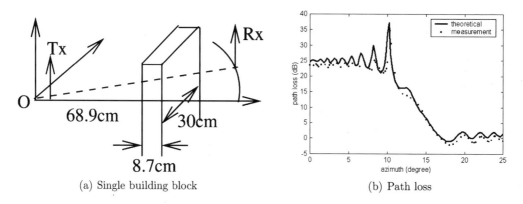

(a) Single building block   (b) Path loss

Figure 9: Path loss computation for a 3D environment obtained with the Fresnel method and comparison with measurements

the reflection coefficient of the roof and vertical walls is +1 and -1. The reflection from the ground is neglected. At frequency of 50GHz, theoretical prediction of the path loss at the receiver (solid line) is compared with the measurement (dashed line) made in [32] as in Figure 9(b). The agreement between the prediction and the measurement is very good.

5   Conclusion

Our efforts in the development of the two-dimensional polygonal line simulator lead to very successful comparisons with measurements. Additionally, our efforts in developing 3D propagation simulators show promising results.

6   Acknowledgement

The authors would like to acknowledge Umberto G. Crovella, Giuseppe D'Elia, Dr. T. Monte, the UIC machine shop, the support of the Andrew Foundation, and the U.S. National Science Foundation that sponsored part of this research under grant ECS 9979413.

References

[1] W. C. Jakes, *Microwave mobile communications*, IEEE Press, New York, 1974

[2] L. D. Parsons, *The mobile radio propagation channel*, Pentech Press, 1992

[3] T. S. Rappaport, *Wireless communications: principles and practice*, Prentice Hall, 1996

[4] H. L. Bertoni, *Radio propagation for modern wireless systems*, Prentice Hall, 2000

[5] M. F. Levy, *Parabolic equation methods for electromagnetic wave propagation*, Inspec/Institution of Electrical Engineers, 2000

[6] N. Blaunstein, *Radio propagation in cellular networks*, Artech House, 2000

[7] R. Janaswamy, *Radiowave propagation and smart antennas for wireless communications*, Kluwer Academic Publishers, 2001

[8] R. G. Kouyoumjian and P. H. Pathak, "A uniform geometrical theory of diffraction for an edge in a perfectly conducting surface," *Proc. IEEE*, vol. 62, no. 11, pp. 1448–1461, Nov 1974

[9] D. Erricolo and P. L. E. Uslenghi, "Knife edge versus double wedge modeling of buildings for ray tracing propagation methods in urban areas," in *Digest of National Radio Science Meeting*, Boulder, CO, USA, Jan 1998, p. 234

[10] D. Erricolo, "Experimental validation of second order diffraction coefficients for computation of path-loss past buildings," *IEEE Transactions on Electromagnetic Compatibility*, vol. 44, no. 1, pp. 272–273, Feb. 2002

[11] M. Albani, F. Capolino, S. Maci, and R. Tiberio, "Diffraction at a thick screen including corrugations on the top face," *IEEE Trans. Antennas Propagat.*, vol. 45, no. 2, pp. 277–283, Feb. 1997

[12] M. I. Herman and J. L. Volakis, "High frequency scattering by a double impedance wedge," *IEEE Trans. Antennas Propagat.*, vol. 36, no. 5, pp. 664–678, May 1988

[13] D. Erricolo, G. D'Elia, and P.L.E. Uslenghi, "Measurements on scaled models of urban environments and comparisons with ray-tracing propagation simulation," *IEEE Trans. Antennas Propagat.*, vol. 50, no. 5, pp. 727–735, May 2002

[14] D. Erricolo, U. G. Crovella, and P.L.E. Uslenghi, "Time-domain analysis of measurements on scaled urban models with comparisons to ray-tracing propagation simulation," *IEEE Trans. Antennas Propagat.*, vol. 50, no. 5, pp. 736–741, May 2002

[15] J. Walfisch and H. L. Bertoni, "A theoretical model of UHF propagation in urban environments," *IEEE Trans. Antennas Propagat.*, vol. 36, no. 12, pp. 1788–1796, Dec 1988

[16] S. R. Saunders and F. R. Bonar, "Explicit multiple building diffraction attenuation function for mobile radio wave propagation," *Electron. Lett.*, vol. 27, no. 14, pp. 1276–1277, July 1991

[17] M. J. Neve and G.B. Rowe, "Assessment of GTD for mobile radio propagation prediction," *Elec. Lett.*, vol. 29, no. 7, pp. 618–620, Apri 1993

[18] Wei Zhang, "A wide-band propagation model based on UTD for cellular mobile radio communications," *IEEE Trans. Antennas Propagat.*, vol. 45, no. 11, pp. 1669–1678, Nov 1997

[19] J. H. Whitteker, "A generalized solution for diffraction over a uniform array of absorbing half screens," *IEEE Trans. Antennas Propagat.*, vol. 49, no. 6, pp. 934–938, June 2001

[20] D. Erricolo. and P.L.E. Uslenghi, "Comparison between ray-tracing approach and empirical models for propagation in urban environments," *IEEE Trans. Antennas Propagat.*, vol. 50, no. 5, pp. 766–768, May 2002

[21] A. M. Watson D. Har and A. G. Chadney, "Comment on diffraction loss of rooftop-to-street in cost-231 walfisch-ikegami model," *IEEE Trans. Veh. Technol.*, vol. 48, no. 5, pp. 1451–1542, Sept. 1999

[22] M. Hata, "Empirical formula for propagation loss in land mobile radio services," *IEEE Trans. Veh. Technol.*, vol. VT-29, no. 3, pp. 317–325, Aug 1980

[23] Y. Okumura, E. Ohmori, T. Kawano, and K. Fukuda, "Field strength and its variability in VHF and UHF land-mobile radio service," *Rev. Electr. Commun. Lab. (Tokyo)*, vol. 16, no. 9 and 10, pp. 825–873, 1968

[24] D. Erricolo and P. L. E. Uslenghi, "Graph-theoretical approach to multiple scattering by polygonal cylinders," in *Intl. Conf. on Electromagnetics in Advanced Application*, Turin, Italy, Sept 1999, pp. 155–159

[25] T. Whitthed, "An improved illumination model for shaded display," *Commun. ACM*, vol. 23, no. 6, pp. 343–349, June 1990

[26] J. Arvo and D. Kirk, "A survey of ray tracing acceleration techniques," in *An introduction to ray tracing*, A. S. Glassner, Ed. Academic Press, 1989

[27] F.A. Agelet, F. P. Fontan, and A. Formella, "Fast ray tracing for microcellular and indoor environments," *IEEE Trans. Magn.*, vol. 33, no. 2, pp. 1484–1487, 1997

[28] Z. Q. Yun, Z. J. Zhang, and M. F. Iskander, "A ray-tracing method based on the triangular grid approach and application to propagation prediction in urban environments," *IEEE Trans. Antennas Propagat.*, vol. 50, no. 5, pp. 750–758, May 2002

[29] L. Saini and U. Casiraghi, "A 3D fourier split-step technique for modeling microwave propagation in urban areas," in *Fourth European Conference on Radio Relay Systems*, 1993, pp. 210–214

[30] J. H. Whitteker, "Fresnel-kirchhoff theory applied to terrain diffraction problems," *Radio Sci.*, vol. 25, no. 5, pp. 837–851, Sept-Oct 1990

[31] T. A. Russell, C. W. Bostian, and T. S. Rappaport, "A deterministic approach to predicting microwave diffraction by buildings for microcellular systems," *IEEE Trans. Antennas Propagat.*, vol. 41, no. 12, pp. 1640–1649, Dec 1993

[32] G. A. J. van Dooren and M. H. A. J. Herben, "Polarisation-dependent site-shielding factor of block-shaped obstacle," *Electron. Lett.*, vol. 29, pp. 15–16, 1993

Wave Propagation,Scattering and Emission in Complex Media
Edited by Ya-Qiu Jin
Science Press and World Scientific,2004

367

An Overview of Physics-based Wave Propagation in Forested Environment

K.Sarabandi, I.Koh

EECS Dept., University of Michigan, Ann Arbor, MI 48109-2122
E-mail: saraband@eecs.umich.edu

Abstract *In this paper recently developed physics-based models for wave propagation in a forested area are briefly described. These models are applicable to various communication configurations, and can predict any statistical properties of the channel.*

1. Introduction

For accurate assessment or design of various wireless communication schemes/systems, accurate characterization of wireless communication channels plays an essential role. Various prediction models have been developed for the different environments that are entirely or partially based on measurements, such as the Hata model. Therefore their prediction has little dependence on a specific regional planning such as environmental or location information. As a result, the predictions given by these models at an individual point in a particular environment may have large errors. Only physics-based predictions that can account for the specific propagation environment hold the possibility of reducing this prediction error.

The physical process of wave propagation through random media like a forest has been extensively studied, but not reduced to attenuation rates through vegetation. It has been found that fields penetrating the leaves and branches from an overhead source initially decay at a high rate as a result of absorption and scattering. After decaying by around 15 dB, the decay rate is much smaller, and the signal is carried via a diffusion process by multiply scattered fields. The effective decay rate to a terminal at ground level will therefore depend on the path length in the vegetation, as well as the frequency. The variation of attenuation rate with frequency is known in principle for random media, but not in detail for forests. This attenuation for a forest has been studied recently. In this paper path-loss, field fluctuation, impulse response, and other channel parameters for a forest using physics-based models are presented.

2. All Terminals inside Forest

When both a receiver and a transmitter are embedded in a forest canopy and the operating frequency is below UHF band, the forest may simply be modeled by a homogeneous dielectric slab that was first introduced by Tamir [1]. A flat boundary between the air and the canopy is assumed and effective foliage permittivity is calculated using a mixing formula or measurement. It is well known that inside a stratified lossy media with smooth boundaries, a lateral wave is the dominant component of the field at a distant observation point. In real world however, the interface between the air and the canopy is not smooth. To investigate the roughness effect on the wave propagation in a forest environment, an analytical model is developed by Sarabandi *et al.* [2] using distorted Born approximation.

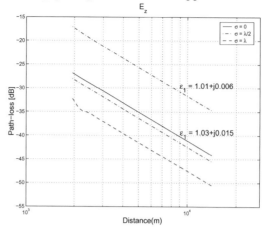

Figure 1: Path-loss of a vertical dipole as function of a distance and surface roughness.

Figure 1 shows the magnitude of the electric field as a function of distance between the receiver and the transmitter, and of also the surface roughness. It is observed that the surface roughness can slightly (5dB) reduce a level of the electric field for a moderate surface roughness. The above explained model can only estimate the mean field. To predict other statistical characteristics of the forest channel, the field fluctuation should also be accurately estimated. At VHF and below, only tree trunks can significantly scatter the field within the medium. Other canopy constituents such as leaves, twigs, and branches are (much) smaller than the wavelength, and therefore diffuse scattering is negligible. Hence these components may mainly contribute to only attenuation.

A comprehensive wave propagation model for a forest environment that accounts for multiple scattering among the tree trunks and includes the interaction between this scattered field and lateral waves is proposed by Sarabandi *et. al.* [3]. The geometry of the simplified forest model is shown in Fig. 2.

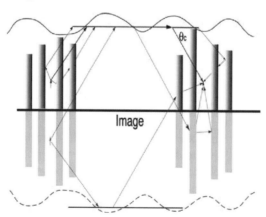

Figure 2: Scattering mechanisms resulted from wave interaction with tree trunks in a forest environment.

The multiple scattering among the tree trunks is exactly calculated using the method of moment (MoM). To obtain statistical properties of a channel, a Monte Carlo simulation is used.

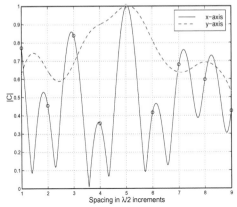

Figure 3: Magnitude of spatial correlation of a vertical dipole.

For example, spatial field correlation (magnitude) at an observation point 1km away from a vertical dipole is calculated and shown in Fig. 3. For this calculation, 150 scatterers near the source, and 50 scatteres near the observation points are retained. Using the same model, time domain response of the channel is also computed from the calculation of the frequency domain response over a relatively wide band. Figure 4 shows the time domain response reconstructed from frequency response of the channel from 10MHz to 90MHz. Using the Gaussian quadrature numerical integration technique instead of regular FFT the required number of frequency points is drastically reduced.

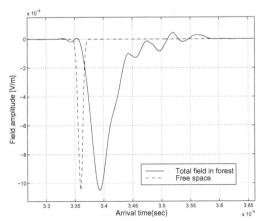

Figure 4: Time domain response when a receiver and transmitter are inside a forest.

3. One Terminal outside Forest

For other important scenarios, the case where a receiver or a transmitter is outside a forest environment, a different channel model must be considered since the wave propagation mechanism is totally different from for the previous cases. For this case where signal propagation is possible for the entire range of EM spectrum, not only the attenuation due to leaves, branches, twigs, and trunks is important but also the scattering from each constituents has to be accounted for because the receiver or the transmitter may be in the close proximity of these scatterers. Hence scattered field from each segment of the trees must be accurately calculated. For frequencies up to 3GHz, it is known that the structural information of trees should be preserved to accurately predict the coherence in the scattered field [4]. To construct a realistic tree species conveniently, a fractal algorithm known as Lindenmayer system can be used. Using these artificial tree structures in conjunction with auxiliary botanical and statistical information. a desired forest can be simulated as shown in Fig.5.

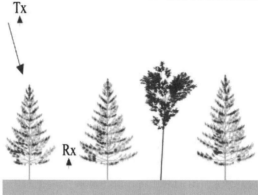

Figure 5: A receiver inside a forest is communicating with a transmitter above.

Simple building blocks for the fractal trees are cylinders representing branches, needles, and tree trunks, or small disks for flat leaves.

Analytical coherent scattering model based on the single scattering theory is developed to reduce complexity [5]. To estimate attenuation, and phase shift that take place inside a forest, Foldy's approximation is used. For scattering from a finite cylinder, an accurate analytical closed-form expression of the scattered field is formulated assuming the induced current on the finite cylinder surface is the same as the induced current on the infinite cylinder surface with the same dielectric constant, and radius as those of the finite one. For scattering from a leaf, a Rayleigh-Gans PO solution can be used because of very small thickness of a leaf.

Using Monte Carlo simulation, statistical channel characteristics such as mean field, field fluctuation, depolarization, etc. are estimated. Figures 6, and 7 show performance assessment of a GPS receiver under a deciduous and coniferous forest with the density of $0.1/m^2$.

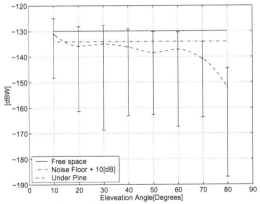

Figure 6: Plot of the received signal power under a coniferous forest.

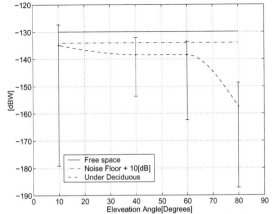

Figure 7: Plot of the received signal power under a deciduous forest.

For this simulation, 400 realizations are performed, and 10 trees are kept around the receiver. The GPS receiver is assumed to have noise figure of 1.5dB, and a threshold signal-to-noise-ratio of 10dB is imposed. Using the same model, a time domain response of the channel is also computed using a wide-band frequency response (1GHz) as shown in Fig. 8.

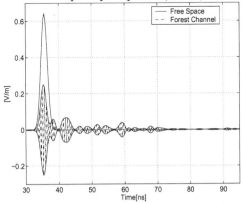

Figure 8: Time domain response under a coniferous forest when a transmitter is above a forest.

Reference

[1]   T. Tamir, "On Radio-Wave Propagation in Forest Environments," *IEEE Trans.Antennas Propagat.,* vol. AP-15 pp.806-817, Nov. 1967

[2]   K. Sarabandi, and I. Koh, "Effect of Canopy-Air Interface Roughness on HF-UHF Wave Propagation in Forest," *IEEE Trans. Antennas Propagat.* Vol. 50, No. 2, pp.111-121, Feb. 2002

[3]   K. Sarabandi, and I. Koh, "A Complete Physics-Based Channel Parameter Simulation for Wave Propagation in a Forest Environment,". *IEEE Trans. Antennas Propagat.* vol.49, No.2 pp.260-271, Feb. 2001

[4]   Y. Lin, "A Fractal-Based Coherent Scattering and Propagation Model for Forest Canopies," Ph.D Thesis, The University of Michigan, 1997

[5]   I. Koh, and K. Sarabandi, "Polarimetric Channel Characterization of Foliage for Performance Assessment of GPS Receivers Under Tree Canopies," *IEEE Trans. Antennas Propaga.* Vol. 49, No. 2, pp.260-271, Feb. 2002

Wave Propagation,Scattering and Emission in Complex Media
Edited by Ya-Qiu Jin
Science Press and World Scientific,2004

373

Angle-of-arrival Fluctuations due to Meteorological Conditions in the Diffraction Zone of C-band Radio Waves, Propagated over the Ground Surface

T.A.Tyufilina, A.A.Mescheryakov, M.V.Krutikov

Tomsk State University of Control Systems and Radioelectronics

40 Lenin av., Tomsk, 634050, Russia.

T/F: +7-3822-413949. E-mail: rwplab@orts.tomsk.ru

Results of direct measurements of slowly fluctuating C-band radio wave angles-of-arrival are described. The measurements were done using a conical scan radar over paths about 36 km long during one winter month near Tomsk. As a target there was used a corner reflector installed at the height of 48 m on a wood tower. Meteorological data and data concerning earth's surface are used to analyze mean angles of elevation and azimuth, observed at 15 min intervals. These results are used to predict radar tracking errors by angle of elevation and azimuth caused by earth's surface and troposphere refraction.

The accuracy of radar angular measurements at low grazing angles is not good due to reflections from rough surfaces, diffraction on obstacles and radio wave refraction. The angular measurement errors have quick and slow fluctuations. The quick fluctuations are about zero as a result of averaging on the intervals of measurement. The slow fluctuations are not averaged on the intervals of measurement and provide displacements of angle estimates. These errors are important for radar tracking in case of far away targets and obstacles near the radar.

In this paper the results of direct measurements of angles-of-arrival of slowly fluctuating C-band radio waves are described. As a target there was used a corner reflector. The measurements were done using conical scan radar over paths about 36 km long during one winter month near Tomsk, West Siberia.

Brief description of the investigated radio path.

The radar (the first point of the path) was located on the river bank at the height of 120 m over the sea level. The radar antenna was at the height of 4 m over ground. The corner reflector (the second point of the path) was installed at the height of 162 m over the see level on a wood tower. The height of the wood tower was 48 m over ground. The radio path went mainly over irregular terrain covered with deciduous. A terranian path profile is shown in Fig. 1. The profile was built on the concept of equivalent earth's radius:

$$H_i = h_i + \frac{R_i(R - R_i)}{2a_{\vartheta}},$$

h_i -current value of the profile without taking earth's curvature and refraction into consideration; R - path length; R_i - path length from the receiver to a current point of the profile; a_{ϑ} - equivalent earth's radius.

As following from Fig.1 the height of the path over terrain was 2 m for normal conditions of refraction at the distance of 8 km from radar, because the first frenel zone radius is 26 m and the path is of diffraction type. A cross section of a terranian obstacle profile is shown in Fig. 2 for the lowest height of the path. As a whole the obstacle is represented by a flat area with left down inclination relatively to the path.

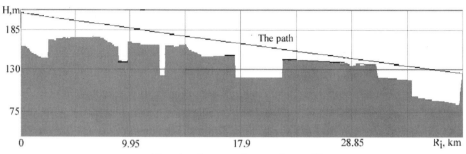

Fig. 1. A terranian path profile

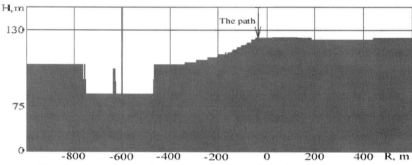

Fig. 2. A cross section of a terranian obstacle profile

Principles of instrumentations and angles-of-arrival processing.

The measurements were done using a conical scan radar. As a target there was used a corner reflector. Angles-of-arrival were measured using signals reflected by the corner reflector. The main lobe width antenna was of 3.5 degree. Angles-of-arrival were measured relatively to the direction of the corner reflector with accuracy of 0.1 angle minute. For measurements the automatic mode of radar operation was switched off in the process of radar tracking. The errors of elevation and azimuth estimation were recorded by the analog data recorder. The radar antenna was moved by 10 angle minute steps in order to calibrate the radar tracking errors. The output voltage of the angle discriminator were also recorded by the analog data recorder for determination of the current angles-of-arrival by interpolation. Later angles-of-arrival were averaged during 15 minutes. After averaging the errors of angles-of-arrival estimation in elevation and azimuth lied within the limits of 0.2 angle minutes.

Contiguous meteorologically data and weather patterns.

Parameters of meteorological elements (pressure, temperature, humidity) were registered by a weather-station at 01, 04, 07, 10, 13, 16, 19, 22 o'clock of local time.
The temperature condition was distinguished, first, by big 24 hour amplitudes; second, by scaled up of the season temperature towards the end of March. These results are: The lowest temperature -26 C, the highest temperature -3 C, the mean temperature -13.4 C.
The humidity condition was distinguished by the same big 24 hour amplitudes (85-100% night, 50 % day).

The wind was distinguished mainly quarters of the West. Exeption was observed on the 20, 21, 22 of March, when the East wind took place.

The weather patterns were determined by a powerful anticyclone with the centre over the   North Ural. The investigated path was located in the anticyclone East periphery. Thefore there were mainly observed there Western and North-western streams, small field gradient of pressure and small nebulosity. Exeption was observed on the 20-21, 24-25, 27-29 of March, when cold fronts took place.

The refractive index N was calculated using pressure, temperature and humidity data.

The general description of experimental data.

The realizations of angle-of-arrival   azimuth, cross path wind and refractive index N are shown in Fig.3 for the period of observation. The general description of experimental data is shown in the Table 1. The additional characteristics of angles-of-arrival by elevation and azimuth are given in Fig. 4 and Fig. 5.

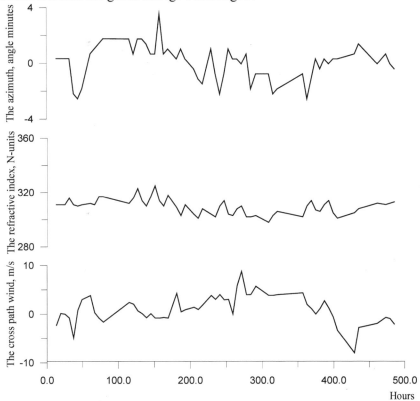

Fig. 3. The realizations of angle-of-arrival   azimuth, cross path wind and refractive index N

Table1
The general description of experimental data

	Means	Minimum	Maximum	The mean square values		
				The full measurements	22, 01,04 o'clock	10, 13,16 o'clock
The azimuth, Angle minutes	0,049	-2.57	3.55	1.24	1.22	1.23
The elevation, Angle minutes	0,016	-5.12	2.80	1.66	1.92	1.37
The refractive index, N-units	309,52	298.0	325.0	5.59	5.09	4.49
The cross path wind speed, m/s	1,03	-7.99	8.77	2.81	2.71	2.85
The along path wind speed, m/s	-1,07	-10.14	6.31	2.60	2.12	2.87
The humidity	74,3	38.0	100.0	6.84		
The temperature, C	-13,4	-26	-3.0	6.61		

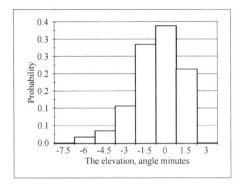
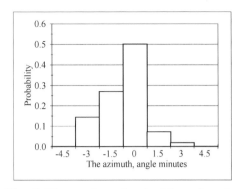

Fig. 4. The of angles-of-arrival by elevation   Fig. 5. The angles-of arrival by azimuth

As follows from Fig.4 and Fig.5 the distribution of angles-of-arrival by elevation and azimuth is asymmetrical relatively the initial direction. The probability of arrival radio waves from the back (negative) direction is higher.

The mean squares of the measured magnitudes in the Table 1 are given for the period of observations as well as for night hours (01 and 07) and day hours (13 and 19). As follows from the Table1 the fluctuations of angles-of-arrival by elevation and azimuth are comparable. The fluctuations of angles-of-arrival by azimuth are not changing during 24 hours whereas the fluctuations of angles-of-arrival by elevation has maximum at night. Fluctuations of refractive index N are about (4-5) N-units and has maximum at night. At the same time fluctuations of wind speed has maximum at the day time.

The correlation of the measured magnitudes is shown in the Table 2. As follows from the Table 2 the correlation coefficient of the angles-of-arrival by elevation and azimuth are -0.49 and is higher at night than at the day time. The correlation coefficient of the angles-of-arrival by azimuth with the refractive index N and the cross path wind is higher at night, when the correlations of the mentioned variable elements are 0.6 and 0.4 accordingly. At the same time the correlation of the refractive index N and angles-of-arrival by elevation does not depend on the time during the day and it equals -0.52.

In such a way estimations of azimuth increase when angles-of-arrival by elevation decrease and when refractive index N increases. The mentioned relationship shows the effect of earth's surface reflection.

Table 2
The correlation coefficients

The measured magnitudes	The azimuth			The elevation		
	The full measu-rements	22, 01,04 o'clock	10, 13,16 o'clock	The full measu-rements	22, 01,04 o'clock	10, 13,16 o'clock
The azimuth, Angle minutes	1.00	1.00	1.00			
The elevation, Angle minutes	-0.41	-0.37	-0.49	1.00	1.00	1.00
The refractive index, N-units	0.34	0.61	0.41	-0.44	-0.52	-0.52
The cross path wind speed, m/s	-0.28	-0.48	-0.07	0.17	0.27	0.05
The along path wind speed, m/s	-0.01	0.20	-0.08	0.07	-0.02	0.15
The humidity	0.07	0.38	0.09	-0.19	-0.20	-0.32
The temperature, C	-0.16	-0.33	-0.26	0.38	0.41	0.52

The relationship between weather patterns and experimental data.

The experimental data in Fig. 6 - Fig.11 show the the dependences of angles-of-arrival by elevation and azimuth on the weather patterns such as anticyclone and cold fronts along the path at different time of day and night. Every picture is presented three dependences, such as dependence azimuth and refractive index, dependence elevation and refractive index, dependence elevation and azimuth. All dependences are presented by two-dimensional pictures of the mentioned pairs of variable elements measured simultaneously. Number and mean square values of the measured azimuth are shown in the Table 3.

As it was expected, dependence between azimuths and refractive index is weak in case of anticyclone and cold fronts, fully described in Fig.6 and Fig 9, while increase refractive index results in decrease elevations.

Mentioned above dependences of azimuth on the refractive index, elevation on the refractive index elevation on the azimuth at some particular time of day and night show multipath propagation in case of investigation. Fig. 7, drawn for measurements at 22, 01, 04 o'clock in case of anticyclone shows multipath propagationbest of all. As follows from Fig.7 estimates of angles-of-arrival elevation divide into too groups. The first group contains estimates, which correspond the straight ray and have elevation 1...2 angle minutes. The second group containes estimates, which corresponds the earth's surface reflected ray and have elevation – (1.5...4) angle minutes. The estimate of angle-of-arrival depends on the ratio of these rays amplitudes. The ray having higher amplitude imposes the direction of arrival for interference amount with some errors. The first group of estimates localizes uniformly gaunt above the null-elevation along by azimuth. The second group of estimates localizes uniformly - like a small cloud below null-elevation. So in case of measuring the straight ray elevation azimuth is measured with bigger error then in case of measuring the Earth's surface reflection ray elevation.

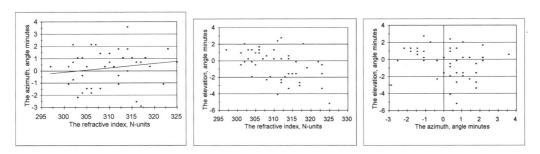

Fig. 6. The dependences of angle-of-arrival in case of anticyclone at the full measurements

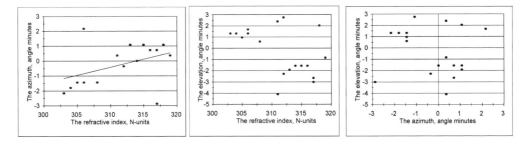

Fig. 7. The dependences of angle-of-arrival in case of anticyclone at 22, 01, 04 o'clock

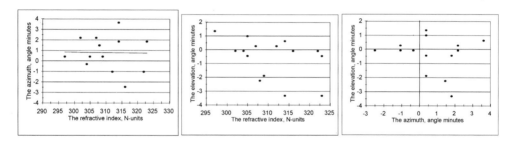

Fig. 8. The dependences of angle-of-arrival in case of anticyclone at 10, 13, 16 o'clock

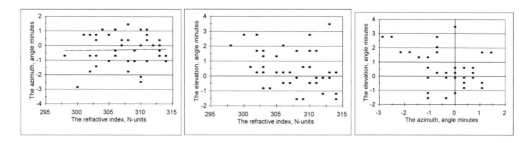

Fig. 9. The dependences of angle-of-arrival in case of cold fronts at the full measurements

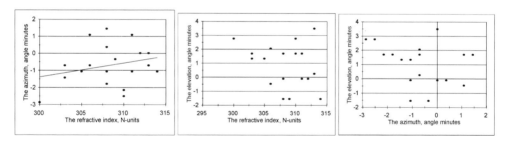

Fig. 10. The dependences of angle-of-arrival in case of cold fronts at 22, 01, 04 o'clock

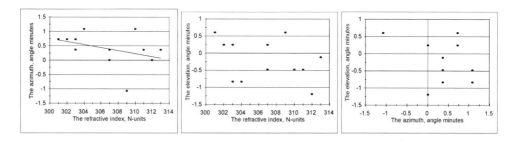

Fig. 11. The dependences of angle-of-arrival in case of cold fronts at 10, 13, 16 o'clock

The dependences of elevation on azimuth at 10, 13, 16 o'clock are illustrated by Fig. 8 in case of anticyclone. As follows from Fig.8 azimuth does not depend on the refractive index. Nevertheless, dependence of elevation on the refractive index exists, similarly to the Fig.7. When the refractive index is higher than 305 N-units elevation estimates change in two ways: the first way corresponds slow changes near null-elevation, the second way describes sharp decrease to value – (3...4) angle minutes. Similarly to Fig.7 dependence of elevation on azimuth is regular are also in this case. This regularity indicates that the estimates divides in two groups, corresponding the straight rays and by earth's surface reflected rays. Thus the azimuth estimates change at large-scale both day as night as s result of straight ray azimuth alternation.

Table 3
The mean square values of azimuth

| | The full measu-rements | The weather patterns | | | | | |
| | | The case of anticyclone | | | The case of cold fronts | | |
		The full measu-rements	10, 13, 16 o'clock	22, 01, 04 o'clock	The full measu-rements	10, 13, 16 o'clock	22, 01, 04 o'clock
Number	89	46	15	18	43	12	19
Mean square values of azimuth, angle minutes	1,26	1.41	1.64	1.30	1.04	0.57	1.19

As follows from Fig.10 and Fig.11 the cold fronts' conditions at different time of the day and night are similar to those in Fig.7. The regular dependence of elevation on azimuth is also observed.

Summary.

So, the azimuth errors depend on fluctuations of the angles-of-arrival by the azimuth of the straight ray and do not depend on meteorological elements. The only suitable case to predict the azimuth errors is to make measurements at 22, 01, 04 o'clock for anticyclone when the dispersion of estimates is minimal.

Simulating Radio Channel Statistics Using Ray Based Prediction Codes

Henry L. Bertoni

Center for Advanced Technology in Telecommunications
Polytechnic University
6 MetroTech Center
Brooklyn, NY 11201, USA

Abstract Advance radio system designs have been studied that are intended to overcome, and even take advantage of the multipath nature of the radio channel in urban environments. The efficient design and evaluation of such systems requires knowledge of various higher order channel statistics. Measurements of the channel statistics are expensive to carry out, and are usually done in only a few locations. To overcome these limitations we have used a 3D ray tracing code to carry out Monte Carlo simulations of various channel statistics.

I. Introduction

Starting with the concept of frequency reuse for cellular mobile radio, designers of modern radio system have developed innovative methods to greatly increase the capacity of communications systems. The subsequent use of digital transmission opened the door for advanced techniques such as CDMA, OFDM smart antennas and MIMO for increasing system capacity. Each advance of communication technology brings with it the need for additional information on the radio channel. For example, the bit error rate of digital systems depends on the symbol period as compared to the delay spread (reverberation time) of the channel.

The choice of parameters used to represent the radio channel typically views the received signal in terms of multiple copies of the transmitted signal. Because the communication systems will operate at many locations over a wide range of building environments, the description of the parameters are necessarily statistical in nature. This view of the channel is consistent with the use of ray models of propagation, which can be used to generate Monte Carlo simulations of the statistical distribution of channel parameters.

It is of course possible to determine the statistical nature of desired channel parameters from measurements. Measurements are in fact essential for evaluating system concepts and validating simulated channel parameters. However, measurements are expensive and time consuming. They are typically made in locations convenient for the experimenters, leaving the question of how well the measurements will apply to building environments in other cities.

In contrast to measurements, ray based Monte Carlo simulations are relatively inexpensive to carry out. Operating conditions, such as antenna height and frequency, are easily varied, and there is no difficulty in running simulations for different cities. In essence, validating the results of ray simulations through a set of measurements increases the value of those measurements by allowing them to be translated to other building environments and other operating conditions.

II. Computer Codes for Ray Tracing

Several ray-tracing codes have been developed specifically for prediction of radio propagation in urban environments. These codes account for specular reflection from the walls of buildings and diffraction around the vertical and horizontal edges of the walls. The simplest codes apply to base station antennas that are much lower than the surrounding buildings. Under these circumstances the propagation takes place around the sides of the buildings, so that computations use only a two dimensions (2D) database of the building footprints [1, 2, 3, 4]. In some codes the contributions from propagation around the sides of the buildings is augmented by the single contribution from a ray lying in the vertical plane that is diffracted over the tops of the buildings between the transmitter and receiver [5]. Such codes can give relatively good predictions, even though they omit rays that travel around some buildings and over other buildings.

Ray codes that fully account for the three dimensional (3D) shape of the buildings have been written [6], they run extremely slowly since they must break each building edge into segments, and then run a secondary ray trace from each segment. Because of the large number of segments that must be accounted for, the codes are limited to a maximum of two diffraction events per ray.

Recognizing that the sides of the buildings are almost always vertical, leads to a much faster running 3D ray code. This approach makes the single approximation that the rays diffracted at the horizontal edges of buildings lie either in the vertical plane of the incident ray or the vertical plane containing the reflected ray. In effect, the Keller diffraction cone at horizontal edges is replaced by two tangent planes. The code correctly treats diffraction at vertical building corners. With this approximation, allows the rays in 3D to be found by combining a 2D ray trace for propagation in the horizontal plane with

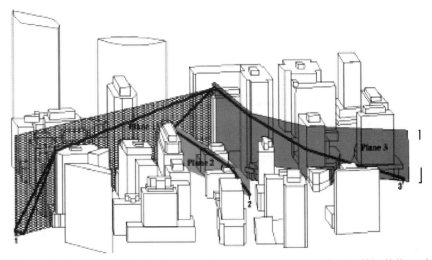

Figure 1. VPL method traces vertical planes as they pass over and reflect off buildings in Rosslyn, VA. Ray paths in the vertical planes are found by analytic methods.

analytic treatment of propagation over the buildings. This approach, which we call the Vertical Plane Launch (VPL) method was first suggested by Rossi [7], but has been independently implemented by others [8].

Figure 1 shows the VPL method, as applied to the buildings in Rosslyn, Virginia. When view from above, the lines in the 2D trace represent vertical planes containing the 3D rays. Figure 2 shows a comparison of predicted and measured small area average path gain (ratio of received to transmitted power) for 900 MHz at mobile locations along the streets of Rosslyn, VA when the base station antenna is place on a rooftop. The mean error of these predictions is 0.75 dB and the standard deviation is 5.4 dB. Including diffuse scattering from the buildings in the prediction code, as shown in Figure 2, usually has little effect for high antennas. Ray codes typically give predictions with average error of about 1 dB and standard deviations of 6 – 10 dB.

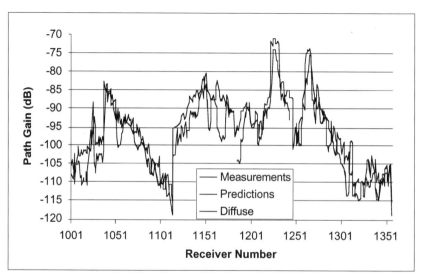

Figure 2. Comparison of predicted and measured small area average power at 900 MHz for low receivers located along the streets and a rooftop base station antenna. Predictions (lighter curve) made with and without diffuse scattering overlap.

III. Predictions of Delay and Angle Spread

In addition to giving the amplitude of the ray field, ray codes also give the ray directions at the transmitter and the receiver, and the total length of the ray path. From this information it is possible to compute the power delay profile, delay spread, angle spread, and other channel parameters. Recent simulations show good agreement between the measurements of power delay profile [9] and delay spread [10] on a point-by -point basis.

Delay spread (*DS*) and angle spread (*AS*) are defined as the second-order moments of the average impulse power response. Delay spread is the same if measured at either end of the link, but angle spread will be different at each end. Rays typically arrive at the mobile from all directions in the horizontal plane, while rays at an elevated base station arrive from a limited wedge of angles. For a given mobile location, let the subscript m denote the different rays from the mobile to the base station. Let A_m be the ray amplitude and let $\tau_m = L_m/c$ be the time delay of the ray. With these definitions, the square of *DS* is given by

$$DS^2 = \left\{ \sum_m (\tau_m - \langle \tau \rangle)^2 |A_m|^2 \right\} \Big/ \left\{ \sum_m |A_m|^2 \right\} \tag{1}$$

where $\langle \tau \rangle$ is the average excess delay of the rays arriving from the mobile.

If ϕ_m is the azimuth angle of arrival of the ray, then it is common to define the *AS* using a formula similar to (1). While this approach is valid for small angle spreads, it is not coordinate independent. The correct definition of *AS* is given in terms of the unit vector $\vec{v}_m$ pointing towards the direction of the arriving ray [11], and is given by

$$AS = \frac{180}{\pi} \sqrt{1 - |\langle \vec{v}_m \rangle|^2} \tag{2}$$

The factor $180/\pi$ is included in (2) to compare with other results expressed in degrees, while the average $\langle \vec{v}_m \rangle$ is found from

$$\langle \vec{v}_m \rangle = \left\{ \sum_m \vec{v}_m |A_m|^2 \right\} \Big/ \left\{ \sum_m |A_m|^2 \right\} \tag{3}$$

Using the VPL ray code, *DS* and *AS* have been computed for many mobile locations using databases of buildings for several cities [12]. As an example, Figure 3 shows scatter plots of computed *DS* and *AS* versus distance from the base station for 900

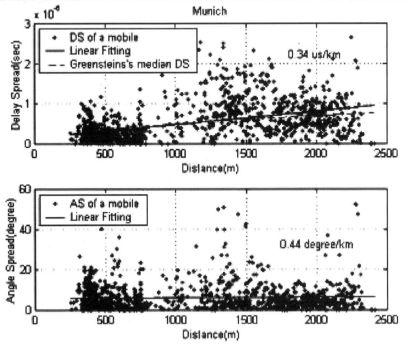

Figure 3. Scatter plots of DS and AS versus distance of the mobile from the base station for 900 MHz in Munich, Germany. Solid lines are linear regression fits.

MHz in Munich. The solid lines in these plots are linear regression fit to the scatter plots. The lines indicate that the median delay spread increases with distance, but that the median angle spread is independent of distance the mobile from the base station. Measurements of *DS* and *AS* indicate a systematic increase of delay spread with distance, but no systematic increase in angle spread has been found.

Reviewing many measurements of *DS*, Greenstein *et al.* [13] have concluded that its statistical properties can be modeled by the expression

$$DS = T\sqrt{R}\xi \tag{4}$$

where R is the distance in km, T is the mean delay spread at $R = 1$ km with value in the range of

0.4 - 1.0 μs, and ξ is a random variable such that $10\log\xi$ is has a normal distribution. In *DS* plot of Figure 3 we have plotted the Greenstein's distance variation for the mean delay spread using $T = 0.5$ μs. This variation is close to that of the regression fit, in Figure 3, and is consistent with the simulations.

The nature of the statistical distribution of *DS* about the regression fit lines in Figure 3 has been examined by plotting the cumulative distribution function of $10\log(DS - \langle DS \rangle)$. The results are shown in Figure 4 for Seoul, Korea and Munich. Germany. The vertical scale in Figure 4 is distorted such that a normal distribution has a straight line plot. The distributions generated using the simulation results are seen to be close to normal, especially for Seoul. The standard deviations of 3.37 and 3.73 for Seoul and Munich, respectively, are in the range 2 – 6 indicated for the Greenstein model [13].

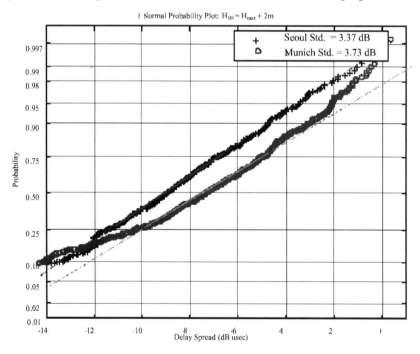

Figure 4.   CDF plotted on a normal scale for values of $10\log(DS - \langle DS \rangle)$ obtained from simulations in Seoul and Munich using a normal scale.

Simulation can easily be used to evaluate the dependence of *DS* and *AS* on antenna height H_{bs} and building height distributions. Simulations have been carried out for the original building database of Seoul and for various modifications of the building height distribution. Table 1 shows the mean (50%) and 90% values of *DS* and *AS* distributions. Here H_m is the maximum building height, while H_{95} and H_{80} represent the heights corresponding to the 95% and 80% points in the building height distribution. From this table it is seen that the delay spread is not sensitive to the height of the base station antenna, or to the distribution of building heights. However, the angle spread changes dramatically when the base station antenna is made lower that the maximum building height. Moreover, angle spread is dependent on the distribution of building heights, as found by comparing simulations for different cities [12]. The simulations can also examine higher order statistical parameters. For example, DS and AS are found to be weakly correlated with a correlation coefficient around $0.5 - 0.6$ [12].

Table 1.   DS (μs) and AS (degrees) for different antenna heights and different distributions of building height.

Building Height	Hbs	DS (50%)	DS (90%)	AS (50%)	AS (90%)
Original Database	Hm + 5 m	0.13	0.35	6.9	16.1
	Hm +2 m	0.14	0.37	7.5	16.3
	H95	0.18	0.46	15.4	33
	H80	0.19	0.48	17.5	35.8
12 - 21 m Uniform	Hm + 5 m	0.17	0.43	11.9	26.5
Distribution	Hm +2 m	0.18	0.38	16.7	34.5
	H95	0.15	0.42	22.8	43
	H80	0.17	0.44	32.7	46
12 - 21 m Rayleigh	Hm + 5 m	0.17	0.45	11.7	26
44.6	44.6 + 5 m	0.14	0.35	13.2	32.3
	44.6 - 5 m	0.12	0.59	24.4	51.4
6 - 9 m Uniform	2 m	0.23	0.62	34.7	54.2

IV. Finger Statistics for CDMA Rake Receivers

Another example of the use of ray simulations to predict channel characteristics is provided by the statistics associated with reception of individual fingers by CDMA rake receivers. In designing the rake receiver it is important to have information on the number of available fingers, which are separated by a minimum of one chip period, and on the time available for the receiver acquire and receive the fingers. These values will have statistical distributions that depend on the chip rate or chip period of the receiver, which defines the minimum time separation of the fingers, and the range of range of finger amplitudes that are detected.

A simulated impulse response consisting of the individual ray arrivals derived from the VPL ray tracing in Soul, Korea is depicted in Figure 5.a [14]. The ray amplitudes are shown relative to the strongest ray, and only those within 20 dB of the strongest are retained. The band limited received signal shown in Figure 5.b is derived from the impulse response of Figure 5.a for a system that in free space would have a sinc function response having full width between zeros of 0.2 μs, corresponding to a bandwidth of 5 MHz [14]. To arrive at Figure 5.b, we sum the sinc functions, taking the voltage amplitude, phase and time delay of individual sinc functions to be those of the individual rays.

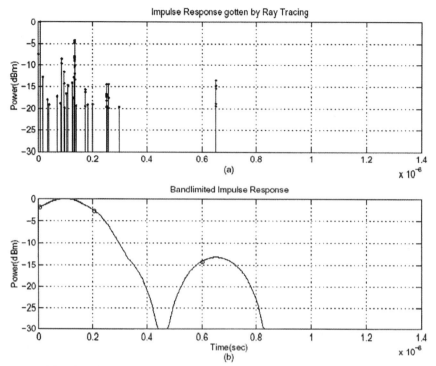

Figure 5.   Simulated ray arrivals at a receiver in Seoul, Korea and the predicted
power delay profile for a system having 5 MHz bandwidth.

For the received signal of Figure 5.b, three fingers at most can be detected under the condition that the threshold for additional fingers is 15 dB below the strongest. The foregoing simulation of the number of usable fingers has been carried out for many locations in Seoul. The resulting histogram of percent of locations at which N fingers are usable is shown in Figure 6 for line-of-sight (LOS) and non-LOS propagation conditions. By tracking the distance along the streets over which N fingers are usable, it is possible to identify the distribution of finger life distance, which translates to finger lifetime when the speed of the mobile is known. The median life distance for the second finger is found to depend strongly on base station location, and to lie in the range $10 - 30$ m. The life of the third finger is less than about 10 m.

High-resolution measurements of the pulse response of the radio channel have been made for many mobile locations in Helsinki, Finland. From these measurements, the number of usable fingers has been computed assuming a 5 MHz bandwidth and a 13 dB threshold [15]. The histogram of locations at which N fingers are usable is shown in Figure 7. The measurement base histogram of Figure 7 is seen to be very similar to that obtained from simulations. In both figures, $N = 3$ fingers are usable at about 60% of the locations, while $N = 5$ fingers are usable at about 30% of the locations.

V. Conclusions

The simulations described above show the power of using ray tracing techniques to generate Monte Carlo simulations of a verity of radio channel statistics. The ray representation of radio wave propagation is consistent with the delay line model of the radio communication cannel used for system design and evaluation. Simulations of the statistical distribution of various channel parameters are found to be in good agreement with measurements. The simulations are substantially easier and less expensive to carry out than are measurements. As a result, simulations are well suited to generating the statistical data needed to compare system performance in different building environments.

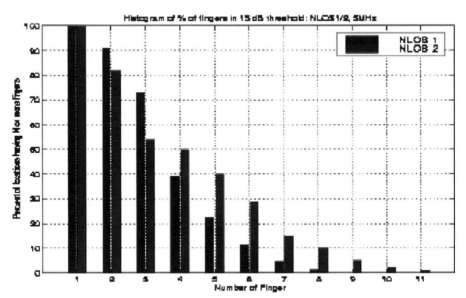

Figure 6. Percent of locations in Seoul at which ray simulations give *N* usable fingers under LOS (left member of each pair of bars) and non-LOS (right member of each pair of bars) conditions.

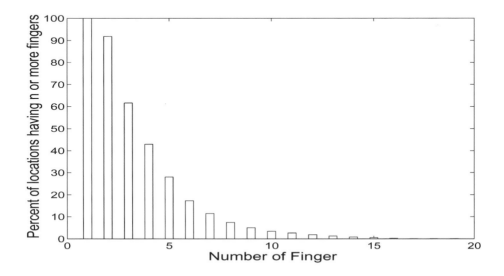

Figure 6. Percent of locations in Helsinki where measurement indicate *N* fingers are usable.

VI. References

[1]   M. C. Lawton and J. P. McGeehan, "The application of a deterministic ray launching algorithm for the prediction of radio channel characteristics in small-cell environments," *IEEE Trans. Veh. Technol.*, vol. 43, pp. 955-969, 1994

[2]   S. Y. Tan and H. S. Tan, "A microcellular communications propagation model based on the uniform theory of diffraction and multiple image theory, "*IEEE Trans. Antennas Propagat.*, vol. AP-44, pp. 1317-1325, 1996

[3]   K. Rizik, J-F. Wagen, and F. Gardiol, "Two-dimensional ray-tracing modeling for propagation prediction in microcellular environments, "*IEEE Trans. Veh. Technol.*, vol. 46, pp. 508-517, 1997

[4]   A. G. Kanatas, I. D. Kountouris, G. B. Kostraras, and P. Constantinou, "A UTD propagation model in urban microcellular environments," *IEEE Trans. Veh. Technol.*, vol. 46, pp. 185-193, 1997

[5]   T. Kurner, D. J. Cichon, and W. Wiesbeck, "Evaluation and verification of the VHF/UHF propagation channel based on a 3-D wave propagation model," *IEEE Trans. Antennas Propagat.*, vol. 44, pp. 393-404, 1996

[6]   S. C. Kim, B. J. Guarino, Jr., T. M. Willis III, V. Erceg, S. J. Fortune, R. Valenzuela, L. W. Thomas, J. Ling, and J. D. Moore, "Radio propagation measurements and prediction using three dimensional ray tracing in urban environments at 908 MHz and 1.9 GHz," *IEEE Trans. Veh.. Technol.*, vol. 48, pp. 931-946, 1999

[7]   J.P. Rossi, J.C. Bie, A.J. Levy, Y.Gabillet and M. Rosen, "A ray-launching method for radio-mobile propagation in urban area," *Digest of the IEEE APS Symposium* (London, Ontario, June 1991), pp. 1540-1543

[8]   G. Liang and H. L. Bertoni, "A new approach to 3-D ray tracing for propagation prediction in cities," *IEEE Trans. Antennas Propagat.*, vol. AP-46, pp. 853-863, 1998

[9]   C. Oestges B. Clerckx, L. Raynaud and D. Vanhoenaker-Janvier, "Deterministic channel modeling and performance simulation of microcellular wide-band communications systems," *IEEE Trans. Veh. Technol.*, vol. 51, pp. 1422-1430, 2002

[10]   H. El-Sallabi, G. Liang, H.L. Bertoni and P. Vainikainen, "Influence of diffraction coefficient and database on ray prediction of power and delay spread in urban microcells," *IEEE Trans. Antennas Propagat.*, vol. 50; No. 5; pp.703-712, 2002

[11]   B.H. Fleury, "Direction, dispersion and space selectivity in the mobile radio channel," *Proc. Fall 2000 IEEE VTC*, Boston, MA, Sept. 2000

[12]   C. Cheon, G. Liang and H.L. Bertoni, "Simulating radio channel statistics for different building environments," *IEEE Journal on SAC*, vol. 19, pp. 2191-2199, 2001

[13]   L.J. Greenstein, V. Erceg, Y.S. Yeh and M.V. Clark, "A new path-gain/delay-spread propagation model for digital cellular channels," *IEEE Trans. Veh. Technol.*, vol. 46, pp. 477-485, 1997

[14]   C. Cheon and H.L. Bertoni, "Fading of Wide Band signals Associated with displacement of the Mobile in Urban Environments," *Proc. Spring 2002 IEEE VTC*, Birmingham, AL, May. 2002

[15]   H. El-Sallabi, H.L. Bertoni and P. Vainikainen, "Experimental Evaluation of Rake Finger Life Distance for CDMA Systems," *IEEE Trans. Antennas Propagat. Lett.*; to be published

Wave Propagation,Scattering and Emission in Complex Media
Edited by Ya-Qiu Jin
Science Press and World Scientific,2004

389

Measurement and Simulation of Ultra Wideband Antenna Elements

Werner Sörgel, Werner Wiesbeck

Institut für Höchstfrequenztechnik und Elektronik, Universität Karlsruhe (TH),
Kaiserstr. 12, 76128 Karlsruhe, Germany. E-Mail: werner.wiesbeck@ihe.uka.de

Abstract   Antennas are mandatory devices for ultra wideband systems. A comprehensive characterization and its application are demonstrated for examples of different antenna types for the frequency range from 3.1 GHz to 10.6 GHz. Investigations of the transient radiation behavior of the logarithmic periodic dipole array, the Vivaldi antenna and the Spiral antenna are presented.

Introduction

Since approximately 20 years ultra wideband communication and radar systems are subject to research [12,13]. The interest was in the beginning mainly focused on military applications with the goals of high resolution, high information rate and low detectability. In the beginning of the nineties several conferences and workshops had been held on these subjects. As far as information is available ultra wideband systems still lack a wide application in the areas of initial interest.

In 2002 the United States Federal Communication Commission (FCC) has ruled that among others the frequency range from 3.1 GHz to 10.6 GHz is license free available for applications in communications and radar with a maximum output power (EIRP) of minors -41.25 dBm per MHz [1]. This announcement stimulated research in ultra wideband technology since the beginning of 2002 significantly all over the world, in the hope that a license free operation will be available in many other countries also, research especially in the areas of near range communications, especially for indoor high data rates mushroomed. UWB technology has also been suggested for near range radar applications for example for the detection of anti personnel landmines or test of road conditions started. From the earlier research work it is known that the excitation of RF front ends with ultra short pulses can not be handled in mathematical modeling, measurements and parametric description like narrow band or single carrier systems. Especially the antennas exhibit a completely different and embarrassing behavior, when they are excited with ultra short pulses. No longer the standard return loss and gain are the only relevant parameters for the characterization of their behavior. The ultra wideband short pulse excitation can be better described by the frequency dependent transfer function and the related time dependent transient response [5], [6], [7]. The access to this transfer function is most readily available from numerical electromagnetic modeling of these antennas, especially with the finite difference time domain codes (FDTD) [14]. The FDTD can take into account pulse lengths shorter than the antenna size, moving phase centers, dispersive radiation and ringing of the antennas. In the following this unconventional characterization of ultra wideband antennas for communications together with results for different types of ultra wideband antennas are presented.

Antenna Model

In general the electrical properties of antennas are characterized by input impedance, efficiency, gain, effective area, radiation pattern and polarization properties [2], [3], [4]. For narrow band applications it is possible to analyze these at the centre frequency of the system. For larger bandwidths all of them become strongly frequency dependent, but the straightforward evaluation of the named parameters as functions of frequency is not sufficient for the characterization of the transient radiation behavior. One proper approach to take transient phenomena into account is to model the antenna as a linear time invariant transmission system with the exciting voltage V_{exc} at the connector as input parameter and the radiated electrical far field E_{rad} as output parameter. This system can be fully characterized by its transient response. Assuming free space propagation this can be written like equation (1) as shown in [5], [6], [7], [8]. The dimension of the antennas normalized time and angle dependent transient response $h_n(\tau,\theta,\psi)$ is [m], which relates to the meaning of an effective antenna height, which is similar to the more common effective antenna area.

$$\frac{\vec{E}_{rad}(t,r,\theta,\psi)}{\sqrt{Z_0}} = \frac{1}{2\pi r c} \vec{h}_n(\tau,\theta,\psi) * \frac{1}{\sqrt{Z_C}} \frac{dV_{exc}(t)}{dt} * \delta(\tau - \frac{r}{c}) \tag{1}$$

The radiated far field is given by the convolution of the antennas normalized transient response h_n and the time derivative of the driving voltage at the feeding port. The derivative character of the antenna model can be explained by the fact that there has to be a capacitive or inductive coupling of the source voltage to the radiated wave. The coupling characteristics are covered by the properties of the transient response h_n. Z_0 denotes the characteristic impedance of free space, Z_c is the characteristic impedance of the antenna connector (assumed to be frequency independent), r is the distance from the antenna. The convolution with the Dirac function $\delta(\tau-r/c)$ represents the time retardation due to the finite speed of light c. The antennas transient response depends on the regarded direction (Θ,ψ) of radiation and is a vector according to the polarization vector properties (co-polarization and cross-polarization) of the modeled antenna.

Assuming an incident plane wave out of the direction (Θ_w,ψ_w) with the polarized field strength $(E_{inc,\Theta Pol}, E_{inc,\psi Pol})$ the given model fulfils the reciprocity theorem [6] and the output voltage of the antenna in receive mode can be characterized by

$$\frac{V_{rec}(t)}{\sqrt{Z_C}} = \vec{h}_n(\tau,\theta_w,\psi_w) * \frac{\vec{E}_{inc}(t)}{\sqrt{Z_0}} \tag{2}$$

This unconventional antenna model enables the description of the radiation of arbitrary waveforms like Gaussian pulses or chirps etc. The model covers all dispersive effects that result from a particular antenna structure (eg. the influence of coupled resonators and the related varying group delay due to nonlinear phase response). The influence of frequency dependent matching and ohmic losses are also covered. Thus quality measures for the effectiveness of a particular more or less dispersive UWB antenna under test (AUT) can be derived directly from the antennas transient response. Important quality measures of the transient response are the peak value, the envelope width (defined as full width at half maximum, FWHM), the duration of the oscillation ringing (defined as the duration until the oscillations envelope has fallen below a certain lower bound).

Measurement of Transient Response

Measurements have to be performed in order to verify the model and to get realistic values for the transient gain. The measurement of the antennas transient response can be performed either in time domain or in frequency domain. Measurements in time domain using very short pulses or steep step functions as driving voltage can potentially be faster but have then a lower dynamic range than measurements in frequency domain using a vector network analyzer in combination with a suited numerical transformation into the time domain. Here the sharp limitation of the measurement bandwidth can cause unrealistic overshoots and ambiguities of the regarded time interval. This can be handled by choosing a measurement bandwidth that is large compared to that of the waveform to be radiated in combination with a proper frequency resolution.

The measurements presented here have been performed using a HP8530A vector network analyzer and a PHYTRON positioner supporting the antenna under test (AUT) within an anechoic chamber. As reference antenna an ultra wide band TEM horn antenna is used. This measurement system is controlled by self-developed control and processing software. The measurement frequency range is 400 MHz to 20 GHz (25 MHz resolution). The direct result of the measurement is the transmission coefficient between the ports of the AUT and the reference antenna, which can be determined by combining eq. (1) and (2) and transforming the result in frequency domain [8],[11]:

$$s_{21}(f) = \frac{U_{rx}(f)}{U_{tx}(f)} = H_{tx}(f)\, H_{rx}(f)\, \frac{j\,f}{Rc}\, e^{j2\pi f R / c} \tag{3}$$

Using two identical UWB horn antennas the complex transfer function $H_{tx}(f)$ of the reference antenna can be calculated from (3) since the distance R between the two antennas is known. With the known reference $H_{tx}(f)$ the transfer functions of the AUT is easily calculated solving (3) for $H_{rx}(f)$. This has to be done for all relevant cut planes of the antenna radiation sphere with a set of two independent polarizations (co- and cross-polarization).

The resulting complex transfer function is transformed into the time domain using the discrete Fourier transformation. As shown in [7] the familiar CW gain can be calculated from the transfer function:

$$G(f,\theta,\psi) = \frac{4\pi\, f^2}{c^2} \left| \vec{H}_n(f,\theta,\psi) \right|^2 \tag{4}$$

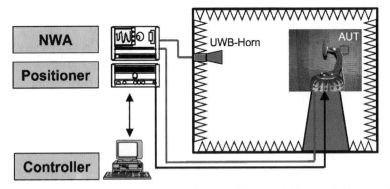

Fig. 1.   Measurement set up with antenna under test (AUT) within anechoic chamber

UWB Antenna Concepts for Communications

Based on the FCC regulations [1] the regarded frequency range is 3.1 GHz to 10.6 GHz. This fractional bandwidth B_f = 109% can be basically achieved with different broadband antenna concepts. The investigations here focus on Logarithmic Periodic Dipole Arrays (LPDA), Vivaldi antennas and Spiral antennas. LPDA and Vivaldi antenna are apt to produce some gain over the regarded bandwidth. Since antenna size is a hard constraint for portable wireless communication devices, these antennas have been optimized for minimum size. Spiral antennas can be set up as conformal antennas and thus are likely to meet the requirements for a suitable form factor. For each type of antenna simulations and measurements have been performed with special interest on the transfer function within the FCC frequency range.

Vivaldi Antenna

The Vivaldi antenna consists of an exponentially tapered slot etched onto the metallic layer on a Duroid 5880 (ε_r = 2,2) substrate as shown in fig. 3. The narrow side of the slot is used for feeding the antenna, the opening of the taper points to the direction of the main radiation. The Vivaldi antenna presented here is designed with a Marchand balun feeding network, which has been developed for the FCC frequency range (return loss better −10 dB). The geometric dimensions of this antenna are 78x75x1.5 mm (fig. 2 left side). The radiation is supported by the travelling wave structure formed by the slot. The antennas characteristics exhibit low dispersion. This can be seen best from the Vivaldis high peak value ($h_{n,max}$=0.017 m) and the short FWHM (115 ps) of the transient responses envelope (ref fig. 2 right side). An explanation for the low dispersion of the Vivaldi antenna is that the phase velocity on the tapered structure is nearly constant over the regarded frequency range. The ringing of the antenna is due to higher modes supported by the rectangular edges of the antenna. Enlarging the transverse dimension of the antennas metal flares the ringing can be reduced. The measured patterns for the complex frequency domain transfer function and the related CW gain are shown in fig. 3. The maximum CW gain is 7.9 dBi at 5.0 GHz.

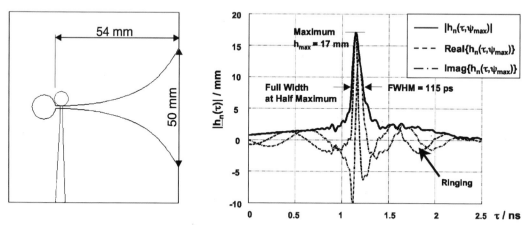

Fig. 2.   **Left side:** structure of the regarded Vivaldi antenna, **Right side:** measured transient response h(τ,ψ_{max}) in main beam direction (E-plane, copolarization)

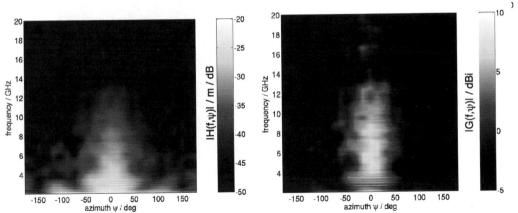

Fig. 3.   **Left side:** Vivaldi antenna, magnitude of complex transfer function (E-Plane, co-polarization). **Right side:** resulting gain pattern.

Logarithmic Periodic Dipole Array (LPDA)

The LPDA in general consists of a number of adjoining unit cells (dipoles), with the dimensions of adjacent cells scaled by a constant factor [9]. In figure 1 the geometric structure of the regarded printed LPDA is shown. Each dipole is etched with one half on the top layer and the other half on the bottom layer of the substrate (Duroid 5880, $\varepsilon_r = 2,2$). The antenna is fed by a triplate balun as proposed in [10]. This structure has been optimized for low return loss (better –10 dB) in the frequency range from 3.1–10.6 GHz. The width of dipole i is denoted as w_i, the distance to its next neighbour as d_i. The length of the feed line is S=45 mm (see fig. 4). The length ratio of neighbouring dipoles is L_{i+1}/L_i=0.8. The ratio of half distance to length of the dipoles is d_i/L_i=0.22. These parameters underline the compact design of the antenna (60x50x2 mm) which result in a HPBW of 65° in the E-plane and 110° in the H-plane. These values are quite constant over the regarded frequency range. FDTD simulations and NWA based measurements (meas. BW 20 GHz @ 25 MHz resolution) of the transient response show that the LPDA exhibits a strong oscillation (fig 5, right side). This can be explained by the excited ringing of coupled resonating dipole elements. Due to the ringing the FWHM of the transient responses envelope is 805 ps long and has a peak value of only 0.007 m. A comparison of the LPDA with the Vivaldi antenna shows that the slightly smaller LPDA has a comparable CW gain over the regarded bandwidth. The maximum CW gain is 7.4 dB at 3.4 GHz. The LPDA exhibits a very dispersive behavior, which has to be taken into account for the design of a proper modulation scheme.

In fig. 5 the results from a FDTD simulation and the linear antenna model (eq. 1) are compared. The transient response data for the model has been measured as described above. As excitation for simulation and measurement based model a Gaussian pulse with FWHM = 88 ps has been chosen. The results for the time dependent co-polarized electric field have each been normalized to their maxima, because the used FDTD simulation tool is not able to handle absolute values for the input power level. The comparison shows a good agreement of the resulting waveforms. This indicates that the linear antenna model is well suited for further optimization of UWB system parameters like modulation schemes.

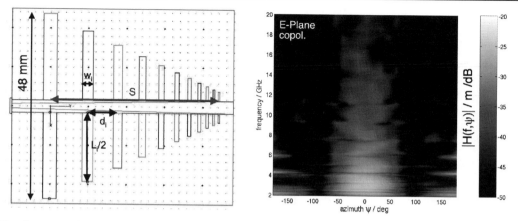

Fig. 4.   **Left side:** structure of compact LPDA. **Right side:** magnitude of measured complex transfer function.

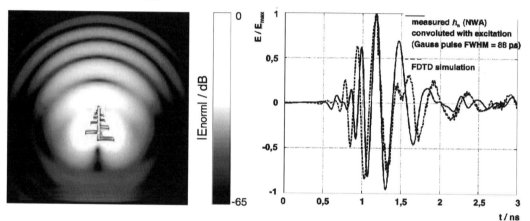

Fig.5.   **Left side:** FDTD simulation magnitude E-field, snapshot after 1100 ps.. **Right side:** comparison of FDTD simulation and antenna model for the E-field (main beam, co-polarization).

Spiral Antenna

Since antenna size is an important constraint for communication devices conformal antennas have to be considered. As an example a two-arm spiral antenna with a diameter of 10 cm has been simulated with FDTD. The archimedic spiral structure has been modeled with a stair case approximation as shown in fig. 6. It is excited with a 188 Ω Voltage gap source located between the inner ends of the perfectly electric conducting two arms. The excitation voltage function is a Gaussian Pulse (FWHM = 88 ps). The maximum of the excitation voltage is reached after 160 ps. The pulse propagates on the spiral structure and is subsequently radiated as shown in fig. 6. Reflections from the open ends cause some ringing (fig.7, left side).

In fig. 7 the resulting electric field in main beam direction 240 mm above the center of the spiral is shown. On the left side of fig. 7 the time dependence of the components of the electric field vector are plotted. As can be seen from the diagram the time delay between 0 ns and the first rise of the electric field strength correlates well with the distance of the simulated field probe (c_0/240 mm=800 ps). A varying delay between the E_x and the E_y component is observed besides the E_z component (component in direction of propagation) is zero. That indicates, that the propagating TEM Mode has already formed.

On the right side of fig. 7 the locus of the electric field vector is given in a polar plot where the radius marks the linear magnitude and the angle marks the temporary polarization of the electric field strength. The observer is looking along the direction of propagation (into the chart) conformal with the IEEE definition of polarization. The diagram shows that within the time of high magnitude only a partial turn of the electric field vector occurs. That illustrates that for UWB polarization aspects like axial ratio can become dependent on the applied modulation. Using independent transfer functions for independent polarization axes, this effect is covered by the presented model.

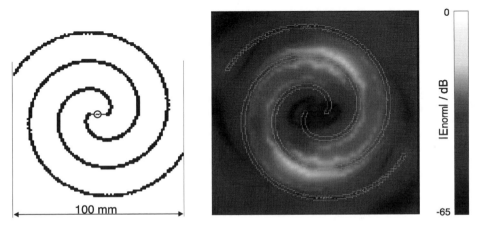

Fig. 6.   **Left side:** simulation model of two-arm spiral as perfect electric conducting arms in air. **Right side:** snapshot of the electric field magnitude distribution after 430 ps.

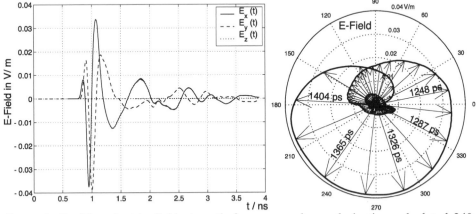

Fig. 7.   **Left side:** electric field strength for x-,y- and z- polarization calculated 240 mm above the center of the spiral antenna. **Right side:** time dependent polarization, locus of the electric field vector.

Summary

Ultra wideband antennas support pico second pulse applications in communications and radar. For the overall system specification in these techniques the proper knowledge of the antenna behavior and the dedicated parameters is of highest importance. In this paper it has been demonstrated that the electromagnetic characterization of ultra wideband antennas allows the proper access to the transfer function. This transfer function can together with the convolution of the exciting ultra short pulse be used to derive the known antenna gain. In addition this characterization gives valuable a priori knowledge for communication channel modeling. This will lead to a joint optimization of antenna and modulation properties in order to exploit the capacity of UWB communications. Ultra wideband systems with a low level radiation allow an efficient use of the limited spectrum by an overlay to present applications. The final determination of the interference to different other active and passive applications of the microwave spectrum have to be experienced in the near future.

References

[1]   Federal Communications Commission (FCC), "Revision of Part 15 of the Commission's Rules Regarding Ultra Wideband Transmission Systems,,, First Report and Order, ET Docket 98-153, FCC 02-48; Adopted: February 14, 2002; Released: April 22, 2002

[2]   M. Thumm, W. Wiesbeck, S. Kern, Hochfrequenzmeßtechnik, B.G.Teubner Stuttgart, 1998

[3]   C.A. Balanis, Antenna Theory, Analysis and Design, Wiley & Sons, Inc. New York, 1997

[4]   Meinke; Gundlach, Taschenbuch der Hochfrequenztechnik, Springer Berlin, 1992

[5]   C. E. Baum, General Properties of Antennas, Sensor and Simulation Notes, Note 330, Air Force Research Laboratory, Directed Energy Directorate, Kirtland, New Mexico, 1991

[6]   E. G. Farr, C. E. Baum, Extending the Definitions of Antenna Gain and Radiation Pattern into the Time Domain, Sensor and Simulation Notes, Note 330, Air Force Research Laboratory, Directed Energy Directorate, Kirtland, New Mexico, 1992

[7]   L. Bowen, E. Farr, W. Prather, A Collapsible Impulse Radiating Antenna, Ultra-Wideband Short-Pulse Conference, Tel Aviv 2000

[8]   B. Scheers, M. Acheroy, A. Vander Vorst, Time Domain simulation and characterisation of TEM horns using a normalised impulse response, IEE Proc.-Microw. AP, Vol. 147, No. 6, pp. 463-468 Dez. 2000

[9]   P.E.Mayes, G. A. Deschamps and W. T. Patton, Backward-Wave Radiation from Periodic Structures and Application to the Design of Frequency-Independent Antennas, Proceedings of the IEEE, Vol. 80, pp. 103-112, January 1992

[10]   R. Pantoja, A. Sapienza, F. Medeiros Filho, A Microwave Printed Planar Log-Periodic Dipole Array Antenna, IEEE Transactions on antennas and propagation, Vol. AP35, No. 10, pp .1176-1178, Oct. 1987

[11]   R. Clark Robertson, M. A. Morgan, Ultra-Wideband Impulse Receiving Antenna Design and Evaluation, Ultra-Wideband, Short-Pulse Electromagnetics 2, Plenum Press, New York, pp. 179-186, 1995

[12]   Bennett, C.L. and Ross, G.F., Time-domain Electromagnetics and its Applications, Proceedings of the IEEE, Vol. 66, No. 3, pp. 299-318, 1978

[13]   History of UWB Technology, http://www.multispectral.com/history.html

[14]   A. Tavlove, Computational Electrodynamics: The Finite Difference Time Domain Method, Artech House, Boston, MA, 1995

[15]   T. B. Hansen, A.D. Yaghjian, Planar Near Field Scanning in the Time Domain, IEEE Transactions on Antennas and Propagation, Vol. 42, No. 9, pp. 1280-1291, Sept. 1994

Wave Propagation, Scattering and Emission in Complex Media
Edited by Ya-Qiu Jin
Science Press and World Scientific, 2004

The Experimental Investigation of a Ground-placed Radio Complex Synchronization System

V. P. Denisov, V. Yu. Lebedev, D. E. Kolesnikov, V. G. Kornienko, M. V. Krutikov

Tomsk State University of Control Systems and Radio Electronics

40 Lenin av., Tomsk, 634050, Russia

T/F: +7-3822-413949. E-mail: rwplab@orts.tomsk.ru

The operation principles and the field test results of the synchronization system are stated in the paper. The system is intended for ground-placed sites time-scale synchronization. The system being an essential part of an active or passive multiposition radio complexes has a mainly influence on objects location determination precision. The time-scale synchronization error of the ground-placed sites must be smaller then σ_i/c, where σ_i - coordinate metering error and c - velocity of light. So the time-scale synchronization error must be smaller then 10 nsec when the coordinates metering error smaller then 10 meters is required.

The possibility of an object location determination precision smaller then 10 meters is investigated in the paper. Such high precision may be provided by UHF synchronization signal transmission if the conditions are favourable [1]. In the investigation it was necessary to define the possible equipment synchronization precision in a ground-placed sites with a priori unknown coordinates. Therefore the synchronization equipment is to provide the signal propagation time measurement from one ground-placed site to other and to make universal time-scales in the sites.

The time-scale is a binary sequence synchronized by a reference signal. The reference time-scale is formed by the leading site. The reference signal synchronizes the pseudorandom m-sequence with the period of 1023 symbols and the partial symbol duration of 100 nsec. The UHF synchronization signal is the phase-shift keyed UHF carrier with the frequency of 1065 MHz modulated by the pseudorandom signal. The UHF signal main lobe bandwidth is 20 MHz. The leading site transmitter has a turnstile antenna with a circular pattern in a horizontal plane. In the peripheral site the m-sequence is synchronized by the signal received from the leading site and modulates the UHF carrier with the frequency of 625 MHz. The phase-shift keyed carrier is emitted by the near-omnidirectional antenna. The received signal in a leading site synchronizes the leading site receiver m-sequence. The clock pulses of transmitted and received sequences form the input signals for the intervalometer. The measured time delay consists of signal propagation delay and equipment delays. The half of total delay is a time-scale error for two ground-placed sites.

The time-scale error consists of two main components. The first component is a equipment faultiness and the second one depends on a propagation conditions.

The main factors of a radio-wave propagation time error for a near-surface radio path are the radio-wave velocity inconstancy and the radio scattering by the relief heterogeneousnesses. The equation for a radio-wave velocity in a space with the index coefficient n is given by $c = c_0/n$, where c_0 - radio-wave velocity for a free space.

The refractive index depends on an atmosphere pressure, a temperature and an air moisture. The equation for the refractive index is given by $n = 1 + N \cdot 10^{-6}$, where N = 200...400 for near-surface area.

The equation for a radio-wave propagation time delay is given as [1]:

$$\tau_d = \frac{1}{c_0} \cdot \int_L n(S)dS \qquad (1)$$

where the integration is performed for a radio path corresponding the minimum propagation time (see the Ferma`s principle). It is necessary to know n as a coordinate function in detail for the computation of (1). It is a problem in most cases because the index coefficient depends on both the synoptic conditions and a radio path relief. If the mean value of N is estimated by weather conditions for a some space point and a mean value deviation is smaller than ΔN then using (1) the propagation time fluctuations for a radio-wave path length L are smaller than

$$\Delta \tau_d = \Delta N \cdot L / C_0 \qquad (2)$$

From this it follows that

$$\Delta \tau_d / \tau_d = \Delta N$$

where τ_d - free space propagation time.

The experimental fluctuation appraisals of the refraction coefficient difference for the near-surface radio paths at the different space points are given by O. N. Kiselev [2]. Measuring the temperature, atmosphere pressure and air moisture at the same time, the mean-square difference of the refraction coefficients $\sigma_{\Delta N} = \sqrt{\overline{\Delta N^2}}$ for the space points with the distance of 103 km is from 2 to 8,5. The experimental measurements were made in the Kemerovo and Tomsk regions (Russia).

If $\sigma_{\Delta N} = 10$, $L = 40$ km then: $\sigma_{\tau_d} = 1,5$ nsec. So it is possible to calculate how the propagation time depends on a radio wave velocity deviation using the mean value.

Let`s consider how the synchronization precision depends on a radio-waves scattering due to a relief. The correlation receiver block diagram in a simplified version is

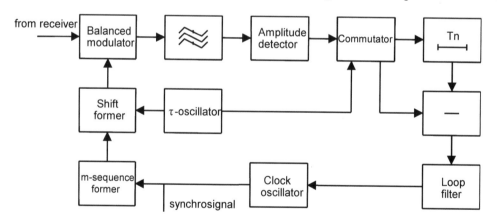

Fig. 1 - Correlation receiver block diagram

given in Fig. 1. The correlation receiver has the m-sequence former similar the transmitter one. For the forming signal of the time error between the input signal and the reference one the reference m-sequence is time shifted by $T_1/2$ or $-T_1/2$ every m-sequence period. T_1 is a sequence partial symbol duration. τ -former sets a time-shift value and controls electronic commutator. The electronic commutator switches the correlation function signal either directly or through the time-delay line with the time-delay T_1 to the subtraction circuit.

Let $x(t) = a \cdot u_0(t) + u_s(t) + n(t) = \operatorname{Re} x(t) \cdot e^{j \cdot \omega_0 \cdot t}$

where $a \cdot u_0(t)$ - useful signal, $u_s(t)$ - relief scattering signal, $n(t)$ - receiver self-noise. And let

$$a \cdot u_0(t) = a \cdot \mathrm{Re}\, A(t) \cdot e^{j \cdot \varphi(t)} \cdot e^{j \cdot \omega_0 \cdot t} = a \cdot \mathrm{Re}\, U(t) \cdot e^{j \cdot \omega_0 \cdot t} \tag{3}$$

where the equation for useful signal complex amplitude is

$$U(t) = \sum_{i=1}^{N} U_{0i}(t - (i-1) \cdot T_1) * $$

and $U_{0i} = \pm 1$. The complex envelope correlation of an input signal and a reference m-sequence is performed by the function multiplier, narrow band filter and amplitude detector. Therefore the equation for the subtraction circuit output signal voltage is:

$$u_{discr}(\tau) = \int_0^T x(t - \tau) \cdot U_0(t - \frac{T_1}{2}) dt - \int_0^T x(t + T - \tau) \cdot U_0(t + T + \frac{T_1}{2}) dt \tag{4}$$

Since the m-sequence is periodic it is possible to transform (4):

$$u_{discr}(\tau) = \int_0^T x(t - \tau) \cdot U_0(t + \frac{T_1}{2}) dt - \int_0^T x(t + T - \tau) \cdot U_0(t + \frac{T_1}{2}) dt \tag{5}$$

If there are no relief scattering signals and receiver self-noises are epsilon squared then:

$$u_{discr}(\tau) = a_0 \cdot \int_0^T U_0(t - \tau) \cdot U_0(t - \frac{T_1}{2}) dt - a_0 \cdot \int_0^T U_0(t - \tau) \cdot U_0(t + \frac{T_1}{2}) dt +$$

$$+ a_1 \cdot \int_0^T U_0(t - \tau_1 - \tau) \cdot U_0(t - \frac{T_1}{2}) dt - a_1 \cdot \int_0^T U_0(t - \tau_1 - \tau) \cdot U_0(t + \frac{T_1}{2}) dt = \tag{6}$$

$$= a_0 \cdot [K(\tau - \frac{T_1}{2}) - K(\tau + \frac{T_1}{2})] + a_1 \cdot [K(\tau + \tau_1 - \frac{T_1}{2}) - K(\tau + \tau_1 + \frac{T_1}{2})]$$

where $K(\tau)$ - m-sequence autocorrelation function

Eq. (6) is illustrated by Fig. 2. The correlation function main lobes are showed. The time discriminator characteristic is showed in Fig. 2. The characteristic is linear if $|\tau| < T_1 / 2$.

$$u_{discr}(\tau) = \frac{2 \cdot a_0 \cdot \tau}{T_1} \tag{7}$$

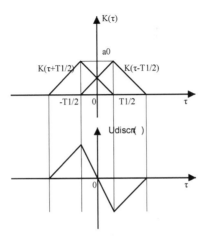

Fig. 2 Time discriminator characteristic

The time discriminator zero value voltage means the input and reference m-sequences time matching in a noise-free conditions.

If there are no receiver self-noises but there is relief scattering signal then using Eq. (5)

$$u_{discr}(\tau) = a_0 \cdot \int_0^T U_0(t-\tau) \cdot U_0(t-\frac{T_1}{2})dt - a_0 \cdot \int_0^T U_0(t+T-\tau) \cdot U_0(t+\frac{T_1}{2})dt +$$
$$+\int_0^T U_0(t-\tau) \cdot U_0(t-\frac{T_1}{2})dt - \int_0^T U_0(t+T-\tau) \cdot U_0(t+\frac{T_1}{2})dt$$

$$(8)$$

Two last terms of Eq. (8) are the error signal minimized by the follow-up system. The time error depends on scattering signal statistics and follow-up system dynamic performance.

Generally, the scattering signal is a sum of reflected rays [3]:

$$u_s(t) = \sum_{i=1}^N a_i u_0(t-\tau_i)$$

$$(9)$$

where a_i - amplitude factor, τ_i - useful and reflection signal propagation time difference

The form of signal $u_s(t)$ depends on radio path type, leading and peripheral sites antenna heights directional patterns. According to the conditions the reflection signals number in (9) may variate from 0 to high value. Generally, signal $u_s(t)$ is a stationary normal random process.

For follow-up system operation analyses the reflection signals with time delays smaller than follow-up system discriminator characteristic width relatively useful (direct) signal are most interesting. Let`s use the discrete-time conception of $u_s(t)$ to understand the synchronization error rise mechanism. Let only reflection signal with the amplitude factor a_1, time delay $\tau_1 < T_1$, and the carrier phase shift $\varphi_1 = 0$.

Eq. (8) is transformed to

$$u_{discr}(\tau) = a_0 \cdot \int_0^T U_0(t-\tau) \cdot U_0(t-\frac{T_1}{2})dt - a_0 \cdot \int_0^T U_0(t-\tau) \cdot U_0(t+\frac{T_1}{2})dt +$$
$$+a_1 \cdot \int_0^T U_0(t-\tau_1-\tau) \cdot U_0(t-\frac{T_1}{2})dt - a_1 \cdot \int_0^T U_0(t-\tau_1-\tau) \cdot U_0(t+\frac{T_1}{2})dt =$$

$$(10)$$

$$= a_0 \cdot [K(\tau-\frac{T_1}{2})-K(\tau+\frac{T_1}{2})] + a_1 \cdot [K(\tau+\tau_1-\frac{T_1}{2})-K(\tau+\tau_1+\frac{T_1}{2})]$$

Therefore (taking into consideration (7))

$$u_{discr}(\tau) = \frac{2 \cdot a_0}{T_1} \cdot \tau + \frac{2 \cdot a_1}{T_1} \cdot (\tau+\tau_1), \quad |\tau| < \frac{T_1}{2}, \quad |\tau-\tau_1| < \frac{T_1}{2}$$

$$(11)$$

Let $\frac{a_1}{a_0} = K_1$. From Eq. (11), the steady-state follow-up system time error due to reflection signal is expressed as

$$\tau = -\frac{\tau_1 \cdot K_1}{1+K_1}, \quad |\tau| < \frac{T_1}{2}$$

$$(12)$$

The negative sign of the equality right part means reference sequence lag relatively input useful signal.

If there are N reflection signals Eq. (12) is transformed to

$$\tau = -\sum_{i=1}^{N} \tau_i \cdot K_i / (1 + \sum_{i}^{N} K_i), \qquad |\tau| < T_1/2 \tag{13}$$

Let's clear up how the synchronization precision depends on phase difference of direct signal and reflection one. In diagram (see Fig. 1) every input signal forms harmonic oscillation at the detector input. The harmonic oscillation amplitude is proportional to cross-correlation function of an input signal complex envelope and a reference m-sequence. Let there are three signals at the receiver input: direct signal and two reflection signals. The cross-correlation function of a total input signal and reference sequence is calculated as:

$$K_\Sigma(\tau) = \sqrt{\left(K_0(\tau) + K_1(\tau) \cdot \cos\phi_1 + K_2(\tau) \cdot \cos\phi_2 \right)^2 + \left(K_1(\tau) \cdot \sin\phi_1 + K_2(\tau) \cdot \sin\phi_2 \right)^2}$$

where $\phi_1 = \omega\tau_1 - \varphi_1$ - the phase difference of direct signal and first reflection, $\phi_2 = \omega\tau_2 - \varphi_2$ - the phase difference of direct signal and second reflection, $K_1(\tau)$ - cross-correlation function of reference and direct signals, $K_2(\tau)$ - cross-correlation function of reference signal and second reflection, τ_1 - radio-wave propagation time difference of first reflection and direct signal, τ_2 -radio-wave propagation time difference of second reflection and direct signal, φ_1 - the first reflection signal phase, φ_2 - the second reflection signal phase.

 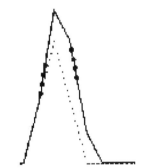

Fig. 3 - Cross correlation function of input and reference signals
$k_1 = a_1 / a_0 = 0.5, a_2 = 0, \tau_1 = 100n\sec,$
$\varphi_1 = \pi, time\,error\,-10n\sec$

Fig. 4 - Cross correlation function of input and reference signals
$k_1 = a_1 / a_0 = 0.5, a_2 = 0, \tau_1 = 50n\sec,$
$\varphi_1 = 0, time\,error\,-17n\sec$

Eq. (14) is too difficult to use so there is expediency of applying the computer simulation for synchronization error analysis. For forming the m-sequence with the length of 1023 symbols it is necessary to simulate the 10-bit shift register with the feed-backs. Every m-sequence symbol consists of 20 discrete values. The discrete value number determines the simulation precision 100/20=5 nsec. There are differences in a phases and amplitudes of $K_0(\tau)$, $K_1(\tau)$, $K_2(\tau)$ but there is no difference in a correlation function form. The input signal consists of direct signal and only reflection one. In the search mode of a follow-up system the correlation function maximum is defined and the correlation function maximum time location is fixed. In the autotrack mode the values of $K_\Sigma(\tau - 50)$ and $K_\Sigma(\tau + 50)$ are compared. Depending on compare result the increase by 1 or decrease by 1 of τ is performed while $K_\Sigma(\tau - 50)$ is not equal $K_\Sigma(\tau)$. The simulation results for different reflection signal parameters are showed in Figs. 3,4. The dotted line is a cross-correlation function of a reference signal and input one in a reflection-free condition. The points of a diagram mean the follow-up system operation in an autotrack mode. The simulation precision is smaller than 5 nsec.

The full-scale test results of the synchronization system are placed in the table (see below).

Radio path number	Measured delay, nsec	RMS. nsec	True delay, ncec	Sinchronization system data, m	GPS data, m	Measured error, m
1	54710,02	0,80	54411,48	8167,13	8154,38	4,14
2	72649,50	2,85	72330,76	10853,96	10839,55	5,81
3	55128,19	1,53	54662,89	8204,82	8191,09	5,13
4	55217,42	1,68	54752,12	8218,20	8187,03	11,57**
5	55222,40	0,49	54724,04	8213,99	8187,03	18,36
6	55222,10	0,54	54723,67	8213,94	8187,03	18,30
7	55226,30	2,03	54727,9	8214,58	8187,03	18,94
8	55232,09	3,70	54733,69	8215,44	8187,03	19,81
9	55239,52	1,98	54741,12	8216,55	8187,03	20,92
10	55239,53	1,10	54741,13	8216,55	8187,03	20,92
11	55236,57	0,96	54738,17	8216,11	8187,03	20,48
12	55236,52	0,89	54738,12	8216,10	8187,03	20,47

In the 6-th column there are data of GPS manufactured by Ashtech with the potential accuracy smaller than 1 m. In the 7-th column there are the differences of distances which are calculated using the radio-waves propagation times and GPS data. The time dependence of a minute mean radio-wave propagation delays for two different radio paths are showed in Figs. 5,6.

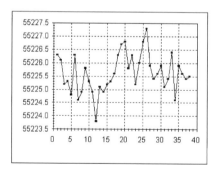

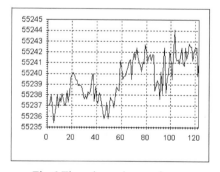

Fig.5 Time dependence of a minute mean radio-wave propagation delay for a radio path 7

Fig.6 Time dependence of a minute mean radio-wave propgation delay for a radio path 8

The full-scale test result analysis:
1. The time-scale error for two ground-placed sites consists of:
       quick fluctuations (up to 3.7 nsec);
       slow fluctuations (up to 8 nsec);
       time displacement (up to 60 nsec).

2. The time-scale error is about 60 nsec using the tested equipment without an additional correction in a condition of measured and allowed equipment errors.

3. For minimize the time scale error it is necessary to make additional correction. The additional correction value may define for example by comparing the radio-wave propagation time from one ground-placed site to other and GPS data. The time-scale error in this case is smaller than 8,8 nsec including the slow and quick fluctuations.

References

1. Winkler G.M. Path Delay, Its Variations, and Some Implications for the Field Use of Precise Frequency Division Proceeding of IEEE, v60, №5 (may 1972)

2. Киселев О.Н. Статистика неоднородностей коэффициента преломления в приземном слое атмосферы // 9 Всесоюзная конференция по распространению радиоволн. Тезисы докладов. Ч.1, Харьков, 1969 г

3. Turin G.L. Introduction to Spread – Spectrum Antimultipath Techniques and Their Application to Urban Digital Radio. Proceedings of IEEE, 1980, v68, №3, p30

VII. Computational Electromagnetics

Wave Propagation,Scattering and Emission in Complex Media
Edited by Ya-Qiu Jin
Science Press and World Scientific,2004

Analysis of 3-D Electromagnetic Wave Scattering with the Krylov Subspace FFT Iterative Methods

R.S.Chen, Z. H. Fan, D.Z.Ding

Department of Communication Engineering,

Nanjing University of Science and Technology, Nanjing, 210094, China

eerschen@mail.njust.edu.cn

Edward.K.N.Yung

Department of Electronic Engineering,

City University of Hong Kong, Kowloon, Hong Kong, China

Abstract   In this paper, the electromagnetic wave scattering is formulated in terms of the electric field integral equation (EFIE) for a dielectric body of general shape, inhomogeneity, and anisotropy. Applying the pulse-function expansion and the point-matching technique, the integral equation can be solved by the efficient Krylov subspace iterative fast Fourier transform (FFT) technique. Several typical Krylov subspace iterative FFT methods are used to solve the problems and their convergence behaviors are compared.

Key Words   Electromagnetic wave scattering, Krylov subspace iterative methods, fast Fourier transform

1. Introduction

The scattering of electromagnetic waves by penetrable inhomogeneous objects of arbitrary shape is an important research subject recently. They are found to have a wide application in communications, target identification, nondestructive testing, geophysical exploration, and microwave imaging. The pioneering works treating 3D arbitrary dielectric bodies were mainly based on the method of moment (MoM) [1]. The conventional method of moment (MoM) for this problem involves discretizing the vector volume electric field integral equation (EFIE) and then solving a linear system with N unknowns. Therefore, the computer memory requirement is $O(N^2)$ and the CPU requirement is $O(N^3)$ if the matrix equation is solved by the direct solvers, or $O(KN^2)$ by a Krylov iterative solver, where K is the number of iterations. Unfortunately, for volumetric inhomogeneities, even a moderate size volume will make this computation intractable. For example, an inhomogeneous volume of dimension $4\lambda \times 4\lambda \times 4\lambda$ will roughly result in a system with 190000 unknowns, requiring 147 GB memories to store the elements of the dense MoM matrix. The conjugate gradient method (CGM) was introduced in the early 1980s as an efficient and accurate technique. It can treat large problems with thousands of tens of thousands of unknowns by avoiding the storage of the MoM matrices. This was accomplished by re-computing the elements of the MoM matrix in the iteration. The FFT technique is used to speed up the matrix-vector multiplication in the iterative methods to solve integral-differential equations and this technique became the CG-FFT method [2] when the CG algorithm was used as the iterative method. Using such a solution technique, the efficiency of the integral equation approach is greatly improved, in that the requirements of the primary memory and the computation time are reduced to those proportional to N and $N \log N$, respectively. However, except for two-dimensional transverse magnetic scattering case, the conventional procedure for solving the EFIE using the pulse-basis block model has one fatal drawback: the relative permittivity of the scatterer must be kept small. Otherwise, the iterative CGM converges very slowly or may even stagnate [2]. Even if a solution is obtained using the CGM or other methods, it may contain serious errors. These errors are due to a term in the EFIE representing the effect of induced polarization charge. Because of such a charge term, the magnitudes of off-diagonal elements and the condition numbers of the resulting matrices increase as the permittivities increase. To attack such a trouble, a new procedure is proposed to use the face-centered node points [3-4]. Regardless of the magnitudes of the permittivity, the condition numbers of the corresponding matrices can be kept small and the polarization charge induced at the cell surfaces can be modeled more accurately. In fact, the CG algorithm is one of Krylov subspace iterative methods. The other algorithms of Krylov subspace iterative methods for solving general, large linear systems have been used widely in many areas of scientific computing [5]. In this investigation, the convergence feature of several typical Krylov subspace iterative methods, such as BCG [6], CGS [7], Bi-CGSTAB [8], TFQMR [9], BCGSTAB2 [10], GPBiCG [11], GMRES [12] and FGMRES[13] is compared when they are applied for the analysis of electromagnetic wave scattering from three-dimensional dielectric bodies.

2. Theory Analysis

To calculate the scattering from a dielectric body with arbitrary shape, inhomogeneity, and anisotropy, the method based on the electric field integral equation is widely employed. From the magnetic vector potential $\overline{A}$ and the electric scalar potential $\overline{\Phi}$, the electric filed $\overline{E}$ due to time-harmonic current and charge sources radiating in free space can be expressed as

$$\overline{E}(\overline{r}) = \overline{E}^i(\overline{r}) - j\omega\overline{A}(\overline{r}) - \nabla\Phi(\overline{r}) \tag{1}$$

where $\Phi(\overline{r}) = \dfrac{1}{\varepsilon_0}\iiint G(k_0 R)\rho(\overline{r}')d\overline{r}'$, $A(\overline{r}) = \mu_0\iiint G(k_0 R)J(\overline{r}')d\overline{r}'$, the free-space Green's function is $G(k_0 R) = e^{-jkR}/(4\pi R)$, here $R = |\overline{r} - \overline{r}'|$, $k_0^2 = \omega^2\mu_0\varepsilon_0$ and $\overline{J}$ and ρ denote, respectively, the electric current and charge density distributions, excluding the sources generating the incident electric field $\overline{E}^i$.

It is known that the Green's function satisfies

$$\nabla^2 G(k_0 R) + k_0^2 G(k_0 R) = -\delta(\overline{r} - \overline{r}') \tag{2}$$

Consider the scattering problem of a dielectric body with a general tensor permittivity distribution $\varepsilon_0\overline{\overline{\varepsilon}}(\overline{r})$. The sources induced in this dielectric scatterer are polarization current $\overline{J}$ and polarization charge ρ. It is known that the electric field integral equation:

$$\overline{E}(\overline{r}) = \overline{E}^i(r) + k_0^2\iiint_\infty \overline{\overline{G}}(k_0 R)\cdot\left\{\overline{\overline{\chi}}(\overline{r}')\cdot E(\overline{r}')\right\}d\overline{r}' \tag{3}$$

where $\overline{\overline{G}}(k_0 R) = \left\{\overline{\overline{I}} + \dfrac{\nabla\nabla}{k_0^2}\right\}G(k_0 R)$, $\overline{\overline{\chi}}(\overline{r}') = \overline{\overline{\varepsilon}}(\overline{r}') - \overline{\overline{I}}$.

Scattering from an arbitrary body is identical to that from an orthorhomboid composed of the body and the surrounding space with $\overline{\overline{\chi}} = 0$. Divide the orthorhomboid into $N = m_1 \times m_2 \times m_3$ identical orthorhombic cells with the volume $\Delta v = \Delta x \times \Delta y \times \Delta z$ (Fig.1) which are centered at (x_i, y_j, z_k), $i = 0, 1, ..., m_1 - 1$, $j = 0, 1, ..., m_2 - 1$, and $k = 0, 1, ..., m_3 - 1$. Suppose the cells are small enough that every component in $\overline{\overline{\varepsilon}}(x, y, z)$ can be treated as constant over each of the cells. Then, on applying pulse-function expansion and point matching technique, the integral equation reduces to a set of $3N$

simultaneous linear equations in terms of $3N$ unknowns $\overline{E}(x_i, y_j, z_k)$ as

$$-\sum_{i=0}^{m_1-1}\sum_{j}^{m_2-1}\sum_{k}^{m_3-1}\sum_{v} g_{uv}(i-i', j-j', k-k') \cdot \sum_{\zeta}\chi_{v\zeta}(i', j', k')E_{\zeta}(i', j', k')$$

$$+ E_{\mu}(i, j, k) = E_{\mu}^{i}(i, j, k), \qquad \begin{aligned} &\mu = x, y, z \\ &i = 0, 1, ..., m_1 - 1 \\ &j = 0, 1, ..., m_2 - 1 \\ &k = 0, 1, ..., m_3 - 1 \end{aligned} \qquad (4)$$

where the summation indices $v, \zeta = x, y, z$, and $f(i, j, k)$ ($f = E_{\mu}, E_{\mu}^{i}, or \ \chi_{\mu\upsilon}$) denotes the associated functional value at points (x_i, y_j, z_k). The above equations can be efficiently solved by combined the Krylov subspace iterative methods with FFT.

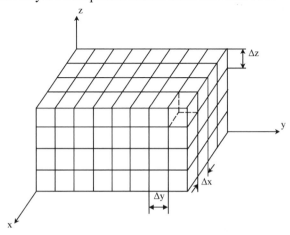

Fig.1 An orthorhomboid composed of $m_1 \times m_2 \times m_3$ identical, orthorhombic cells with volume $\Delta v = \Delta x \times \Delta y \times \Delta z$ encloses the entire scattering body.

Once the field $\overline{E}^{i}(x_i, y_j, z_k)$ throughout the dielectric body is solved, the scattered field $\overline{E}^{s}(x, y, z)$ at any position can be calculated by evaluating the integral in (4). In this section, we assume that incident field to be a uniform plane wave, that is

$$\overline{E}^{i}(x_i, y_j, z_k) = (\hat{x}A_x + \hat{y}A_y + \hat{z}A_z)\exp\left[-j(k_x x + k_y y + k_z z)\right]$$

Where $k_x^2 + k_y^2 + k_z^2 = k_o^2$ and A_{μ} are constants denoting the corresponding amplitudes and satisfying the relation $A_x k_x + A_y k_y + A_z k_z = 0$. Far away from the dielectric body, the scattered field propagates in the direction of $\overline{r}$, and its polarization is orthogonal to this direction. Then, we have

$$\sigma(\theta, \phi) = \lim_{r \to \infty} 4\pi r^2 \left\{ \left|E_{\phi}^{s}(r, \theta, \phi)\right|^2 + \left|E_{\theta}^{s}(r, \theta, \phi)\right|^2 \right\} \Big/ \left|\overline{E}^{i}\right|^2 \qquad (5)$$

where

$$E_{\phi}^{s}(r, \theta, \phi) = -E_x^{s}(r, \theta, \phi)\sin\phi + E_y^{s}(r, \theta, \phi)\cos\phi$$

and

$$E_{\theta}^{s}(r, \theta, \phi) = E_x^{s}(r, \theta, \phi)\cos\phi\cos\theta + E_y^{s}(r, \theta, \phi)\sin\phi\cos\theta - E_z^{s}(r, \theta, \phi)\sin\theta$$

The most powerful iterative algorithm of these types is the conjugate gradient algorithm for solving positive definite linear system [2]. Although the BiCG [6], a variation of the CG algorithm, sometimes converges much faster than CG, however, it has several well-known drawbacks. Among these are (i) the need for matrix-vector multiplications with A^T, (ii) the possibility of breakdowns, and (iii) erratic convergence behavior. Various attempts have been made in the last fifty years to extend the highly successful CG algorithm to the nonsymmetric case. Many recently proposed methods could be viewed as improvements over some of these drawbacks of BiCG. The most notable of these is the ingenious CGS method proposed by Sonneveld [7], which cures the first drawback mentioned above by computing the square of the BiCG polynomial without requiring A^T. Hence, when BiCG converges, CGS is an attractive, faster converging alternative. However, this relation between the residual polynomials also causes CGS to behave even more erratically than BiCG, particularly in near-breakdown situations for BiCG. These observations led to the development of Bi-CGSTAB algorithm [8], a more smoothly converging variant of CGS. Nevertheless, although the Bi-CGSTAB algorithms were found to perform very well compared to CGS in many situations, there are still some cases where convergence is still quite erratic. Therefore, a new version of CGS, called TFQMR [9], was proposed to "quasi-minimizes" the residual in the space spanned by the vectors generated by the CGS iteration. Numerical experiments show that in most cases TFQMR retains the good convergence features of CGS and improve on its erratic behavior. The transpose-free nature, low computational cost, and smooth convergence behavior make TFQMR to be an attractive alternative to CGS. On the other hand, since the square of the residual polynomial of BiCG is still in the space being quasi-minimized, the CGS and TFQMR sometimes converge in about the same number of steps in many practical examples. It is also known that the CGS residual polynomial can be quite polluted by round-off error. QMRBICGSTAB [10] method is proposed and its smoothed convergence is observed when the quasi-minimization principle is applied to the Bi-CGSTAB method. By defining a new three-term recurrence relation modeled after the residual polynomial of BiCG, a diverse collection of generalized product-type methods based on BiCG is proposed including CGS, Bi-CGSTAB, GPBiCG [11] and BICGSTAB2, without the disadvantage of storing extra iterates like that in GMRES [12]. The residual norm of GMRES (∞) can converge to zero optimally among all Krylov subspace methods and it is guaranteed to converge in at most N steps where N is unknowns number, but this algorithm would be impractical if many steps are required for convergence, since the computation and, in particular, the memory requirements for this method are prohibitively high. Therefore some measurement has to be taken to remove this disadvantage for this algorithm. The simplest remedy is periodically to restart the algorithm. The Flexible GMRES [13] is a variant of the GMRES algorithm that allows changes in the preconditioning at every step. Any iterative method can be used as a preconditioner due to its flexibility. Here the standard GMRES algorithm is used. Thus we have GMRES for the preconditioner, or inner iterations, and FGMRES as the (outer) flexible method, denoted as FGMRES-GMRES (m), where m is dimension numbers of Krylov subspaces. In the next section, the convergence of Krylov subspace iterative FFT methods is discussed in detail for several typical scattering bodies.

3. Numerical Results and Discussions

A. Cube

As shown in Fig. 2(a), a homogeneous cube with the side length L and relative permittivity $\varepsilon_r = 4.0$ is considered. The scattering cross section $\sigma(\theta,\phi)/\lambda_0^2$ is shown in Fig.2 (b), when the incident field $\overline{E}^i$ is horizontally polarized with $k_z = k_o$

$(A_z = A_y = 0, A_x \neq 0)$ (where E-plane denotes parallel to the polarized direction, that is $\phi = 0°$, and H-plane, vertical $\phi = 90°$). The length L is chosen as $L_1 = 0.2\lambda_0$ and $L_2 = 0.5\lambda_0$ where λ_0 is wavelength in free space.

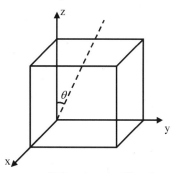

Fig. 2 (a) a homogeneous dielectric cube illuminated by a plane wave with side length L and relative permittivity ε_r

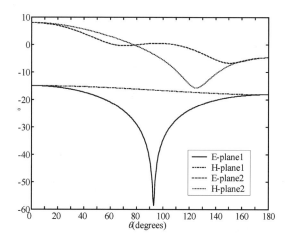

Fig.2 (b) Scattering cross section per square wavelength $\sigma(\phi = 0°, 90°, \theta)/\lambda_0^2$ as a function of the scattering angle θ, from homogeneous, isotropic cube for two cases (1) $L = 0.2\lambda_0$ (2) $L = 0.5\lambda_0$

As given in Fig. 3, the residual norm history is plotted for the analysis of cube dielectric body with $N = 32 \times 32 \times 32$, $\varepsilon_r = 4.0$, $L = 0.6\lambda_0$. In order to compare the convergence influenced by the dielectric permittivity, the residual norm history of Krylove subspace iterative FFT methods is given in Fig.4 and Fig.5 for the cube dielectric bodies between $\varepsilon_r = 14.0$ and $\varepsilon_r = 40.0$, where $N = 16 \times 16 \times 16$, $k_0 L = 2$, and their stop precision is both 10^{-3}. When the relative permittivity increases, for example $k_0 L > 2$, $\varepsilon_r = 40$, the iteration numbers of Krylove subspace iterative FFT methods becomes large enough. It is observed that the CG method is robust for both high permittivity and electrically large size of dielectric scattering bodies. It is expected that the preconditioning techniques can be used to accelerate the convergence of Krylove subspace iterative FFT methods.

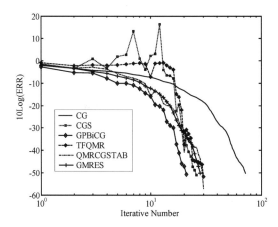

Fig.3 The residual norm history for the cube dielectric body with $N = 32 \times 32 \times 32$, $\varepsilon_r = 4.0, L = 0.6\lambda_0$

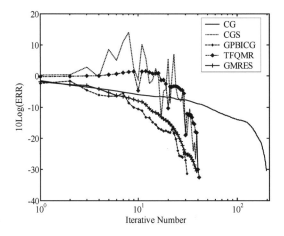

Fig.4 The residual norm history for the cube dielectric scatterer
with $N = 16 \times 16 \times 16$, $\varepsilon_r = 14.0, k_0 L = 2$

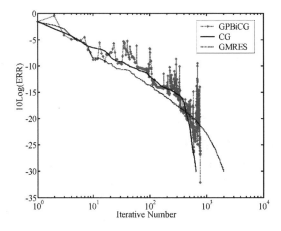

Fig.5 The residual norm history for the cube dielectric body
with $N = 32 \times 32 \times 32$, $\varepsilon_r = 40.0, k_0 L = 2$

B. Sphere

As shown in Fig. 6 (a), an isotropic, homogeneous, sphere dielectric body with diameter L is considered. The relative permittivity is taken to be $\varepsilon_r = 4$, the frequency operation is taken to be 100 MHz, the sphere is put into a cube with size of $L \times L \times L$, and the cube is subdivided with (1) $N = 16 \times 16 \times 16$ and (2) $N = 32 \times 32 \times 32$, with incident E-filed parameters $k_z = k_o$, $A_y = A_z = 0$, $A_x \neq 0$. The corresponding scattering cross section $\sigma(\theta, \phi) / \lambda^2$ is shown in Fig.6 (b).

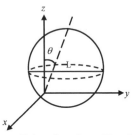

Fig. 6 (a) a homogeneous dielectric sphere illuminated by a plane wave
with the diameter L and relative permittivity ε_r

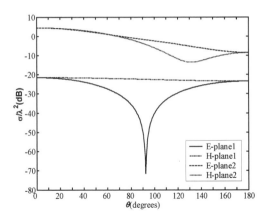

Fig.6 (b) Scattering cross section per square wavelength $\sigma(\phi = 0°, 90°, \theta) / \lambda_0^2$ as a function of the scattering angle θ, from homogeneous, isotropic sphere for two cases (1) $L = 0.2\lambda_0$ (2) $L = 0.5\lambda_0$.

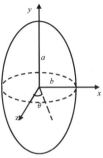

Fig. 7 (a) a lossy prolate spheroid with major axes a and minor axes b illuminated by a plane wave with relative permittivity ε_r.

As shown in Fig.7(a), a lossy prolate spheroid with relative permittivity ε_r is analysed when it is illuminated by a plane wave. The E-plane and H-plane bistatic pattern is given in Fig.7 (b) of for the lossy prolate spheroid having permittivity of $\varepsilon_r = 4 - j1$ and electric size $k_0 a = \pi / 2$ and $a/b = 2$, where a and b are the major and minor axes of the spheroid, respectively. The prolate spheroid is subdivided with $N = 16 \times 32 \times 16$ and with incident E-filed parameters $k_z = k_o$, $A_y = A_z = 0, A_x \neq 0$.

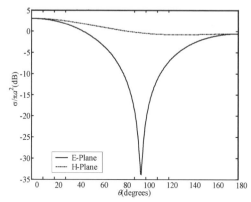

Fig.7 (b) E-plane and H-plane bistatic scattering pattern of a lossy prolate spheroid having $\varepsilon_r = 4 - j1, k_0 a = \pi / 2$ and $a/b = 2$.

C. Finite-length Cylinder

As shown in Fig.8, the structure is a finite-length dielectric rectangle cylinder with a smaller hollow elliptic cylinder at center. The dielectric body is homogeneity, anisotropy, and loss. The parameters is as follows: dielectric rectangle cylinder x-length = $2 \lambda_0$, y-length = $2.5 \lambda_0$, z-length = $3 \lambda_0$; the hollow elliptic cylinder x half-axis length = $0.5 \lambda_0$, y half-axis length = $0.625 \lambda_0$, z-length = $3 \lambda_0$, where λ_0. is wavelength of incident field. The permittivity tensor is

$$\overset{=}{\varepsilon_r} = \begin{bmatrix} 3.5 - 2j & 0.5 & 0.5 \\ 0.5 & 2.5 - 2.5j & 0.5 \\ 0.5 & 0.5 & 3.0 - 1.5j \end{bmatrix}$$

with the $N = 16 \times 16 \times 32$. Here the incident electric field satisfies $A_x = A_y = 0, A_z = 1$, $k_0 = k_x, k_y = k_z = 0$.

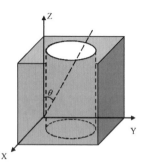

Fig.8 Finite-length dielectric cylinder illuminated by a plane wave with relative permittivity tensor ε_r.

From the above analysis, it can be seen that when the relative permittivity become large (say $\varepsilon_r > 20$), the convergence rates of Krylov subspace iterative methods all become slow. In order to speed up the convergence, the FGMRES-GMRES (m) iterative method is used to accelerate iteration. As shown in Fig.2 (a), a homogeneous, isotropic cube is considered with the relative permittivity $\varepsilon_r = 50, L = 0.23\lambda_0$. The subdivided orthorhombic cells number is taken to be $16 \times 16 \times 16$ and the frequency operation 100 MHz. The incident field $\overline{E}^i$ with $k_z = k_o$ is horizontally polarized $(A_z = A_y = 0, A_x \neq 0)$. To check the computational efficiency of the proposed method, we compare the convergence history of FGMRES-GMRES (30) and restarted GMRES (30) methods combined with FFT technique. In the inner iteration of FGMRES-GMRES, the stop precision is 1.E-3. The residual norm history for the cube dielectric body is given in Fig 11. It can be seen that the FGMRES-GMRES method can considerably improve its convergence speed when the relative permittivity becomes large.

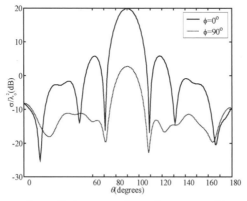

Fig.9 Bistatic scattering pattern of a homogeneity, anisotropy and lossy finite length cylinder with $N = 16 \times 16 \times 32$.

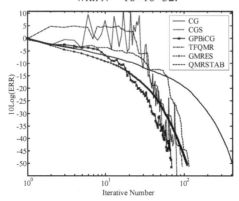

Fig.10 The residual norm history for a homogeneity, anisotropy and lossy finite length cylinder with $N = 16 \times 16 \times 32$

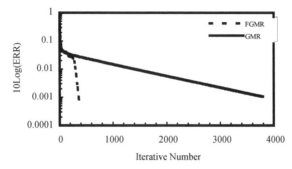

Fig.16 The residual norm history for the cube dielectric body with $N = 16 \times 16 \times 16$, $\varepsilon_r = 50, L = 0.23\lambda_0$.

4. Conclusions

The electric field integral equation has been investigated by employing the pulse-function expansion and the point-matching technique. Several Krylov subspace iterative methods are combined with FFT technique to solve the equation. It can be found that GPBiCG is a competitive method in solving this problem when the relative permittivity is small, since its convergence rate is faster with less memory need than others. It can be seen that when the relative permittivity become large (say $\varepsilon_r > 20$), the convergence rates of Krylov subspace iterative methods are all slowed down. It is also concluded that the FGMRES-GMRES is robust for the analysis of high permittivity and electrically large size of dielectric scattering bodies. In the future, the more suitable preconditioning technique is proposed to accelerate their convergence.

Acknowledgement

This work is partially supported by Young Scholar Foundation of Nanjing University of Science & Technology, the Excellent Young Teachers Program of Moe, PRC, and Natural Science Foundation of China under Contract Number 60271005.

Reference

1. J. H. Wang, *Generalized Moment Methods in Electromagnetics,* John Wiley & Sons, Inc., 1991
2. C.C.Su, "Electromagnetic scattering by a dielectric body with arbitrary inhomogeneity and anisotropy," *IEEE Trans. on Antennas and propagation,* 1989, 37(3): 384-389
3. C. C. Su, "A procedure for solving the electric field integral equation for a dielectric scatterer with a large permittivity using face-centered node points", *IEEE Trans. on Microwave Theory and Techniques*, 1991, 39(6): 1043-1048
4. C. C. Su, "The three-dimensional algorithm of solving the electric field integral equation using face-centered node points, conjugate gradient method, and FFT", *IEEE Trans. on Microwave Theory and Techniques*, 1993, 41(3): 510-515
5. Y. Saad, *Iterative Methods for Sparse Linear Systems*, PWS Pub. 1996
6. C. Lanczos, "Solution of systems of linear equations by minimized iterations," *J. Res. Natl. Bur. Stand.*, 1952, 49: 33-53
7. P. Sonneveld, "CGS, a fast Lanczos-type solver for nonsymmetric linear systems," *SIAM J. Sci. Statist. Comput.*, 1989, 10: 36-52
8. H. A. Van Der Vorst, "BI-CGSTAB: A fast and smoothly converging varient of BI-CG for the solution of nonsymmetric linear systems," *SIAM J. Sci Statist. Comput.*, 1992, 13: 631-644
9. R.W. Freund, "A transpose-free quasi-minimal residual algorithm for non-Hermitian Linear systems," *SIAM J. Sci. Comput.*, 1993, 14(2): 470-482
10. T. F. Chan, *et al.*, "A quasi-minimal residual variant of the Bi-CGSTAB algoithm for nonsymmetric systems," *SIAM J. Sci. Comput.*, 1994, 15(2): 338-347
11. S. L. Zhang, "GPBi-CG: Generalized product-type methods based on Bi-CG for solving nonsymmetric linear systems", *SIAM J. Sci. Comput.*, 1997, 18(2): 537-551
12. Y. Saad and M. Schultz, "GMRES: A generalized minimal residual algorithm for solving non- symmetric linear systems," *SIAM J. Sci. Stat. Comput.*, 1986, 7: 856-869
13. Y. Saad, "A Flexible inner-outer preconditioned GMRES algorithm," *SIAM J. Sci. Comput.,* 1993, 14(2): 461-469

Wave Propagation,Scattering and Emission in Complex Media
Edited by Ya-Qiu Jin
Science Press and World Scientific,2004

Sparse Approximate Inverse Preconditioned Iterative Algorithm with Block Toeplitz Matrix for Fast Analysis of Microstrip Circuits

L. Mo, R.S. Chen, E. K. N. Yung

Department of Communication Engineering

Nanjing University of Science & Technology, Nanjing 210094, China

eerschen@mail.njust.edu.cn

Abstract   In this paper, several preconditioning technique such as symmetric successive overrelaxation (SSOR), block diagonal matrix, sparse approximate inverse and wavelet based sparse approximate inverse are applied to conjugate gradient (CG) method for solving the dense matrix equations from the mixed potential integral equation (MPIE). Our numerical calculations show that the PCG-FFT algorithms with these preconditioners can converge much faster than the conventional one to the MPIE for the analysis of microwave circuits. Some typical microstrip discontinuities are analyzed and the good results demonstrate the validity of these proposed algorithms.

Key Words   mixed potential integral equation, CG-FFT, SSOR, sparse approximate inverse preconditioning, microwave circuits

1. Introduction

Integral equation method has been proved to be a powerful alternative to finite element method (FEM) and finite difference method (FDM) for solving a class of partial differential equations (PDE). In contrast to the sparse matrices associated with FDM and FEM, discretization of the integrate equation usually leads to a fully populated matrix. The resultant matrix equation is then solved, for example, by the conjugate gradient (CG) iterative method, requiring $O(N^2)$ operations for the matrix-vector multiplication in each iteration, where N is the number of unknowns [1-3]. A number of techniques have been proposed to speed up the evaluation of the matrix-vector multiplication [1,2]. However, it has been known that the eigenvalues of these full matrices approximate those of its underlying integral operator. As far as the iterative method is applied for solutions, its iteration number largely depends on the spectral properties of the integral operator or the matrices of discrete linear systems. This observation naturally suggests that we should study operator preconditioning in connection with matrix preconditioning [4]. To reduce the number of iterations, various preconditioning techniques have been used [2-6]. The SSOR preconditioner can be directly derived from the coefficient matrix without additional cost, but can lead to significant convergence improvement for sparse linear systems. Another advantage of the SSOR as preconditioner is that it contains more information of the coefficient matrix when compared with block diagonal matrix [16,17]. Therefore, SSOR preconditioner can speed up CG algorithm more efficiently. But FFT technique can not be applied to CG algorithm if SSOR preconditioner is used. Recently, one of the significant advances in direct methods for sparse matrix solution is the development of the multifrontal method [10,18,19]. The method organizes the numerical factorization into a number of steps, and each involves the formation of a dense smaller frontal matrix followed by its partial factorization. There are several advantages of this approach over other factorization algorithms: effective vector processing on dense frontal matrices, better data locality, efficient out-of-core scheme, and its natural adaptability to parallel computations. The price paid is the extra working storage required for the frontal matrices and the extra arithmetic needed to perform the assembly operations.

2. Theory Analysis

Consider a general microstrip structure on an infinite substrate having relative permittivity ε_r and thickness h. The microstrip is in the $x-z$ plane and excited by an applied field $\vec{E}^a$. The induced current on the microstrip can be found by solving the following MPIE:

$$j\omega\mu_0\hat{y}\times[\vec{A}(\vec{r})+\frac{1}{k_0^2}\nabla\Phi(\vec{r})]=\hat{y}\times\vec{E}^a(\vec{r}) \tag{1}$$

where the vector and scalar potentials can be expressed as

$$\vec{A}(\vec{r})=\iint_S \ddot{G}_A(\vec{r},\vec{r}')\cdot J(\vec{r}')ds' \tag{2}$$

$$\Phi(\vec{r})=\iint_S G_q(\vec{r},\vec{r}')\nabla'\cdot J(\vec{r}')ds' \tag{3}$$

here $\ddot{G}_A$ and G_q denote the Green's functions for the magnetic vector and electric scalar potentials, respectively.

To solve the integral equation (1), we place the conducting surface of the microstrip in a rectangular cell, which is then divided into $M\times N$ small rectangular cells along x and z directions, respectively. Assume that $\vec{f}_{m,n}^x=f_{m,n}^x\hat{x}$ and $\vec{f}_{m,n}^z=f_{m,n}^z\hat{z}$ are vector basis functions in both x and z directions, respectively, where $f_{m,n}^{x,z}$ can represent the two-dimensional pulse functions or the rooftop functions. Applying the Galerkin's method to equation (1) results in the following matrix equation

$$\begin{bmatrix} Z_{xx} & Z_{xz} \\ Z_{zx} & Z_{zz} \end{bmatrix}\begin{bmatrix} I_x \\ I_z \end{bmatrix}=V \tag{4}$$

The impedance matrix Z obviously is a dense and symmetric complex coefficient matrix. When the direct solution methods such as LU decomposition are used the computational complexity is $O(N^3)$ operations. As the problem size increase, the computational expense of these operations becomes prohibitive. This has led to the development of CG technique that solves for the surface current with the computational complexity of $O(N^2{}_{iter})$, where N_{iter} denotes the number of iterations. To reduce the number of iterations with the preconditioning technique, the following explicitly preconditioned CG algorithm has to be used. Given an initial electrical current guess I_0 and a symmetric positive definite matrix M, we generate a sequence $\{I_k\}$ of approximations to the solution I^* of equation (4) as follows:

$r_0=V-AI_0$, Solve $Md_0=r_0$ and $p_0=d_0$ for k=0,1,... until convergence do

$$\alpha_k=(d_k^T\cdot r_k)/(p_k^T\cdot Ap_k), \quad I_{k+1}=I_k+\alpha_k p_k, \quad r_{k+1}=r_k-\alpha_k Ap_k, \quad \text{solve } Md_{k+1}=r_{k+1}$$

$$\beta_k=(d_{k+1}^T\cdot r_{k+1})/(d_k^T\cdot r_k), \quad p_{k+1}=d_{k+1}+\beta_k p_k \tag{5}$$

The basic conjugate gradient (CG) method can be obtained by choosing M to be the identity matrix. The preconditioning matrix M^{-1} is never computed explicitly and for any vector r, the linear system $Md=r$ can be solved by the iterative methods. It should be noted that the above algorithm is only suitable for real coefficient matrix equations and the complex matrix equation (4) can be transferred into real coefficient matrix equation so that the above preconditioned CG algorithm is applied for solution [12]. Since the impedance matrix Z is in the form of blocks and subblocks, which are themselves multilevel block-Toeplitz matrices, the $O(N\log N)$ FFT-based method to expedite matrix-vector multiplications involving multilevel block-Toeplitz matrices is used. The method is also a minimal memory method with $O(N)$ memory requirements because only nonredundant entries of the multilevel block-Toeplitz matrices are stored [11].

Although FFT can greatly enhance the computational efficiency of matrix-vector multiplication, it can't reduce the iteration number of CG method which is largely depends on the spectral properties of the integral operator or the matrices of discrete linear systems. In this paper, the following several preconditioning schemes are chosen to accelerate the convergence of the PCG algorithms.

A. Block Diagonal Matrix Preconditioner

An important variation of preconditioned conjugate gradient algorithms is inexact preconditioner implemented with inner-outer iterations [19], where the preconditioner is solved by an inner iteration to a prescribed precision. The block diagonal matrix preconditioner M is chosen to only have diagonal elements of sub-matrices Z_{xx}, Z_{xz}, Z_{zx}, and Z_{zz}, so that only $O(N)$ computation cost is required in the inner iteration.

B. SSOR preconditioning

For SSOR preconditioning scheme [16], the preconditioner M is chosen as follows
$$M = (\widetilde{D} + L)(\widetilde{D})^{-1}(\widetilde{D} + U) \tag{6}$$
where $Z=L+D+U$ in equation (4), L the lower triangular matrix, D the positive diagonal matrix, U the upper triangular matrix, and $\widetilde{D} = D/\omega$, $0 < \omega < 2$. It is said that the value of ω doesn't have great influence on the convergence of SSOR preconditioned CG algorithm [6] and we simply choose its value to be one. Since the convergence speed of CG method largely depends on the condition number of coefficient matrix Z, the linear system equation (4) could be scaled by a preconditioner M and then be transformed into the system given by
$$\hat{Z}\hat{I} = \hat{V} \tag{7}$$
where
$$\hat{Z} = \widetilde{D}(\widetilde{D} + L)^{-1}Z(\widetilde{D} + U)^{-1}, \ \hat{V} = \widetilde{D}(\widetilde{D} + L)^{-1}V, \ \hat{I} = (\widetilde{D} + U)I \tag{8}$$

If $NZ(Z)$ denotes the number of nonzero entries in matrix Z, this straightforward implementation of PCG would require $6N+4NZ(Z)$ multiply-adds per iteration. However, Considering that $(\widetilde{D} + L)$, $(\widetilde{D} + U)$ the part of coefficient matrix Z, matrix-vector multiplication in peconditioned CG algorithm can be computed efficiently. The efficient implementation of SSOR-PCG would require $10N+2NZ(Z)$ multiply-adds per iteration. This cost of computation is asymptotically half as many as the straightforward implementation if $NZ(Z)$ is large enough and is almost the same as the direct CG method.

C. Sparse Approximate Invert Preconditioner

The sparse preconditioner M is chosen through keeping the larger elements of the origin matrix and dropping out those lower than threshold. Then the multifrontal method is used to solve this resultant sparse preconditioning matrix equation so that the sparse approximate inverse operation can be completed during the iterations.

D. Wavelet Based Sparse Approximate Inverse Preconditioner

Start with a moment method matrix equation (4) which results from some commonly used local expansion and testing functions. This may be transformed into the following equation [18]:
$$WZW^T J = WV \tag{9}$$
The matrix W performs a discrete wavelet transform. Fast computational method for performing the multiplication by W uses its factorization into many matrices, each involving functions with a different characteristic length scale. With the following definitions
$$W^T J = I, \ \widetilde{Z} = WZW^T, \ U = WV \tag{10}$$

We have the following transformed equation:

$$\widetilde{Z}J = U \tag{11}$$

To reduce the number of iterations with the wavelet based sparse approximate inverse preconditioned conjugate gradient technique, M is used as a preconditioner to the equation $\widetilde{Z}J = WV$, where M is a sparse matrix taken from $\widetilde{Z}$ by use of a threshold. Then the matrix-vector multiplication may only take N operations if M has N nonzeros where $N=O(n\log n)$. Therefore, the efficiency gain is obtained due to matrix compression if the iterative solver is convergent.

3. Results and Discussions

Although the approximate inverse preconditioning scheme is widely used for CG algorithm, it relies on finding a sparse preconditioning matrix M, which minimizes the Frobenius norm of the residual matrix. The construction of M is made usually in column by column manner in order to minimize the construction time. An important aspect of this approach is that only few columns of M need to be constructed and these columns are typically chosen to refer to rows of the original coefficient matrix. It can be inferred that the approximate inverse preconditioning matrix contains more of the global information of coefficient matrix when compared with block diagonal preconditioner. However, much time is spent in the construction phase and the example given in [7] shows that approximate inverse preconditioning scheme doesn't lead to a significant improvement in CPU time consuming.

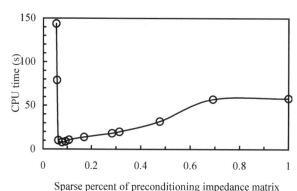

Figure 1 CPU time versus the sparse percent of preconditioning impedance matrix

To investigate the influence of the sparse percent of preconditioning matrix on PCG-FFT algorithm, Figure 1 gives CPU time curve versus the sparse percent of preconditioning impedance matrix for multifrontal method preconditioned CG-FFT algorithm when the discretization is taken in two dimensions (31 longitudinal grids and 9 transverse grids). It is observed from Figure 1 that 7.86% is an optimal sparse percent and the CPU time requirement for solution is shortest. At this optimal sparse percent, the multifrontal method preconditioned CG-FFT algorithm even achieves 7.275 times faster than the multifrontal method to reach -130dB residual errors $R=|b-Zx|/|b|$. As plotted in Figure 2, CPU time versus the unknown number for different solvers is displayed. It can be found from Figure 2 that the multifrontal method preconditioned CG-FFT algorithm can be 956.5 times faster than the conventional CG-FFT algorithm to reach -130dB residual errors (177.8 times for -40dB residual errors).

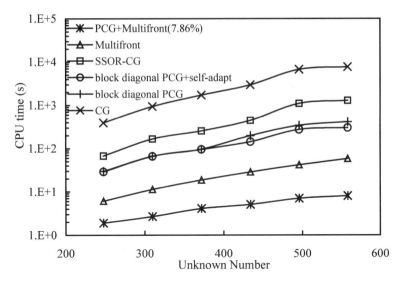

Figure 2 CPU time versus the unknown number for different solvers to reach −130dB residual error.

As shown in Figure 3, CPU time is plotted for multifrontal method preconditioned CG-FFT algorithm versus the unknown number for different sparse percent of preconditioning impedance matrix. It is observed that 7.86% is an optimal sparse percent for different unknowns and the sparse percent has a large influence on the convergence speed of the preconditioned CG-FFT algorithms. To investigate the efficiency of the multifrontal method preconditioned CG-FFT algorithm for the solution of a dense matrix equation from the integral equation, residual errors for both conventional CG-FFT and multifrontal method preconditioned algorithms with different sparse percent are illustrated versus iteration number in Figure 4. It can be noted that multifrontal method preconditioned CG-FFT algorithms with the suitable sparse percent can reach the convergence in significantly less iteration number than the conventional CG-FFT algorithm.

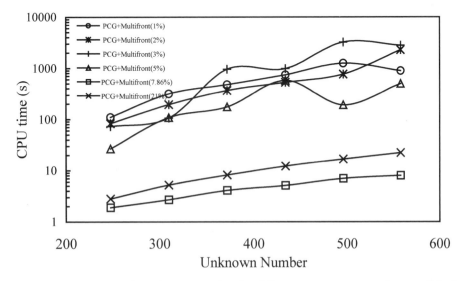

Figure 3 CPU time versus the unknown number for different sparse percent of preconditioning impedance matrix to reach −130dB residual error.

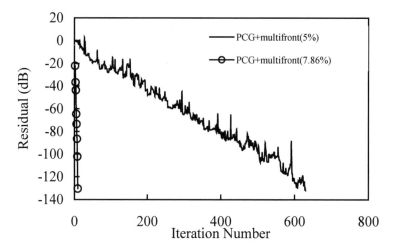

Figure 4 Residual errors versus iteration number for both conventional CG-FFT and multifrontal method preconditioned algorithm for different sparse percent of preconditioning impedance matrix

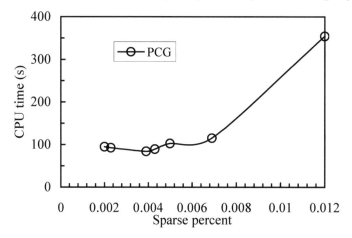

Figure 5 CPU time versus the sparse ratio of preconditioning impedance matrix
for wavelet based approximate inverse preconditioned CG algorithm to reach –60dB residual error.

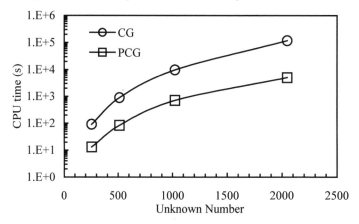

Figure 6 CPU time versus the unknown number for different solvers to reach –60dB residual error.

Considering the great consumption of the memory by the multifrontal method, the wavelet transform is used to compress the impedance matrix, which is first used to transform the impedance matrix Z into $\widetilde{Z}$. Then the sparse preconditioner M is chosen through keeping the larger elements of the matrix $\widetilde{Z}$ and dropping out those lower than threshold. After that the multifrontal method is used to solve this resultant sparse preconditioning matrix equation so that the sparse approximate inverse operation can be completed during the iterations.

To investigate the influence of the sparse ratio of preconditioning matrix on PCG algorithm, Figure 5 gives CPU time curve versus the sparse ratio of preconditioning impedance matrix for multifrontal method preconditioned CG algorithm when the discretization is taken in two dimensions (64 longitudinal grids and 4 transverse grids). It is observed from Figure 5 that the $0.2\% - 0.4\%$ sparse ratio dominates elements of the transformed impedance matrix $\widetilde{Z}$ is enough and the CPU time requirement for solution is shortest. It is also noted that the sparse ratio has a large influence on the convergence speed of the preconditioned CG algorithms.

As plotted in Figure 6, CPU time versus the unknown number is displayed. It can be found from Figure 6 that the multifrontal method preconditioned CG algorithm can be 23.43 times faster than the conventional CG algorithm for 2048 unknowns to reach –60dB residual errors. To investigate the efficiency of the wavelet based approximate inverse preconditioned CG algorithm for the solution of a dense matrix equation from the integral equation, residual errors for both conventional CG and wavelet based approximate inverse preconditioned algorithms are illustrated versus iteration number in Figure 7. It can be noted that the preconditioned CG algorithms with the suitable sparse ratio can reach the convergence in significantly less iteration number than the conventional CG algorithm.

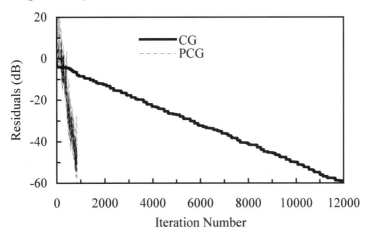

Figure 7 Residual errors versus iteration number

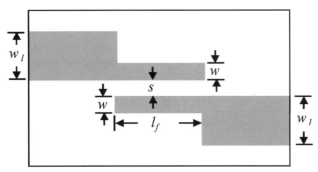

Figure 8 A parallel coupled line filter and its dimension: $\varepsilon_r = 2.37$, $w_1 = 2.37mm$, $w = s = 0.24mm$, $h = 0.79mm$, $l_f = 10mm$

Numerical computations have been carried out for some microstrip discontinuities. As shown in Figure 8, the scattering parameters of a parallel-coupled line filter are calculated. Figure 9 shows very good agreement between our method and measurement [14]. In particular, there is no frequency shift in the stop band between them. Another example of a microstrip stub filter is analyzed. A comparison of the scattering parameters S_{12} is made with those [15]. It is observed from Figure 10 that they are in a good agreement with each other.

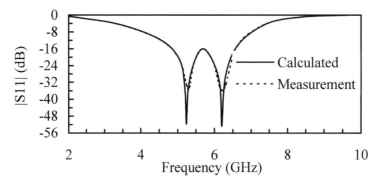

Figure 9 Magnitude of the scattering parameters of the parallel-coupled line filter versus frequency

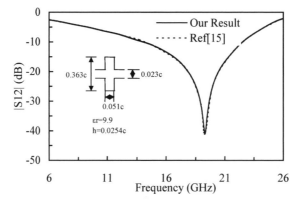

Figure 10 Microstrip stub discontinuity and its insertion loss versus frequency

4. Conclusions

In this paper, several kinds of preconditioners are adapted to CG-FFT algorithm to reduce the condition number of the operator equations. The SSOR preconditioned CG algorithm converges at about as high as 7 times faster than the conventional one, but FFT technique can't be used. When the block diagonal preconditioning scheme is applied. This PCG algorithm with sparse preconditioner in inner iteration converges at nearly 30 times faster than the conventional one for the discretization in two dimensions. Finally, the Sparse Approximate Inverse preconditioning is introduced. Our numerical results demonstrate that more than hundreds of times convergence speed can be achieved for microstrip circuits. Considering the great consumption of the memory by the multifrontal method, the wavelet transform is used to compress the impedance matrix. The numerical results show that the wavelet based sparse approximate inverse PCG algorithm with the suitable sparse ratio can reach the convergence in significantly less iteration number than the conventional CG algorithm.

Acknowledgement

This work was supported by Young Scholar Foundation of NUST, the Excellent Young Teachers Program of Moe, and Natural Science Foundation of China (60271005).

References

[1] F.X. Canning, " Improved impedance matrix localization method," *IEEE Trans. Antennas Propagat.*, Vol.41, No.5, May 1993, pp.659-667

[2] F.X. Canning, and Iames F.Scroll, " Diagonal preconditioners for the EFIE using a wavelet basis," *IEEE Trans. Antennas Propagat.*, Vol.44, No 9, Sept. 1996, pp.1239-1246

[3] R.S. Chen, E.K.N. Yung, C.H.Chan, and D.G.Fang, "Application of preconditioned CG-FFT technique to method of lines for analysis of the infinite-plane metallic grating," *Microwave and Optical technology letters*, Vol.24, No.3, pp.170-175, Feb 5 2000

[4] J.C. West, "Preconditioned iterative solution of scattering from rough surfaces", *IEEE Trans. Antennas Propagat.*, Vol.48, No.6, June 2000 pp.1001-1002

[5] R.S. Chen, E.K.N.Yung, C.H.Chan, and D.G.Fang, "Application of preconditioned conjugate-gradient algorithm to the edge FEM for electromagnetic boundary-value problems," *Microwave and Optical technology letters*, Vol.27, No.4, pp.235-238, Nov 20 2000

[6] O.Axelsson, L.Yu.Kolotilina, "Preconditioned conjugate gradient methods", Proceedings 1989, in Lecture Notes in Mathematics Vol.1457, Edited by A.Dold, B.Eckmann and F.Takens, Springer-Verlag 1990

[7] Y.Y. Botros and J. L. Volakis, "Precoditioned Generalized Minimal Residual Iterative Scheme for Perfectedly Matched Layer Terminated Application", *IEEE Microwave and Guided Wave Letters*, Vol. 9, No.2, Feb. 1999, pp.45-47

[8] R.S. Chen and K.F. Tsang, E.K.N.Yung "An Effective Multigrid Preconditioned CG algorithm For Millimeter Wave Scattering by an Infinite Plane Metallic Grating", *International Journal of Infrared and Millimeter Waves*, Vol.21, No.6, June. 2000, pp.945-963

[9] R.S. Chen, D.G. Fang, K.F. Tsang, E.K.N. Yung, "Analysis of Electromagnetic Wave Scattering by an Electrically large Metallic Grating Using Wavelet-based Algebraic Multigrid Preconditioned CG method", *Journal of Electromagnetic Waves and Applications*, Vol. 14, Nov. 2000, pp.1559-1561

[10] R.S.Chen, D.X.Wang, and E.K.N.Yung and J.M. Jin "Application of the multifrontal method to the vector FEM for analysis of microwave filters", Microwave and Optical Technology Letters, Vol.31, No. 6, December 20 2001

[11] B.E. Barrowes, F.L. Teixeira, and J.A. Kong, "Fast algorithm for matrix-vector multiply of asymmetric multilevel block-toeplitz matrices in 3-D scattering," *Microwave and Optical technology letters*, Vol.31, No.1, pp.28-32, Oct 5 2000

[12] D. Day and M.A. Heroux, "Solving complex-valued linear systems via equivalent real formulations," *SIAM J. Sci. Comput.*, Vol.23, No.2, 2001, pp.480-498

[13] K. Wu, R.Vahldieck, "A new method of modeling three dimensional MIC/MMIC circuits:The space-spectral domain approach" , *IEEE Trans. Microwave Theory and Tech.*, vol. MTT-38, no. 9, Sept 1990, pp. 1309-1318

[14] S.B. Worm, "Full-wave analysis of discontinuities in planar waveguides by the method of lines using a source approach", *IEEE Trans. on Microwave Theory and Technique*. Vol. 38, No.10, Oct. 1990, pp.1510-1513

[15] F. Giannini, G. Bartolucci and M. Ruggieri, "Equivalent circuit models for computer aided design of microstrip rectangular structures", *IEEE Trans. on Microwave Theory and Tech*. Vol. 40, No. 2, Feb. 1992, pp378-388

[16] K.F.Tsang, R.S.Chen, Mo Lei, E.K.N. Yung "Application of the Preconditioned Conjugate Gradient Algorithm to the Integral Equations for Microwave Circuits", *Microwave and Optical Technology Letters*, Vol. 34, No.4, August 20, 2002, pp.266-270

[17] K.F. Tsang, M. Lei, R.S.Chen, "Application of the Preconditioned Conjugate-Gradient Algorithm to the Mixed Potential Integral Equation", *IEEE AP-S Int Antenna and Propagation Symposium* Vol 4, pp.630-633, June 16-21, 2002, San Antonio, Texas, USA

[18] R.S. Chen, K.F. Tsang, L. Mo, "Wavelet Based Sparse Approximate Inverse Preconditioned CG Algorithm for Fast Analysis of Microstrip Circuits", *Microwave and Optical Technology Letters*, Vol. 35, No. 5, December 5, 2002

[19] R.S. Chen, E.K.N.Yung, K.F.Tsang, L.Mo, "The Block-Toeplitz-Matrix-Based CG-FFT Algorithm With An Inexact Sparse Preconditioner for Analysis of Microstrip Circuits", *Microwave and Optical Technology Letters*, Vol. 34, No. 5, September 5, 2002, pp347-351

Wave Propagation,Scattering and Emission in Complex Media
Edited by Ya-Qiu Jin
426   *Science Press and World Scientific,2004*

An Efficient Modified Interpolation Technique for the Translation Operators in MLFMA

Jun Hu, Zaiping Nie, Guangxian Zou
School of Electrical Engineering,
University of Electronic Science and Technology of China
Chengdu 610054, China
E-mail: scatterhu@263.net,   Phone:028-83202056

Abstract   Based on multilevel fast multipole algorithm (MLFMA), large target-scale problems can be solved efficiently now. The translation operators are amenable for the efficiency of MLFMA. In this paper, an efficient modified interpolation technique is developed for fast evaluation of translation operators. Different form other efficient techniques for translation operators in which only the evaluation of multiple angles is performed, the space-distance interpolation is implemented by introducing a modified factor. Two-dimensional Lagrange interpolation technique is used to realize fast computation of the operators in case of multiple space-distance and angles at a given level. Low computer storage for translation operators is required. The present technique is suitable for distributed memory machines.

Key Words   multilevel fast multipole algorithm, translation operator, interpolation

1.Introduction

Recently, fast multipole method (FMM) [1] and multilevel fast multipole algorithm (MLFMA) [2] have been pay much attention. As two powerful integral equation solvers, FMM and MLFMA reduce the complex complexity to $O(N^{1.5})$, $O(N \log N)$ respectively from $O(N^3)$ of traditional Gaussian elimination method. The memory requirement is also reduced to the same order. A parallel MLFMA called ScaleME has been developed [3]. Based on MLFMA, many large target-scale problems can be solved efficiently now.

In MLFMA, the matrix-vector multiplication is implemented in a multilevel multistage fashion [4,5]. Diagonal translation operators are amenable for the efficiency of MLFMA. In general, the translation operators are precomputed at the matrix filling stage and stored. The operators are only dependent of relative position of field-point groups and source-point groups because of the translational invariance. If the translation operators are evaluated directly, the complexity is $O(L^3)$, where L is the mode term at each level. For surface integral equation, L is proportional to $\sqrt{N}$ at the coarsest level, where N is the number of unknowns. So the computation complexity of each translation operator is $O(N^{3/2})$. Obviously, the expensive cost can not be afforded for large scale problems especially for distributed numerical computations. To achieve fast evaluation and low storage for translation operators, some efficient techniques have been proposed [6,7]. A fast Legendre expansion algorithm (FLEA) developed by Alpert and Rokhlin [6] has an asymptotic complexity of $O(L \log^2 L / \log \log L)$. Velamparambil and Chew precompute the values of the operators at a suitably chosen set of nodes in [-1,1], then evaluate the interpolation by using a slightly modified version of one-dimensional FMM and fast polynomial interpolation algorithms [7]. This algorithm has a memory requirement of only $O(L)$, a complexity of $O(\log 1/\varepsilon)$ for evaluating single point, where ε is the required precision. The error can be controlled and is independent of L. In above two algorithms, the space-distance between field-point groups and source-point groups in the translation operators is assumed. Repeated computations including the setup stage are required when the space-distance varies. Although the number of the operators at a given level is finite, this still limits the speed-up of efficiency of MLFMA. In this paper, a modified interpolation technique is developed for fast evaluation of translation operators, in which both the angle and the space-distance can be arbitrary. Two-dimensional Lagrange interpolation technique is used to realize fast computation of the operators in multiple space-distance and angles case at a given level. The space-distance interpolation is implemented by introducing a modified factor, the angle interpolation is still based on fast Legndre expansion algorithm (FLEA). Fast evaluation and low storage of the operators can be achieved. The present technique is efficient and simple, easy to implement, suitable for distributed memory machines.

2. Fast Legendre Expansion Algorithm (FLEA)

FLEA is used to realize fast evaluation of a translation operator for a given space-distance between field-point groups and source-point groups. For a translation operation

$$\alpha_{mm'}(\hat{r}_{mm'} \cdot \hat{k}) = \sum_{l=0}^{L-1} i^l (2l+1) h_l^{(1)}(kr_{mm'}) P_l(\hat{r}_{mm'} \cdot \hat{k}) \tag{1}$$

where $r_{mm'}$ is the space-distance between field-point groups and source-point groups, $\hat{r}_{mm'}$ and $\hat{k}$ are unit vectors. $h_l^{(1)}$ are the spherical Hankel functions of order l and P_l are Legendre polynomials of order l.

Considering $\cos\theta = \hat{r}_{mm'} \cdot \hat{k}$, we attain a finite Legendre expansion of the form

$$\alpha_{mm'}(\cos\theta) = \sum_{l=0}^{L-1} a_l P_l(\cos\theta) \tag{2}$$

$$a_l = i^l (2l+1) h_l^{(1)}(kr_{mm'}) \tag{3}$$

Eq.(2) can be replaced with a Chebyshev expansion of the same length

$$\alpha_{mm'}(\cos\theta) = \sum_{l=0}^{L-1} b_l T_l(\cos\theta) \tag{4a}$$

$$T_l(\cos\theta) = \cos(l\theta), l \geqslant 0 \tag{4b}$$

where a_l, b_l are related by the equation

$$b_i = \sum_{j=0}^{L-1} M_{ij} a_j, \quad i = 0,1,\cdots L-1 \tag{5}$$

The formula on the M_{ij} is given in [6].

Because the M_{ij} is only dependent of the index i, j, it can be precomputed and stored. The Chebyshev expansion can be performed by fast cosine transform. To attain the values of $\alpha_{mm'}(\hat{r}_{mm'} \cdot \hat{k})$ at $2L^2$ points, the local Lagrange interpolation is used. To assure the interpolation accuracy, enough interpolation nodes are required. For evaluation of translation operators, the number of interpolation nodes will be much larger than L. Numerical results show that excellent accuracy can be achieved by using $16L$ interpolation nodes and only two-point Lagrange interpolation.

3. Modified Lagrange Interpolation Technique (MLIT)

Modified Lagrange interpolation technique (MLIT) is used to rapidly evaluate a translation operator when the space-distance between field-point groups and source-point groups $r_{mm'}$ varies. For $h_l^{(1)}(kr_{mm'})$ is highly oscillatory, an accurate and rigorous interpolation technique for $\alpha_{mm'}(\hat{r}_{mm'} \cdot \hat{k})$ is difficult to find. But for solution of scattering from electrically large targets, most of the interactions between two groups occurs in the main beam of the translation operator. For a certain accuracy of computation, many side lobes can be ignored [8]. When two groups are in the far-field region, the translation operator can be further simplified using the fast far-field approximation (FAFFA) in which only the $\hat{r}_{mm'}$ direction is considered [9]. The translation operator in FAFFA is [9]

$$\alpha_{mm'}^{far} = \frac{e^{ikr_{mm'}}}{r_{mm'}} \tag{6}$$

Based on the FAFFA idea, a modified factor is introduced into local interpolation technique. We define the modified translation operator as

$$\tilde{\alpha}_{mm'}(\hat{r}_{mm'} \cdot \hat{k}) = \frac{kr_{mm'}}{e^{ikr_{mm'}}} \alpha_{mm'}(\hat{r}_{mm'} \cdot \hat{k}) \tag{7}$$

The modified translation operator can be evaluated by traditional Lagrange interpolation

$$\tilde{\alpha}_{mm'}(\hat{r}_{mm'} \cdot \hat{k}) = \sum_{i=1}^{n} W_i \, \tilde{\alpha}_{mm'}^i(\hat{r}_{mm'} \cdot \hat{k}) \tag{8}$$

$$W_i = \prod_{\substack{j=1 \\ j \neq i}}^{n} \frac{r_{mm'} - r_{mm'}^j}{r_{mm'}^i - r_{mm'}^j} \tag{9}$$

where n is the number of local Lagrange interpolation, $r_{mm'}^j$ is the space-distance at the j-th interpolation node, W_i is the weighting coefficient.

Eq.(7) is substituted into (8) to construct the modified Lagrange interpolation

$$\alpha_{mm'}(\hat{r}_{mm'} \cdot \hat{k}) = \frac{e^{ikr_{mm'}}}{kr_{mm'}} \sum_{i=1}^{n} \tilde{W}_i \, \alpha_{mm'}^i(\hat{r}_{mm'} \cdot \hat{k}) \tag{10}$$

$$\tilde{W}_i = \frac{kr_{mm'}^i}{e^{ikr_{mm'}^i}} W_i \tag{11}$$

where $\alpha_{mm'}^i(\hat{r}_{mm'} \cdot \hat{k})$ is the translation operator at the i-th interpolation node.

For arbitrary space-distance $r_{mm'}$ and the angle $\cos^{-1}(\hat{r}_{mm'} \cdot \hat{k})$, two-dimensional Lagrange interpolation based on Eq.(10) is used to realize fast computation of the operators.

4.Numerical Results and Discussions

In this section, some results are given to indicate the validity and efficiency of the present technique. Only two interpolation nodes are used in both the space-distance interpolation and the angle interpolation.

In the classical MLFMA using an oct-tree, the number of translation operators needed to store in the direct method is 343, the one in the FLEA is 64, the one in the present technique is only 5. The comparison of total memory requirements of translation operators at each level (corresponding to different L) for the direct method, FLEA, the present technique (MLIT) is listed in Table I. We see that the improvement especially for large L is very remarkable.

Table I   Comparison of memory requirements for three techniques
(single precision)

L	Direct method	FLEA	MLIT
8	351kb	65.5kb	5.12kb
16	1.4Mb	131.0kb	10.24kb
32	5.6Mb	262.0kb	20.48kb
64	22.4Mb	524.0kb	40.96kb
128	89.6Mb	1048.0kb	81.92kb
256	358.4Mb	2096.0kb	163.84kb

To verify the validity of the MLIT, two translation operators at different levels are investigated. The results evaluated by the direct method, traditional Lagrange interpolation technique (LIT) are also given for comparison. At the fine level, $L = 8$, $r_{mm'} = \hat{x}D_{cube} + \hat{y}2D_{cube}$. D_{cube} is the size of cube, $D_{cube} = 0.6\lambda$. At the coarse level, $L, r_{mm'}$ and D_{cube} are amplified by factor of 16. Fig.1 (a), Fig.1 (b) respectively shows the real part of the $\alpha_{mm'}(\hat{r}_{mm'} \cdot \hat{k})$ at the fine level in case of $\theta = 79.43°$ and $\theta = 16.2°$. The corresponding imaginary part of the $\alpha_{mm'}(\hat{r}_{mm'} \cdot \hat{k})$ is also shown in Fig.1(c), Fig.1(d) respectively. For the case of $\theta = 79.43°$, the $\hat{k}$ lies in the near region of the $\hat{r}_{mm'}$, a good agreement between the direct result and the result calculated by the MLIT can be achieved. For the case of $\theta = 16.2°$, the result of real part of $\alpha_{mm'}(\hat{r}_{mm'} \cdot \hat{k})$ calculated by the MLIT agrees not well with the direct result. But this is not important, for the $\hat{k}$ lies in the far region of the $\hat{r}_{mm'}$. A good agreement between the direct result and the result calculated by traditional LIT can not be achieved even in case of $\theta = 79.43°$.

Fig.2 (a), (b) shows the real part and imaginary part of the $\alpha_{mm'}(\hat{r}_{mm'} \cdot \hat{k})$ at the coarse level in case of $\theta = 89.3°$. We see that all the results calculated by the MLIT agree with the one by the direct method much better than the traditional LIT especially when the $\hat{k}$ lies in the near region of the $\hat{r}_{mm'}$. This is to say that most of the interactions between two groups has been accurately took into account in the MLIT.

Fig.3 shows the normalized bistatic RCS of conducting sphere with electrical size $ka = 35.0$, a is the radius of sphere. The number of unknowns is 39800, the conforming rooftop basis functions and line-matching method are used. It is clearly indicated that an excellent agreement between the MIE series result and the result calculated by the MLFMA based on the MLIT can be achieved. But we can not attain a satisfactory result if the MLFMA based on traditional LIT is used.

Fig.4 shows the normalized bistatic RCS of conducting sphere with electrical size $ka = 60.0$. The number of unknowns is 140481, 6-level MLFMA based on the MLIT is used. The result calculated by the present technique agrees very well with the MIE series result.

The above numerical results clearly demonstrate that the present technique is valid and efficient.

5. Conclusions

In this paper, a novel modified interpolation technique is developed for fast evaluation of translation operators. The interpolation on the space-distance between field-point groups and source-point groups $r_{mm'}$ is implemented by introducing a modified factor, the angle interpolation is based on the fast Legendre expansion algorithm (FLEA). The present technique is efficient, simple and easy to implement. Very low computer storage for translation operators is required. It is suitable for distributed memory machines.

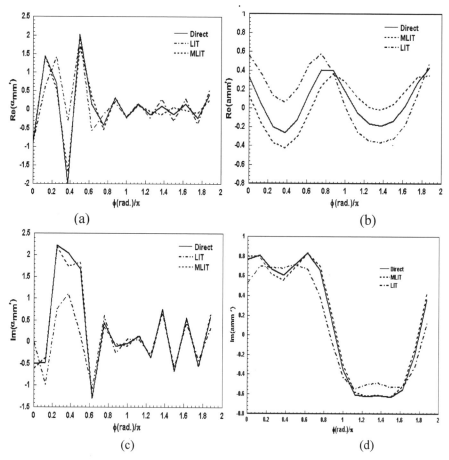

Fig.1   Comparison of the translation operator computed by direct method and two interpolation techniques.(a),(c) $\theta = 79.43°$; (b),(d) $\theta = 16.2°$

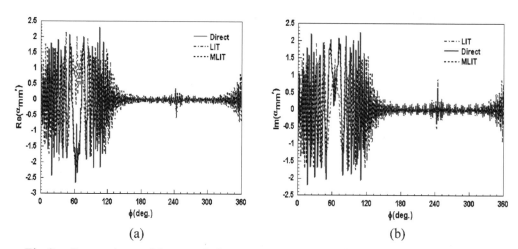

Fig.2   Comparison of the translation operator computed by direct method and two interpolation techniques in which $\theta = 89.3°$

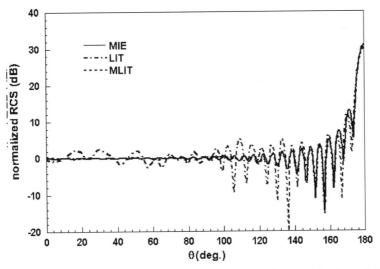

Fig.3   Normalized bistatic RCS of conducting sphere with $ka = 35.0$

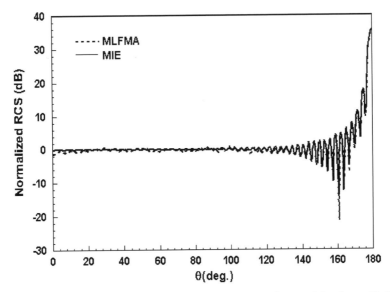

Fig.4   Normalized bistatic RCS of conducting sphere with $ka = 60.0$.

References

[1] Rokhlin,V, "Rapid solution of integral equations of scattering theory in two dimension "J.Comput.Phys.1990,86(2),pp.414-439

[2] J.M.Song, C.C.Lu, and W.C.Chew, "Multilevel fast multipole algorithm for electromagnetic scattering by large complex objects", IEEE Trans., AP-45, No.10, Oct., 1997, pp.1488-1493

[3] S.V. Velamparambil, J.M. Song, and W.C. Chew, "A portable parallel multilevel fast multipole solver for scattering from perfectly conducting bodies", IEEE Antennas Propaga. Symp., Vol.1, 1999, pp. 648-651

[4] W. C. Chew, etc. "Fast and efficient algorithms in computational electromagnetics", Artech House Publishers, 2001

[5] HU Jun, "The efficient method for vector electromagnetic scattering from complex target- fast multipole method and its applications", The Dissertation [PHD.], University of

Electronic Science and Technology of China, Cheng du, 2000, Jun

[6] B. K. Alpert and V. Rokhlin, "A fast algorithm for the evaluation of Legendre expansions", SIAM J. Sci. Statist. Comput., 12, 1991, pp.158-179

[7] S.V. Velamparambil, W.C. Chew, "A fast polynomial representation for the translation operators of an MLFMA", Microwave and Optical Technology Letters, Vol.28, No. 5, Mar., 2001, pp.298-303

[8]R. L. Wagner, W. C. Chew, " A ray propagation fast multipole algorithm ", Microwave and Optical Technology Letters, Vol.7, No.10,   1994, pp.435-438

[9]C. C. Lu, J.M.Song, and W.C.Chew, " The application of far-field approximation to accelerate the fast multipole method", IEEE Antennas Propaga. Symp., 1996, pp. 1738-1741

Wave Propagation,Scattering and Emission in Complex Media
Edited by Ya-Qiu Jin
Science Press and World Scientific,2004

Efficient Solution of 3-D Vector Electromagnetic Scattering by CG-MLFMA with Partly Approximate Iteration

Jun Hu, Zaiping Nie
School of Electrical Engineering,
University of Electronic Science and Technology of China
Chengdu 610054, China
E-mail: scatterhu@263.net,   Phone:028-83202056

Abstract In the conjugate gradient iteration-multilevel fast multipole algorithm (CG-MLFMA), the matrix-vector multiplication can be implemented by the interactions from nearby region and non-nearby region. Based on CG-MLFMA, an efficient partly approximate iteration technique is developed for solving the scattering from the object with large electrical sizes in this paper. When the iterative error is low than the critical iteration error (CIE), determined by the contributions of nearby region and the accuracy required, the matrix-vector multiplication can be computed approximately by the interactions from nearby region. Compared to the interactions from non-nearby region, the interactions from nearby region only possess a very small portion of total complexity. So the total CPU time can be reduced greatly, although the total number of iteration needed in the present method is nearly same to the one in CG-MLFMA. Numerical results are given to demonstrate the validity and efficiency of the present method.

Key Words   multilevel multipole, conjugate gradient, matrix-vector multiplication

1. Introduction

Fast multipole method (FMM) developed by Rokhlin [1] is an efficient method to expedite matrix-vector multiplication in the iterative method. Compared with the complexity of matrix-vector multiplication of $O(N^3)$ of Gaussian elimination method, the complexity of matrix-vector multiplication of FMM is only $O(N^{1.5})$, N is the number of unknown. A multilevel fast multipole algorithm (MLFMA) developed by Song etc. [2], attaining the complexity of $O(N \log N)$, is especially suitable to solve scattering from the target with large electrical sizes.

Although MLFMA can solve scattering from the object with very large electrical size, the required computer storage and CPU time are still expensive. To attain a faster algorithm, many improvements on the efficiency of iteration solution have been developed [3-4]. A block diagonal preconditioner [3] is used to attain much faster convergence. FMM and CG iteration with physical optics initial guess (FMM-POI) has been developed for efficient solution of scattering from conductive object with large electrical sizes [4]. These methods can reduce dramatically the total number of iteration, but cannot reduce the complexity of matrix-vector multiplication in each iteration. In this paper, MLFMA with partly approximation iteration (MLFMA-PAI) is developed to attain a faster MLFMA. In fact, this work in the paper is the extension of FMM-PAI [5]. Different from the above methods, the present method does not reduce the total number of iteration, but can reduce the total complexity of matrix-vector multiplication. To assure the stability of the partly approximation iteration, the conjugate gradient (CG) iteration is applied in this paper. In CG iteration, the iterative error is reducing with increase of number of iteration. To implement the present method, a critical iterative error (CIE) is set, which is dependent on the contributions of nearby region and the required accuracy. After the iterative error has satisfied CIE, an approximation formula is adopted to calculate the matrix-vector multiplication. In the formula, only the contributions from nearby groups are considered. It needs much less CPU time than the conventional MLFMA needs, so the total CPU time can be reduced greatly. The present method is very efficient for solving EM scattering from the object with large electrical sizes.

2.   Multilevel Fast Multipole Algorithm (MLFMA)

For 3-D conductive object, electrical field integral equation (EFIE) is given by

$$\hat{t} \cdot \int_S \overline{G}(r,r') \cdot J(r')dS' = \frac{4\pi i}{k\eta} \hat{t} \cdot E^i(r) \tag{1}$$

$$\overline{G}(r,r') = \left[\overline{I} - \frac{1}{k^2}\nabla\nabla'\right]\frac{e^{ik|r-r'|}}{|r-r'|} \tag{2}$$

where $\hat{t}$ is any unit tangent vector on S, E^i is the illuminating field.

The basic mathematics tool in FMM is the addition theory. By applying the theory, we attained finally [6]

$$\overline{G}(r_j,r_i) = \frac{ik}{4\pi}\int d^2\hat{k}(\overline{I} - \hat{k}\hat{k})e^{ik\cdot(r_{jm}-r_{im'})}\alpha_{mm'}(\hat{r}_{mm'}\cdot\hat{k}), \quad |r_{mm'}| > |r_{jm} - r_{im'}| \tag{3}$$

where r_j,r_i is the field point vector and source point vector respectively, r_m, $r_{m'}$ is the group center vector of the field points group and the source points group respectively.

Choosing appropriate basis function j_i and testing function t_j, linear algebraic equations is attained from (1)

$$\sum_{i=1}^N A_{ji}a_i = b_j \quad , \qquad j = 1,2,\cdots N \tag{4}$$

where

$$A_{ji} = \int_S dst_j(r) \cdot \int_S ds'\overline{G}(r,r') \cdot j_i(r') \tag{5a}$$

$$b_j = \frac{4\pi i}{k\eta}\int_S dst_j(r) \cdot E^i(r) \tag{5b}$$

Equation (3) is applied to (5a) to attain the computation representation of FMM,

$$\sum_{i=1}^N A_{ji}a_i = \sum_{m'}\sum_{i\in G_{m'}} A_{ji}a_i + \frac{ik}{4\pi}\int d^2\hat{k}\, V_{fmj}(\hat{k})$$

$$\times \sum_{m'}\alpha_{mm'}(\hat{k}\cdot\hat{r}_{mm'})\sum_{i\in G_{m'}} V_{sm'i}^*(\hat{k})a_i, \quad j \in G_m \tag{6}$$

In Equation (6), the first term represents the contributions from nearby groups, the second term represents the contributions from the non-nearby groups. $V_{sm'i}(\hat{k})$, $V_{fmj}(\hat{k})$ and $\alpha_{mm'}$ represents aggragation, disaggregation and translation term respectively,

$$V_{sm'i}(\hat{k}) = \int_S ds'\, e^{ik\cdot r_{im'}}\left[\overline{I} - \hat{k}\hat{k}\right] \cdot j_i(r_{im'}) \tag{7a}$$

$$V_{fmj}(\hat{k}) = \int_S ds\, e^{ik\cdot r_{jm}}\left[\overline{I} - \hat{k}\hat{k}\right] \cdot t_j(r_{jm}) \tag{7b}$$

$$\alpha_{mm'}(\hat{r}_{mm'}\cdot\hat{k}) = \sum_{l=0}^L i^l(2l+1)h_l^{(1)}(kr_{mm'})p_l(\hat{r}_{mm'}\cdot\hat{k}) \tag{7c}$$

In the above, L is the number of multipole terms, dependent on the electrical size of the divided group. $h_l^{(1)}$ are the spherical Hankel functions of order l, P_l are Legendre polynomials of order l.

MLFMA is the multilevel extension of FMM. First, one cube that contains the scatterer is divided into eight sub-cube, then each sub-cube is again divided into eight smaller sub-cube. Repeat this process until the size of cube in the finest level is about 0.5 wavelength. For the *l*-level, total number of cubes is 8^l. MLFMA only treats the cube that contains current elements. In the finest level, if the current element *i* is far from the current element *j*, the field due to element *i* at element *j* is computed in a multilevel fashion [7]. The field due to the near neighbor elements is computed directly by MoM. The far current elements calculation is divided into the aggregation process, the translation process, and the disaggregation process. The aggregation process is to compute the outgoing waves of the sources at different levels recursively starting from the finest level by interpolation technique. The translation process converts the outgoing waves into the incoming waves, the disaggregation process converts the incoming waves from a coarser level into incoming waves at a finer level by anterpolation technique. Detailed formula is shown in references [6,7]. For a perfectly conducting scatterer, the complexity of matrix-vector multiplication and the storage requirement in MLFMA are $O(N \log N)$, *N* is the number of unknowns.

3. Partly Approximate Iteration (PAI) Technique

MLFMA can expedite matrix-vector multiplication because $\overline{A} \cdot a$ is decomposed into two parts: $\overline{A}^{near} \cdot a$ and $\overline{A}^{far} \cdot a$. The former term is calculated directly, and the latter on the interaction between field point and source point is calculated through their groups in a multilevel multistage fashion. In MoM, each current element is one subscatterer, total $O(N^2)$ complexity is required for the interactions between all current elements. But in MLFMA, the group contains current elements is the subscatterer. So the total number of the subscatters is reduced dramatically. Only $O(N \log N)$ of total complexity is required for the interactions between all current elements. Actually, $\overline{A}^{near} \cdot a$ is dominant among $\overline{A}^{near} \cdot a$ and $\overline{A}^{far} \cdot a$ because the elements of $\overline{A}^{near}$ concentrates the strongest contribution of the sources. Based on this fact, MLFMA with partly approximation iteration is developed to improve the efficiency of MLFMA in this paper. To assure the stability of the partly approximation iteration, conjugate gradient (CG) iterative method is used to solve $\overline{A} \cdot a = b$, where $[\overline{A}]_{ji} = A_{ji}$, $[a]_i = a_i$ and $[b]_j = b_j$.

In CG-MLFMA, the iterative error, defined by $\left\| \overline{A} \cdot x - b \right\|_2 / \left\| b \right\|_2$, monotonously reduces as the iteration is going on. The modified unknown vector $s^{(i+1)} = x^{(i+1)} - x^{(i)}$ also reduces as the iteration is going on. $\left\| \cdot \right\|_2$ represents the 2-norm. When the iterative error reduces to a certain value, the matrix-vector multiplication $\overline{A} \cdot s$ can be computed approximately by [5]

$$\overline{A} \cdot s^{(i+1)} \approx \overline{A}^{near} \cdot s^{(i+1)}, \quad i > n \tag{8}$$

where *n* is the iteration number needed to reach this certain value. This value is called critical iterative error (CIE), which is mainly determined by the contributions of $\overline{A}^{near}$ and the required accuracy. The efficiency will be speed up but the accuracy will be lose when a larger CIE is adopted in MLFMA-PAI.

The physical basis of the present method is that the main contributions from non-nearby groups have been almost accounted before the iterative error reduces to CIE and the contributions of non-nearby groups is not important compared with the one of nearby groups in later iteration. Compared to the interactions from non-nearby groups, the interactions from nearby groups only possess a very small portion of total complexity. On the other hand, the total number of iteration needed in the present method is nearly same to the one in CG-MLFMA. So compared to conventional CG-MLFMA, this method is more efficient and very suitable to solve EM scattering from the object with large electrical sizes.

4. Numerical Results and Discussions

In this section, some typical numerical results are given to prove the validity and efficiency of the present method. In the following examples, CIE is the critical iterative error, Tol is the iterative error for convergence.

Fig.1 shows the bistatic normalized RCS of conducting sphere with radius of a. $ka = 15.0$, horizontal polarization case. The sphere is modelled by parametric quadratic surface [8] with 9120 unknowns. The results calculated by MLFMA-PAI with CIE=0.08, Tol=0.01 and MLFMA with Tol=0.01 respectively are given for comparison. It is seen that these results agree well with each other and all agree with the Mie series very well. The agreement degree can not be attained for conventional MLFMA with Tol=0.08. The results calculated by MLFMA-PAI with CIE=0.5, Tol=0.01 and MLFMA with Tol=0.5 respectively are also given in Fig.2 to illustrate the validity of MLFMA-PAI ulteriorly. It is shown the result calculated by MLFMA-PAI with a large CIE still can attain a relatively reasonable accuracy, which can not be realized by conventional MLFMA with Tol=0.5. The comparison of efficiency for MLFMA-PAI and MLFMA is listed in Table I.

Table I Comparison of MLFMA-PAI and conventional MLFMA
Solving scattering from conducting sphere.

$ka = 15.0$	Number of iteration	CPU time for iteration	Root Mean Square (RMS) error (dB)
Tol=0.01	41	25.62 minute	0.68
Tol=0.08	12	7.50 minute	1.0
CIE=0.08	44	7.88 minute	0.66
Tol=0.50	3	1.88 minute	6.52
CIE=0.50	43	2.68 minute	1.0

Fig.3 shows the bistatic vertical polarization RCS of conducting cube with length of $a = 3.0\lambda$. The number of unknowns is 3888. The results calculated by MLFMA-PAI with CIE=0.5, Tol=0.01, MLFMA with Tol=0.01 and MLFMA with Tol=0.50 respectively are given for comparison. It is seen that the result calculated by MLFMA-PAI with a CIE=0.5 large error still can agree with the one calculated by MLFMA with Tol=0.01, which can not be realized by conventional MLFMA with Tol=0.5. The comparison of efficiency for MLFMA-PAI and MLFMA is also listed in Table II. It is shown that MLFMA-PAI expedite the iteration process greatly and do not sacrifice the accuracy greatly.

Table II Comparison of MLFMA-PAI and conventional MLFMA
Solving scattering from conducting cube.

$a = 3.0\lambda$	Number of iteration	CPU time for iteration
Tol=0.01	14	4 minute
Tol=0.50	2	0.6 minute
CIE=0.50	10	0.76 minute

To prove the efficiency of MLFMA-PAI for solving scattering for the object with large electrical sizes, the bistatic normalized RCS of conducting sphere with $ka = 35.0$ is calculated. The number of unknowns is 39800. The results calculated by MLFMA-PAI with CIE=0.05, Tol=0.01 and CIE=0.5, Tol=0.01 and MLFMA with Tol=0.01 respectively are given in Fig.4. It is seen that the results by MLFMA-PAI with CIE=0.05, Tol=0.01 and MLFMA with Tol=0.01 all agree with the Mie series very well. The result calculated by MLFMA-PAI with CIE=0.5, Tol=0.01 can also attain a reasonable accuracy. But, the CPU time for iteration in MLFMA is 76 minutes, the one in MLFMA-PAI with CIE=0.05 is only 33 minutes, less than half of the CPU time required in MLFMA. The CPU time for iteration will be less when a larger CIE is adopted. only 10 minutes CPU time is required for iteration in MLFMA-PAI with CIE=0.5.

MLFMA-PAI can solve efficiently not only bistatic scattering but also monostatic scattering. The monostatic horizontal polarized RCS results of benchmark target-almond is shown in Fig. 5. The illuminating frequency is 3GHz. The nmuber of unknowns is 2380. To calculate 46 points different azimuth RCS, the CPU time for iteration in MLFMA with Tol=0.01 is 50 minutes, the one in MLFMA-PAI with CIE=0.08, Tol=0.01 is only 24 minutes. Less CPU time for iteration is needed for a larger CIE. only 15 minutes CPU time is required for iteration in MLFMA-PAI with CIE=0.5. It is seen that the results by MLFMA-PAI with CIE=0.08, Tol=0.01 and MLFMA with Tol=0.01 all agree with the measure result very well. The result calculated by MLFMA-PAI with CIE=0.5, Tol=0.01 can also attain a reasonable accuracy except plane wave illuminating towards two ends of almond.

5. Conclusions

In this paper, CG-MLFMA with partly approximate iteration (MLFMA-PAI) is developed for efficient solution of EM scattering from the object with large electrical sizes. The MLFMA-PAI is a simple and efficient method. It can expedite the iteration process greatly and do not sacrifice the accuracy when an appropriate critical iterative error (CIE) is adopted. Numerical results show that MLFMA-PAI with CIE=0.05-0.08 can attain a good accuracy for a general 3D object, only less than half of the CPU time for iterations in conventional MLFMA is required. The present method can solve efficiently not only bistatic RCS of object but also monostatic RCS of object. It is also a general method, can be implemented easily into other extended FMM such as fast far-field approximation (FAFFA), ray propagation fast multipole method (RPFMA) etc. It can also be combined with other efficient technique such as block diagonal preconditioner technique, POI etc. to achieve a more faster algorithm.

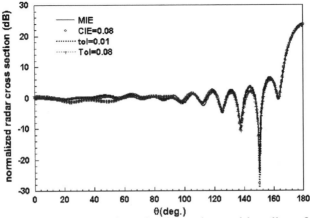

Fig.1   Bistatic Normalized RCS of conducting sphere with radius of a. $ka = 15.0$, HH-pol. The solid: MIE result; the circlet: the results calculated by MLFMA-PAI with CIE=0.08, Tol=0.01. The slotted: the results calculated by MLFMA with Tol=0.01. The +-slotted: the results calculated by MLFMA with Tol=0.08.

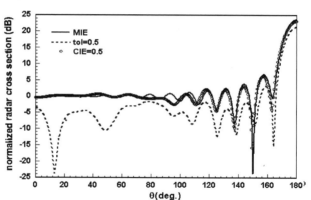

Fig.2   Bistatic Normalized RCS of conducting sphere with radius of a. $ka = 15.0$,
HH-pol. The solid: MIE result; the circlet: the results calculated by MLFMA-PAI with
CIE=0.5, Tol=0.01. The slotted: the results calculated by MLFMA with Tol=0.5.

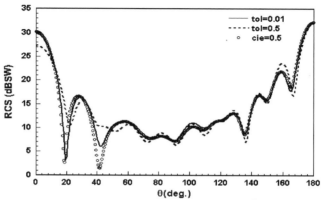

Fig.3   Bistatic Normalized RCS of conducting cube with length of a. $a = 3\lambda$, VV-pol.
The solid: the results calculated by MLFMA with Tol=0.01; the circlet: the results
calculated by MLFMA-PAI with CIE=0.5, Tol=0.01. The slotted: the results calculated by
MLFMA with Tol=0.5.

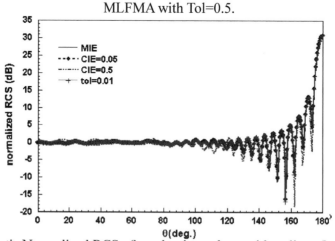

Fig.4   Bistatic Normalized RCS of conducting sphere with radius of a. $ka = 35.0$,
HH-pol. The solid: MIE result; the dot-slotted: the results calculated by MLFMA-PAI with
CIE=0.05, Tol=0.01. The slotted: the results calculated by MLFMA-PAI with CIE=0.5,
Tol=0.01. The +-solid: the results calculated by MLFMA with Tol=0.01.

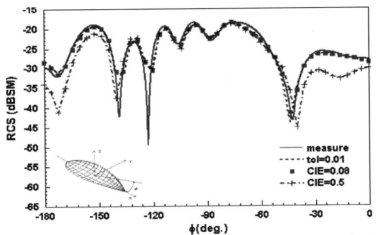

Fig.5   Monostatic RCS of benchmark target- almond. The illuminating frequency is 3GHz, HH-pol. The solid: the measure result; the dotted: the results calculated by MLFMA-PAI with CIE=0.08, Tol=0.01. The +-slotted: the results calculated by MLFMA-PAI with CIE=0.5, Tol=0.01. The slotted: the results calculated by MLFMA with Tol=0.01.

References

[1] Rokhlin,V, "Rapid solution of integral equations of scattering theory in two dimension "J.Comput.Phys.1990,86(2),pp.414-439

[2] J.M. Song, W.C. Chew, "Multilevel fast-multipole algorithm for solving combined field integral equations of electromagnetic scattering", Microwave Opt.Tec.Lett. Vol.10, No.1, pp.14-19,1995

[3] J.M. Song, C.C. Lu, and W.C. Chew, "Multilevel fast multipole algorithm for electromagnetic scattering by large complex objects", IEEE Trans., AP-45, No.10, Oct., 1997, pp.1488-1493

[4] Z. Nie, J. Hu, "New iterative methods in efficient numerical solution of 3-D vector electromagnetic scattering", Chinese Journal of Radio Science, Vol.14, sup., Apr.,1999

[5] J. Hu, Z. Nie, "Efficient Solution of 3-D vector Electromagnetic Scattering by FMM with partly approximate iteration", IEEE Antennas and Propagation Symposium, pp.656-659, Orlando, Florida, 1999

[6] J. HU, "The efficient method for vector electromagnetic scattering from complex target-fast multipole method and its applications", PhD. Dissertation, University of Electronic Science and Technology of China, Cheng du, Jun., 2000

[7] W. C. Chew, et al., "Fast and efficient algorithms in computational electromagnetics", Artech House Publishers, 2001

[8] J.M. Song and W.C. Chew, "Fast multipole method solution using parametric geometry", Microwave and Opt. Tech. Lett. Vol.7, No.16, Nov.1994, pp.760-765

The Effective Constitution at Interface of Different Media

Longgen Zheng, Wenxun Zhang
State Key Lab. of Millimiter Waves, Southeast University
Nanjing 210096, China

Abstract This paper presents a universal way to estimate the effective constitutive parameters for a grid across an interface between two different bi-anisotropic media. Its degenerated cases of bi-isotropic, anisotropic and isotropic interface are discussed respectively in detail.

1. Introduction

In many cases, the spatial domain of unknown electromagnetic field in a boundary value problem is discretized into a number of cells characterized by grids(see Fig.1). Most grids are filled with homogeneous medium, but some run across interfaces of different media and need to be homogenized with effective constitutive parameters. For isotropic medium,

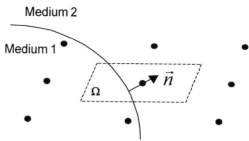

Fig.1 Spatial discretization: a grid across the boundary

the following relations exist: $\vec{D} = \varepsilon \vec{E}$ and $\vec{B} = \mu \vec{H}$. The question is how to define the effective ε_{eff} and μ_{eff}. We must do discrimination between the tangential and normal components of the field. For the former, $E_{t,\tau}$ and $H_{t,\tau}$ are continuous across the interface, thus $D_{t,\tau}$ and $B_{t,\tau}$ should be calculated from $E_{t,\tau}$ and $H_{t,\tau}$ (see Fig.2. D_t and D_τ are tangential components of $\vec{D}$). This results in $D_{t,\tau} = \varepsilon_{eff} E_{t,\tau}$ and $B_{t,\tau} = \mu_{eff} H_{t,\tau}$; for the latter, D_n and B_n are continuous, thus both E_n and H_n should be calculated from D_n and B_n. This results in $E_n = (\varepsilon^{-1})_{eff} D_n$ and $H_n = (\mu^{-1})_{eff} B_n$. In [1], similar formulae to compute effective parameters for isotropic and anisopropic media are suggested but the expression for anisotropic media lacks of a clear analytical foundation. In this paper, a universal way to estimate the effective constitutive parameters for grids across a interface between two different media is proposed. The effective parameters for an interface of bi-anisotropic media are derived in dètail, and then their degenerated formulae for bi-isotropic, anisotropic and isotropic cases are given respectively.

2. General steps to compute the effective medium

We suggest the following 6 steps to compute the parameters of an effective homogeneous medium. The basic rule is that we should always express non-continuous components across the boundary in terms of continuous ones, no matter how many field

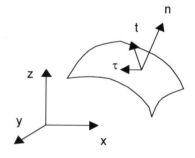

Fig.2 Local coordinate system

components are involved.

(1) Establish a local coordinate system with three orthonormal unit vectors $\hat{t}, \hat{\tau}$ and $\hat{n}$ as its bases by rotating unit vector $\hat{z}$ to the direction $\hat{n}$ and taking the rotated unit vectors $\hat{x}$ and $\hat{y}$ as $\hat{t}$ and $\hat{\tau}$ respectively (Fig.2). The coordinate-transform matrix $[T]$ satisfies $(\hat{t} \quad \hat{\tau} \quad \hat{n}) = (\hat{x} \quad \hat{y} \quad \hat{z})[T]$ with $[T]^T[T] = [T]^{-1}[T] = [I]$.

(2) Decompose fields into normal and tangential components, and transform the medium parameters from $\hat{x}\hat{y}\hat{z}$ system to $\hat{t}\hat{\tau}\hat{n}$ system;

(3) Exchange field components in the constitutive formulae so that all the non-continuous components are expressed in terms of continuous ones;

(4) Define the effective medium parameters in $\hat{t}\hat{\tau}\hat{n}$ system based on individual averaging procedure of these non-continuous components over the grid;

(5) Rearrange field components so that the flux density are expressed in terms of field strength or vice versa;

(6) Inverse coordinate-transform the resulted effective medium parameters from $\hat{t}\hat{\tau}\hat{n}$ system to $\hat{x}\hat{y}\hat{z}$ system, or integrate them to form a dyad.

3. Effective medium parameters in a grid across two bi-anisotropic media

For bi-anisotropic medium, the constitutive relations are formulated as:

$$\begin{pmatrix} \vec{D} \\ \vec{B} \end{pmatrix} = \begin{pmatrix} \overline{\overline{\varepsilon}} & \overline{\overline{\xi}} \\ \overline{\overline{\zeta}} & \overline{\overline{\mu}} \end{pmatrix} \begin{pmatrix} \vec{E} \\ \vec{H} \end{pmatrix} \tag{1}$$

In $\hat{t}\hat{\tau}\hat{n}$ system, the matrix form of Eq.(1) is:

$$\begin{pmatrix} \begin{pmatrix} D_t \\ D_\tau \\ D_n \end{pmatrix} \\ \begin{pmatrix} B_t \\ B_\tau \\ B_n \end{pmatrix} \end{pmatrix} = \begin{pmatrix} [\varepsilon_T] & [\xi_T] \\ [\zeta_T] & [\mu_T] \end{pmatrix} \begin{pmatrix} \begin{pmatrix} E_t \\ E_\tau \\ E_n \end{pmatrix} \\ \begin{pmatrix} H_t \\ H_\tau \\ H_n \end{pmatrix} \end{pmatrix} \tag{2}$$

where $[\varepsilon_T] = [T]^T[\varepsilon][T]$, $[\xi_T] = [T]^T[\xi][T]$, $[\zeta_T] = [T]^T[\zeta][T]$ and $[\mu_T] = [T]^T[\mu][T]$; blocks $[\varepsilon]$, $[\xi]$, $[\zeta]$ and $[\mu]$ are described in $\hat{x}\hat{y}\hat{z}$ system.

In order to exchange D_n with E_n and B_n with H_n, we expand the 2 by 2 blocks into 2 by 4 blocks (actually we get six homogeneous equations with twelve components as unknowns):

$$[A] = \begin{pmatrix} -[I] & [0] & [\varepsilon_T] & [\xi_T] \\ [0] & -[I] & [\zeta_T] & [\mu_T] \end{pmatrix} \tag{3}$$

By making column exchanges(namely 3rd with 9th and 6th with 12th in Exp.(3)), and reducing the left 4 blocks to their original states(2 zero blocks and 2 negative identities respectively), one gets another $[A']$ for rearranged field components as :

$$[A'] = \begin{pmatrix} -[I] & [0] & [\varepsilon_{T'}] & [\xi_{T'}] \\ [0] & -[I] & [\zeta_{T'}] & [\mu_{T'}] \end{pmatrix} \tag{4}$$

where blocks $[\varepsilon_{T'}]$, $[\xi_{T'}]$, $[\zeta_{T'}]$ and $[\mu_{T'}]$ satisfy:

$$
\begin{pmatrix}
\begin{pmatrix} D_t \\ D_\tau \\ E_n \end{pmatrix} \\
\begin{pmatrix} B_t \\ B_\tau \\ H_n \end{pmatrix}
\end{pmatrix}
=
\begin{pmatrix}
[\varepsilon_{T'}] & [\xi_{T'}] \\
[\zeta_{T'}] & [\mu_{T'}]
\end{pmatrix}
\begin{pmatrix}
\begin{pmatrix} E_t \\ E_\tau \\ D_n \end{pmatrix} \\
\begin{pmatrix} H_t \\ H_\tau \\ B_n \end{pmatrix}
\end{pmatrix}
\tag{5}
$$

These two operations can be done conveniently using matrix multiplication.

Now the averaging procedure can be done directly over the grid, which yields:

$$
\begin{pmatrix}
\begin{pmatrix} \widetilde{D}_t \\ \widetilde{D}_\tau \\ \widetilde{E}_n \end{pmatrix} \\
\begin{pmatrix} \widetilde{B}_t \\ \widetilde{B}_\tau \\ \widetilde{H}_n \end{pmatrix}
\end{pmatrix}
=
\begin{pmatrix}
[\widetilde{\varepsilon}_{T'}] & [\widetilde{\xi}_{T'}] \\
[\widetilde{\zeta}_{T'}] & [\widetilde{\mu}_{T'}]
\end{pmatrix}
\begin{pmatrix}
\begin{pmatrix} E_t \\ E_\tau \\ D_n \end{pmatrix} \\
\begin{pmatrix} H_t \\ H_\tau \\ B_n \end{pmatrix}
\end{pmatrix}
\tag{6}
$$

where notation $\widetilde{\rho}$ means the average value of ρ over the grid: $\widetilde{\rho} = \dfrac{1}{\Omega}\displaystyle\int_\Omega \rho(\vec{r})\,d\Omega$.

Moving D_n and B_n back to the left hand side of the equations using the same technique above yields:

$$
\begin{pmatrix}
\begin{pmatrix} \widetilde{D}_t \\ \widetilde{D}_\tau \\ D_n \end{pmatrix} \\
\begin{pmatrix} \widetilde{B}_t \\ \widetilde{B}_\tau \\ B_n \end{pmatrix}
\end{pmatrix}
=
\begin{pmatrix}
[\widetilde{\varepsilon}_{T''}] & [\widetilde{\xi}_{T''}] \\
[\widetilde{\zeta}_{T''}] & [\widetilde{\mu}_{T''}]
\end{pmatrix}
\begin{pmatrix}
\begin{pmatrix} E_t \\ E_\tau \\ \widetilde{E}_n \end{pmatrix} \\
\begin{pmatrix} H_t \\ H_\tau \\ \widetilde{H}_n \end{pmatrix}
\end{pmatrix}
\tag{7}
$$

The matrix form of effective constitutive parameters in xyz system can be written as $[\widetilde{\varepsilon}] = [T][\widetilde{\varepsilon}_{T''}][T]^T$, $[\widetilde{\mu}] = [T][\widetilde{\mu}_{T''}][T]^T$, $[\widetilde{\xi}] = [T][\widetilde{\xi}_{T''}][T]^T$ and $[\widetilde{\zeta}] = [T][\widetilde{\zeta}_{T''}][T]^T$.

The respective dyadic form can be obtained formally in the following way:

$$
\begin{pmatrix}
\overline{\overline{\varepsilon}} & \overline{\overline{\xi}} \\
\overline{\overline{\zeta}} & \overline{\overline{\mu}}
\end{pmatrix}_{eff}
=
\begin{pmatrix} \hat{t} & \hat{\tau} & \hat{n} \end{pmatrix}
\begin{pmatrix}
[\widetilde{\varepsilon}_{T''}] & [\widetilde{\xi}_{T''}] \\
[\widetilde{\zeta}_{T''}] & [\widetilde{\mu}_{T''}]
\end{pmatrix}
\begin{pmatrix} \hat{t} \\ \hat{\tau} \\ \hat{n} \end{pmatrix}
\tag{8}
$$

4. Effective medium parameters in a grid across two bi-isotropic media

For bi-isotropic medium, the constitutive relations are formulated as:

$$
\begin{pmatrix} \vec{D} \\ \vec{B} \end{pmatrix}
=
\begin{pmatrix} \varepsilon & \xi \\ \zeta & \mu \end{pmatrix}
\begin{pmatrix} \vec{E} \\ \vec{H} \end{pmatrix}
\tag{9}
$$

Blocks $[\varepsilon_{T'}]$, $[\xi_{T'}]$, $[\zeta_{T'}]$ and $[\mu_{T'}]$ in Eq.(5) can be simply derived as:

$$\left(\begin{matrix}[\varepsilon_{T'}] & [\xi_{T'}] \\ [\zeta_{T'}] & [\mu_{T'}]\end{matrix}\right) = \left(\begin{matrix}\begin{pmatrix}\varepsilon & 0 & 0 \\ 0 & \varepsilon & 0 \\ 0 & 0 & \overline{\mu}\end{pmatrix} & \begin{pmatrix}\xi & 0 & 0 \\ 0 & \xi & 0 \\ 0 & 0 & -\overline{\xi}\end{pmatrix} \\ \begin{pmatrix}\zeta & 0 & 0 \\ 0 & \zeta & 0 \\ 0 & 0 & -\overline{\zeta}\end{pmatrix} & \begin{pmatrix}\mu & 0 & 0 \\ 0 & \mu & 0 \\ 0 & 0 & \overline{\varepsilon}\end{pmatrix}\end{matrix}\right) \tag{10}$$

where $\overline{\mu} = \dfrac{\mu}{\varepsilon\mu - \xi\zeta}$, $\overline{\xi} = \dfrac{\xi}{\varepsilon\mu - \xi\zeta}$, $\overline{\zeta} = \dfrac{\zeta}{\varepsilon\mu - \xi\zeta}$ and $\overline{\varepsilon} = \dfrac{\varepsilon}{\varepsilon\mu - \xi\zeta}$.

The effective medium parameters, namely blocks $[\widetilde{\varepsilon}_{T''}]$, $[\widetilde{\xi}_{T''}]$, $[\widetilde{\zeta}_{T''}]$ and $[\widetilde{\mu}_{T''}]$ in Eq.(7) are:

$$\left(\begin{matrix}[\widetilde{\varepsilon}_{T''}] & [\widetilde{\xi}_{T''}] \\ [\widetilde{\zeta}_{T''}] & [\widetilde{\mu}_{T''}]\end{matrix}\right) = \left(\begin{matrix}\begin{pmatrix}\widetilde{\varepsilon} & 0 & 0 \\ 0 & \widetilde{\varepsilon} & 0 \\ 0 & 0 & \widetilde{\varepsilon}_n\end{pmatrix} & \begin{pmatrix}\widetilde{\xi} & 0 & 0 \\ 0 & \widetilde{\xi} & 0 \\ 0 & 0 & \widetilde{\xi}_n\end{pmatrix} \\ \begin{pmatrix}\widetilde{\zeta} & 0 & 0 \\ 0 & \widetilde{\zeta} & 0 \\ 0 & 0 & \widetilde{\zeta}_n\end{pmatrix} & \begin{pmatrix}\widetilde{\mu} & 0 & 0 \\ 0 & \widetilde{\mu} & 0 \\ 0 & 0 & \widetilde{\mu}_n\end{pmatrix}\end{matrix}\right) \tag{11}$$

with

$$\widetilde{\varepsilon}_n = \widetilde{\overline{\varepsilon}} \Big/ \left(\widetilde{\overline{\varepsilon}}\widetilde{\overline{\mu}} - \widetilde{\overline{\xi}}\widetilde{\overline{\zeta}}\right), \quad \widetilde{\mu}_n = \widetilde{\overline{\mu}} \Big/ \left(\widetilde{\overline{\varepsilon}}\widetilde{\overline{\mu}} - \widetilde{\overline{\xi}}\widetilde{\overline{\zeta}}\right), \quad \widetilde{\xi}_n = \widetilde{\overline{\xi}} \Big/ \left(\widetilde{\overline{\varepsilon}}\widetilde{\overline{\mu}} - \widetilde{\overline{\xi}}\widetilde{\overline{\zeta}}\right) \text{ and } \widetilde{\zeta}_n = \widetilde{\overline{\zeta}} \Big/ \left(\widetilde{\overline{\varepsilon}}\widetilde{\overline{\mu}} - \widetilde{\overline{\xi}}\widetilde{\overline{\zeta}}\right).$$

The respective dyadic form can be written as:

$$\left(\begin{matrix}\overline{\overline{\varepsilon}} & \overline{\overline{\xi}} \\ \overline{\overline{\zeta}} & \overline{\overline{\mu}}\end{matrix}\right)_{eff} = \left(\begin{matrix}\widetilde{\varepsilon}(\hat{t}\hat{t} + \hat{\tau}\hat{\tau}) + \widetilde{\varepsilon}_n\hat{n}\hat{n} & \widetilde{\xi}(\hat{t}\hat{t} + \hat{\tau}\hat{\tau}) + \widetilde{\xi}_n\hat{n}\hat{n} \\ \widetilde{\zeta}(\hat{t}\hat{t} + \hat{\tau}\hat{\tau}) + \widetilde{\zeta}_n\hat{n}\hat{n} & \widetilde{\mu}(\hat{t}\hat{t} + \hat{\tau}\hat{\tau}) + \widetilde{\mu}_n\hat{n}\hat{n}\end{matrix}\right) \tag{12}$$

5. Effective dyadic permittivity in a grid across two anisotropic media

For anisotropic media, only $\vec{D} - \vec{E}$ relation should be considered:

$$\vec{D} = \overline{\overline{\varepsilon}} \cdot \vec{E} \tag{13}$$

The general formulae in Section 3 can be trivially adapted to this case. To view the details of $\overline{\overline{\varepsilon}}_{eff}$, let us take a 2-D PBG structure as an example.

Let a cylinder parallel to $\hat{z}$, its dyadic permittivity has the matrix form as:

$[\varepsilon] = \begin{pmatrix}\varepsilon_{11} & \varepsilon_{12} & 0 \\ \varepsilon_{21} & \varepsilon_{22} & 0 \\ 0 & 0 & \varepsilon_{33}\end{pmatrix}$ and the unit vectors of the local

coordinate system are $\hat{t} = -\sin\phi\,\hat{x} + \cos\phi\,\hat{y}$, $\hat{\tau} = \hat{z}$, $\hat{n} = \cos\phi\,\hat{x} + \sin\phi\,\hat{y}$, then $[\varepsilon_T]$ in Eq.(2) can be written as

$[\varepsilon_T] = \begin{pmatrix}\varepsilon_{T11} & 0 & \varepsilon_{T13} \\ 0 & \varepsilon_{T22} & 0 \\ \varepsilon_{T31} & 0 & \varepsilon_{T33}\end{pmatrix}$ with

$\varepsilon_{T11} = \varepsilon_{11}\sin^2\phi + \varepsilon_{22}\cos^2\phi - (\varepsilon_{21} + \varepsilon_{12})\sin\phi\cos\phi$

$\varepsilon_{T33} = \varepsilon_{11}\cos^2\phi + \varepsilon_{22}\sin^2\phi + (\varepsilon_{21} + \varepsilon_{12})\sin\phi\cos\phi$

$\varepsilon_{T13} = -\varepsilon_{12}\sin^2\phi + \varepsilon_{21}\cos^2\phi + (\varepsilon_{22} - \varepsilon_{11})\sin\phi\cos\phi$

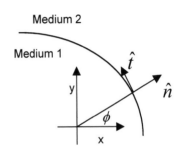

Fig. 3 Local coordinates

$$\varepsilon_{T31} = \varepsilon_{12} \cos^2 \phi - \varepsilon_{21} \sin^2 \phi + (\varepsilon_{22} - \varepsilon_{11}) \sin \phi \cos \phi$$

$$\varepsilon_{T22} = \varepsilon_{33}$$

$[\tilde{\varepsilon}_{T''}]$ in Eq.(7) and $\bar{\bar{\tilde{\varepsilon}}}_{eff}$ in Eq.(8) can be derived as:

$$[\tilde{\varepsilon}_{T''}] = \begin{pmatrix} \tilde{\varepsilon}_{T11} & 0 & \tilde{\varepsilon}_{T13} \\ 0 & \tilde{\varepsilon}_{T22} & 0 \\ \tilde{\varepsilon}_{T31} & 0 & \tilde{\varepsilon}_{T33} \end{pmatrix} \tag{14}$$

$$\bar{\bar{\tilde{\varepsilon}}}_{eff} = \tilde{\varepsilon}_{T11}\hat{t}\hat{t} + \tilde{\varepsilon}_{T13}\hat{t}\hat{n} + \tilde{\varepsilon}_{T22}\hat{\tau}\hat{\tau} + \tilde{\varepsilon}_{T31}\hat{n}\hat{t} + \tilde{\varepsilon}_{T33}\hat{n}\hat{n} \tag{15}$$

where

$$\tilde{\varepsilon}_{T11} = \text{ave}\left(\varepsilon_{T11} - \frac{\varepsilon_{T31}\varepsilon_{T13}}{\varepsilon_{T33}} \right) + \frac{\text{ave}(\varepsilon_{T31}\varepsilon_{T33}^{-1})\text{ave}(\varepsilon_{T13}\varepsilon_{T33}^{-1})}{\text{ave}(\varepsilon_{T33}^{-1})}$$

$$\tilde{\varepsilon}_{T13} = \frac{\text{ave}(\varepsilon_{T13}\varepsilon_{T33}^{-1})}{\text{ave}(\varepsilon_{T33}^{-1})}$$

$$\tilde{\varepsilon}_{T31} = \frac{\text{ave}(\varepsilon_{T31}\varepsilon_{T33}^{-1})}{\text{ave}(\varepsilon_{T33}^{-1})} \qquad \tilde{\varepsilon}_{T22} = \text{ave}(\varepsilon_{33}) \qquad \tilde{\varepsilon}_{T33} = \frac{1}{\text{ave}(\varepsilon_{T33}^{-1})}$$

Here notation "ave(A)" also means average of A (*i.e.* $\tilde{A}$).

Since both $\tilde{\varepsilon}_{T13}$ and $\tilde{\varepsilon}_{T31}$ do not vanish in general, Eq.(15) is likely different from the one in [1] which has a incomplete form:

$$\bar{\bar{\tilde{\varepsilon}}}_{eff} = \frac{1}{2}\left(\text{ave}(\bar{\bar{\varepsilon}}) \cdot (\bar{\bar{I}} - \hat{n}\hat{n}) + (\bar{\bar{I}} - \hat{n}\hat{n}) \cdot \text{ave}(\bar{\bar{\varepsilon}}) \right) + \frac{1}{2}\left(\text{ave}^{-1}(\bar{\bar{\varepsilon}}^{-1}) \cdot \hat{n}\hat{n} + \hat{n}\hat{n} \cdot \text{ave}^{-1}(\bar{\bar{\varepsilon}}^{-1}) \right)$$

6. Effective permittivity in a grid across two isotropic media

When both media are isotropic, the medium relation can be simplified to:

$$\vec{D} = \varepsilon \vec{E} \tag{16}$$

then Exp.(11) can be adapted as:

$$[\tilde{\varepsilon}_{T''}] = \begin{pmatrix} \tilde{\varepsilon} & 0 & 0 \\ 0 & \tilde{\varepsilon} & 0 \\ 0 & 0 & (\text{ave}(\varepsilon^{-1}))^{-1} \end{pmatrix} \tag{17}$$

and Exp.(12) as

$$\bar{\bar{\tilde{\varepsilon}}}_{eff} = \tilde{\varepsilon}(\bar{\bar{I}} - \hat{n}\hat{n}) + (\text{ave}(\varepsilon^{-1}))^{-1}\hat{n}\hat{n} \tag{18}$$

7. Conclusion

A general rule for calculation of the effective medium parameters at the interface of different media is established, and its derived process is simple in concept with a clear analytical foundation.

References

[1] R. D. Meade, A. M. Rappe, K. D. Brommer, J. D. Joannopoulos, and O. L. Alerhand, "Accurate theoretical analysis of photonic band-gap materials," Phys. Rev. B vol.48, pp.8434-8437 (1993).Erratum: S. G. Johnson, ibid vol.55, pp.15942(1997)

Wave Propagation,Scattering and Emission in Complex Media
Edited by Ya-Qiu Jin
Science Press and World Scientific,2004

445

Novel Basis Functions for Quadratic Hexahedral Edge Element

Peng Liu [1], Jia-Dong Xu [2], Wei Wan [2], Ya-Qiu Jin [1]
1. Center for Wave Scattering and Remote Sensing,
Fudan University, Shanghai 200433, China
2. Dept. of Electronic Engineering,
Northwestern Polytechnical University, Xi'an 710072, China

Abstract Two sets of basis functions are presented for the second-order hexahedral edge element of Serendipity type. These basis functions improve the orthogonality to facilitate the enforcement of Dirichlet boundary conditions and speed up the convergence rate of the iterative solver. Some examples of three-dimensional (3-D) scattering solution by using the finite element method with these basis functions show good accuracy in handling the field singularities around the edge and tip of structures.

Key Words   edge element,finite element method, electromagnetic scattering.

Edge elements are very useful tool in solving vector electromagnetic problems by the finite element method (FEM). The three-dimensional (3-D) vector electric fields in edge elements are expanded in terms of vector basis functions. These bases enforce the tangential continuity of the electric fields between adjacent elements while allow the abrupt change of the field's normal component, which is preferred in handling dielectric interface or field singularities around the edge and corner regions of physical structure. Edge elements of different shape and order have been developed [1~6]: low-order edge elements are simple and convenient to use [1,2], while high-order elements [3~6] are complicated, but may provide more accuracy and efficiency. In [3], Kameari developed the quadratic hexahedral edge element of Serendipity type from the linear 12-edge brick element.

In this paper, two sets of new basis functions are presented by biasing the plane associated with the polynomials in Kameari's basis functions. An orthogonality improvement of these new bases can be made and is discussed in detail by analyzing diagonal dominance of the elemental stiffness matrix. Fast convergence rate of iterative solver is demonstrated by solving 3-D scattering problem in the FEM approach. Moreover, the second set of the bases facilitates the implementation of Dirichlet boundary condition.

1 Kameari's vector basis functions

Consider a curvilinear hexahedral brick shown in Fig.1. 36 degrees of freedom (DOFs) are on a 20-node brick, as shown in Figs. 2(a,b), 24 out of these are associated with the element's edges and 12 are associated with the element's surfaces.

The electric fields $\mathbf{E}$ in the edge element is expanded as

$$\mathbf{E} = \sum_{i=1}^{36} \mathbf{N}_i E_i \tag{1}$$

where E_i is the expansion coefficient, and $\mathbf{N}_i$ are the edge basis function.

For the edges in ξ direction of the local coordinates (ξ, η, ζ), Kameari's original basis functions are written as [3]:

$$\mathbf{N}_i = \frac{1}{8}(1 + \eta\eta_i)(1 + \zeta\zeta_i)(\xi\xi_i + \eta\eta_i + \zeta\zeta_i - 1)\nabla\xi \qquad (2)$$

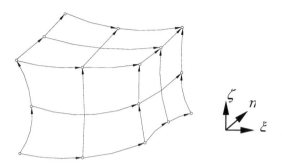

Fig. 1. Quadratic hexahedral edge element.

for DOFs on 8 edges and

$$\mathbf{N}_i = \frac{1}{4}(1 + \zeta\zeta_i)(1 - \eta^2)\nabla\xi \qquad (3)$$

for DOFs on 2 surfaces $\zeta = \pm 1$ and

$$\mathbf{N}_i = \frac{1}{4}(1 + \eta\eta_i)(1 - \zeta^2)\nabla\xi \qquad (4)$$

for DOFs on 2 surfaces $\eta = \pm 1$. Here ξ, η and ζ rang from -1 to 1; $\xi_i = \pm 1/2$, $\eta_i = \pm 1$ and $\zeta_i = \pm 1$ are the center of the i-th edge. Similar expressions can be obtained for basis functions in η and ζ direction.

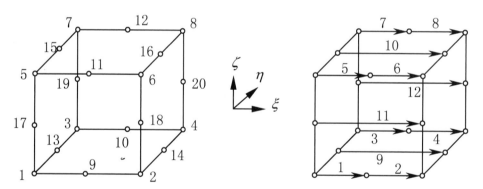

a. 20-node brick element   b. edge vectors in ξ direction

Fig. 2. Edge definition in ξ direction.

Definition of the element edge in ξ direction is shown in Fig.2b. Each DOFs i in Fig.2b is associated with source node i_1 and target node i_2 in Fig.2a, such as DOFs 1 is with source node 1 and target node 9.

Kameari's basis functions are mutually orthogonal in the integral of the tangential component along hexahedral edges and mid-surface lines

$$\int_{l_j} \mathbf{N}_i \cdot \mathbf{t}_j \, ds = \delta_{ij} = \begin{cases} 1 & i = j \\ 0 & i \neq j \end{cases} \qquad (5)$$

here σ_{ij} is Kronecker's σ, l_j is the length of j-th edge or mid-surface line of the Hexahedron, and $\mathbf{t}_j$ is the unit tangential vector. Hence for DOFs sharing a hexahedral edge, the corresponding equations must be solved to determine the unknown expansion coefficients for Dirichlet boundary condition.

2 New vector basis functions

Note that every linear polynomial in each basis function may be regarded as the left-hand member of a plane equation. On these planes, the basis function decays to zero. For example, the first polynomial in (2) is the left-hand member of the plane $\eta - 1 = 0$ when $\eta_i = -1$.

We biased the plane associated with the third polynomial of (2), so that it passes across the midpoint of the other element edge residing at the same hexahedral edge, as illustrated in Fig.3 (here edge vector $\overline{1}$ is as an example with parameters $\xi_i = -1/2, \eta_i = -1$, and $\zeta_i = -1$).

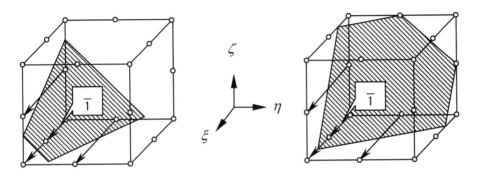

a. Kameari's basis function   b. New basis function

Fig. 3. The plane associated with the third polynomial in local coordinates.

For the edges in ξ direction, the new basis functions are given by

$$\mathbf{N}_i = \frac{1}{8}(1 + \eta\eta_i)(1 + \zeta\zeta_i)(4\xi\xi_i + \eta\eta_i + \zeta\zeta_i - 1)\nabla\xi \qquad (6)$$

As a result, Eqs. (6) and (3), (4) support a new set of basis functions, as named as the set **I**. This set of bases ensures the integral orthogonality in (5), and further more, an element edge level integral orthogonality is achieved: the line integration of the tangential component of $\mathbf{N}_i$ along the i-th edge of edge element equals one, and its integration along other edges, including the edge sharing a hexahedral edge, are equal to zero.

$$\int_{s_j} \mathbf{N}_i \cdot \mathbf{t}_j \, ds = \delta_{ij} \qquad \begin{cases} s_j = \dfrac{1}{2}l_j & j = 1,2,\cdots,8 \\ s_j = l_j & j = 9,10,11,12 \end{cases} \qquad (7)$$

If the magnitudes of Eqs. (3) and (4) are doubled as:

$$\mathbf{N}_i = \frac{1}{2}(1 + \zeta\zeta_i)(1 - \eta^2)\nabla\xi \qquad (8)$$

$$\mathbf{N}_i = \frac{1}{2}(1 + \eta\eta_i)(1 - \zeta^2)\nabla\xi \qquad (9)$$

then at the midpoint of j-th element edge, it yields

$$\mathbf{N}_i \cdot \mathbf{t}_j = \sigma_{ij} \tag{10}$$

Consequently, Eqs. (6) and (8), (9) support another set of basis functions, as named as the set **II**. They have the following properties:

a) The integral orthogonality of the element edge level is maintained (although not in a normalized form).

b) Dirichlet boundary condition can be easily imposed and computation load is reduced (because from (9), there is no need to evaluate the expansion coefficient through numerical integration).

3 Orthogonality analysis

In the wave propagation problem, the entries of the FEM element stiffness matrix $[K^e]$ contains the inner product of $\mathbf{N}_i^e$ and its curl [7]

$$A_{ij}^e = \iiint_{v^e} \left(\nabla \times \mathbf{N}_i^e\right) \cdot \left(\nabla \times \mathbf{N}_j^e\right) dv \tag{11}$$

$$B_{ij}^e = \iiint_{v^e} \mathbf{N}_i^e \cdot \mathbf{N}_j^e dv \tag{12}$$

with

$$K_{ij}^e = \alpha A_{ij}^e + \beta B_{ij}^e \tag{13}$$

here α and β are determined by the media property and wave number. A_{ij}^e and B_{ij}^e must calculate numerically for the quadratic hexahedral edge elements.

The ideal orthogonal property for K_{ij}^e should be

$$K_{ij}^e = \sigma_{ij} \tag{14}$$

from which, after the FEM assembling procedure, an diagonal FEM mass matrix is generated, and the matrix solution is obviated. In [8], the orthogonality of B_{ij}^e was achieved for the tetrahedral edge elements, and the matrix solution in each time step is eliminated in a time domain FEM. However, in the case of frequency domain FEM, as far as we know, it is impossible to achieve the orthogonality for both A_{ij}^e and B_{ij}^e.

An alternative way is to increase the diagonally dominant trend of the elemental stiffness matrix $[K^e]$. After the FEM assembling procedure, the condition number of the mass matrix of FEM can be reduced.

To clarify this, as an example, a quadratic hexahedral edge element with the shape of the mother element, i.e., a cubic with edge length 2, is discussed. A diagonally dominant analysis of this cubic edge element is enough, since curvilinear hexahedral edge element with different shape can be viewed as transforming from this mother element. The vector basis functions in $\nabla\xi, \nabla\eta$ and $\nabla\zeta$ directions have been orthogonal to each other in the mother element, so only the 12 vector basis functions in the ξ direction are discussed as an example.

Assuming that local coordinates ξ, η, and ζ in the mother element coincide with the $\mathbf{a}_x$, $\mathbf{a}_y$, and $\mathbf{a}_z$ directions of the global Cartesian coordinates separately, then basis function $\mathbf{N}_i^e$ could be written as

$$\mathbf{N}_i^e = \phi_i^e\left(\xi,\eta,\zeta\right)\nabla\xi = \phi_i^e\left(\xi,\eta,\zeta\right)\mathbf{a}_x \tag{15}$$

here $\phi_i^e(\xi,\eta,\zeta)$ is the scalar component of the vector basis function. For bases of Kameari's and our new sets **I** and **II**

$$\phi_i^e = \frac{1}{8}(1+\eta\eta_i)(1+\zeta\zeta_i)(n\xi\xi_i + \eta\eta_i + \zeta\zeta_i - 1) \tag{16}$$

where $n=1$ or $n=4$, it yields

$$B_{ij}^e = \iiint_{v^e} \phi_i^e \cdot \phi_j^e \, dv \tag{17}$$

The magnitude of ϕ_i^e decays from the edge or mid-surface line it resides. Therefore, two neighboring ϕ_i^e and ϕ_j^e tends to produce a big B_{ij}^e to destroy the diagonally dominant property of K_{ij}^e, such as the edge 1 and edge 2 in Fig. 2b. As illustrated in Fig. 3a and Fig 3b, the primary difference between Kameari's and our basis function is the volume above or below the plane associated with term $(n\xi\xi_i + \eta\eta_i + \zeta\zeta_i - 1)$. In Kameari's basis function, the overlapping volume of edge 1 and edge 2 where $(n\xi\xi_i + \eta\eta_i + \zeta\zeta_i - 1)$ taking the same sign (+ or -) are larger and results a bigger B_{12}^e.

The magnitude of B_{1j}^e and B_{9j}^e in the aforementioned mother element is plotted in Fig.4. From the viewpoint of diagonal dominant, our bases set **I** and **II** perform better than Kameari's basis function for B_{1j}^e, and the set **II** is superior to **I** for B_{1j}^e, $j>9$ and B_{9j}^e.

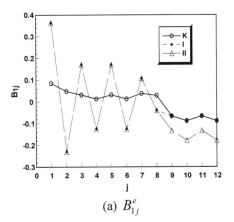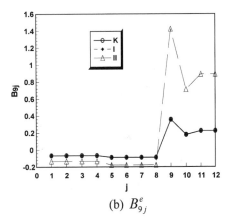

(a) B_{1j}^e (b) B_{9j}^e

Fig. 4. The magnitude of B_{ij}^e in element with the shape of the mother element.

The analysis of A_{ij}^e is more complicated. First, we can show that [9]

$$\nabla \times \mathbf{N}_i^e = \frac{1}{|J|}\left(\frac{\partial \phi_i^e}{\partial \zeta}\frac{\partial \mathbf{r}}{\partial \eta} - \frac{\partial \phi_i^e}{\partial \eta}\frac{\partial \mathbf{r}}{\partial \zeta}\right) \tag{18}$$

here $|J|$ is the Jacobi matrix, and

$$A_{ij}^e = \iiint_{v^e}\left(\frac{\partial \phi_i^e}{\partial \zeta}\cdot\frac{\partial \phi_j^e}{\partial \zeta} + \frac{\partial \phi_i^e}{\partial \eta}\cdot\frac{\partial \phi_j^e}{\partial \eta}\right)dv \tag{19}$$

For the bases $1\sim 8$ of both Kameari's and our new sets **I** and **II**, $\partial\phi_i^e/\partial\zeta$ has the form

$$\frac{\partial \phi_i^e}{\partial \zeta} = \frac{1}{8}(1+\eta\eta_i)\zeta_i\left[(n\xi\xi_i + \eta\eta_i + \zeta\zeta_i - 1)+(1+\zeta\zeta_i)\right] \tag{20}$$

Since for the edges 1 and 2, $(1+\eta\eta_i)$ and $(1+\zeta\zeta_i)$ take the same value, then it is noticed once again that the property of A_{12}^e is determined by $(n\xi\xi_i + \eta\eta_i + \zeta\zeta_i - 1)$, i.e. the third term in (2), (6), and (13).

The magnitude of A_{1j}^e and A_{9j}^e in the edge element with the same shape as the mother element is plotted in Fig.5. From the viewpoint of diagonal dominant, Kameari's bases perform badly, while our new set **I** perform much better and set the **II** performs as the best one. The set **II** is superior to Set **I** for A_{1j}^e, $j > 9$ and A_{9j}^e.

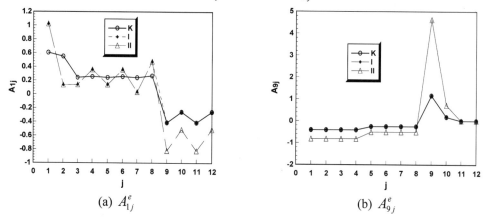

(a) A_{1j}^e (b) A_{9j}^e

Fig. 5. The magnitude of A_{ij}^e in the element with the shape of the mother element.

4 Numerical validation

Scattering from a PEC sphere is applied to testing both the Kameari's basis functions and two sets of our new basis functions. The sphere is 0.4λ in diameter and is enclosed by cubical perfectly matched layer (PML) absorber [10]. The new basis functions provide an overall better accuracy as shown in Fig.6, here the H plane case is considered and $\theta = 0°$ is backscattering. The relative bistatic RCS error is defined as

$$\delta_{RCS} = (\sigma_{numerical} - \sigma_{analytic})/\sigma_{analytic} \tag{21}$$

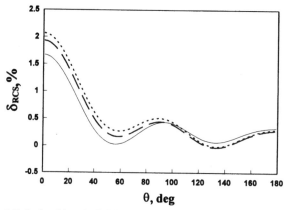

Fig.6 Relative bistatic RCS errors for different basis functions.
----- Kameari's basis functions, —·—·· new bases set **I**, —— new bases set **II**.

We have applied the Conjugate Gradient (CG) [4] method to solve the system equations. The total number of unknowns for the finite element mesh is 67264. When using 10^{-4} as the termination criteria, the iteration steps for Kameari's basis functions are 15086 while the new bases **I** are 10575 and **II** are 7450. It is evident that due to the orthogonality improvement of the new basis functions, the better-conditioned finite element matrices are obtained, and the iterative solver converges much faster.

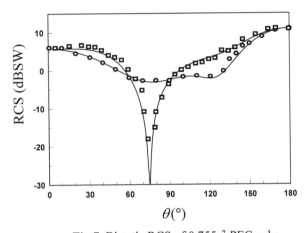

Fig.7. Bistatic RCS of 0.755 λ PEC cube.
$\square$ Measurement $\varphi = 0°$, $\bigcirc$ measurement $\varphi = 90°$, ——— new bases set II.

Another example is a 0.755 λ PEC cube. The surrounding cubical PML is 0.2 λ thick and 0.2 λ away from it. Here $\theta = 0°$ is backscattering and the new basis functions of the set **II** is used. The good agreement in Fig.7 with measured data [11] demonstrates the power of edge elements in handling field singularities around edge and vertex of the conducting structure.

Next, consider the NASA almond of one λ – length. Figure 8(a) is the quadrilateral mesh on the surface of the almond, and Figure 8(b) is the hexahedral edge element mesh of the entire computational domain. This example has been published in [12] with emphasis on the use of conformal PML (CPML) absorbing boundary.

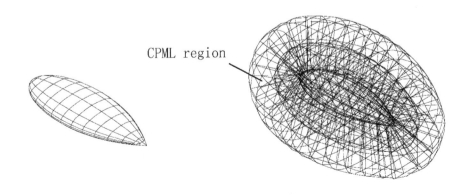

(a)   (b)
Fig. 8. The mesh of the NASA almond. (a) Quadrilateral mesh on the surface of the scatterer; (b) hexahedral mesh conformal to the scatterer (except for the tip).

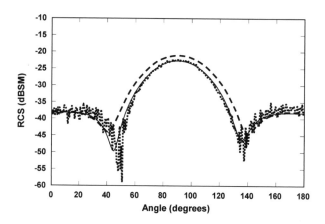

Fig. 9. Monostatic (VV) RCS of the NASA almond.
┈┈┈Measured, ──── FERM, ──── FEM-CPML

Fig.9 shows the computed results at 1.19 GHz using the new basis functions of the set **II**, and comparison with measured values [13] and a 3D MoM code FERM [13]. The FEM-CPML result is in better agreement with the measured data. The advantage of edge elements in handling field singularities around tips of a conducting structure is validated. Here the flat plane of almond coincides with the horizontal plane, and the tip of the almond corresponds to zero degree.

5 Conclusions

In this paper, by biasing the plane associated with the polynomials in Kameari's basis functions, two sets of new basis functions are developed. An orthogonality improvement of these basis functions is proved by the diagonal dominance of the FEM elemental stiffness matrix. Convergence rate of iterative solver in solving some 3-D scattering problems by using the FEM is demonstrated. Moreover, the second set of basis functions facilitates the implementation of Dirichlet boundary condition.

Acknowledgments

This work was supported by the China National Key Basic Research Program 2001 CB309401 and NSFC 60171009.

References

[1] J.P. Webb, "Edge elements and what they can do for you", IEEE Trans. Magn., vol. 29, pp.1460-1465, Jan. 1993

[2] J.C. Nedelec, "Mixed finite elements in R3", Numer. Math., vol 35, pp. 315-341, 1980

[3] A. Kameari, "Calculation of transient 3D eddy current using edge-elements", IEEE Trans. Magn., vol. 26, pp. 466-469, Mar. 1990

[4] A. Kameari, "Symmetric second order edge elements for triangles and tetrahedral", IEEE Trans. Magn., vol. 35, pp. 1394-1397, May. 1999

[5] J.P. Webb, "Hierarchal vector basis functions of arbitrary order for triangular and tetrahedral finite elements", IEEE Trans. Antennas Propagat., vol. 47 , pp.1244-1253, Aug. 1999

[6] V. Yioltsis and T.D. Tsiboukis, "Development and implementation of second and third order vector finite elements in various 3-D electromagnetic field problems", IEEE Trans. Magn., vol. 33, pp. 1812-1815, Mar. 1997

[7] J.M. Jin, The Finite Element Method in Electromagnetics, New York: Wiley, 1993

[8] D. Jiao, J.M. Jin, "Three-dimensional orthogonal vector basis functions for time-domain finite element solution of vector wave equations", IEEE Trans. Antennas Propagat., vol. 51 , pp.59-66, Jan. 2003

[9] Ida, Nathan, Jaão, Electromagnetics and Calculation of Fields, New York: Springer-Verlag, 1997

[10] J.Y. Wu, D.M. Kingsland, J. F. Lee, and R. Lee, "A comparison of anisotropic PML to Berenger's PML and its application to the finite-element method for EM scattering," IEEE Trans. Antennas Propagat., vol. 45, pp. 40~50, Jan. 1997

[11] R.P. Penno, G.A. Thiele, and K.M. Pasala, "Scattering from a perfectly conducting cube", Proceedings of the IEEE, vol. 77, pp. 815-823, May 1989

[12] P. Liu, J.D. Xu, W. Wan, "A finite-element realization of 3D conformal PML", Microwave Opt. Technol. Lett., vol. 30, pp.170－173, 2001

[13] A.C. Woo, H.T.G. Wang, M.J. Schuh, and M.L. Sanders, "Benchmark radar targets for the validation of computational electromagnetics programs", IEEE Antennas Propagat. Mag., vol. 35, no.1, pp. 84-89, 1993

Wave Propagation,Scattering and Emission in Complex Media
Edited by Ya-Qiu Jin
Science Press and World Scientific,2004

A Higher Order FDTD Method for EM Wave Propagation in Collision Plasmas

Shaobin Liu [1, 2], Jinjun Mo [1], Naichang Yuan [1]

1) Institute of Electronic Science and Engineering,
National University of Defense Technology, Changshai 410073, China
2) Basic Courses Department, Nanchang University, 330029, China
Email: liushaobin273@sohu.com

Abstract A fourth-order in time and space, finite-difference time-domain (FDTD) algorithm is applied to study the electromagnetic propagation in homogeneous, collision and warm plasma. The approach can significantly minimize the dispersion errors while still maintaining minimal memory requirement. For the problem of three-dimensional propagation, the scheme requires only three additional memory cells per Yee cell over and above that of the generic FDTD scheme. To investigate the validation of the fourth-order FDTD algorithm, the reflection coefficient of a slab of non-magnetized collision plasma is calculated. Comparisons between the accurate data and the results of second-order or fourth-order FDTD methods are discussed.

Key Words Fourth-order, FDTD, electromagnetic wave, plasma

1. Introduction

In recent years, the finite-difference time-domain (FDTD) algorithm has received considerable attention. The popularity of this algorithm stems from the fact that it is not limited to a specific geometry. And researchers can use the method for various materials (for example: isotropic material, anisotropic material [1,2], linearly dispersive media [3,4], and time-varying plasma media [5,6]). Furthermore, it provides a direct solution to problems with transient illumination. One of linear dispersive media is the isotropic collision plasma. For EM wave propagation in this medium, Young [7,8] recently proposed a direct integration finite-difference time-domain (DI-FDTD) algorithm that requires a minimal amount of memory and is second-order accurate. However, to obtain accurate solutions one must set the time step to be some arbitrarily small value (how small depends on the accuracy needed). Hence, Young [9] presented a fourth-order in time and space, finite-difference time-domain (FDTD) algorithm for radio-wave propagation in collisionless plasmas. The approach can significantly minimize the dispersion errors while still maintaining minimal memory requirements.

In this paper, we extended the fourth-order in time and space, FDTD algorithm to radio-wave propagation in collision, warm plasmas. One significant advance over the previous paper is that the fourth-order FDTD method is applied to loss plasmas. Thus, the attenuation of radio-wave by plasmas can be simulated using the fourth-order FDTD technique.

2. The FDTD Algorithm

A one-component, fluid model for electromagnetic wave propagation in a warm, isotropic, collision, plasma medium is assumed in the following discussion. And the motion of ions is neglected due to the ion' larger mass. Then the Maxwell's equations and the constitutive relation are given by (1)-(4)

$$\nabla \times \mathbf{H} = \varepsilon_0 \frac{\partial \mathbf{E}}{\partial t} + \mathbf{J} \tag{1}$$

$$\nabla \times \mathbf{E} = -\mu_0 \frac{\partial \mathbf{H}}{\partial t} \tag{2}$$

$$\frac{\partial \mathbf{J}}{\partial t} + v\mathbf{J} = \varepsilon_0 \omega_p^2 \mathbf{E} \tag{3}$$

Here, $\mathbf{E}$ is the electric field, $\mathbf{H}$ is the magnetic intensity, $\mathbf{J}$ the polarization current density, ε_0 the permittivity of free space, μ_0 the permeability of free space, $\omega_p^2 = n_e e^2 / m\varepsilon_0$ the square of plasma frequency, n_e is the electron concentration, v the electron collision frequency, and e and m are the electric charge and mass of an electron, respectively.

For the ionosphere, the magnitude of the electron collision frequency for momentum loss was determined by the formula [10]

$$v = 8.3 \times 10^3 \, \pi a^2 \sqrt{T} N_m \tag{4}$$

where a is the radius of the molecule (the molecule is assumed to be rigid sphere. For air, $a = 1.2 * 10^{-10} m$ [10]), N_m is the number density of the molecules, T the plasma temperature.

The temporal derivatives and spatial derivatives (spatially-centered approximation) are discretized using the following fourth-order approximation [9]

$$\frac{\partial f(x,t)}{\partial t} \approx \frac{f(x,t+\Delta t/2) - f(x,t-\Delta t/2)}{\Delta t} - \frac{(\Delta t)^2}{24} \frac{\partial^3 f(x,t)}{\partial t^3} \tag{5}$$

$$\frac{\partial f(x,t)}{\partial x} \approx \frac{9}{8} \frac{f(x+\Delta x/2,t) - f(x-\Delta x/2,t)}{\Delta t} - \frac{1}{24} \frac{f(x+3\Delta x/2,t) - f(x-3\Delta x/2,t)}{\Delta x} \tag{6}$$

For the one-dimensional case, the electromagnetic wave is assumed to be propagation in x-direction, and consider only E_y and H_z components. Then the substitutions of (5)-(6) in (1)-(3), and using leap-frog integration give the following equations

$$\varepsilon_0 E_y\big|_i^{n+1} = \varepsilon_0 E_y\big|_i^n - \Delta t \left[\frac{9}{8} \left(\frac{H_z\big|_{i+1/2}^{n+1/2} - H_z\big|_{i-1/2}^{n+1/2}}{\Delta x} \right) - \left(\frac{H_z\big|_{i+3/2}^{n+1/2} - H_z\big|_{i-3/2}^{n+1/2}}{24\Delta x} \right) \right]$$

$$- \Delta t J_y\big|_i^{n+1/2} + \varepsilon_0 \frac{(\Delta t)^3}{24} \frac{\partial^3 E_y\big|_i^{n+1/2}}{\partial t^3} \tag{7}$$

$$\mu_0 H_z\big|_{i+1/2}^{n+1/2} = \mu_0 H_z\big|_{i+1/2}^{n-1/2} - \Delta t \left[\frac{9}{8} \left(\frac{E_y\big|_{i+1}^n - E_y\big|_i^n}{\Delta x} \right) - \frac{1}{24} \left(\frac{E_y\big|_{i+2}^n - E_y\big|_{i-1}^n}{\Delta x} \right) \right]$$

$$+ \mu_0 \frac{(\Delta t)^3}{24} \frac{\partial^3 H_z\big|_{i+1/2}^n}{\partial t^3} \tag{8}$$

$$J_y\big|_i^{n+1/2} = J_y\big|_i^{n-1/2} + \Delta t \varepsilon_0 \omega_p^2 E_y\big|_i^n - v\Delta t J_y\big|_i^n + \frac{(\Delta t)^3}{24} \frac{\partial^3 J_y\big|_i^n}{\partial t^3} \tag{9}$$

Here, n signifies the time $n\Delta t$, Δt is the time step, i signifies the space $i\Delta x$, Δx is the space step. After repeated use of (1)-(3), the terms of threes differentiations in (7)-(9) can be written as

$$\mu_0 \frac{\partial^3 \mathbf{H}}{\partial t^3} = -c^2 \nabla^2 (\nabla \times \mathbf{E}) + \omega_p^2 \nabla \times \mathbf{E} - \frac{v}{\varepsilon_0} \nabla \times \mathbf{J} \tag{10}$$

$$\varepsilon_0 \frac{\partial^3 \mathbf{E}}{\partial t^3} = c^2 \nabla \times \nabla^2 \mathbf{H} + v\varepsilon_0 \omega_p^2 \mathbf{E} - \omega_p^2 \nabla \times \mathbf{H} + \omega_p^2 \mathbf{J} + c^2 \nabla \times \nabla \times \mathbf{J} - v^2 \mathbf{J} \tag{11}$$

$$\frac{\partial^3 \mathbf{J}}{\partial t^3} = -c^2 \varepsilon_0 \omega_p^2 \nabla \times \nabla \times \mathbf{E} - \varepsilon_0 \omega_p^4 \mathbf{E} + v\omega_p^2 \mathbf{J} \tag{12}$$

Using (10)-(12), we generate the following FDTD equations for (7)-(9) as

$$E_y\Big|_i^{n+1} = \frac{48\varepsilon_0 + \varepsilon_0 (\Delta x)^3 v\omega_p^2}{48\varepsilon_0 - \varepsilon_0 (\Delta x)^3 v\omega_p^2} E_y\Big|_i^n$$

$$-\frac{48\Delta t}{48\varepsilon_0 - \varepsilon_0 (\Delta x)^3 v\omega_p^2}\left[\frac{9}{8}\left(\frac{H_z\big|_{i+1/2}^{n+1/2} - H_z\big|_{i-1/2}^{n+1/2}}{\Delta x}\right) - \left(\frac{H_z\big|_{i+3/2}^{n+1/2} - H_z\big|_{i-3/2}^{n+1/2}}{24\Delta x}\right) + J_y\big|_i^{n+1/2}\right]$$

$$+\frac{2(\Delta t)^3}{48\varepsilon_0 - \varepsilon_0 (\Delta x)^3 v\omega_p^2}\left[\begin{array}{l} \dfrac{\omega_p^2}{\Delta x}\left(H_z\big|_{i+1/2}^{n+1/2} - H_z\big|_{i-1/2}^{n+1/2}\right) - \dfrac{c^2}{(\Delta x)^2}\left(J_y\big|_{i+1}^{n+1/2} - 2J_y\big|_i^{n+1/2} + J_y\big|_{i-1}^{n+1/2}\right) \\ -\dfrac{c^2}{(\Delta x)^3}\left(H_z\big|_{i+3/2}^{n+1/2} - 3H_z\big|_{i+1/2}^{n+1/2} + 3H_z\big|_{i-1/2}^{n+1/2} - H_z\big|_{i-3/2}^{n+1/2}\right) - v^2 J_y\big|_i^{n+1/2} \end{array}\right] \tag{13}$$

$$\mu_0 H_z\Big|_{i+1/2}^{n+1/2} = \mu_0 H_z\Big|_{i+1/2}^{n-1/2} - \frac{9\Delta t}{8}\left(\frac{E_y\big|_{i+1}^n - E_y\big|_i^n}{\Delta x}\right) + \frac{\Delta t}{24}\left(\frac{E_y\big|_{i+2}^n - E_y\big|_{i-1}^n}{\Delta x}\right)$$

$$+\frac{(\Delta t)^3}{24}\left(\begin{array}{l} \omega_p^2 \dfrac{E_y\big|_{i+1}^n - E_y\big|_i^n}{\Delta x} - \dfrac{v}{\varepsilon_0}\dfrac{J_y\big|_{i+1}^{n+1/2} - J_y\big|_i^{n+1/2}}{\Delta x} \\ -c^2 \dfrac{E_y\big|_{i+2}^n - 3E_y\big|_{i+1}^n + 3E_y\big|_i^n - E_y\big|_{i-1}^n}{(\Delta x)^3} \end{array}\right) \tag{14}$$

$$J_y\Big|_i^{n+1/2} = \frac{48 - 24v\Delta t + v\omega_p^2 (\Delta t)^3}{48 + 24v\Delta t - v\omega_p^2 (\Delta t)^3} J_y\Big|_i^{n-1/2} + \frac{48\Delta t\varepsilon_0 \omega_p^2}{48 + 24v\Delta t - v\omega_p^2 (\Delta t)^3} E_y\Big|_i^n$$

$$+\frac{2(\Delta t)^3}{48 + 24v\Delta t - v\omega_p^2 (\Delta t)^3}\left(\begin{array}{l} c^2 \varepsilon_0 \omega_p^2 \dfrac{E_y\big|_{i+1}^n - 2E_y\big|_i^n + E_y\big|_{i-1}^n}{\Delta x^2} \\ -\varepsilon_0 \omega_p^4 E_y\big|_i^n \end{array}\right) \tag{15}$$

Here, the second-order accuracy spatial derivatives in (10)-(12) are computed due to the presence of the multiplicative term $(\Delta t)^3$, and shown as follows.

$$\frac{\partial f(x,t)}{\partial x} \approx \frac{f(x + \Delta x/2, t) - f(x - \Delta x/2, t)}{\Delta x} \tag{16}$$

$$\frac{\partial^2 f(x,t)}{\partial x^2} \approx \frac{f(x + \Delta x, t) - 2f(x,t) + f(x - \Delta x, t)}{(\Delta x)^2} \tag{17}$$

$$\frac{\partial^3 f(x,t)}{\partial x^3} \approx \frac{f(x + 3\Delta x/2, t) - 3f(x + \Delta x/2, t) + 3f(x - \Delta x/2, t) - f(x - 3\Delta x/2, t)}{(\Delta x)^3} \tag{18}$$

3. Numerical Verification

To investigate the accuracy of the fourth-order FDTD method we compute the reflection coefficient of an unmagnetized collision plasma slab

$(\omega_p = 2\pi * 10 * 10^6 \, rad/s, \quad v = 100 * 10^6 \, rad/s = 1.59 * 10^7 \, Hz)$

with a thickness of 12.5 m. (In the D layer of the isothere, the electron collision frequency for momentum loss is $v \sim 6 * 10^6 - 6 * 10^7$ while the number density of the air molecules is $N_m \sim 10^{15} - 10^{16}$ and the plasma temperature is $T = 300K$). In addition, we also compute the same coefficients using the second-order FDTD method. The computational domain is 125 m long and the plasma slab is defined by the region [56.25, 68.75] m. The reflection coefficient were computed by simulating the transient response of a normally incident plane wave on the plasma slab. The incident wave used in the simulation is a Gaussian pulse whose frequency spectrum peaks at 50 MHz and is 10dB down from the peak at 100MHz. For FDTD parameters, the spatial discretization, Δx, used in the simulations is 0.125 m (i.e., 48 points per free-space wavelength at $f = 50MHz$) and the time step $\Delta t (= 0.5\Delta x/c)$ is 0.208 ns. Then, the computational domain is subdivided into 1000 cells, plasma slab occupies cells 450 to 550, free space from 0 to 500 and 550 to 1000.

Figures 1 compare the reflection coefficients computed using the second-order FDTD and the fourth-order FDTD methods with those of the analytical solution for a plasma with collision frequency $v = 100 * 10^6 \, rad/s$. The Fourier transforms (FFT) were used when the reflection coefficients of the plasma slab from the air-to-plasma and plasma-to-air interfaces were computed using the second-order or fourth-order FDTD. And the second-order or fourth-order perfectly match layers [11,12] were used to absorb the electromagnetic waves traveling towards boundaries, respectively. The exact solution of reflection coefficients is the following [13]:

$$R = \frac{\left(\dfrac{n-1}{n+1}\right)\left[1 - \exp\left(-2j\left(\dfrac{\omega}{c}\right)nd\right)\right]}{1 - \left(\dfrac{n-1}{n+1}\right)^2\left[1 - \exp\left(-2j\left(\dfrac{\omega}{c}\right)nd\right)\right]} \tag{19}$$

where R is the reflection coefficient of EM wave, ω is the incident EM wave frequency, d is the thickness of the plasma slab, n is the refractive index of the plasma and has the form

$$n = \sqrt{1 - \frac{\omega_p^2}{\omega^2 + v^2} - j\frac{v}{\omega}\frac{\omega_p^2}{\omega^2 + v^2}} \tag{20}$$

Upon the comparison of those two plots we observe that the bandwidth associated with the fourth-order scheme is wider than its second-order counterpart. Furthermore, at the higher frequencies the frequency shift [9] in the nulls of the reflection coefficient doesn't occur.

In the F layer of the isothere, the electron collision frequency for momentum loss is $v \sim 10^4 \, Hz$ while the number density of the air molecules is $N_m \sim 10^6$ and the plasma temperature is $T = 300K$ [10]. Hence, we computed the reflection coefficients of a plasma slab with collision frequency $v = 2\pi * 10^4 \, rad/s$ and $v = 10^9 \, rad/s$ using the fourth-order FDTD technique and exact solution (Figure 2). Figures 1-2 show that the magnitude of the reflection coefficient can be decreased by increasing the plasma collision frequency. Upon the comparison of Fig. 2, we can observe that the bandwidth associated with the higher collision frequency is wider than its low collision frequency counterpart.

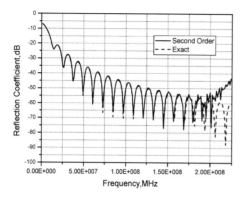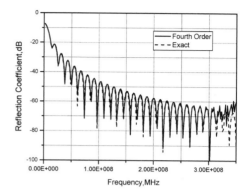

Fig. 1. A frequency domain plot of the reflection coefficient for $v = 100 * 10^6 \, rad/s$. Comparison is between the exact data and the data obtained from the second-order FDTD (fourth-order FDTD).

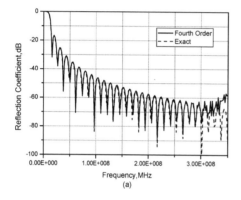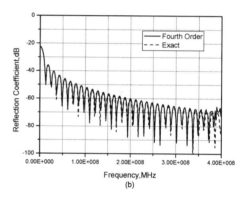

Fig. 2. A frequency domain plot of the reflection coefficient for (a) $v = 2\pi * 10^4 \, rad/s$, (b) $v = 10^9 \, rad/s$. Comparison is between the exact data and the data obtained from the fourth-order FDTD.

4. Summary

In this paper, we have introduced a fourth-order FDTD formulation for predicting the propagation characteristics of an electromagnetic wave in collision plasma. For three-dimensional propagation, the scheme requires only three additional memory cells per Yee cell over and above that of the generic FDTD scheme. With respect to accuracy, the scheme is accurate to fourth-order in both time and space. And the approach can significantly minimize the dispersion errors while still maintaining minimal memory requirements. In particular, the fourth-order FDTD algorithm is well fit for simulating the electromagnetic problems of those devices with bigger structures

The study of the reflection coefficient of collision plasmas is of practical significance. It is well known that collision plasmas can attenuate the energy of incident EM wave. Hence, in some specific cases collision plasmas can be used as EM-wave absorbers [13-14, 16]. The reduction of the radar cross section (RCS) of a target surrounded by collision plasma is an example.

References

[1] J. Schneider and S. Hudson. IEEE Trans. A. P. 1993, 41(7): 994-999
[2] S. D. Gedney. IEEE Trans. A. P. 1996, 44(12): 1630-1639
[3] R. J. Luebbers, F. Hunsberger and K. S. Kunz. IEEE. Trans. A. P. 1991, 39(1): 29-34
[4] J. L. Young and R. O. Nelson. IEEE Antennas and Propagation Magazine, 2001,

43(1): 61-77

[5] J. H. Lee, D. K. Kalluri IEEE Trans. A. P. 1999, 47(7): 1146-1151

[6] J. H. Lee, D. K. Kalluri and G. C. Nigg J. of Infrared and Millimeter Waves, 2000, 21(8): 1223-1253

[7] J. L. Young. IEEE Trans. A. P. 1995, 43(3): 422-426

[8] J. L. Young. Radio Science, 1994, 29(6): 1513-1522

[9] J. L. Young. IEEE Trans. A. P. 1996, 44(9): 1283-1289

[10] V. L. Ginzburg. The Propagation of Electromagnetic Waves in plasmas, 2nd ed., Pergamon, 1970

[11] J. P. Berenger. J. of Computational Physics, 1994, 114: 185-200

[12] A. Taflove. Advances in Computational Electrodynamics: The Finite-Difference Time-Domain Method, Artech House, Boston, London. 1995: 63-105

[13] J. H. Yuan and D. H. Mo, Waves in Plasmas (in Chinese), UESTC press, 1990: 153-156

[14] J. J. Mo, S. B. Liu and N. C. Yuan Chinese Journal of Radio Science (in Chinese), 2002, 17(1): 69-73

[15] S. B. Liu, J. J. Mo and N. C. Yuan Chinese Journal of Radio Science (in Chinese), 2002, 17(2): 134-137

[16] S. Liu, J. Mo and N. Yuan. J. of Infrared and Millimeter Waves, 2002, 23(12)

Attenuation of Electric Field Eradiated by Underground Source

Jiping Dong[1], Yougang Gao[2]

[1] P.O.Box 9628-8, Chinese Academy of Space Technology, Beijing, China 100086
email: dong.j.p@163.com

[2] P.O.Box 171, Beijing University of Posts and Telecommunications, Beijing, China
100876, email: lisf@bupt.edu.cn

Abstract   In the paper, the expressions of subaerial electric field (lower frequency) eradiated by a unit electric current source horizontally placed underground is derived for half–space using the method of image charges and numerical calculations are also performed for various epicentral distances ($\leqslant$500Km),several different frequencies (1Hz, 10Hz,100Hz,1KHz) and depths of the source ($\leqslant$30Km). It is found that attenuation of electric field is much smart with depth of electric current source, frequency, and epicentral distance increasing. According to actual threshold for observation of electric field, the condition and possibility for detection of SES is also discussed.

Key Words   seismo electric signal; electric current source; the method of image charges.

1. Introduction

There are a lot of reports concerning about electromagnetic signal observed in a wide frequency range from ULF to HF before earthquakes occur [1]. Electromagnetic radiation during experiments of rock fracture also convinces us that seismo electromagnetic signal comes directly from seismic source, considering that earthquakes occur when rock body fractures suddenly due to high stress in crust. An important phenomenon is that electromagnetic radiation happens simultaneously with acoustic emission [2], so it seems that seismic electromagnetic signal is closely related to fracturing process. The process is like that a dipole is being charged and recharged, thus electromagnetic wave is eradiated [3].

Seismo electromagnetic signal usually takes on various forms and doesn't always appear before any earthquakes for each station. A noticeable result found by some reports is the so-called SES selectivity effect, which is caused by inhomogenous medium according to some discussions [1,4]. The observational result of electromagnetic phenomena associated with earthquakes should be related to the three factors: radiation mechanism, propagation and attenuation of electromagnetic wave and observational situation [5]. The paper mainly deals with the attenuation of SES in crust for a half-space.

2. LF Electric Field eradiated by a unit electric current source underground

The coordinates O-xyz are as showed as Figure.1, with axis z in vertical direction, while axis x and axis y in horizontal direction. z>0,z=0 and z<0 separately represent atmosphere, the earth's surface and crust. Supposing that a unit electric current source Idl is placed along axis x with its depth h, underground in the crust which is linear, homogenous and isotropic. $\varepsilon_1, \mu_1, \sigma_1$ are electromagnetic constants for atmosphere, and $\varepsilon_2, \mu_2, \sigma_2$ for crust. When $\omega\varepsilon \ll \sigma$,i.e. for LF SES, we have

$$\varepsilon_1 \approx \varepsilon_0, \quad \sigma_1 \approx 0, \quad \mu_1 \approx \mu_2 \approx \mu_0 = 4\pi \times 10^{-7} H/m;$$

$$k_1^2 = \omega\mu_1(\omega\varepsilon_1 - j\sigma_1) \approx 0, \quad k_2^2 = \omega\mu_2(\omega\varepsilon_2 - j\sigma_2) \approx -j\omega\mu_0\sigma_2.$$

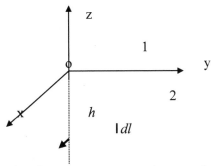

Figure.1 An unit electric current source underground for a half-space

Using the method of image charges, the electric field for electric current source Idl is $(z<0)$ [6]

$$E_{2x}^i = \frac{Idl}{4\pi\sigma_2}(k_2^2 + \frac{\partial^2}{\partial x^2})(\frac{e^{-jk_2R}}{R} - \frac{e^{-jk_2R_s}}{R_s}) + \frac{Idl}{2\pi\sigma_2}(s_0 - s_1 + s_2\cos^2\phi) \tag{1}$$

$$E_{2y}^i = \frac{Idl}{4\pi\sigma_2}\frac{\partial^2}{\partial x\partial y}(\frac{e^{-jk_2R}}{R} - \frac{e^{-jk_2R_s}}{R_s}) + \frac{Idl}{2\pi\sigma_2}s_2\cos\phi\sin\phi \tag{2}$$

$$E_{2z}^i = \frac{Idl}{4\pi\sigma_2}\frac{\partial^2}{\partial x\partial z}(\frac{e^{-jk_2R}}{R} + \frac{e^{-jk_2R_s}}{R_s}) \tag{3}$$

where $\cos\phi = x/\rho$; $\sin\phi = y/\rho$; $\rho = \sqrt{x^2 + y^2}$

$$R = \sqrt{\rho^2 + (z-h)^2}\ ; R_s = \sqrt{\rho^2 + (z+h)^2}$$

and $s_0 = \int_0^\infty \lambda(\lambda - u_2)J_0(\lambda\rho)e^{u_2(h+z)}d\lambda\ ; s_1 = \int_0^\infty \frac{\lambda}{\rho}J_1(\lambda\rho)e^{u_2(h+z)}d\lambda$

$$s_2 = \int_0^\infty \lambda^2 J_2(\lambda\rho)e^{u_2(h+z)}d\lambda\ ; u_2 = \sqrt{\lambda^2 - k_2^2}$$

s_0, s_1, and s_2 are three generalized Sommerfeld integrals (GSI), can be substituted by analytical expressions as some elementary functions and modified Bessel functions[7]. when $z \to 0$, we get

$$E_x = \frac{Idl}{2\pi\sigma_2}(s_0 - s_1 + s_2\cos^2\phi) \tag{4}$$

$$E_y = \frac{Idl}{2\pi\sigma_2}s_2\cos\phi\sin\phi \tag{5}$$

$$E_z \to 0 \tag{6}$$

3. Attenuation of Electric Field

The absolute value of subaerial electric field for the unit electric current source Idl is

$$E = \sqrt{E_x^2 + E_y^2} = \frac{Idl}{2\pi\ \sigma_2}\sqrt{(s_0 - s_1 + s_2\cos^2\phi)^2 + (s_2\cos\phi\sin\phi)^2} \tag{7}$$

Take $Idl = 1Am$, $\sigma_2 = 0.01 S/m$, the amplitude of electric field at origin is calculated (Table.1) for 1Hz, 10Hz, 100Hz, and 1KHz in deferent depths. More detailed working out is as shown as Fig.1. Thus we can see that attenuation of electric field is much smart with depth of the source h or frequency f increasing. From the top down are sequntly 1Hz, 10Hz,100Hz, and 1KHz.

Attenuation of electric field with epicentral distance increased are as shown as Fig.2 — Fig.8 for different frequencies and depths of the source. The electric field corresponds to maximum for each epicentral distance (ρ). From the top down are sequentially 1Hz, 10Hz,100Hz, and 1KHz.

Table.1 Electric field E (V/m) at point $(0,0,0)$

h \ f	1Hz	10Hz	100Hz	1KHz
1m	1.59×10^{1}	1.59×10^{1}	1.59×10^{1}	1.59×10^{1}
1Km	1.60×10^{-8}	1.73×10^{-8}	1.95×10^{-8}	2.23×10^{-9}
5Km	1.52×10^{-10}	1.10×10^{-10}	1.14×10^{-12}	5.39×10^{-21}
10Km	1.95×10^{-11}	2.23×10^{-12}	2.77×10^{-17}	6.18×10^{-35}
15Km	4.36×10^{-12}	6.35×10^{-14}	8.99×10^{-22}	9.42×10^{-49}
20Km	1.16×10^{-12}	2.05×10^{-15}	3.28×10^{-26}	1.61×10^{-62}
25Km	3.37×10^{-13}	7.10×10^{-17}	1.28×10^{-30}	2.94×10^{-76}
30km	1.03×10^{-13}	2.56×10^{-18}	5.18×10^{-35}	5.57×10^{-90}

4. Discussion

Electric field attenuates more acutely versus h than versus f, so it seems easier for us to detect seismo electric signal from shallower sources than from deeper ones. Electric field radiated by the source should counteract both of the attenuation in crust and the observational threshold, and how far SES can propagate rests with intensity of the source and how deep the source is. Considering that the observational sensitivity for electric field is generally $10^{-7} V/m$, so there should be $Il \cdot E \geqslant 10^{-7} V/m$, if SES could be detected somewhere. Here Il means the intensity of the source, while E means the electric field for a unit electric current source ($1Am$). A reasonable radiation mechanism of seismic electromagnetic signal is proposed by Guo et al. and supported and demonstrated by Dong [7] from the viewpoint of fractal geometry. According to the mechanism, it is quite possible for lots of fracture events associated with earthquake to cause observable electromagnetic radiation.

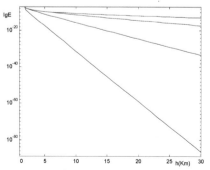

Fig.1 Attenuation of electric field
at origin versus h.

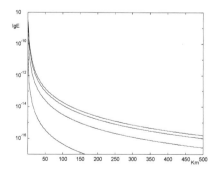

Fig.2 Attenuation of electric field versus
epicentral distance ρ with h=1km

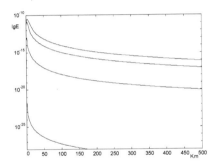

Fig.3 Attenuation of electric field versus ρ with h=5Km.

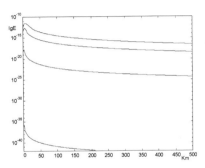

Fig.4 Attenuation of electric field vs ρ with h=10Km.

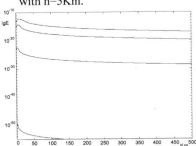

Fig.5 Attenuation of electric field vs ρ with h=15Km.

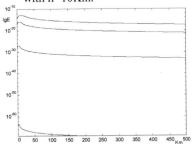

Fig.6 Attenuation of electric field vs ρ with h=20Km.

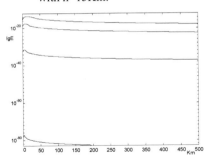

Fig.7 Attenuation of electric field vs ρ with h=25Km.

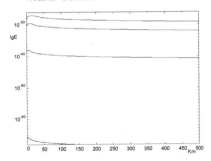

Fig.8 Attenuation of electric field vs ρ with h=30Km.

References：

[1] M. Hayakawa，Atmospheric and Ionospheric Electromagnetic Phenomena Associated With Earthquakes, Terra Scientific Publishing Company, Tokyo,1999

[2] А.И.ГОНЧОРОВ et al., Acoustic Emission and Electromagnetic Radiation under Uniaxial Pressure, World Earthquake Translations,1983,No.3

[3] J. Dong, Y. Gao, Electromagnetic Radiation of the Finite Moving Sources, AP-RASC'01, Tokyo Japan, August 1-4,2001

[4] International Workshop on Seismo Electrmagnetics, okyo,Japan, September 19-22, 2000

[5] J. Dong, Y. Gao, Analysis on Propagation of the Electromagnetic Waves Associated with Earthquakes, 2002 3rd International Symposium on Electromagnetic Compatibility, May 21-24, 2002 Beijing, China

[6] Y. Lei, the Analytical Methods for Time Harmonic Electromagnetic Field (in Chinese), Scientific Publishing Company,2000

[7] J. Dong, Y. Gao, Radiation Mechanism of SEMS (in Chinese), China EMC'2002, Tianjin China, October 22-25,2002

Author Index